TROPICAL FORAGE LEGUMES

FAO Plant Production and Protection Series 6082717 No. 2

Tropical forage legumes

Second edition

Revised and expanded

by
P.J. Skerman
D.G. Cameron
F. Riveros

**Food and
Agriculture
Organization
of the
United Nations**

Rome, 1988

P-10
ISBN 92-5-102190-2

David Lubin Memorial Library Cataloguing in Publication Data

Skerman, P.J.
   Tropical forage legumes. 2nd ed.
   (FAO Plant Production and Protection Series, no. 2)

1. Feed legumes.   2. Tropical zones.
I. Cameron, D.G.   II. Riveros, F.   III. Title.   IV. Series.

FAO code: 11   AGRIS: F01   1988
ISBN 92-5-102190-2

Printed in Italy

Foreword

Use of forage legumes in the tropics for animal feeding has been limited, mainly because of the lack of knowledge and of inadequate economic incentives. However, advances in improving varieties and in developing suitable agronomic and management practices are promising and legume-based forage deserves consideration.

The increasing costs of nitrogen fertilizer and the need to improve soils and conserve natural resources create a certain urgency to realize the possibilities offered by tropical forage legumes.

Increasing demands for food and feed have in the past been met largely through expanding the area of cultivated land. But as uncultivated arable lands become scarce, this solution is progressively less viable and the emphasis must be shifted to raising the productivity of land now in cultivation and maintaining or improving its potential.

The better arable lands will need to be used more and more for food and cash crops. These crops will also receive priority in the use of fertilizer, particularly nitrogen. It is also essential to improve productivity of land not suitable for cultivation and to develop complementary land-use systems so as to raise and diversify total production. Ruminant breeding stock will depend increasingly upon the grazing lands not suitable for crop production, while their progeny are grown out on improved and planted pastures, fodder crops and agricultural by-products. Overall economic benefits of pastures and fodder crops will not only be their yield of animal products, but will largely depend on how well they supplement natural grazing lands and how effectively they maintain soil fertility. How well these forages may be substituted for protein concentrates is also important. In these circumstances, it is clear that high-protein, soil-enriching forage legumes deserve a special place in future agricultural development.

Much remains to be learned about tropical forage legumes. There is a need to explore and use the immense gene pool, which is still largely untapped.

However, existing knowledge in synthesized form could go far toward meeting immediate needs and increase the likelihood of success, especially in developing countries. This is the *raison d'être* for this handbook.

The manuscript was prepared by Dr Percy Skerman of the Department of Agriculture, University of Queensland, St. Lucia-Brisbane. As a teacher, research worker and ranching consultant, Dr Skerman has been closely associated with research on tropical legumes and land development in Australia. He has also become well acquainted with the problems of the developing countries, having worked in the tropics for FAO on various United Nations Development Programme projects and for the World Bank. Dr Skerman has produced a handbook which should prove useful to everyone concerned with improving tropical pastures and livestock production especially in the developing countries. FAO is grateful to Dr Skerman and his colleagues who contributed with special chapters or in other ways.

Wider use of tropical forage legumes, particularly in developing countries, will generate valuable new information. This will undoubtedly call for revision of this handbook from time to time.

Roald A. Peterson
Former Chief, Crop and Grassland Production Service,
Plant Production and Protection Division,
Food and Agriculture Organization
of the United Nations, Rome

Contents

Illustrations

Colour plates

Introduction

This handbook is designed to fill a need for factual and up-to-date information on the use of forage legumes in livestock production, with particular relevance for developing countries.

Most tropical countries have used leguminous pulse crops for centuries as staple food to supply some or all of their protein requirements and their experience with such crops predates that of the more technically advanced countries. In countries whose religious traditions have precluded the use of animal protein the pulse crops have been widely developed, and these areas provide a background of legume culture in some depth.

The need to protect the soil between rows of plantation crops led to a technique of intercropping with trailing leguminous species which would provide a dense and quick vegetative cover both to nullify the effects of raindrop-splash erosion on bare soil and to negate the erosive effect of rapidly flowing water across the soil surface. In both of these practices the beneficial effect of a leguminous crop upon yields of succeeding annual and companion plantation crops was early appreciated and a system of rotation with leguminous crops in an annual cropping programme developed to control this advantageous effect.

The practice of shifting cultivation, traditional in many countries, also depends upon some fertility contribution from leguminous components of primary and secondary forest cover.

The use of legumes in pastures sown for the grazing animal is of more recent history. The early development and use of temperate legumes such as clovers, trefoils and lucerne took place close to the centres of origin of these plants in Europe and around the Mediterranean region. This experience showed that leguminous plants were useful both as forage species and for their beneficial effect on the growth of associated grasses. A vast knowledge of the performance of these species has accumulated and has been exploited

1

in restricted high-altitude areas of tropical countries where climatic conditions can be satisfactorily equated.

The use of tropical legumes as sown species in tropical countries received little attention until some 15 to 20 years ago, when cash cropping for sale began to attract capital investment from overseas. The role of livestock in the national economy had been a feature of the subsistence economy on the part of nomadic herdsmen who used arid land unsuited to arable agriculture for their purposes. In these areas range management as a science, developed in North America, is being applied; and in the process, the importance of leguminous browse for livestock is appreciated, so that preservation of the range will of necessity be combined with improvement of sown pastures where the environment is favourable.

The development of high-quality tropical pastures is of comparatively recent date. It was first based on the use of better quality grasses, because of the belief that tropical legumes would not survive the rigours of the tropical environment — excessive rainfall or drought, high temperatures and, in general, highly leached soils of low fertility or heavy clays too difficult to harness. Mainly due to the insight and enthusiasm of the late Dr John Griffiths Davies and his team of scientists at the CSIRO Cunningham Laboratory in Queensland, Australia, coupled with strong support from the Queensland Department of Primary Industries, it has been shown that legumes can be successfully introduced into and maintained in tropical pasture mixtures in northern Australia. Dr Jorge Ramos de Otero pioneered some valuable species evaluation in Brazil.

Quality in livestock products from the developing countries is now a possibility, and in some cases a reality, as a result of better breeding, informed husbandry and adequate feeding. Improved legume-based pastures appear to be the most economical approach to this latter objective.

With the knowledge that adaptation of legumes to their ecological niches by adjustment of soil nutrients, by matching *Rhizobium* affinities, and by balancing management to integrate the growth cycle of the plant with animal requirements, it is possible for all tropical countries to embark upon a stage of rapid forage development.

This handbook, designed to assist them toward this goal, should provide a ready reference for workers in the field of tropical forage legumes and for students in the general field of agrostology or pasture agronomy.

For the purposes of this discussion, the tropics are defined as all land lying between latitudes 30°N and 30°S of the equator. Owing to changes of climate with latitude, with rainfall, with the duration of the wet season, and with

2

altitude, it is difficult to delimit the tropics precisely as an ecological unit.

Although the handbook is concerned mainly with tropical legumes, some utilization of temperate species is necessary in certain high-altitude areas in tropical countries. As legumes are almost universally associated with grasses in forage mixtures, discussion of suitable companion grasses has been undertaken and the performance of these grass/legume mixtures recorded.

I wish to acknowledge the initiative and help of Dr Roald A. Peterson, former Chief of the Crop and Grassland Production Service of FAO in arranging for this book to be written, of Dr J.J. Norris, and of Dr P. Brumby and Mr P. Reid of the World Bank for encouragement in its production.

The help of my colleagues, Mr D.R. Bailey, Mr F.H. Kleinschmidt and Drs E.F. Henzell, E.M. Hutton and D.J. Minson, in providing specialist chapters is gratefully acknowledged.

Taxonomy of the forage legumes is still under investigation. A visit was made to the Royal Botanical Gardens at Kew, United Kingdom, to obtain opinions on some uncertain names. I am grateful to Messrs R. Podhill and B. Verdcourt for their help in this regard.

The author would also like to express appreciation to the many other individuals and organizations who helped with information and photographic material.

P.J. Skerman
FAO, Rome, 1976

1. The role of legumes in agriculture

Survival of an agricultural civilization depends upon its ability to retain soil fertility at a level enabling continuous productivity of sufficient land to meet subsistence needs for food and clothing. Growth of such a civilization depends upon a continual improvement of soil productivity.

Once primeval forest is removed to expose soil to the combined influences of erosion, leaching, insolation and radiation, fertility decline is rapid, and without some injection of fertility successive crops dwindle quickly until it is not worth while expending effort to plant further crops. Such crises are well documented in the history of agriculture.

In an effort to postpone fertility drain and improve the productive capacity of the soil, several systems of agriculture have developed. They include the periodic abandonment of cropped land to allow regeneration of natural vegetative cover and then reutilization for a cropping cycle (shifting cultivation); the use of animal and vegetable waste as compost to augment fertility; the culture of green manure crops to be ploughed in as part of a rotation; the growth of inter-row cover crops in plantation crops; and the development of short-term pasture leys or permanent pastures using the grazing animal to recycle the nutrients. In each of these systems legumes usually play an important part.

Legumes in shifting agriculture

Shifting agriculture represents an intelligent attempt by early agriculturists — who had no access to or knowledge of fertilizers — to prolong soil productivity. In the regenerating "bush", leguminous species generally play a part in fertility building.

In the "goz" sands of the Kordofan Province of the Sudan, where gum arabic is an important article of commerce, the use of the gum-producing

5

shrub *Acacia senegal* in a gum-cropping cycle represents a planned fertility-building exercise. After a cycle under gum production, the trees are coppiced and a cropping cycle is taken of groundnuts, sorghum (dura), bulrush millet (dukn) and possibly sesame (sim-sim) until soil fertility is exhausted as indicated by declining yields. The *Acacia senegal* trees are then allowed to regenerate. Being deciduous, the leaf canopy falls to the ground in November; this material is mineralized and, with the added nitrogen fixed by symbiotic fixation, fertility is gradually restored sufficiently to support another cropping cycle. Thus an ancient agricultural civilization has survived. This legume-based land-use system is sound but nonintensive, and population pressure can lead to short legume-producing cycles and consequent lower crop yields.

Legumes as green manure

The use of green manure crops in a cultivation cycle is aimed at replenishing organic matter, at the same time adding any available nitrogenous material, and protecting soil during periods of high erosion hazard. It is not necessary to use legumes, but because of their multiple advantages they are generally preferred. However, where a legume may harbour an insect or disease affecting the main crop, a heavy-yielding grass or gramineous forage crop may be used (e.g. in a commercial bean-growing venture a leguminous green manure crop may increase bean fly damage). In the culture of sugar cane the use of cowpeas, velvet beans or some other leguminous green manure crop is common practice. Such a crop can protect the soil during the monsoonal rains and, if ploughed in some six weeks before planting the succeeding sugar cane, will reduce the amount of fertilizer needed to grow the plant crop.

The amount of nitrogen present in a hectare of the common green manure crops used for sugar cane is presented in Table 1.1.

Legumes as cover crops

The need to protect the soil from the erosive effects of heavy rainfall and to assist in maintaining fertility led plantation agriculture to introduce a system of planting a fast-growing perennial leguminous cover crop between the rows of the main plantation crop. Besides protecting the soil, this crop increased fertility by mineralization of the annual leaf drop from the legume and

6

TABLE 1.1 **Nitrogen present in 1 hectare of the common green manure crops used for sugar cane**

Variety	Yield (tonnes/ha)	Nitrogen (kg/ha)	Sulphate of ammonia (equivalent kg/ha)
Cowpea	11.4	80.7	393
Poona pea	11.8	113.3	558
Mauritius bean	8.6	123.3	602
Gambia pea	28.6	232.0	1 132
Velvet bean	17.4	339.4	1 646

Source: King, Mungomery and Hughes, 1965

accumulation of combined nitrogen through symbiotic nitrogen fixation by nodules growing on the roots of the leguminous plants.

Such cover crops were introduced to young plantations of rubber, oil palm, coconut and sisal. The persistence of the legume depended largely on its shade tolerance and its ultimate competition with the main plantation crop for water and nutrients.

The requirements of a cover crop are that it should be easy to establish, cover the ground quickly and well, be aggressive enough to exclude weed growth — but not so aggressive as to cover its companion crop or compete adversely with it for light, water and nutrients — and actually to improve soil fertility and hence crop yields.

Several successes with cover crops have been recorded in the relevant literature. At the Sisal Research Station at Tanga in the United Republic of Tanzania (latitude 5°S) a cover crop of tropical kudzu or puero (*Pueraria phaseoloides*), growing as an inter-row crop with sisal on a red loam soil over gneiss, increased the yield of sisal by 48 percent above that of a clean-weeded, inter-row cultivated crop. In addition to protecting the soil from erosion, the legume encouraged earlier unfurling of the sisal leaves and added 635 kg of nitrogen per hectare to the soil, equivalent to the application of 1.4 tonnes of urea per hectare (Rijkebusch, 1967). After 74 months the soil organic carbon had increased from 1.85 to 2.96 percent under the *Pueraria* cover, and the total nitrogen from 0.163 to 0.248 percent. At the end of the period the difference between the soil nitrogen under *Pueraria* and under the clean-cultivated soil was 2 195 kg per hectare of cover (Hopkinson, 1968, unpublished).

7

Legumes in short-term pasture leys

"Ley-farming" was developed as a soil-renovating system in British agriculture, the ley either being grazed by livestock or harvested to provide hay or silage for feeding during the winter months.

The system was introduced into the southern Australian wheat belt in about 1850. From 1870 the Australian wheat yield began to decline as a result of falling soil phosphorus and nitrogen levels. The introduction in 1900 of a rotation of wheat and bare fallow and fertilization with superphosphate arrested this decline by mineralization and accumulation of nitrogen from the soil organic matter by microbial activity during the fallow, and by the provision of readily available phosphorus. By 1950, yields were back to the 1870 average figure. The introduction of superphosphate meanwhile had encouraged the growth of the accidentally introduced subterranean clover (*Trifolium subterraneum*) which became the basis for improved pastures. When these clover-based pastures were introduced as leys into the wheat-growing cycle, grain yields rose by 28 percent in the next decade and a prosperous wheat and fat lamb industry developed as a sound land-use system with marked improvement in wheat flour quality. Carter (1966) has estimated that in southern Australia there are 16 million hectares of sown pasture based on legumes belonging to the genera *Trifolium* and *Medicago* which are top-dressed annually with an average of 121 kg of superphosphate per hectare. The annual cost of this fertilizer is estimated at $A270 million and the cost of pasture seed at $A10 million. The fixed nitrogen was valued at $A400 million.

Schofield (1945), who was the first to introduce tropical leguminous cover crops into Queensland, Australia, for use as pastures, compared the nitrogen accession in a soil under a seven-month bare fallow with that added by ploughing in tropical legumes after 18 months of growth. The nitrogen level in the soil under the bare fallow treatment was 34.4 parts per million (ppm), while that in the legume-incorporated soil was 66.7 ppm for *Calopogonium mucunoides*; 71.7 ppm for *Centrosema pubescens*; 171.8 ppm for *Pueraria phaseoloides*; and 54.5 ppm for *Stylosanthes guianensis* (see Figure 1). *Pueraria phaseoloides* was particularly effective in raising soil fertility.

At Serere Research Station, Uganda (lat. 1°32′N), with an annual bimodal rainfall of 1 365 mm, the early research on leys was conducted with pure Rhodes grass (*Chloris gayana*). During a three-year ley, mean animal production from this unfertilized grass was 209 kg per hectare per year. The inclusion of the legumes *Stylosanthes guianensis* and *Centrosema pubescens* increased animal production by 11 to 49 percent (Stobbs, 1969c). Recently

8

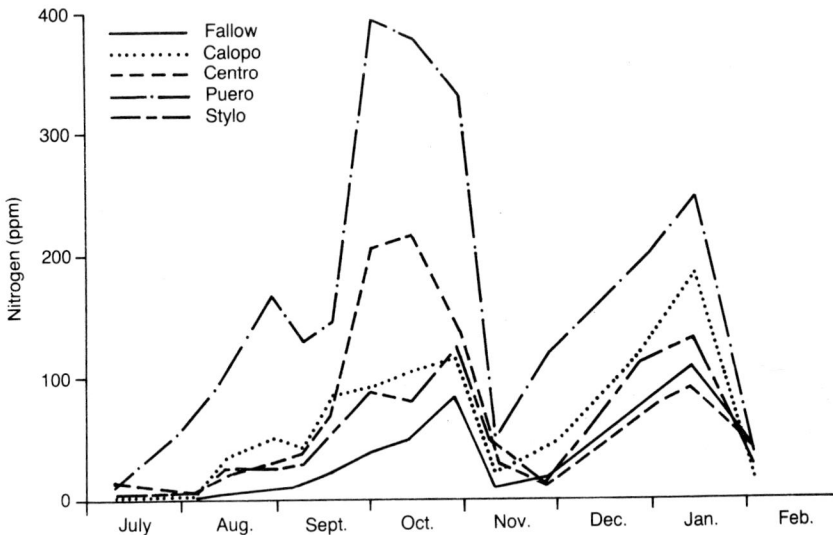

Figure 1. Nitrate nitrogen in the 0- to 15-cm layer of soil, July 1941 - February 1942 (**Source:** Schofield, 1945)

Panicum maximum/Desmodium intortum swards have given 517 kg per hectare per year live-weight gain.

Pasture composed of grass alone gave a per acre (0.405 hectare) return over costs of U Sh 130 ($A14.00), an unfertilized grass/legume mixture U Sh 275 ($A29.00) and a grass/legume mixture fertilized with superphosphate a return of U Sh 475 ($A50.00) over fixed costs. This fertilized mixed grass/legume ley also considerably increased the yield of the subsequent crops.

At Ibadan, Nigeria, the inclusion of *Centrosema pubescens* in a giant star grass (*Cynodon plectostachyus*) pasture growing on a red podzolic soil resulted in significantly higher levels of organic matter and total nitrogen. The nitrogen content of the soil under the pasture containing the legume was 275 kg per hectare-33 cm per year higher than under the pasture of pure grass. The associated legume raised the nitrogen content of the grass from 1.8 to 2.4 percent. Under grazing with White Fulani cattle the grass/legume pasture gave a daily live-weight gain over 12 months of 0.37 kg compared with 0.30 kg from pure grass (Moore, 1962).

In Puerto Rico, the introduction of tropical kudzu (*Pueraria phaseoloides*) into pastures of molasses grass (*Melinis minutiflora*) in increasing percentages

9

raised the protein level from 4.2 percent in the pure grass pasture to over 8 percent with 60 percent of legume by weight in the mixture (Vicente-Chandler, Caro-Costas and Figarella, 1953).

In Brazil, a cotton crop following four years of *Neonotoria wightii* pasture ley yielded 3 026 kg per hectare of seed cotton. The yield of the crop grown with a complete fertilizer mixture consisting of 40:40:30 NPK applied at 440 kg/ha without a previous legume crop was 3 048 kg/ha, and the unfertilized and unrotated control crop yielded only 1 097 kg/ha (Menegario and Richi, 1970, unpublished).

The short-term ley has a rejuvenating effect on soil structure and fertility, but involves periodic capital outlay in ploughing out, cultivation and reseeding to crops.

Survival under wet conditions

Long-term or permanent pastures are, therefore, preferable where they can be used for profitable animal production while at the same time improving soil fertility.

Bruce (1965) examined the soil fertility under two *Panicum maximum/ Centrosema pubescens* pastures established 11 and 16 years earlier on unfertilized red loam (latosol) at Utchee Creek, north Queensland (lat. 17°30'S, rainfall 3 750 mm/year), which had previously carried dense tropical rain forest. In comparing the data with those obtained under adjacent rain forest, he found that the percentage of total nitrogen in the 0 to 7.5-cm layer of soil under the tropical rain forest was 0.423 percent and under the 11-year and 16-year grass/legume pasture was 0.406 and 0.432 percent respectively. It can thus be seen that, given good management, an effective grass/tropical legume pasture can maintain fertility even under intensely wet conditions.

The low-lying infertile coastal sandy country in southeastern Queensland, popularly known as the *wallum* (an aboriginal word for *Banksia aemula*, a small proteaceous tree inhabiting this ecological niche), was virtually unoccupied by livestock prior to 1954 because of low productivity.

A research team of soil scientists, agronomists and ecologists initiated investigations into establishing pastures for beef and fat lamb production in this area in 1953. Soil fertility studies of these low-humic gleys, ground-water podzols, humic gleys and relict sandy lateritic soils revealed general gross deficiencies of nitrogen, phosphorus, potassium, calcium, copper and sulphur, and deficiency of zinc and molybdenum. The fertilizer needed for pasture establishment was determined as 616 kg single superphosphate, 616 kg lime, 123 kg muriate of potash, 7.7 kg copper sulphate, 7.7 kg zinc sulphate and 56 g elemental molybdenum per hectare, with a subsequent annual maintenance requirement of 246 kg superphosphate and 123 kg potassium chloride per hectare. Grass and legume species were found which could utilize the new environment and the *Rhizobium* requirements of the legumes were ascertained.

The initial pasture so established carried 7.5 sheep per hectare; this figure rose in successive years to 11.25, 11.25, 15.0, 17.5, 18.75, 25.0 and 30.0 sheep per hectare. At this stage, cattle were introduced into the management system. Mixed grass/legume pastures carrying 1.2 beasts per hectare in 1958 improved to handle 2.2 beasts per hectare from 1962 and the live-weight gain per hectare rose from 150 kg in 1958 to 220 kg in 1959, 330 kg in 1961, and reached 569 kg in 1967 (Bryan, personal communication).

The improvement in soil fertility during a 15-year period is shown in Table 1.2.

TABLE 1.2 **Improvement in soil fertility over a 15-year period in the *wallum* country, southeastern Queensland**

	pH	Organic carbon	N	Available P	Total K	Total Ca
		(%)			(ppm)	
Virgin soil	5.2	0.84	0.047	4	31	30
After 15 years' pasture	5.0	1.56	0.103	40	202	320

Source: Bryan, personal communication

It can thus be seen that adequately fertilized, selected legume/grass pasture mixtures can exert a profound improvement in both plant and animal performance in an area of unproductive but climatically favoured land.

Is legume introduction economic?

Three case histories are presented to answer this question:

Zaire. Risopoulos (1966) has shown the following results obtained with pasture improvement and supplementary feeding in the Gandajika area of the southern savannas. The traditional extensive system of livestock husbandry, using natural *Hyparrhenia* pastures burnt annually, yields only 10-15 kg/ha annually in live-weight gain in cattle. Improved practices can increase production five times and give a useful profit over expenditure. System (*c*) is economically viable in the first year, and from the second year returns a profit. (See below for relationship between husbandry practised and live-weight gains.)

Husbandry practised:	Live-weight gains (kg/ha/yr)
a) Burned day pastures, traditional method of stock raising	10-15
b) Day pastures, burned once every 4 years; night pastures improved by sowing *Stylosanthes guianensis* and cutting with a rotary cutter. Fodder crops and distribution of silage (a *Stylosanthes*-sorghum mixture) during the last three months of the dry season at 10 kg/head/day	46
c) Unburned natural pastures wholly treated mechanically to over-sow *Stylosanthes guianensis* at 2.5 kg/ha: distribution of silage during the last three months of the dry season at 10 kg/head/ day. No fertilizers used	76

Serere, Uganda. Stobbs (1969a) evaluated the economics of improved pastures at Serere, Uganda, the results of which are shown in Figure 2. Nitrogen applications produced highly significant increases in herbage and beef production, but the excellent pastures obtained, although they produced up to 550 kg/ha live-weight gain, were uneconomic when the price of beef was low. The grass/legume treatment provided the greatest margin of profit.

Dairy farm, coastal Queensland. Farm area - 100 hectares. Number of milking cows - 61. Area of improved legume-based pastures - 8 increasing to 22 hectares.

Over a five-year period, production increased from 3 909 to 6 773 kg butterfat with the same number of cows.

Pasture establishment costs - US$70 per hectare.

Annual maintenance costs - $17.50 per hectare.

Total additional cost - $2 930.

Total additional returns - 7 910 kg butterfat over five years at 47 cents = $5 472.

Total increase in net return - $2 542 (average $508 per year).

Note: All figures given refer to costs at the time of publication of the studies quoted.

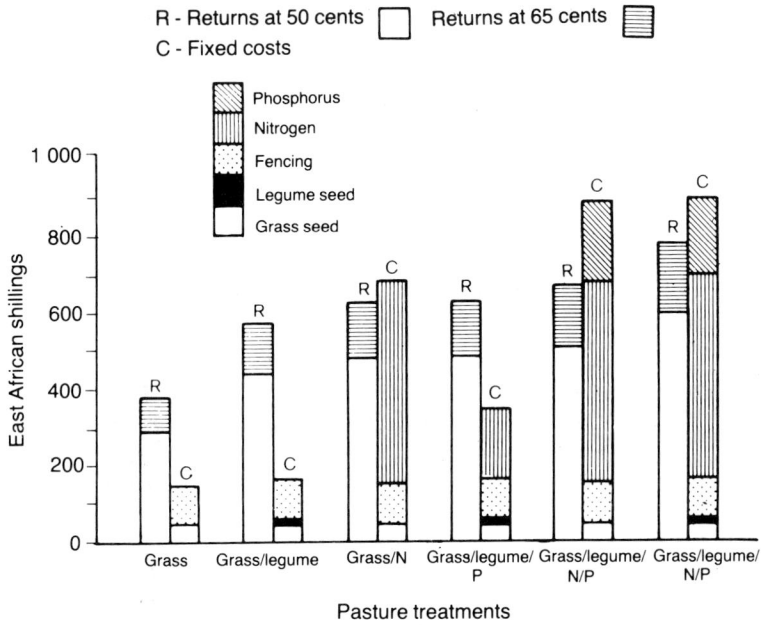

Figure 2. Histograms showing estimated returns from fertilized and unfertilized grass and grass/legume pastures, with approximate fixed costs (**Source:** Stobbs, 1969a)

2. Evaluation of natural resources for pasture production

One of the most important requirements in the sound development of the land of any country is an inventory of its natural resources. In the developed countries such inventories were often carried out too late to make full use of the information and some resources were wasted for lack of planning. In the developing countries there is still time for initial inventories, especially in pasture development and livestock production, so that planning can be instituted and development programmes implemented.

The natural resources in land use are climate, land form, soil vegetation, fauna and water (Stewart, 1968).

Climate and plant adaptation

The macroclimate of any particular locality is basically a fixed resource which depends on geographical location, although it is still generally beyond human capacity to manipulate. The microclimate immediately surrounding a plant, however, is often subject to human control, and a study of the micro-environment is important in gaining knowledge of a plant's behaviour.

The mapping of the world's climates by Köppen (1931) and Thornthwaite (1948) include broad classifications useful for comparative purposes, but not detailed enough for the assessment of plant performance. Several others have tried to derive more meaningful agrobioclimatic indices, and readers are referred to a summary of these attempts by J.M. Henry in the book by Frankel and Bennett (1970).

The major climatic elements for plant growth are solar radiation (energy input), temperature and precipitation.

14

Solar radiation. The energy for plant growth is derived directly or indirectly from the energy of sunlight through the process of photosynthesis. In the photosynthetic process, light energy taken up by the plant is transformed and stored in the form of energy-rich carbon compounds which are then themselves sources of energy both to the plant itself and to animals which use the plant as food. The amount of light energy or solar radiation is influenced primarily by latitude, being comparatively uniform near the equator and varying as distance from the equator increases. In the tropics, average daily radiation is highest during the summer months (600 cal/cm^{-2} in January in Australia) but the degree of cloudiness associated with summer rainfall conditions reduces radiation input in the coastal areas (by 50 to 150 cal/cm^{-2}/day in Australia). The number of wet days and hence cloudiness influence the rate of sugar cane production in the tropics and higher sugar production is usually accomplished by irrigating the crop in an area where there is more sunlight and fewer days of rainfall. However, radiation is seldom a limiting factor for high pasture production in the tropics.

The day length, or photoperiod, which varies with latitude, influences daily energy input and controls flowering in plants. It is thus an important factor in selecting pasture legumes to fit into the various ecological niches found in the tropics and also has a strong effect on seed production.

Atkinson (1970) has listed the tropical legumes in three groups according to latitude:

— legumes with wide latitudinal range in both northern and southern subtropics (*Rhynchosia, Stylosanthes, Trifolium repens, Trifolium, Vicia, Lupinus, Phaseolus, Medicago, Melilotus, Cassia* and *Desmodium*);

— an intermediate group with a distribution north and south of the respective tropics, but with an apparent equatorial "gap" of 10 to 20° latitude across the equator (e.g. *Zornia, Galactia, Indigofera, Crotalaria, Aeschynomene,* and *Astragalus*); and

— an obligatory equatorial group (*Alysicarpus, Calopogonium, Centrosema, Canavalia, Pueraria* and *Teramnus*).

Atkinson states that further plant exploration work may modify this generalization. The author has found *Aeschynomene* species at Pucallpa, Peru, in the upper Amazon basin at latitude 8°S.

Plants may be either short-day plants which flower only when the daily

15

period of illumination is shorter than a particular critical length; long-day plants which flower only when the daily period of illumination exceeds some critical duration; or day-neutral plants which flower over a wide range of day lengths. Not only genera differ in flowering response to day lengths, but also species and cultivars.

'T Mannetje (1965) found that photoperiod effects differed in four species of *Stylosanthes*: *S. guianensis* and *S. humilis* were short-day plants; *S. montevidensis* was a long-day plant, *S. mucronata* tended to be day-neutral, flowering better on short days. Under photoperiods of 8 and 10 hours *S. humilis* had a prostrate habit and *S. montevidensis* was stunted, while under photoperiods of 12 and 14 hours both species grew erect. Dry-matter yields for the 12- and 14-hour photoperiods were significantly higher than for the 8- and 10-hour photoperiods for all species except *S. mucronata*, which gave a significantly higher yield only under the 14-hour photoperiod.

Cameron (1967b) studied natural selections among short-day *S. humilis* in the field. Within a selection, a southward (increasing latitude) progression of flowering occurred, that is, a variety which flowered in 86 days at lat. 19°40'S took 97 days to flower at lat. 24°S and 108 days to flower at lat. 27°22'S. A day length of 13 hours was reached at lat. 19°40'S on 1 February, at lat. 24°S on 10 February, and at lat. 27°22'S on 16 February. By late March all day lengths were shorter and equal. The extreme variation in flowering times between selections at one site was 60 days.

Light is a potent factor in competition among pasture plants (Donald, 1963). Beneath a dense pasture the light is of very low intensity and species of dwarf stature suffer seriously from shading by taller species. Sillar (1967) showed that, under 0.74 daylight, Townsville stylo (*Stylosanthes humilis*) markedly declined in dry-matter production in both top growth and roots. The successful establishment of this species is achieved only by drastic reduction of natural grass competition before seeding (grazing, disc harrowing or burning) and by heavy stocking after sowing. The successful introduction of a sod-seeded legume into a pasture also depends on reduction of grass competition for water, light and nutrients.

Where water, nutrients and temperature are not limiting factors, the rate of growth of a sward of particular genetic composition will depend on the amount of light alone. The leaf canopy of a pasture intercepts more and more light until interception is complete. Donald (1963) suggests that each species has a "ceiling yield" of dry matter in a particular environment governed by the amount of incident light. In New Zealand, Brougham (1956) noted in a pasture of perennial ryegrass and white clover that when a canopy of leaves

16

had developed equal to five times the surface area of the ground covered (leaf area index or LAI = 5.0), then light values at the ground surface were almost zero and the rate of dry-matter increment became constant. This index will differ somewhat with species and location, but the principle is a useful concept in pasture studies. Defoliation at the appropriate frequency and intensity to maintain a satisfactory LAI gives a leafy herbage without loss of yield.

Temperature. Temperature has a major effect on the growth pattern of plants in the tropics. In subtropical areas with a hot, wet summer growth period and a dry, cold winter period, winter production of tropical species is severely restricted. They are usually supplemented by the more widely adaptable temperate legumes (e.g. *Trifolium repens*) if growth conditions are suitable; otherwise supplementation by conserved fodder or concentrates is necessary for continuous livestock production. Selection of tropical species for cold tolerance is extremely important in such an environment (Atkinson, 1970).

Temperature has a controlling effect on flowering in some legume species. Siratro (*Macroptilium atropurpureum*) flowers well under day-night temperature regimes of 24/19°C, 27/22° and 30/25°, for example, but not at 18/13° (Davies and Hutton, 1970).

Extremely high temperatures in the semi-arid areas and for short periods in the subhumid climates can cause establishment failures, seedling mortality (Skerman, 1958a), and they have a serious effect on the survival and activity of the associated *Rhizobium* (Bowen and Kennedy, 1959). High temperatures are also associated with high evapotranspiration figures.

One particular aspect of high temperatures in pasture performance is resistance to fire. Fire has a drastic effect on ecology and annual burning tends to control the type of vegetation (e.g. *Eucalyptus* species in Australia, *Imperata* grasslands in New Guinea, *Pennisetum polystachyon* fire subclimax in Fiji). Lamprey (1967) has shown that if fire can be kept out of the grassland on the Serengeti plain in northwest Tanzania, *Acacia drepanolobium* is able to regenerate for wildlife browse. Passage of fire is often a prerequisite for the germination of many *Acacia* seeds.

Many of the fires occurring in the subhumid areas cause serious damage to pasture. It is important to know which species can withstand the passage of fire. Siratro (*Macroptilium atropurpureum*) will rapidly regenerate; *Stylosanthes guianensis* is often killed outright, though if the ground is moist the damage may not be so severe.

17

Under lowland tropical conditions a steady mean temperature of between 25 and 30°C is often maintained, with a comparatively small daily range. Distance from the sea and increasing altitude produce greater extremes in the diurnal range, which have a marked effect on plant performance.

Fitzpatrick and Nix (1970) have shown that for maximum dry-matter production, the tropical legumes fall between the group of temperate grasses and legumes and the tropical grasses and reach highest production in the temperature range of 26.5 to 32.2°C, with little growth below 10°C.

In frost-free environments, the tropical legumes have a much greater potential for dry matter than the temperate legumes.

Seasonal temperatures are modified by altitude. Atkinson (1970) has listed the occurrence of tropical legumes in relation to altitude as follows:

Sea level to 915 m — *Alysicarpus, Calopogonium, Centrosema*, some large-leaved *Desmodium* species.
915 m to 1 840 m — *Aeschynomene, Desmodium canum, D. scorpiurus, Indigofera, Phaseolus, Rhynchosia, Stylosanthes guianensis* and *Zornia*.
1 840 m to 3 810 m — *Vicia, Trifolium, T. repens, Melilotus, Medicago, M. lupulina, Lupinus, Astragalus* and *Adesmia*.

Some species are able to span greater differences in altitude: *Macroptilium atropurpureum* (siratro) is found from 600 to 1 200 m and *Stylosanthes* from sea level to 1 800 m.

Species within genera and cultivars within species also show differences (Britten, 1961, has demonstrated the great divergence in flowering times with altitude of white clover, *Trifolium repens*, in Hawaii).

A screen temperature of 0°C is taken as the limit of the growing period due to frost, but a light ground frost may occur with a screen (1.2 m above ground) temperature of 2.2°C. However, growth may cease at a temperature well above 0°C. Savage (1970) found that *Neonotonia wightii* growth slowed down at 16°C at Redland Bay, Queensland, Australia, (lat. 27°40'S) and ceased at 13°C. Allen (1960) reported leaf shedding in Queensland at a temperature of 1.5°C. Jones (1969) took as the criterion of frost tolerance the survival percentage in the spring of the plants which were counted the previous autumn. The lowest terrestrial minimum temperature reached during the winter was −8.3°C. *Teramnus uncinatus, Alysicarpus vaginalis, Lablab purpureus* (L.) Sweet (*Dolichos lablab*), *Macrotyloma uniflorum* (Lam.) Verdc. (*D. uniflorus*), *Macrotyloma africanum* (Wilczek) Verdc. (*D. baumanii*) and

18

Macrotyloma axillare (E. Mey.) Verdc. (*D. axillaris*) did not survive the winter, *Medicago sativa* and *Macroptilium atropurpureum* (DC) Urb. (*Phaseolus atropurpureus*, siratro) survived well and *D. uncinatum*, *N. wightii*, *D. intortum* and *Lotononis angolensis* had about 50 percent survival — mainly due to well-developed root systems.

Some tropical browse species, because of their height, are able to withstand light frosts (e.g. *Leucaena leucocephala*) and this is an advantage in winter survival. The association of legumes with tall-growing grasses also offers them some protection from frost, and mixed pastures of a tall grass and a tropical legume deferred from grazing in summer can carry some green leguminous growth near the base of the grasses into the winter.

Precipitation. Rainfall is the main component of precipitation in the tropics, the relatively few snow-capped peaks having little effect on the moisture regime. Rainfall is extremely variable, ranging from well over 10 000 mm in parts of Hawaii to practically zero in desert regions. Annual rainfall is both variable in amount and erratic in incidence, so that annual totals must be used with care in interpreting probable plant growth response. A knowledge of annual amount and seasonal dominance is, however, a useful guide.

Kendrew (1961) gives details of the climatic influences affecting each individual continent and his book is a ready source of information. He identifies six types of rainfall regimes, three of them tropical.

Equatorial. Where two seasons of heavy rain occur in the course of the year, with no quite dry season. This type occurs within only a few degrees of latitude on each side of the equator (e.g. Cameroon).

Tropical. Between the equatorial area and the neighbourhood of the Tropics of Cancer and Capricorn; most rain in the hottest months when the sun is highest; winter a pronounced dry season. It can be subdivided into:

— inner tropical, with two maxima of monthly rainfall found in some regions between the equatorial zone and the neighbourhood of lat. 10°N and S: it approximates to the equatorial regime, but the two maxima following the overhead sun are closer, and winter is a long dry season (e.g. Juba, the Sudan); and

— outer tropical, poleward of inner tropical; the two maxima coalesce and the dry season is longer (e.g. Khartoum).

19

Monsoonal. Has a marked maximum in summer and a long dry season, much like outer tropical.

More important in plant growth is the water balance determined by precipitation, the evapotranspiration of the plant and the water-holding capacity of the soil. In general, evapotranspiration is closely related to seasonal energy input; unfortunately, high energy input and water deficit often go together.

Prescott (1934) developed a formula relating precipitation to evaporation and defined months where precipitation exceeded a specified fraction (according to locality) of the evaporation from a free water surface as effective months for plant growth, checking his results with pasture growth in a Mediterranean-type climate. This formula does not take into account stored moisture, especially in clay soils.

Thornthwaite (1948) related precipitation to evapotranspiration and developed a concept of water balance. Fitzpatrick and Nix (1970) have developed a moisture index which takes into account stored water and soil-moisture movement in a drying soil. Each new development leads to a better understanding of plant-soil-water interactions.

Regions with a uniform distribution of a higher annual rainfall support a more luxuriant growth than one in which a similar total rainfall is concentrated in wet seasons separated by long dry seasons. However, in tropical forage legumes, luxuriant growth is attractive to numerous leaf-eating insects and plant-sucking bugs which transmit virus diseases, and it also provides favourable conditions for growth of pathogens such as *Rhizoctonia solani.* Hence, very high rainfall is inimical to the growth of some leguminous species, and a dry season or a very cold season is useful in suppressing insect pests and diseases.

High rainfall associated with slow drainage can result in waterlogging, and in many tropical areas (e.g. the flood plain of the Orinoco [Llanos *inundables*], Venezuela) waterlogging is a problem of some magnitude. In sudden downpours of rain, flooding may occur for only a few hours to a few days, whereas waterlogging is of longer duration. Grasses such as rice grass (*Leersia hexandra*) and Para grass (*Brachiaria mutica*) have adapted themselves to waterlogging and it is important to assess pasture legumes in relation to their survival under such conditions. Tropical kudzu (*Pueraria phaseoloides*) and calopo (*Calopogonium mucunoides*) both grow and nodulate under very wet conditions, whereas *Lablab purpureus* and lucerne (*Medicago sativa*) are extremely sensitive to this condition. Leguminous cover crops have been introduced into tropical plantations to present loss of soil caused by erosion after intense summer storms. A pasture sward has a similar effcct.

A uniform medium rainfall regime of 1 250 to 1 750 mm generally leads to the best performance from tropical forage legumes. Bimodal rainfalls (long and short rains) such as those occurring in Kenya, Uganda and India often interfere with overall production and make pasture establishment difficult.

Numerous leguminous species have adapted themselves to arid and semiarid conditions. Adaptation is usually achieved by annuals which germinate, grow and set seed in the brief rainy season. They are thus drought-evading rather than drought-tolerant species (annual *Stylosanthes* and *Alysicarpus* species are typical of such plants). The drought-resistant perennials adapt themselves by becoming deciduous; they shed their leaves at the onset of the dry season (e.g. *Acacia senegal*). Others are deep-rooted and draw their

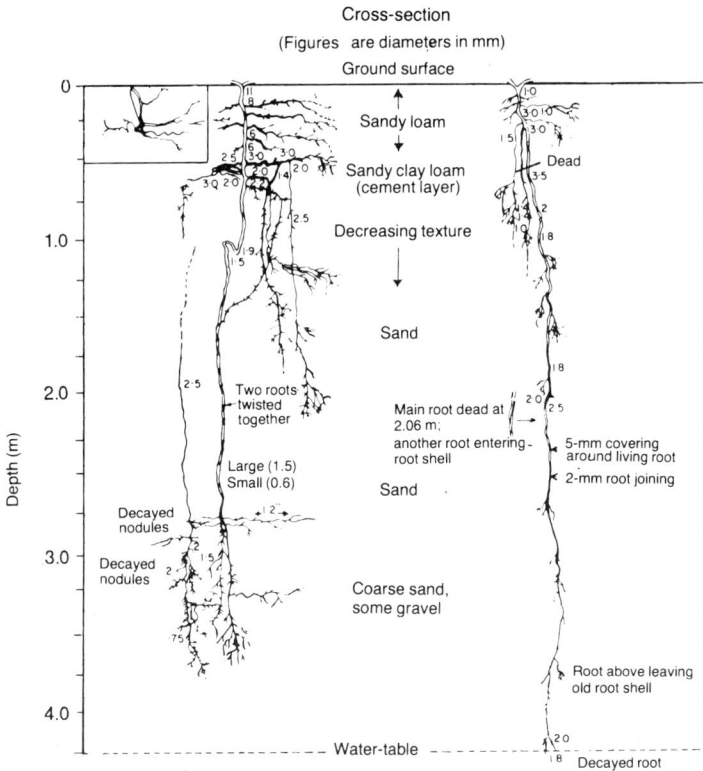

Figure 3. Root systems of *Psoralea patens* growing in deep levee sand at Charleville, Queensland, Australia (**Source:** Kerridge and Skerman, 1968)

21

water from a low-lying water-table, e.g. *Psoralea patens* roots (Figure 3) frequently tap a water-table (Kerridge and Skerman, 1968).

Lucerne (*Medicago sativa*) has a deep, well-developed root system (Paltridge, 1955), as have siratro (*Macroptilium atropurpureum*) and glycine (*Neonotonia wightii*). Leaf adaptations to reduce transpiration are also common in drought-resisting plants.

Land survey, mapping and selection

The evaluation of climate will indicate the extent to which sown pastures can be used to advantage over natural grazing and the climatic regimes within which individual species of legumes and grasses can be expected to perform satisfactorily. The next step is an evaluation of land resources.

A broad reconnaissance survey of the geology, geomorphology (land form), soils, vegetation, water resources and existing land use can be made by aerial inspection and strategic land traverses for a particular developmental project under consideration.

Aerial photography. Piecemeal development may be convenient while awaiting an overall survey, but for long-term planning, a complete aerial-photographic cover should be undertaken. Aerial photogrpahy should be carried out during the season of the year when the absence of cloud and haze will enable clear photographs to be taken. Flying heights for general assessment work are 3 700 to 4 400 metres. For a cheaper cover with less detail, a flying height of 7 400 metres can be adopted. The Queensland brigalow areas were photographed from this height. An overlap of 60 percent is required for the photographs to be examined as stereoscopic pairs.

Preparation of maps. Maps can be prepared to scale from aerial photographs and stereoscopic plotters can draw contours down to an interval of 1.5 m difference in height. A contoured map compiled from aerial photographs checked with existing ground controls of the scale of 1:40 000 to 1:60 000 is a convenient size for transference of detail from the aerial photographs. For more detailed information the aerial photographs can be enlarged to 1:15 000.

Costs of survey. The amount of detail obtained determines the costs of the survey. Each survey should fit into a pattern of expanding detail so that each successive survey enlarges the knowledge of an area outlined in a previous survey and does not render obsolete the existing data.

In general, field costs are proportional to the square of the map scale (the larger the scale, the greater the costs).

Survey methods. Generally, broad-scale reconnaissance surveys should be undertaken to define areas such as those immediately available for development, those unsuitable for early development and those which will require high capital expenditure to bring them into production. The defined areas can then be surveyed in greater detail for particular purposes.

One of the best preliminary methods of assessment and mapping is the "land system" of Christian and Stewart (1952). A land system is "an area, or group of areas, throughout which there is a recurring pattern of topography, soils and vegetation". A change in this pattern determines the boundary of the land system. Land-system mapping is of more direct geographic interest than land use, but it gives a preliminary scientific picture of the landscape and enables land systems to be selected on the basis of potential for development. These selected land systems can then be broken down into smaller land units and subjected to classification as to land capability as proposed by Bennett (1955), who drew up the following classes.

Class I. Land of good productivity, practically free of erosion and suitable for cultivation without special practices. Some areas may need simple preparation to fit them for cultivation, such as clearing of brush or providing simple drainage.

Class II. Land of moderate to good productivity, suitable for cultivation with ordinary or simple practices to prevent erosion or effect satisfactory drainage. Such practices as contouring, growing protective cover crops, and effecting simple drainage operations, as with small ditches, are to be employed where needed.

Class III. Land of moderate to good productivity, suitable for cultivation with intensive practices, such as terracing, strip cropping, heavy fertilization, and installation of extensive drainage facilities. Combinations of these practices may also be used.

Class IV. Land of moderate productivity, primarily suitable for pastures and hay because of steepness of slope and critical erosion hazard. Such land may

23

occasionally be used for cultivated crops, but usually for not more than one or two years out of a 6- to 12-year period. It should be kept in grass most of the time.

Class V. Land not suitable for cultivation, but useful for grazing or forest. Normal grazing or forestry precautions must be taken to ensure sustained land use.

Class VI. Land not suitable for cultivation, but suitable for grazing or the growing of trees. Strict grazing or forestry precautions must be taken for sustained use.

Class VII. Land not suitable for cultivation, but suitable for grazing or forestry when used with extreme care to prevent erosion.

Class VIII. Land not suitable for cultivation, grazing or forestry, although frequently having value for wildlife or recreation. Such land is ordinarily extremely rough, steep, stony, sandy, wet, salty or susceptible to severe erosion.

Bennett's scheme may be modified to adapt it to local conditions: thus, Class I land could be used for high-value irrigated pastures, or a grass/legume pasture ley could beneficially fit into a rotation cycle with crops on Class II and Class III land.

Soil survey and classification

The soil survey can also be done by broad-scale reconnaissance, using aerial photographs and ground traverses to determine soil associations. For pasture development purposes, soil association mapping is usually sufficient until problem soils are revealed or soils are required for development of irrigated pastures when water relationships of the soils will need preliminary investigation. Detailed grid surveys are then required.

As soil is a product of the combined effects of geology (parent material), climate, vegetation, topography and time, it follows that a great diversity of soil types exists. However, there are sufficient similarities to permit the classification of the world's soils into groups which might be expected to behave similarly under similar conditions of land use, and maps of such soil groups are available. Different criteria have been used by various mapping authorities but each soil unit in any country can be distinguished sufficiently to fit it into a classification scheme if sufficient criteria are available for the soil in question. Basic data such as the physical description of the soil profile and some chemical tests for soil acidity, cations and organic matter are needed. Because of the major and minor earth movements over the centuries, folding and subsequent erosion of the original geological strata have produced abrupt changes in soil parent material and in topography even over short distances, so that the reliability of a soil map is only as good as the amount of detail recorded in the soil survey. Hence, published soil maps represent a compromise, usually at the soil group or soil association level, and soil treatment in land use must also involve compromise.

Soil maps on a world basis give an indication of what soil treatment may be necessary in a given country to match results attained on a similar soil elsewhere, but they are not a substitute for local research. Soil maps are also useful in plant exploration, as a similar suite of genera are usually found on similar soils with similar climatic elements elsewhere.

With the development of the modern fertilizer industry and, in the case of legumes, the appreciation of *Rhizobium* behaviour, it is often possible to modify the soil environment to accommodate the desired species. In many countries, however, the cost of modifying the soil environment with the use of fertilizers is at present of unproved economic benefit, so that the use of the soils in their virgin state, or manipulation of them within the bounds of existing economics, is all that can be achieved.

Soils in relation to legume adaptation

A knowledge of soils is necessary for pasture scientists, extension workers and farmers effectively to understand plant, and especially legume, performance. The following outline will describe the soils commonly encountered in tropical countries. The descriptions of the soils are given under the headings used in the Soil map of the world (United States Department of Agriculture). This soil map (see Figure 4) gives only very broad soil groups, and an attempt, therefore, has been made to adapt the groups to a broader range of tropical soils, representatives of which are described in *A handbook of Australian soils* (Stace, Hubble *et al.*, 1968).

25

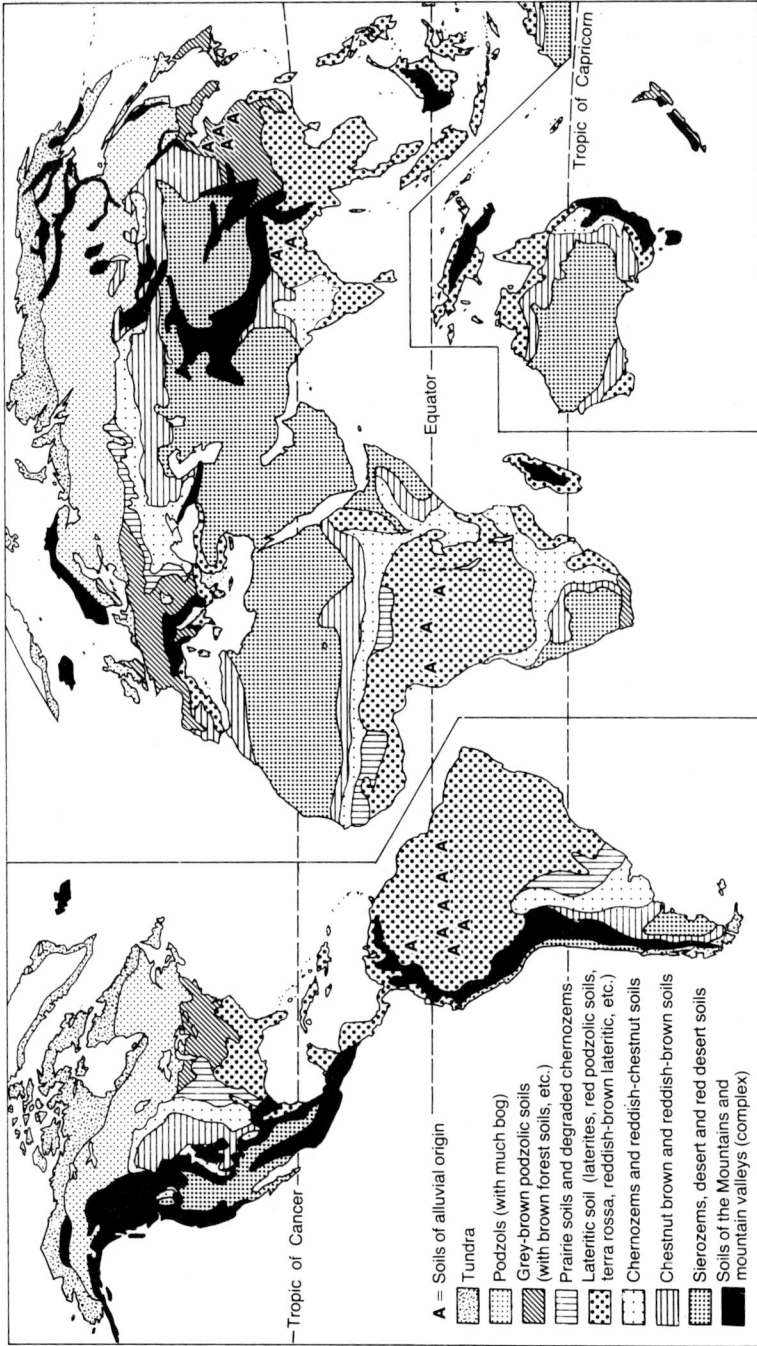

A = Soils of alluvial origin

Tundra

Podzols (with much bog)

Grey-brown podzolic soils
(with brown forest soils, etc.)

Prairie soils and degraded chernozems

Lateritic soil (laterites, red podzolic soils,
terra rossa, reddish-brown lateritic, etc.)

Chernozems and reddish-chestnut soils

Chestnut brown and reddish-brown soils

Sierozems, desert and red desert soils

Soils of the Mountains and
mountain valleys (complex)

Figure 4. Soil map of the world

Soils commonly encountered in tropical countries

Dark grey and black soils of the subtropics and tropics (vertisols). These are heavy clays, which crack when drying and swell when wet, enabling water to enter quickly and wet the subsoil. On wetting, the soil colloids swell, the cracks close and infiltration of water is subsequently slowed down, and either becomes ponded or runs off the land, causing sheet or gully erosion, or evaporates into the atmosphere.

Clay soils, however, can store a large amount of water against gravity and this principle is utilized in the system of bare fallowing for subsquent cropping (e.g. in the wheatlands of Australia and the United States, where rainfall during the growing season is marginal). In tropical countries the grass produced on these heavy soils is reserved for dry-season grazing, because there is usually sufficient soil moisture to support some growth into the dry season. Another factor in this system of grazing is that the heavy clays are often too wet in the rainy season and, if they are grazed, severe pugging and soil compaction adversely affect soil physical condition.

Although the heavy clays can store a good deal of moisture (in the range of 5 cm per metre of soil profile), the percentage of moisture at the wilting point is quite high (15 to 16 percent). Consequently, available moisture for plant growth between field capacity (about 33 to 35 percent water) and wilting point amounts to 18 to 19 percent or less than would be expected.

The heavy clays usually have a surface mulch of loose crumb-structured soil varying from 1 to 7 cm in depth. This ensures a well-aerated and receptive surface layer which quickly dries. It has been found that establishment of small seeded pasture species in such soils is difficult (Leslie, 1965). Rolling immediately after planting improves germination considerably. Larger seeded species can establish quite readily.

This group of soils contains several varieties, which can be placed in two main units: (*a*) grey, brown, red clays; and (*b*) the black earths.

Grey, brown, and red clays. These occur as broad plains or as bottom lands in a soil catena (e.g. the *mbuga* of East Africa). In the more arid regions the brown and red clays predominate.

The better drained types usually carry good-quality natural grasslands (e.g. the *Astrebla* or Mitchell-grass grasslands of central Queensland). They have a very loose surface mulch; the surface soil is either slightly acid or slightly alkaline, becoming more alkaline with depth; and the lower profile contains soft lime below which there is often an accumulation of gypsum. They are

27

usually very fertile soils and in areas of better rainfall are utilized for cropping. Natural legumes belonging to the genera *Rhynchosia, Glycine, Psoralea,* and *Medicago* (in winter rainfall areas) occur, but generally rather sparsely. Improved pastures can be established with legumes requiring or tolerating high levels of calcium such as lucerne or alfalfa (*Medicago sativa*), the annual medics (*Medicago scutellata, M. truncatula, M. polymorpha*) where there is some winter rainfall, as well as *Neonotonia wightii, Lablab purpureus, Macrotyloma africanum* and *Dolichos sericeus (formosus)*.

The wetter type of grey clays often have an acid or slightly alkaline surface soil and are strongly acid at depth. There may be mottling in the subsoil (often a sticky plastic clay). The surface microrelief may be uneven with alternating hummocks or depressions (gilgais).

These soils are brought into cultivation with difficulty, as there is only a short period when the soil is in a satisfactory physical state in which to prepare a seed bed (it is often either too wet or too dry and frequently is deficient in phosphorus). Species tolerating wet conditions (e.g. *Aeschynomene*) may have a place here.

Black earths. These are usually extremely fertile and where rainfall is adequate they are used for extensive cultivation of sorghum and pulse crops, and wheat where the climate is suitable. They occupy large areas of the Sudan, Ethiopia, Kenya and Tanzania, central India, central Queensland and smaller areas in Central America. In Africa they commonly carry leguminous thornbush browse in their native state and many of the game parks have been established on such sites. In other areas grassland is a common feature. Basalt and alluvia are common parent materials.

Timing of cultural operations is important and rolling after seeding is beneficial.

Soil phosphorus, calcium, and potash are usually in good supply and animals grazing on these lands normally develop into big-framed beasts if the land is not overstocked. Deficiencies of zinc may occur under high phosphorus and calcium, and sulphur is sometimes limiting.

Suitable leguminous species. Leguminous species suitable for this soil in the subtropics include lucerne (*Medicago sativa*) and the annual medics if summer temperatures are not too high and there is some winter rain; in the tropics *Neonotonia wightii, Dolichos sericeus (formosus)* and the annuals, *Lablab purpureus, D. uniflorus* and *Cajanus cajan.* Suitable grasses include *Panicum maximum* var. *trichoglume, P. coloratum* var. *makarikariense* and *Chloris gayana.*

28

Under irrigation these soils can produce high-yielding pastures but careful levelling of the soil must be achieved to prevent waterlogging.

Prairie soils, similar to the black earths and sometimes included with them, often occupy the slopes above a black earth plain. They have a mildly acid to mildly alkaline dark friable loamy surface soil with a good crumb or granular structure. Usually not deep, they grade from a darker surface into a yellow-grey, yellow-brown or reddish-black clay subsoil, and do not contain visible lime. They occur frequently on basaltic slopes in southeastern Queensland and in Costa Rica, where they are planted to sugar cane, and in northern Tanzania, where maize and beans are grown.

They are most generally quite fertile, but may give responses to molybdenum, sulphur and zinc.

Rendzina. Similar to the black earth in colour, this soil is a little finer in texture and not so deep; it is developed from limestone or amorphous lime. In northern Tanzania it is highly prized for cotton cultivation. Its shallow depth makes it more susceptible to dry conditions than the black earths, and soil conservation measures are essential. Similar species to those for the black earths are used, although deep-rooted plants such as lucerne would have a short life because of limited soil moisture.

Brown and reddish-brown soils (alfisols). These soils are more common in the Mediterranean type of climate. They comprise large areas of wheat-growing soils in southern Australia, southern Africa, South America, the central United States and the Mediterranean region. Vegetation is commonly savanna woodland.

Red-brown earths. These soils usually consist of slightly acid to neutral reddish-brown loams or clay loams, overlying a sharply defined blocky calcareous clay subsoil. Where they occur on sloping land they are highly susceptible to erosion. On wetting, the soil surface quickly seals, thus preventing further absorption of rainfall.

They are generally deficient in phosphorus and nitrogen, and the addition of superphosphate gives spectacular response in legumes. A large area of subterranean clover (*Trifolium subterraneum*)-based pastures in southern Australia occupies this soil group. Surface soil structure is considerably improved under pasture.

Sierozems, desert and red desert soils (aridisols). These soils occupy arid regions and are not important in sown pastures. In Kenya and the United States, reseeding of the arid ranges is accomplished mainly with drought-resistant grasses.

Sierozems. Heavy grey clay soils which crack extensively on drying; they occupy some drainage lines in arid and semi-arid regions where spasmodic flooding occurs. In southwest Queensland a sierozem-like soil (classified as a grey clay) extremely high in available phosphorus (up to 1 000 ppm) supports an extensive growth of the annual legume *Trigonella suavissima* when adequate autumn rains occur.

Desert loams. Common in the drier tropics of West Africa, the Sudan, Kenya, Tanzania, India and Australia, they usually have a compact, brown to reddish, neutral to alkaline loamy surface grading into a paler, vesicular horizon above the deeper clay; erode easily in heavy storms; in Africa they are vegetated quite often by thorny *Acacia* spp. Management consists of conservation of the natural vegetation, especially by control of stocking and water distribution.

Siliceous sands. Three types are common in the arid and semi-arid regions: a deep sand (locally known as *goz*) derived from sandstone found throughout Africa just south of the Sahara, granitic sands, and red siliceous sands characteristic of the deserts of central Australia.

The *goz* sands of Africa support a natural leguminous flora of various *Acacia* spp. including the gum arabic tree (*A. senegal*), and deep-rooted perennial legumes such as *Stylosanthes mucronata (fruticosa)*, *Indigofera* spp. and annual *Indigofera*, *Tephrosia* and *Alysicarpus* spp. (Skerman, 1966).

The red siliceous sands of central Australia occur in the 250-mm rainfall zone as sand dunes, the mobile crests of which are initially colonized by the legumes *Crotalaria cunninghamii* and *C. novae-hollandiae* (Skerman, 1947). Fertility is low, with deficiencies of nitrogen and phosphorus, but leaching is minimal. These deep sands are capable of absorbing storm rains quite rapidly and most of this water is available to deep-rooting perennials. The percentage of available water is only about 3.5 to 4 percent of the wet soil, but soil moisture at wilting point is only about 2.5 percent, so a deep wet soil is able to contribute considerable moisture to the plant roots.

Decomposition of acid granites gives rise to the coarse gritty sands which

30

are common in the semi-arid areas of the Sudan, central Tanzania and southern India. Water-holding capacity of these soils is low and erosive capacity is high. They are infertile and are normally vegetated by sparse woodland. In the Sudan and Tanzania these sands often carry a leguminous suite of native legumes, the most important of which is *Stylosanthes mucronata (fruticosa)*, a deep-rooting perennial which can extend into areas with a rainfall as low as 275 mm annually, falling over four summer months. Similar siliceous sands and weakly podzolized sands occur in higher rainfall regimes.

Latosols (oxisols), red-yellow podzolic soils (ultisols) — (with inclusions of hydromorphic soil, lithosols and regosols). This group of soils is by far the most important in potential for pasture improvement. Both the latosolic and podzolic soils occupy a huge area in the southeastern United States, Brazil and Venezuela, central Africa, southern India and Sri Lanka, Southeast Asia and Australia. In their natural state many are too infertile for sustained agriculture and remain under forest cover. With the addition of the necessary fertilizers and selection of suitable cultivars of legumes and grasses, excellent pastures can be established, the economic viability of which depends on relative fertilizer and livestock product prices.

Red earths. These occur in all climatic zones from arid to humid, but are used for pasture improvement only where there is sufficient annual rainfall — usually more than 625 mm. Native vegetation is usually woodland or forest; soils are porous red to red-brown sandy to loamy with clay content increasing with depth. Mildly acid, they may contain small amounts of ironstone (ferromanganiferous) nodules on the surface and throughout the profile. A common type, referred to as "lateritic red earth" or "ferruginous latosol", has a dense horizon of ironstone nodules or reticulate sandy laterite at a depth of 30 to 60 cm below the surface. These soils occur extensively in western Kenya, where the nodular layer is referred to as *murram*, and in the southern highlands of Tanzania, in central western Queensland and in Brazil and Venezuela. In Kenya and Tanzania they were used for the production of wattle bark from *Acacia* spp. and in western Queensland they support dense scrub lands of the edible browse, *Acacia aneura* (mulga). Many of them develop a surface crust, especially if subject to wind erosion, which seals the surface and limits water infiltration. They are usually deficient in available phosphorus (less than 20 ppm) and fix a large proportion of the added phosphorus, rendering it unavailable to plants; they are often very acid and manganese toxicity in legumes is common. Manganese-tolerant species such as *Stylosanthes* should be used if soil liming to increase pH is not practicable.

31

Krasnozems (latosols, red loams). These red friable clay soils derived mainly from basalt are deep and well drained with little differentiation of the profile, though the surface few centimetres are dark reddish brown, owing to higher organic matter content. A few ferromanganiferous nodules may be present.

Natural vegetation is frequently tropical rain forest, and immediately after clearing, the soil is very fertile and supports a vigorous weed population. Most of the tropical food crops are grown on this soil type. Fertility decline is rapid and the addition of nitrogen, phosphorus, sulphur and molybdenum is soon required. *Neonotonia wightii* has proved most adaptable to this soil and grows vigorously if supplied with phosphorus and adequate amounts of molybdenum. *Macroptilium atropurpureum*, *Centrosema pubescens*, *Stylosanthes guianensis* and the *Desmodium* spp. also perform well. Companion grasses can be *Setaria sphacelata* and *Panicum maximum* cultivars in the higher rainfall areas, and *Panicum maximum* var. *trichoglume* and *Chloris gayana* in the drier belt.

Deep yellow-earth variants of the krasnozems occur, indicating slightly more restricted internal drainage, but otherwise the soils respond similarly to the krasnozems.

Red-yellow podzolic soils. These are soils with an acid sandy surface soil from 15 to 30 cm deep overlying a friable clay subsoil exhibiting definite structure. There is a clear boundary between the surface and the subsoil and a slightly paler band may occur just above the clay containing small ferromanganiferous nodules.

If the clay is red, the soil is classed as red-podzolic, if yellow, as yellow-podzolic. The subsoil colour is usually determined by drainage: red-podzolics occur on well-drained crests, the yellow-podzolics farther down the slope, and humic gleys may occur at the valley floor. All podzolic soils are deficient in nitrogen and phosphorus and often in potash, copper (Teitzel, 1969a), zinc and boron.

Some weakly podzolic soils with a deep coarse sandy profile and only a small amount of clay above the parent rock are found on granite soils and even on coral in higher rainfall areas near the coastline. They are intermediate between the siliceous sands and the podzolics. In tropical coastal areas these coralline sands, which are actually acid in reaction, are used for the cultivation of coconuts and cashews. In the natural state they have a leguminous component of *Rhynchosia, Dolichos, Stylosanthes, Tephrosia, Lablab* and *Indigofera* (Schmidt, 1967). When phosphorus is supplied, this

32

leguminous flora responds markedly and dominates the vegetation under coconut trees (Schmidt, 1967; Anderson, 1968a).

Lateritic podzolic soils are usually more sandy than the red and yellow ones.

The salt-affected soils. In this group of soils the downward movement of colloidal clay has resulted from dispersion of soil colloids by the sodium ion. They include soloths, solodized solonetz and solodic soils.

The soloths have affinities with the podzolics in that the basic cations have been leached below the profile so that the latter is acid throughout, but the structural effects of the sodium ion still persist.

The solodized solonetz and solodic soils, on the other hand, have not been as severely leached, so that they are acid in the surface soil and the upper subsoil and strongly alkaline, because of the presence of the sodium ion, in the subsoil. They occur as extensive plains, on valley floors and on the lower slopes in rainfall areas ranging from 381 to 1 106 mm per year. They have a moderately acid, grey to grey-brown sandy loam to loamy surface soil with dense grey-brown, yellow-brown or red clay subsoils. Just above the clay is a distinct bleached band at the base of the surface layer, which may contain ferromanganiferous nodules. In the solodic soils the dense blocky clay subsoil has a level surface, whereas in the solodized solonetz soils it is arranged in round-topped prisms. In the deeper subsoil soft lime is usually encountered.

These soils can be difficult to bring into production because of extremely low fertility and bad physical characteristics; in the wet season they become waterlogged above the clay, and in the dry season the surface is very hard, making seed-bed preparation difficult. They seal on wetting. They are markedly deficient in nitrogen and phosphorus and trace elements, and often are of low calcium status.

In frost-free areas, Townsville stylo (*Stylosanthes humilis*) occurs naturally and can be easily established (Figure 5). Performance is enhanced by applications of molybdenized superphosphate.

Coastal saline soils, or solonchaks, have been derived from alluvial material which has been highly salinized by seawater. They usually are devoid of all edible vegetation and support only a sparse population of halomorphic plants.

In the tropics these soils are at present unused, but reclamation by drainage, irrigation and use of gypsum followed by initial use of salt-tolerant plants, including lucerne, as has been developed in the Netherlands, is a possibility for the future.

33

Figure 5. Townsville stylo (*Stylosanthes humilis*) establishing naturally at Townsville, north Queensland, Australia (latitude 19°S)

Hydromorphic soils. These are organic soils restricted to low-lying areas with the groundwater table near the surface. They are common on flat coastal and estuarine areas. They range from sands (dominant) to clays and the native vegetation is low woodland or heath. The soils are usually strongly acid with a dark organic surface layer merging into grey wet subsoils of sandy clay mottled with red and yellow streaks. The deeper subsoil is permanently waterlogged and consists of grey or bluish-grey clay.

These soils are most infertile, but with good rainfall and situated close to coastal areas, they may justify development. The *wallum* country of southeastern Queensland, which has been successfully developed for improved pastures, contains much humic gley soil. CSIRO workers have shown that the initial fertilizer dressing for establishment of legume-based pastures is 550 kg lime, 616 kg single superphosphate, 132 kg potassium chloride, 8 kg copper sulphate, 8 kg zinc sulphate and 55 g elemental molybdenum per hectare (Bryan, personal communication) with a subsequent annual maintenance need of 265 kg single superphosphate and 132 kg potassium chloride per hectare.

With this treatment, white clover (*Trifolium repens*), *Lotononis bainesii*, *Desmodium* spp., *Macroptilium atropurpureum* and *Vigna luteola* are among the pasture legumes successfully established in this frost-prone environment.

Soils of the mountains and mountain valleys (complex). These soils are mapped as a complex and include the vast Andean chain in South America, the pumice mountains in Guatemala, the southern Himalayan chain in the Indian subcontinent, which represent high altitude temperate zones in the tropics, and certain highlands in Papua New Guinea. Some of these soils are covered by rain forest in the tropics and belong to the krasnozem group. In Guatemala, in the southern highlands of Tanzania and in the Nakuru area of Kenya they are pumice soils derived from volcanic products. In most other areas they are shallow, stony soils developed frequently on severely metamorphic rocks.

Lithosols. These stony or gravelly soils are generally shallow with no development of profile. They occupy the upper slopes and mountainous areas. Their fertility depends on the character of the parent material. Under a sound land-use system these soils would be best left under natural vegetation to protect the soil and the watershed generally; however, some have already been cleared for pastoral production. They are unsuitable for cultivation except by extensive terracing.

Alluvial soils. This group of diverse soils, because of their high fertility, are much more important than their limited geographical distribution. They occupy the valley floors within most of the soil groups already discussed. The degree of fertility depends on the types of parent materials from which they are derived and the extent of prior utilization. They are topographically suited to irrigation and mechanization and so usually have a prior demand from food and fibre crop production. To justify their use for pastures, they are devoted to higher value animal production such as fresh milk, fat lamb or veal production under intensive management and heavy use of fertilizer. The cultivation of lucerne or alfalfa (*Medicago sativa*) primarily as a hay crop but with some grazing is a common form of land use.

3. Land clearing

Competition for light, water and nutrients must be eliminated or reduced to a point where productive pastures can be established and can survive with other vegetation. In some types of woodland savanna and open forest it is possible to establish pastures without any clearing. However, clearing of all vegetation from the area is generally recommended to give the new pasture the best conditions possible for successful establishment. Clearing of all vegetation from the grazing area, however, is not recommended: trees should be left for stream-bank protection, shade, windbreaks, fencing material, firewood, and as wildlife sanctuaries as well as for aesthetic effect.

Clearing of the vegetation is done either with hand-tools or by specially equipped tractors. The modern trend is toward large-scale mechanical clearing, but local considerations may decide that hand clearing is more acceptable. In all cases, any millable timber should first be harvested and sold to offset some of the costs of pasture development.

Hand clearing methods

The approach to hand clearing depends upon the type of vegetation to be removed, the time factor and the amount of clearing required. If the time factor is not important, ringbarking or poisoning of the standing timber can precede by some years its subsequent removal, but usually the time factor necessitates the clearing of green timber.

Hand clearing of standing green timber is usually done with primitive tools such as the machete, panga or *hengo* in developing countries, or by axe, chainsaw and winch in other areas. Where shifting agriculture is practised the larger trees are usually left untouched or lopped to provide fuel for the subsequent fire. Stumps are usually not removed.

In tropical rain forests, the undergrowth of shrubs and lianas is first cut down with a brush hook or similar hand instrument and then trees are felled with an axe or a chainsaw. In both cases considerable work can be avoided if trees are felled in a "drive" by cutting about half way through a series of smaller trees in a line and then felling a large tree across these in such a way that the larger tree will complete the felling of the smaller trees.

Trees with large buttressed roots must be cut above the buttresses with the aid of springboards fixed to the trunk to act as a platform for the axeman. In open forest or savanna woodland there is little undergrowth to contend with and the timber can be felled on a face. To produce a satisfactory result, any large shade trees or groups of edible shrubs should be left standing with sufficient cover around them to protect them from the subsequent fire.

The fallen mass is burnt when it is dry enough, but not too soon before the rainy season; otherwise the valuable ash mulch may be blown away before the pasture seed is planted. Usually burning is done one to two weeks before the rainy season is expected. If there is insufficient combustible material, burning can be delayed for a year to enable grass to grow through the fallen timber to provide more bulk.

For pasture establishment there is no need for further clearing of stumps, but if a mechanized cultivation cycle is envisaged, hand picking of fallen timber and stacking it around stumps is the preliminary operation prior to burning, winching or blowing the stumps out of the ground.

Two hand-clearing arrangements are worthy of mention.

By charcoal burners. In parts of Africa, charcoal is a major source of fuel and is sold by the roadside or in the village market. Before any land clearing is undertaken, the possibility of using some of the existing timber for charcoal should be considered.

In Kenya, land clearance has been successfully accomplished by allowing charcoal burners to clear land for pasture with the right of sale of manufactured charcoal. Specifications regarding clearing methods such as removing stumps to a specified depth and hand picking and burning all timber debris were agreed to by both parties under contract. In this way the persistent leleswha bush (*Tarconanthus camphoratus*) was effectively removed and giant star grass (*Cynodon plectostachyus*) established in its stead. Heavy initial stocking of the star grass to recycle and distribute the nutrients previously mobilized in the forest cover gave an excellent establishment.

By otherwise unemployed labour. Some governments marshal otherwise unemployed males, especially during the dry season, to undertake land clear-

37

ance schemes. Under good technical supervision, such schemes can give valuable leads in pasture improvement in the developing countries.

In all such medium-sized schemes for pasture improvement, it is important not to attempt to establish too large an area of pasture at one time; otherwise bush regrowth may be too vigorous to control in subsequent years. It is also important to establish a sufficient area to cater for the minimal needs of the herd. Otherwise in dry periods a good pasture may be irretrievably damaged by overgrazing at this critical time. At least 0.6 ha of improved pasture per head of stock should be sown.

Large-scale mechanized clearing

It is the large-scale land development work which has the greatest appeal for governments and large landholders because it offers a rapid increase in productivity in a relatively short time. Such large-scale schemes also attract finance, in the case of developing countries mainly from overseas, from such institutions as the World Bank.

Attractive as this kind of development is, it must not be lightly or hurriedly undertaken. A full-scale feasibility study of inputs and outputs should precede any on-the-spot activity. Again, care must be taken as to the scale of the operations to ensure maintenance of the resulting pastures.

As mentioned in Chapter 2, a complete series of aerial photographs should be available from which to plan operations.

Contract specifications. Most large-scale land clearance schemes are based on contract arrangements with private operators. If there are sufficient operators available, competition will favour arriving at a realistic cost.

Specifications for land clearing should be carefully and clearly set forth so that the contractor or government project engineer can clearly understand the job's requirements, its limitations, and exactly what is expected of the clearing operator.

Specifications should be exacting, but still allow some tolerance (e.g. rather than say "removal of all woody vegetation", the contract should read "removal of woody material above 10 cm in diameter to a depth of 20.3 cm below the surface"). Specifications should contain: (a) a clear description of the present condition of the land; (b) a clear description of the finished condition required, specifying maximum and minimum; (c) final completion date and method of payment; and (d) provision for a final check to ensure compliance with the contract.

The present condition of the land should be revealed by aerial photographs, maps and drawings showing location; size of the area; contours, if available; types and sizes of vegetation delineated on the maps; soil conditions (types: stony land, gullies, drainage conditions and water-table); location of streams, rivers, present and future roads and present and future drainage works. A rainfall incidence chart indicating average number of wet days is also a help. The most important detail is tree counts. These should be made for each type of vegetative growth present in the area. Besides the average number of trees per hectare or per acre, the count should include the size (diameter at breast height or immediately above buttress) and species of trees and other vegetation, especially vines. This tree count can best be obtained by random sampling of perhaps three areas 100 metres long by 10 metres wide. The "finished condition required" stipulation includes size and percentage of trees to be left standing as individual trees and as belts of trees for shade, for timber, windbreaks and stream-bank protection; wildlife sanctuaries; degree of stump and root removal required; degree of debris removal required; and amount of submarginal land which may be left uncleared. This condition may be difficult to arrive at for pasture establishment. Rain forest may need no more than felling and burning if weather conditions are right. Forest country will usually also require some windrowing, stick-picking, ripping and cultivation to suppress subsequent timber regrowth.

Specification of time should stipulate starting time, time for completion of portions of the work, and final completion of the job. The number of working days required for completion should be gauged so that the contractor is not penalized for adverse weather conditions.

In hiring a clearing operator, the following considerations should also be taken into account:

● The job should be big enough to permit the use of specialized equipment, efficient methods, competent management and supervision. A project large enough to write off a suite of machinery after completion is an advantage. Distances between maintenance facilities also seriously affect costs.
● The contract should be divided, if possible, into specialized sections — that is, clearing, logging, road construction, and so on — with each contractor doing the job he does best.
● The bidder's experience, reputation and financial solvency should be investigated.
● Both vague and unnecessarily strict specifications can cause the contractor to do a less efficient job.

- Unnecessary, submarginal land should not be cleared.
- Land should not be cleared until arrangements are complete for effective utilization and management.
- Wherever possible, a pilot project should initiate the work, to determine the most efficient and economical methods to help with future budgeting.
- Arrangements should be made to harvest all millable timber before clearing commences.

Types of country to be cleared. The types of country most likely to be cleared are:

— dense tropical rain forest, usually with an undergrowth of vines and lianas which, although growing in a very wet environment with frequent rainy days, is moderately well drained;

— monsoon forest, which is still dense, somewhat deciduous; soils are well drained, and rainfall less frequent;

— wet sclerophyll forest where there is little undergrowth but fairly dense timber and where there are swampy areas of various size from large ponds to small gilgais, and frequent rainy days;

— wet heath country, also subject to waterlogging and frequent rain;

— dry sclerophyll forest generally well drained, with infrequent rain to hold up clearing operations, and varying populations of trees;

— savanna woodland, of variable rainfall and tree density; and

— low thornbush savanna in low to medium rainfall areas.

Clearing of wet sclerophyll forest and heath country can be severely interrupted by bogging and isolation, resulting in costly delays. In these types of country it is essential to examine the areas closely before clearing and to lay out a drainage and road system. Road access should receive early attention in any clearing operation. Roads cut by bulldozer can also function as job boundaries and firebreaks in subsequent burning operations.

Clearing sequence. Clearing operations involve all or some of the following operations: pulling; windrowing before burning; burning; windrowing after burning; ripping, stick-picking and cultivation. The number of operations depends on the use to be made of the land, the rapidity with which establishment of pastures is required and the evenness of stand expected. Much topographically suitable land may end up as cropland or as a rotation of crops and pastures. In such cases, complete clearing ready for mechanized seeding is necessary. Even if crop production is not undertaken, forage or seed harvesting may be required and clean land is necessary for this. The presence of unburnt logs and timber regrowth also interferes with stock handling, weed control and seed and fertilizer distribution, so that in most cases it will ultimately pay to clean the land by windrowing at the outset. This also permits the use of a cheaper wheeled tractor instead of the more expensive bulldozer. In some rain-forest areas where a good burn can be achieved and rotting of timber is rapid and complete, windrowing may be eliminated or postponed to a later stage.

Removing standing timber. The modern approach to large-scale clearing involves the use of high-powered crawler tractors or specially built machines such as the electric tree crusher. Two high-powered bulldozers, 180 or 270 hp, working in tandem dragging a 275-m length of heavy anchor chain are a common unit for pulling over a swath of medium or dense forest respectively (Figure 6).

Figure 6. Two crawler tractors pulling a heavy anchor chain felling thick scrub (brigalow) in central Queensland, Australia (second tractor is out of sight to the left)

41

It is a mistake to use smaller machines; they waste considerable time felling the larger trees and cannot work as economically as more highly powered machines. The only advantage of a smaller unit is its manoeuvrability in awkward terrain. When operating bulldozers in tandem in wet sclerophyll or dry sclerophyll forest country where there is some undergrowth, a highball of 3-metre diameter fixed by swivel link to the centre of the anchor chain is necessary to keep the chain near the ground and not allow it to ride up the trunks to the larger trees and miss the smaller ones. In tropical rain forest and monsoon forest, a chain alone is more efficient because of the dense vegetation which is pulled down on a face. A heavy-duty canopy over each tractor driving seat is necessary to protect the driver from falling and springing timber. In all clearing operations, strips 40 to 80 m wide should be reserved intact to act as firebreaks, windbreaks, sources of fencing material, for shade and for wildlife sanctuary. The problem in such pulling operations is the large tree. If logging (or cutting for timber) has been previously carried out, many of these will have been removed, but other trees may not be suitable as timber species and must be cleared. Such trees can be left standing in the pulling operation but this involves some extra and difficult manoeuvring and increases costs. A third bulldozer with a tree-pushing attachment (tree-dozer) can be employed to trail along behind the tandem of tractors and assist where necessary with the larger trees.

Trees which are left standing can be kept for shade, blown out with explosives or later ringbarked and burnt. In the case of the tree-crusher, it is fitted with a tree-pusher front and its electrically driven steel roller drums are fitted with heavy steel blades so that it both pushes the trees and chops up the undergrowth in one operation at the rate of 1 to 2 hectares per hour, depending upon local conditions.

Large areas of brigalow (*Acacia harpophylla*) forest have recently been cleared in the 625- to 750-mm rainfall country in central Queensland, Australia. The dominant brigalow tree is rarely more than 30 cm in diameter and is the largest tree in the forest. There are usually some smaller trees as an understorey growing to a height of 5 to 6 metres with trunks up to 15 cm in diameter. For the tall forest with little undergrowth, two crawler tractors pulling a steel cable clear the land at 8 to 10 ha per hour. If there is also undergrowth, a cable and a chain are used, the shorter cable riding up the trees to pull them down and the larger chain taking care of the undergrowth (Johnson, 1965). Where the forest is light and not tall, the cable alone does the best job.

In areas where there is low thornbush and the country is considered suitable for pasture establishment, a tree-pusher root-plough machine can be used (Figure 7). At the King Ranch in Texas, United States, mesquite country is treated in this way.

Where tree density is low or in clearing small areas, a single bulldozer can be employed for pushing the individual trees. However, it must be realized that there is a lot of wasted time between trees when the bulldozer is not actually working, and it may still pay to employ two tractors and a chain.

In heath country, land clearing can generally be achieved by simply burning where possible and ploughing with heavy-duty stump-jump ploughs such as the shearer "Majestic" or with heavy-duty tandem disc cultivators with angling devices (Figure 8). In any case, final preparation is carried out with disc cultivating machines to chop up the heath vegetation.

Windrowing. Where the fallen material is dense enough to take a running fire, it is usual to burn it where it has fallen and as soon as it is dry enough. However, with some species (e.g. *Acacia harpophylla*, or brigalow), an early burn without subsequent cultivation has often resulted in excessive regrowth and a recommendation has been made not to burn for nine months after pulling (Johnson, 1965). Thus, some knowledge of the species involved is neces-

Figure 7. "Majestic" disc plough working out brigalow (*Acacia harpophylla*) regrowth in central Queensland (**Photo:** Shearer & Sons, Qld)

43

Figure 8. Front oblique view of the Holt tree-pusher root plough in operation at King Ranch, Texas, United States (**Photo:** King Ranch Ltd)

sary for intelligent management. Burning *in situ* prior to windrowing takes advantage of the abundant dead leaf material which may be lost in the windrowing operation, and the fire consumes the great bulk of woody material so there is less to deal with after the burn. However, waiting for a burn delays land preparation, a running fire destroys valuable humus, and some species (such as *Eucalyptus*) are difficult to burn with a running fire. Where these factors are a consideration, windrowing can begin immediately and the material can be burnt in windrows later.

The windrowing operation is best carried out with a heavy tine rake rather than with a bulldozer blade as a rake allows the fertile topsoil to move through the tines and be left in place, and the weight of debris moved is less. These scrub- or bush-rakes are of varying strength and width, and some are designed to carry timber as well as push it together into windrows (Figure 9). However, in heavy, wet clay soils, rakes can clog up; a bulldozer with an angle blade may ride more easily over the land, but it is better to wait until the soil will pass through the rake.

44

The distance between windrows affects costs. It should not exceed 140 metres. Windrows should also be arranged on the contour of the land so that they can protect it from erosion until the pasture cover has established itself.

Two scrub-rakes working side by side usually accomplish windrowing more easily. Dead running can be avoided if windrowing is done in a forward step-wise direction, reversing the movement when the end of the field is reached.

Root-ripping. This operation is mainly for clearing land for cultivation. Tractors fitted with heavy tines rip the soil to a depth of 20 to 45 cm to remove near-surface roots in advance of ploughing.

Root-raking and stick-picking. Following burning and ripping, there will always be short pieces of timber which have escaped all previous treatment and form a hazard for the operation of ploughing, cultivation and seeding machinery. Small pieces can be picked up by hand or raked together with root-rakes and stick-pickers and subsequently burned. The Wake root-rake operates like a hay-tedding machine and sweeps the sticks into windrows with a side delivery action, to be picked up or burnt *in situ*.

Figure 9. Scrub-rake ready to push smaller trees into a windrow. The tines also perform some ground tillage during this operation (**Photo:** Le Tourneau Foundation)

45

The cost of clearing land. From the foregoing discussion, it will be seen that the cost of land clearing is dependent on local conditions and costs at the time, and the soundest assessment can be arrived at by asking several contractors to tender for a given project. The following costs are listed only as a guide and must be considered as a quotation for a particular job in a particular place at a specified time. Inflation, rising costs and increasing wages quickly make such figures out of date.

1. Figures given by R.G. Le Tourneau Inc. for clearing 759 ha of jungle on the Uyacali river frontage at Pucallpa, Peru, during 1967/68 (on-site cost of diesel fuel 9 cents, rainfall 2 250 mm, 10 000 stems per ha).

	$US per hectare
Felling-crushing with Le Tourneau tree-crusher and burning in place, including supervision, administration and support vehicles	72.89
Raking, windrowing, reburning, disposal of root wads and unburned trunks by pushing into soft spots and depressions, and disc harrowing with Le Tourneau tractor, including supervision, administration and support vehicles	153.82
Total clearing costs per hectare	226.71
Road access to individual colonist's allocation, supervision, administration and support vehicles	62.76
Transport of machines to the job	3.10
Total cost of contract per hectare	292.57

2. Brigalow (*Acacia harpophylla*) forest in central Queensland (rainfall, 625-675 mm).

	$US per hectare
Pulling with tractors and cable or chain	5.00-12.50
Burning, and aerial sowing (without seed)	0.75-1.25

Brigalow burns very readily to a white ash and as the trees are not as large as those found in tropical rain forest, pulling with tractor and chain is easy. For pasture purposes the timber is usually pulled, burnt about nine months later, and sown a few days later by air with a pasture mixture. Hence, cost per hectare to establish pasture (including seed) is about $US12.50-20.00.

	$US per hectare
3. Clearing with a single bulldozer (D8-H crawler). Rain forest (north Queensland) pushing and windrowing	70.00
4. Pushing medium bush (Belize) 1967-70 (0.19 ha per hour).	43.15
Pulling, burning and windrowing (Belize)	167.70
5. Total site preparation from cypress/palmetto country in the southern coastal plain of Florida, United States (Yoho, Dutrow and Mark, 1969).	48.20

This cost was made up of 23.9 percent supervision, 65.1 percent equipment and 6.6 percent overhead.

Note: All figures given refer to costs at the time of publication of the studies quoted.

47

Seeding rate is 2.2 to 3.3 kg/ha seed pods. Seed-pod planting ensures sufficient hard seed to take care of any adverse weather conditions. The plant stand achieved in the first year usually increases about eightfold in the second year. An improvement in planting techniques is to fit a V-shaped light bulldozer blade to the front and loader arms of the tractor to push away small logs and other obstacles in the path of the tractor (Figure 10).

Under sound grazing management to keep the grass so controlled that Townsville stylo is not suppressed by shading, this legume will spread to the unplanted strips and a full cover will be established in about five years. Quicker cover can be achieved by planting a larger proportion of the area.

Risopoulos (1966) successfully introduced *Stylosanthes guianensis* into natural pastures of *Hyparrhenia* spp. at Gandajika, Zaire, by broadcasting after light cultivation with a harrow and after a somewhat heavier cultivation by a brush cutter. Stand counts obtained are presented in Table 4.1.

TABLE 4.1 **Stand counts with different land preparation and sowing methods**

Land preparation method	Broadcast	Broadcast after scarification with harrow	Broadcast after passage of brush cutter
	Seedlings/1 000 m^2		
Without burning; immediate grazing	1 565	1 744	2 678
After burning; immediate grazing	1 606	2 538	1 477
After burning; ungrazed for the first year	1 237	2 505	3 474

At Mvuazi, Zaire, Risopoulos (1966) established *Stylosanthes guianensis, Pueraria phaseoloides* and *Centrosema pubescens* successfully in sandy alluvial soils carrying dense *Imperata cylindrica* by burning the *Imperata* at the beginning of the rainy season, giving one single cultivation with a heavy-duty tandem disc "Rome" plough, broadcasting the seed and deferring grazing for eight months.

Strip planting into poisoned green timber. A successful trial with planting siratro (*Macroptilium atropurpureum*) in sandy podzolic soils has recently been concluded in central Queensland. The *Eucalyptus* woodland trees were poisoned with Tordon 50 herbicide and siratro was planted in cultivated strips at 8.8 kg/ha, double the normal planting rate. The poisoning of the trees by

Figure 10. Broadcasting Townsville stylo seed and superphosphate on to strips cultivated with scalloped discs mounted on three-point linkage. A light bulldozer blade is attached to the front-end lift to clear logs from the path of the tractor (**Photo:** Rockhampton Morning Bulletin)

basal bark injection reduced competition for water, light and nutrients and allowed the siratro to establish. It should be noted that Tordon herbicide has a residual toxic effect on legumes for up to six months after application and, therefore, spot treatment of trees and not an overall spray should be attempted. Tordon toxicity from frequent leaf fall is a possibility. With ringbarking and poisoning, there is a risk of heavy regeneration by seedlings in some types of vegetation as, for example, *Eucalyptus* spp.

The distribution of legume seed by the grazing animal. Legume seed frequently remains viable, and germination may even be enhanced after passage through the digestive tract of the grazing animal. Livestock grazing a ripened field of legume usually ensures its spread within and outside the grazing field. Hay containing ripe seed is also a source of seed distribution. White and strawberry clovers (*Trifolium repens* and *T. fragiferum*) are often spread in this way. Siratro and Townsville stylo seedlings are frequently found germinating in dung pats. If seed harvesting machinery is not available, the grazing of a harvest-ripe legume sward in conjunction with an area of native pasture will lead to seed distribution, but for successful establishment the fertility of the native pasture may have to be raised by broadcasting fertilizer prior to grazing the legume sward.

Lamprey (1967) has suggested that the seeds of leguminous browse whose pods do not dehisce naturally usually require passage through the animal gut for germination. These are the very palatable species such as *Acacia albida*, *A. tortilis*, *A. sieberiana* and *A. nilotica*. Sticky seeds pods, such as those of *Desmodium uncinatum*, and spiny pods, such as those of *Medicago polymorpha* and *M. minima*, are spread by adherence to the animal body and subsequent removal.

With partial clearing

Establishment in the ashes following burning of timber. This is a common pasture establishment method where dense rain forest or scrub has been cleared and burnt without subsequent removal of fire debris and stumps. A hot fire will reduce the fallen timber to a white ash; a less intense fire may leave a good deal of debris.

Fire destroys much weed seed, releases nutrients previously mobilized in foliage and generally provides an ashy seed bed suitable for seedling emergence.

The pasture mixture is broadcast by hand or from aircraft or, if the burn is clean enough, by machine. Broadcasting from a height tends to bury the larger seeds in the ash.

Seeding should take place as soon as the ash is cool enough — usually two to three days following the burn — so that a pasture is established before ash is blown or washed away. Burning and seeding should be done just prior to the usual onset of seasonal rain.

Most of the pastures so far established by aerial seeding into ashes have been based on grasses such as *Chloris gayana, Melinis minutiflora, Panicum maximum, Cenchrus ciliaris* and *Hyparrhenia rufa* because of their quick establishment and subsequent suppression of weeds. Few legumes have been tried, but siratro (*Macroptilium atropurpureum*), lucerne (*Medicago sativa*), *Stylosanthes guianensis* and *S. humilis* have succeeded. Where inoculation of the seed is necessary, pelleting after inoculation (Chapter 7) will help to keep the *Rhizobium* organisms viable for some time after seeding into dry ashes, and will protect the seed from harvesting ants. Normally, *M. atropurpureum*, *S. humilis* and *S. guianensis* do not need inoculation.

Weed growth and often bush regrowth are rapid shortly after a burn and it is, therefore, important to obtain an early and dense cover of improved pasture. It is often recommended that windrowing and cultivation follow a burn and that accurate seeding and fertilizing by ground machine be practised. Land so cleared also lends itself to mechanical slashing for weed control.

On a prepared seed bed

The most successful pasture establishment has been on prepared seed beds. Generally, the better the seed-bed preparation, the better the pasture establishment; also, the smaller the seed to be established, the finer the seed bed should be.

The larger seeded legume species such as *Lablab purpureus, Macrotyloma uniflorum, Centrosema pubescens* and siratro can be satisfactorily established with a minimum of seed-bed preparation (in which cases a higher seeding rate is employed). It is important to determine such minimum seed-bed preparation levels, as each successive cultivation of the soil will add to establishment costs. On the other hand, if seed is costly and a permanent pasture is being established, the better establishment and persistence due to improved seed-bed preparation will generally prove worth while.

53

Figure 11. Contour cultivation strips on to which Townsville stylo (*Stylosanthes humilis*) seed was broadcast at Rockhampton, central Queensland, Australia (**Photo:** Rockhampton Morning Bulletin)

54

Establishment on a fully cultivated seed bed on new land

Pasture establishment is usually more successful on new land because of the higher initial fertility (especially nitrogen) after recent clearing and less competition from weed growth.

Assuming the new land previously carried rain or monsoon forest or scrubland, the vegetation is pulled or felled, then pushed into windrows. Windrowing on the contour will ensure subsequent contour cultivation which will protect the soil against erosion. If this windrowing is done with a heavy tined root-rake, considerable surface cultivation will have been done by the tines in the process. If surface or near-surface roots are extensive, deep ripping with heavy tines, preferably in two directions at right angles, will remove the roots, which can then be raked together with a root-rake and placed in the windrow.

Number of operations. The first cultivation should be given by a heavy-duty stump-jump disc plough which can remove the remaining small stumps and roots and prevent a good deal of bush regrowth.

After this treatment, the seed bed is prepared by a set of disc harrows which can chop small pieces of root and stems so that they constitute no hazard to seeding machinery. In the process of disc harrowing, the uneven microrelief common on newly cleared land is gradually levelled. If a finer and more level seed bed is desired, a subsequent harrowing can be given.

Seeding is carried out preferably with a combined seed and fertilizer drill for accurate measurement and placement of seed and fertilizer, a disc drill being preferred to a tine machine at this stage because there is less chance of blockage by small sticks and debris.

Placement of seed and fertilizer at different depths is an advantage, particularly with inoculated seed which is affected by acid fertilizers.

Spinner-type seed and fertilizer broadcasting machines can be used where the ground is too uneven, too steep or where there is too much debris for drill operation.

For small areas, row seeding may be accomplished with a hand planting machine of the "Planet Junior" type. Glycine is often sown in Brazil in this way.

Small spinner-type hand broadcasting machines can also be used for seed fertilizer application.

Canvas bags holding 2 to 4 kg hanging from the shoulder and covering the

55

Figure 12. Planting pasture seed on a fine, well-prepared seed bed with a combined seed and fertilizer drill

chest are attached to hand-operated spinners that broadcast the seed or fertilizer centrifugally.

Depth of planting depends on seed size. Very small seeds such as *Lotononis bainesii* should never be covered but broadcast on the surface; large seeds such as *Lablab purpureus* can germinate from deep planting. The usual planting depth is 1 to 2.5 cm for most species.

For high germination, the soil particles should be packed firmly around the seed. To accomplish this, small press wheels can be attached at the rear of each drilling tube or a rubber-tired roller made of old motor-car tires can be attached behind the drill. A fluted Cambridge-type metal roller can be used in a separate, slightly delayed operation in moist clay soil as it is likely to gum up with clay if used attached to the seeding drill. In tropical areas the time of seeding is important. Most seed is planted dry just ahead of the expected rainy season. Seeding after rain is more successful with large-seeded legumes and when rolling follows seeding.

56

Establishment by full seed-bed preparation of previously cultivated land

The weed problem. Many areas in the tropics have been abandoned after a series of cropping cycles and have become overrun with weedy herbs and grasses which present severe competition to the establishment of a new pasture. The soils are also depleted of nutrients, especially nitrogen, making pasture establishment difficult without the addition of fertilizers. Abandoned cultivations in the tropics are often covered by such low-fertility-demanding species as mat grass (*Axonopus affinis*), carpet grass (*Axonopus compressus*), couch grass (*Cynodon dactylon*), lalang, kunai or blady grass (*Imperata cylindrica*), or a low-producing matted turf of *Paspalum dilatatum*. These persistent grasses are difficult to eradicate without thorough cultivation. There is also a suite of tropical annual and perennial dicotyledonous weeds. Deep ploughing with a disc plough followed by successive treatments with a rotary hoe will give maximum destruction, but small pieces of stolon often survive and compete with the new pasture. The renovation effect of one ploughing will often stimulate the grass (especially *P. dilatatum*) to much more prolific growth, as the soil disturbance usually releases some mineralized nitrogen. On the steeper hillsides, a tractor with caterpillar tracks (crawler type) and a low centre of gravity should be used, working around the contour (to protect both the tractor from overturning and the soil from eroding during heavy rains). Some initial strip seeding — planting the intervening unploughed land at a later date when the first strips are established — is a worthwhile soil protection measure. One or two cultivations, a week after useful falls of rain will help to reduce the weed seedling population.

Use of herbicides. In some cases, a blanket cover of herbicide can be applied to the prepared ground as a preplant weed-killer. Bailey (1967a) showed that better results were obtained with herbicides sprayed on the germinating seedlings rather than on the prepared seed bed. For guinea grass/*Centrosema pubescens* establishment, he used 2.2 kg 2,4-D amine at 2.2 kg acid equivalent per hectare in 90 litres of water at 2.25 kg per cm^2 pressure on a red latosol derived from basalt. The toxic effect on *Centrosema* disappears after seven days and, therefore, planting of the pasture should be delayed for at least seven days and preferably for two weeks after treatment with 2,4-5-T. *Stylosanthes guianensis* seeding should not take place until 33 days after treatment (Bailey, 1965c). In the West Indies (University of the West Indies, 1963), it was shown that 2,4-D, ametryne, linuron and 2,2-DPA were promis-

57

ing preplant herbicides to use before sowing puero or tropical kudzu (*Pueraria phaseoloides*).

As plants react differently to herbicides, it is important to seek the advice of herbicide specialists as the planting season approaches to ascertain the best herbicide to use for a particular species or mixture, the correct dosage required, and the time and method of application. New herbicides are appearing frequently and knowledge of their effects on plants is increasing from day to day.

When using preplant herbicides, it is also advisable to spray around logs, along headlands and other sources of weed seeds to reduce the weed population around, as well as within, the pasture.

Legume establishment in a nurse, cover or companion crop or in stubble

Establishment of pastures with short-term crops has been practised for a considerable time. The use of oats as a nurse crop for lucerne or alfalfa (*Medicago sativa*) is still commonly practised. The idea is to provide a more favourable microenvironment, especially in terms of humidity, for germination and early establishment of the seedling. Other advantages are that, with a fast-growing crop, some feed can be obtained earlier without damaging the legume and without putting land out of production for a season. There is also some protection of the young seedling from extreme heat or frost. On the debit side, there may often be severe competition by the companion crop for water, light and nutrients.

Risopoulos (1966) successfully established *Stylosanthes guianensis* in alternate rows with sorghum (*Sorghum vulgare*) in Zaire for use first as a combined silage crop with the sorghum for one or two years, after which time the sorghum disappeared leaving a pure *Stylosanthes* pasture for grazing, hay or seed production. The rows of sorghum and *Stylosanthes* were 0.60 m apart.

In Brazil, it is common practice to plant *Lablab purpureus* with maize, using two-thirds by weight seed of maize and one-third *Lablab*. *Lablab* is a slow starter, and when the maize is harvested the *Lablab* is seeding. Cattle are turned in to the harvested maize and graze the *Lablab*, tramping a quantity of the seed into the ground, which serves to regenerate the *Lablab* for another year.

The sowing of pastures for livestock is difficult to fit into small subsistence farming. Staples (1938) tried to introduce pastures into a peasant holding of 20 acres, four of which were cultivated for food crops by a man and his wife.

58

He found that the work of clearing, planting, hedging and controlling bush regrowth for the first four years to establish the uncropped area under pasture was beyond the capacity of the family; once established, however, the pasture could thereafter be kept clean by the family.

In such a situation, the most likely means of pasture improvement is to sow a pasture in the final crop before abandonment.

Thomas and Humphreys (1970) successfully planted *Stylosanthes guianensis, S. humilis* and molasses grass with rain-grown rice at 0.3 × 0.3-m centres in the ashes of a forest burn in Laos (lat. 17°N). There was no significant decrease in the yield of rice, and the pasture was well established when the rice was harvested. Almost two years later the legume component was 41 percent and the weed component only 7 percent.

Introduction of a pasture into the final crop in a cropping cycle presents two major problems. By the time fertility has reached the point at which cropping is uneconomical, the weed problem is often severe and the low soil nitrogen status makes establishment and early development of the pasture difficult. Initial help with fertilizer to make up for nitrogen deficiency would be justified if a permanent legume-based pasture could be established to replace the traditional bush regrowth cycle.

McWilliam and Dowling (1970) have shown that the germination of subterranean clover seed is enhanced by sowing it into standing millet straw. The straw increased the relative humidity in the vicinity of the seed by as much as 5 percent. Rickert (1970) has found that pasture establishment under sorghum mulch is better than under a cover crop such as oats. Stubble and mulch both improved the water relations of the seed.

Sod-seeding legumes into existing pastures

The sod-seeding technique was developed by Breakwell and Jenkins (1953) to improve declining *Paspalum dilatatum* pastures on the northeastern coast of New South Wales, Australia. The sod-seeder (Figure 13) is a heavy-tine seeding implement which can cut a furrow into and beneath a matted grass sod and introduce seed and fertilizer. The greatest success has been achieved in introducing winter-growing species into a relatively dormant summer-growing pasture in autumn — for example, crimson clover (*Trifolium incarnatum*) and common vetch (*Vicia sativa*). Large-seeded species establish better than small-seeded ones.

Murtagh (1963) was successful in sod-seeding *Neonotonia wightii* into a

59

Figure 13. "Grasslands" sod-seeder oversowing legume seed into native pasture (**Photo:** Grasslands Farming Equipment Ltd)

pasture of *Paspalum dilatatum* and *Axonopus compressus*, which was suppressed by application of a herbicide mixture containing 27.7 percent acid equivalent (AE) of the sodium salt of 2,2-DPA (2,2-dichloropropionic acid) and 12.5 percent AE amitrole (3-amino-1,2,4-triazole). The inoculated glycine seed was planted at the rate of 5.3 kg/ha drilled with 480 kg/ha of a neutral fertilizer containing equal amounts of molybdenized superphosphate and dolomite, plus 2.2 percent sodium nitrate. Yields of glycine increased in proportion to the rate of herbicide applied up to 8.2 kg AE per hectare. Glycine failed to establish when sown into the grass which was not treated with the herbicide.

Luck and Douglas (1966) recorded success in introducing *Neonotonia wightii* and *Macroptilium atropurpureum* by sod-seeding these legumes into a *Paspalum dilatatum/Axonopus affinis* sward sprayed with 5.5 kg/ha dalapon. Yields approached those resulting from sowing on a prepared seed bed and were better than those recorded from sowing on once rotovated strips.

60

Choice of species

The aim of improved pastures should be to provide an adequate diet for the grazing animal throughout the year, with particular emphasis on the seasonal curve of animal requirements. Only in favourable climatic conditions is this possible. Individual species are usually not able to accomplish this end and so a suite of species is selected, each to fill a niche in a "feed-year" programme. Thus, winter-growing species would follow summer- and autumn-growing types where climatic conditions permitted. In the tropics there is often a pronounced dry season which coincides with a period of cooler weather so that continuity of feed is a problem. Selection within species can prolong growth into the cooler weather; for example, 'Tinaroo' glycine continues growth longer into the autumn than the cultivars Cooper or Clarence, and so is used as "autumn-saved" pasture. *Lablab purpureus* can provide feed in the autumn after the summer species have finished growth and before a winter grazing crop of oats is ready. Nitrogen applied to summer grasses in autumn helps to prolong the period of adequate feed. In other cases, a plant which "hays off" into the winter or dry season can be used to fill the gap; for example, Townsville stylo (*Stylosanthes humilis*) and common stylo (*S. guianensis*) become more palatable as they dry off and can be grazed as standing hay during the dry season. Norman and Stewart (1964) used standover bulrush (*Pennisetum typhoides*) or pearl millet, supplemented with Townsville stylo hay for dry-season feeding at Katherine, Northern Territory, Australia. Risopoulos (1966) used silage from a mixture of *Sorghum vulgare* and *Stylosanthes guianensis* for dry-season feeding in Zaire.

A knowledge of the growth rhythm of each species at a particular site is important so that mixtures can be arranged to give continuous grazing. Lovadini and Miyasaka (1968) have given the performance of *Neonotonia wightii* (Figure 14) at Campinas, Brazil (lat. 23°S), on a red latosolic loam with an annual rainfall of 1 200 mm in a frost-free environment.

Holder, Swain and Colman (1963) developed the "feed-year" concept for dairying at Wollongbar, New South Wales (lat. 29°S). Summer rainfall and temperatures are adequate and useful winter rains occur, but the spring months are dry and cold; the average minimum daily temperature from June to September is below 10°C and rises to over 15.5°C in November. A *Paspalum dilatatum*-based pasture was supplemented by *Neonotonia wightii*, *Trifolium subterraneum* and *Vicia sativa* to give a sequence of feed, the growth patterns of each component being given in Figure 15.

Trifolium subterraneum and *Vicia sativa* were subsequently shown to be

61

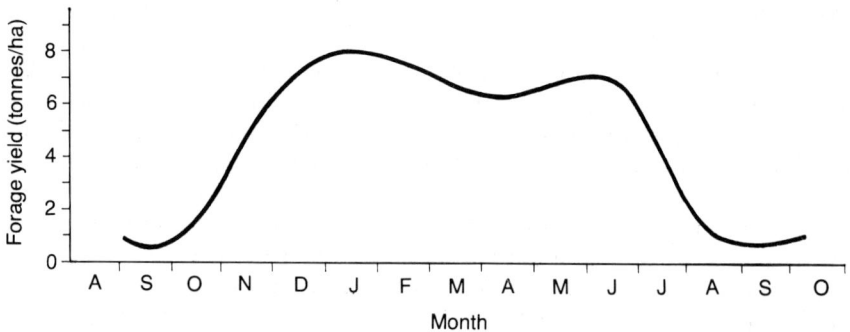

Figure 14. Seasonal production pattern for *Neonotonia wightii* at Campinas, Brazil (latitude 23°S) (**Source:** Lovadini and Miyasaka, 1968)

too unreliable in the drier spring months and were replaced by nitrogen-fertilized oats. Suitable sequences were then developed for the four main soil types in the area (Figure 16) and the economics of the systems examined. Substantial increases in yearly income over the traditional summer-excess-winter-scarcity feed pattern were achieved under the new feed-year systems.

Choice of species according to soil, topography and water relation is also an important on-site decision. The water relations of the soils on the slopes are

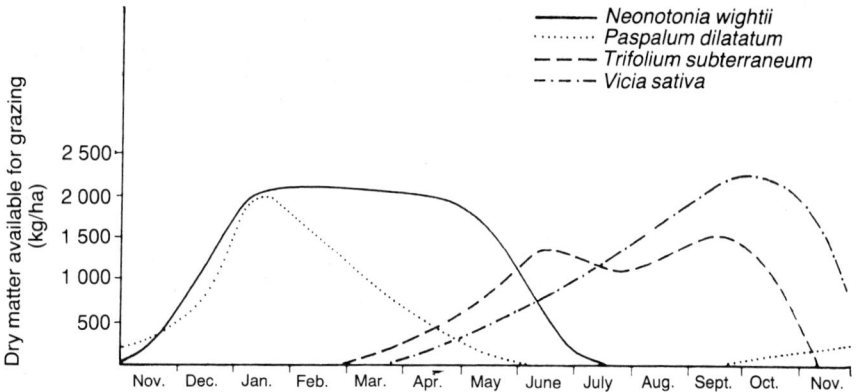

Figure 15. Seasonal production pattern of four pastures at Wollongbar, New South Wales, Australia. Curves represent total production to a particular date if ungrazed from the beginning of the season

62

usually different from those in the valleys; also, in frost-prone areas, the higher slopes are less susceptible to frost influence than the valley floor. In subtropical areas the use of tropical legumes for the slopes, and temperate legumes such as white clover (*Trifolium repens*) for the valley floor, where it is colder, is a sound system. Within this framework decisions must be made. Returns from different grass/legume combinations are given in Table 4.2.

TABLE 4.2 **Varying returns from different grass/legume combinations**

Grass/legume combination	Inputs	Outputs	Return/ha
	(US$)		
7.2 ha pangola grass, white clover; 3.2 to 4 ha sod-seeded oats, ryegrass and clover on heath land	655.84	599.94	7.75
4 ha pangola grass, white clover; 3.2 ha frost-tolerant setaria + nitrogen on heath land	576.84	847.80	37.50
4 ha pangola grass, white clover on heath land; 3.2 ha tropical species on ridge	361.66	760.68	55.42

Source: Wright, 1969

The heath land required more fertilizer and more preparation than the ridge country. The combination of tropical legumes on the ridge and temperature legumes on the valley floor gave the best net returns.

The characteristics of the species of tropical forage legumes are outlined in Chapter 14. This catalogue will be an aid in the selection of suitable combinations. Records of performance in various localities are also given.

Pasture mixtures

Pasture can be either pure swards of grasses or legumes or mixtures of both. In the past, pure grass swards have been widely used and have been successful in time, depending on the fertility of the soil. Pasture swards demand a high level of nitrogen for continuous high production. Henzell (1963) has shown a linear response to 440 kg and a further increase up to 880 kg N/ha with Rhodes grass. At Parada in north Queensland, an irrigated pangola grass (*Digitaria decumbens*) pasture fertilized with 440 kg N/ha/year gave a live-weight yield of 1 980 kg/ha/year.

In areas where a serious weed problem is likely to arise early in the life of the pasture, a species able to give a quick cover to suppress weed growth should be included. In the brigalow (*Acacia harpophylla*) scrublands of central Queensland, Rhodes grass (*Chloris gayana*) is included to give a quick

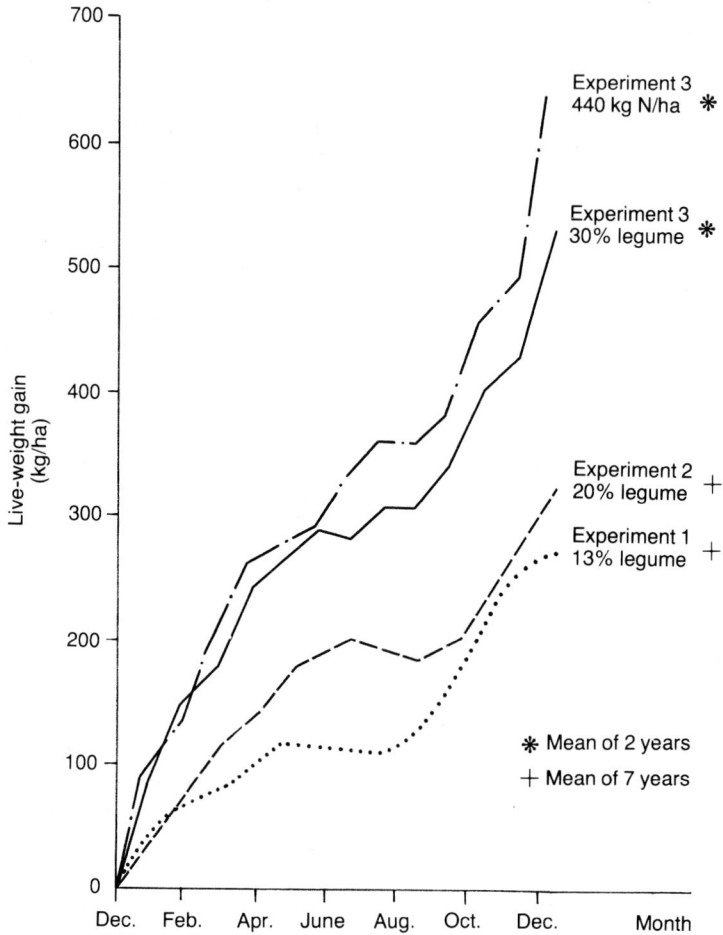

Figure 17. Cumulative live-weight changes in grazing pastures with varying percentages of legumes at Beerwah, southeastern Queensland (**Source:** Bryan and Evans, 1968)

66

cover to suppress sucker regrowth. In the tropical rain forest areas, molasses grass (*Melinis minutiflora*) is included to suppress weed growth and can later be replaced by more productive species. When sowing tropical legume mixtures, it is sound practice to include such a vigorous grass to suppress early weed growth.

Table 4.3 shows some pasture mixtures which might be used.

TABLE 4.3 **Pasture mixes for various environments**

Soil type	Annual rainfall (mm)	Mixture and seeding (rate/ha)
Wet tropics	1 500 or more	2 kg guinea; 3 kg centro; 1 kg stylo
Red latosolic soils	1 125 or more	2 kg guinea or setaria; 3 kg greenleaf *Desmodium* or 3 kg *Glycine* (on basalt)
Podzolic soils	875 or more	3 kg siratro; 1 kg stylo; 2 kg guinea or Rhodes grass
Podzolic or solodic soils with native *Heteropogon, Hyparrhenia* or *Sorghum* spp.	875 or more	3 kg Townsville stylo or 3 kg stylo (in frost-free areas), strip-seeded or aerially sown into native pasture

Seed purchase and preplanting treatment

Seed purchase. The investment in pasture improvement is usually a costly one and is meant to provide long-term benefits; hence, only seed of the highest quality should be planted. Quality in seed means seed with high viability and high purity. Purity is important because contamination with weedy species can cause weed problems for years to come. About 20 years ago, some *Centrosema pubescens* seed imported to Australia from Malaysia was contaminated with Giant Sensitive Plant (*Mimosa invisa*) seeds, and this thorny shrub has been a pest in sugar-cane growing ever since. Most governments have regulations covering the quality of seed presented for sale, including minimum germination percentages, minimum purity percentages, and a list

67

of prohibited seed contaminants. Purchase of government certified seed is, therefore, the first step in sound pasture establishment.

Seed treatment to break dormancy. Most legume seeds have a high percentage of "hard" seeds — that is, seeds which will not germinate even under favourable conditions. This is a protective mechanism to ensure the survival of the species through adverse weather conditions and is especially important in arid regions. In many legumes the percentage of hard seed is between 60 and 90 percent. The dormancy is due mainly to the presence of a seed coat so impermeable to water that imbibition and subsequent germination are delayed. Under natural conditions, the seed coat gradually becomes permeable and a germination of about 10 percent of the seed per year is adequate to ensure survival.

For rapid pasture establishment at the cheapest price (use of minimum amount of seed), it is important to have a high percentage of readily germinable seed; some treatment, therefore, is usually necessary to break the dormancy shortly before planting. Some legume seeds do not require treatment, and mechanically harvested seed is often sufficiently scarified in the process of threshing.

In the following section are listed some treatments to break dormancy.

Scarification. Abrasion of the seed coat will increase permeability to water; this is the common method of treatment for most legume seeds. For small seed lots, scarification can be achieved by rubbing the seed lightly between two pieces of emery or sandpaper, or by treatment in a small hand or electric scarifier. For larger parcels of seed, a mechanical scarifier of larger capacity is needed. Cement mixers to which some gravel has been added with the seed have been successfully used to scarify glycine in Brazil. Commercial rice hullers are also effective.

Chemical treatment. Chemical treatment with alcohol or acetone will dissolve some of the substances inhibiting germination, but treatment with concentrated sulphuric acid is more common. The method used at the Sisal Research Station, Tanzania (Grundy, 1959) is as follows:

68

● Equipment

– 2-4 plastic or enamel jugs with a capacity of about half a litre;
– 1-2 plastic or enamel pails;
– 1-2 plastic or enamel washbasins;
– 1-2 sieves.

The sieves should be about 30 cm square and 10-15 cm deep, with the frames made of wood. The sieving material should be of plastic-coated mosquito gauze and, since this gauze is not very strong, it should be reinforced with three or four wooden battens across the bottom.

– One 200-litre oil drum, open at the top;
– 5-10 kg of lime of a grade suitable for whitewash;
– 4.5 litres concentrated sulphuric acid (24 or 36 Normal) for every 55 kg seed (this acid can be purchased in 22.7-litre carboys).

● Method

The two pails are put in the basins and filled with water from a hose. The 200-litre drum is also filled with water and the lime stirred into it. One jug is half filled with seed and just enough sulphuric acid is added to wet the seed when stirred with a stick. This jug is put aside and a few minutes later acid is added to the seed in another jug. When the seed in the first jug has been in contact with the sulphuric acid for the right length of time (see Chapter 14), the hose is put into one of the pails and the tap is turned on. The gummy mass of seed is transferred from the jug to the pail with the aid of a stick. Any seed that is washed out of the pail will be collected in the basin underneath. When nearly all the seed has been scraped out of the jug (and not before), the rest can be washed out with water into the pail. The diluted acid in the pail and that in the wash-basin together with the seed are poured into a sieve, allowing the acid to run to waste. The hose is turned into the sieve to wash the seed. The sieve is then dipped in the stirred lime and water in the 200-litre drum to wet the seed, after which it is ready for drying prior to planting. By this time, the seed in the second jug should be ready for washing. Before the jug is used again, it should be dried roughly with a cloth.

● Precautions

Concentrated sulphuric acid can cause severe burns; thus, if any comes in

69

contact with the skin, the affected part should be washed at once with water and a little lime to neutralize the acid. The eyes should be protected with goggles, and clothes with an apron. Water should never be poured into sulphuric acid. This is the reason why it is so important to ensure that all the sulphuric acid and most of the seeds are out of the jug before washing the remainder out with water.

While treatment with concentrated sulphuric acid has been very successful with hand-picked seed with a uniform seed coat, it is not a generally safe method for machine-harvested seed. The acid treatment usually increases the germination of siratro (*Macroptilium atropurpureum*) seed from about 10 to 95 percent in hand-picked seed, but with machine-harvested seed the increase is only from about 30 to 60 percent.

Disadvantages of the sulphuric acid method are: (*i*) strict safety precautions must be observed when handling the acid; (*ii*) equipment for treating large quantities of seed is costly; and (*iii*) seed may be damaged by the extreme heat generated when the acid is washed from the seed coat, or the seed coat of machine-harvested seed may not be hard enough to stop acid from penetrating and killing the seed.

Wycherley (1960) found that soaking legume seed in warm glycerine at 50°C broke dormancy in some species, probably by extracting water by its strong osmotic effect. The germination of *Pueraria phaseoloides* seed was increased from 10 to 50 percent and that of *Centrosema pubescens* from 9 to 16 percent by treatment in warm glycerine for one hour. The treatment had no effect on the seed of *Flemingia macrophylla* or *Calopogonium mucunoides*.

Hot water treatment. Soaking the seed in hot water will break the dormancy of some seeds. The seed of *Neonotonia wightii* placed in boiling water which is then allowed to cool overnight will germinate. The dormancy of *Leucaena leucocephala* seeds is broken by heating to 80°C for four minutes or to boiling point for half a minute. Seeds of mulga (*Acacia aneura*) will germinate after boiling for three minutes. The disadvantages of these water treatments is that the absorbed water in the seed allows germination of some of them to take place in the field, even if there is inadequate moisture in the soil. Complete loss may result if rain fails after sowing (Wycherley, 1960).

Infrared lamp irradiation. Wycherley (1960) succeeded in breaking the dormancy of *Calopogonium, Centrosema, Flemingia* and *Pueraria* using Philips Infraphil Type 13373F/479 (150-watt) and Osram I.R.R. 4892 (250-watt) lamps. The germination of *Calopogonium* seed was enhanced by exposure to eight hours under the Philips and 16 hours plus under the Osram; *Centrosema* by 16 hours under the Osram; *Pueraria* by one hour under the Philips and two hours under the Osram; and *Flemingia* by only $7^1/_2$ minutes' exposure under the Philips lamp.

Exposure to intense cold at temperatures of -50 to $-90°C$ (-58 to $-130°F$) and subjecting seed to high pressures will break dormancy in some cases.

Despite a general decline in percentage germination, seeds do not become "hard" again after scarification. It is advisable to use seed soon after treatment, but scarified seed will store for at least two months without deterioration.

Seed inoculation and pelleting

The operations are discussed in Chapter 7.

Seed treatment against insect pests and diseases

Pelleting of seed to protect the *Rhizobium* in the inoculant also protects it from harvesting ants. A dressing of 85 g of 10 percent lindane wettable powder per 10 kg of seed will deter ant attack, but as lindane adversely affects the *Rhizobium* the seed must be pelleted before lindane treatment, and pelleting without lindane treatment is usually effective. Bean fly attack on developing legume seedlings can be controlled by treating the seed before planting with 13 ml of 15 percent dieldrin for each kilogram of seed planted.

Seed treatment for control of seed-borne diseases is not common with legume seed, but dusting with mercury-based dusts such as Ceresan and Agrosan can be undertaken with seed already pelleted.

Seeding rates and seed dilution

The amount of seed to plant for each species is governed by climatic conditions, the complexity of the mixture, the rapidity with which a full pasture sward is required, the competition between plants mainly for available moisture, and the cost of seed.

A higher seeding rate can be adopted in higher rainfall areas than in semi-arid areas and is desirable to give a quick cover to suppress weed growth. In arid areas, a low seeding rate gives the sown species an opportunity to utilize the low soil-moisture status and either deep-rooting perennials or short-term annuals should be selected. Luck (personal communication) recommended increasing the seeding rate as the amount of seed-bed preparation declined, using 9 kg/ha siratro (*Macroptilium atropurpureum*) on roughly prepared seed beds, 4.5 kg on partially prepared seed beds and 1 kg on well-prepared seed beds. Middleton (1970), using *Desmodium intortum, Macroptilium atropurpureum* and *Setaria anceps*, showed that seedling density was proportional to sowing rate and only high legume sowing rates maintained a satisfactory proportion of legume in the pasture. Increasing the sowing rate of the species in the mixture altered the composition but not the final yield of dry matter. His experiments were conducted under a 1 220 mm, 60 percent summer-dominant rainfall on a podzolic soil at latitude 27°44'S.

When seed is broadcast, it is usual to increase seeding rate by up to 50 percent.

To obtain an even distribution, especially of very small seed, it can be mixed with rice hulls, sawdust, sieved cow manure, sand or fertilizer. Where acid fertilizer is used with inoculated seed, the seed should first be pelleted to protect the *Rhizobium*. Neutralized fertilizer can be mixed with inoculated but unpelleted seed. Halving the seed mixture and sowing twice gives a more even distribution of small seed lots.

Fertilizer needs and application

Fertilizer needs. The use of fertilizer for tropical legumes is dealt with more fully in Chapter 5. The legumes will generally respond to dressings of phosphorus, molybdenum and sulphur, and in some soils to potash, copper and zinc. Because legumes have the ability to acquire their needs for combined nitrogen symbiotically, it is usually not added in the fertilizer planting mixture. However, between planting the legume seed and the appearance of the first active nodules, there is a hiatus during which added nitrogen would be beneficial for establishment and even during subsequent growth modest amounts of nitrogen will usually give a response.

At least 11 kg/ha nitrogen should be included with the planting mixture. In mixed swards of pangola grass in Florida (United States), *Stylosanthes humilis* disappeared when 27.5 kg/ha nitrogen were applied, and siratro at about the 66-kg level, due to the increased competition from the stimulated grass.

Legumes and their *Rhizobium* require high levels of phosphorus, but some legumes, especially the tropical species such as the *Stylosanthes*, can extract phosphorus very efficiently from the soil, so that their requirements are not as high as others. Rates of superphosphate from 123 to 493 kg/ha are usually used for establishment.

Potash is not always limiting, but under high production it is usual to add this element at the rate of about 132 kg/ha muriate of potash per year.

It may be necessary to add the trace elements to leached soils. Molybdenum is essential for legume nodulation and is usually applied at 210 to 420 g molybdenum trioxide per hectare, with up to 840 g for *Neonotonia wightii* in latosolic soil. Sulphur may also be deficient, but its deficiency may not be noticed because it is usually corrected by topdressing with superphosphate, which contains about 13 percent sulphur. Copper is supplied as copper sulphate at about 8 kg per hectare, and zinc sulphate at 8 kg per hectare. Any deficiency of boron is corrected with borax at 22 kg per hectare.

In most soils, there is an initial nutrient barrier which must first be broken by relatively heavy applications of fertilizer, but the subsequent yearly application required is much smaller. Thus, there is a special requirement for establishment, and a separate one for yearly maintenance which may be about one-quarter to one-half of the initial application.

Liming may be necessary to raise the soil pH to a satisfactory level for legume growth, but Andrew and Norris (1961) have shown that the needs for nutrition of tropical legumes are rather modest — of the order of 616 kg/ha for extremely acid soils.

A preliminary series of comprehensive soil-fertility studies should be initiated in each locality and on each soil type to determine fertility needs prior to commercial pasture establishment.

Fertilizer application. Fertilizers should be appplied in conjunction with planting if a seed and fertilizer drill is used, and fertilizer placement is more rewarding than blanket dressings. Precautions to protect the *Rhizobium* of inoculated seed should be taken (see Chapter 7).

If liming is to be done, it should be a separate operation carried out ahead of the general fertilizing programme.

Where the land is unsuitable for drilling in the fertilizer, a spinner-type broadcaster, driven by direct coupling to the power takeoff shaft of the tractor, is satisfactory.

5. Mineral nutrition of tropical pasture legumes

In mixed pastures, it has always been a problem to maintain the desired proportion of legumes to grass. A good deal of research has centred on the nutrition of the legume to ensure its survival and maximum performance. The associated grass is usually satisfied nutritionally by adequate fertilization of the legume and it will have some nitrogen (but not usually enough for full potential) from that fixed by the symbiotic fixation of the legume.

Rhizobium complex

It is usually much easier to introduce a grass to a sward of legume than a legume into a grass sward. Phosphorus and nitrogen are two of the main limiting factors in production of pasture from tropical soils. The addition of phosphorus stimulates the growth of legumes, which in turn fix nitrogen. As the cost of nitrogenous fertilizer is usually high in the tropics, nitrogen from legumes is encouraged as the cheapest form available for pasture growth (see Chapter 6).

Research has shown that although phosphorus is the major requirement for adequate legume performance, it is not the whole answer. Potash is also necessary, but the trace elements cobalt and molybdenum are essential for nodule function, and others such as sulphur, zinc, boron, manganese and aluminium may be limiting in excessive amounts. Further, there is need for nutrient balance for full production.

Soil

Under natural conditions, the parent material from which the soils have been derived determines in a large measure the fertility of the soil, especially in comparatively recently developed soils. However, the effects of leaching and land-use systems are also determining factors. Natural vegetation is an indicator of the present nutrient status of the soil, reflecting an equilibrium with the environment reached over the years. To improve land use economically by the introduction of new pasture species, the environment must be modified to satisfy the needs of these species.

Soils derived from sands, sandstones, granite, schists and laterite are usually grossly deficient in plant nutrients.

Soil analyses give a lead to possible nutrient requirements of the plant to be grown thereon, but soil analyses are no substitute for pot and parallel field experiments.

Plant analyses, when the plant is established, serve as a supplementary guide to current soil fertility conditions, but techniques are still being refined to cover a wider spectrum of elements.

The presence of such toxic substances as salt in the soil, and its physical condition (e.g. waterlogging and hardpans), also affect legume growth.

It is necessary to distinguish between the nutritive requirements of the legume host and those of the associated symbiotic rhizobia, although total performance depends on both. Rhizobia nutrition is dealt with in Chapter 7.

The role of mineral elements in the nutrition of the legume host is considered below (nitrogen is discussed in Chapter 6).

Individual mineral elements

Calcium. In the traditional culture of temperate legumes, calcium has been liberally applied and growth response in the legume recorded. The response was usually regarded as being due to raising the pH level of the soil, which in turn enabled successful nodulation of the legume with its associated *Rhizobium*. Andrew and Bryan (1955), growing white clover (*Trifolium repens*) on a tropical low-humic gley soil in southeast Queensland, Australia, were able to show that germination and seedling development of white clover were dependent on elemental calcium being added to the soil as plant food.

Russell (1966) subsequently found that elemental calcium was also necessary for the development of temperate legumes on a solodic soil in southeast Queensland.

The temperate legumes have always been regarded as a group which could not tolerate acidity. Lucerne or alfalfa (*Medicago sativa*) was shown by Spurway (1941) to require for growth a minimum pH of 6.2, McIlvane and

75

Pohlman (1949) arrived at a figure of 6.7, and Bear (1949) at 6.5 for lucerne and 6.0 for white clover. Munns (1968) found that a pH of 4.5 stopped nodulation but did not prevent root growth. Munns (1965), however, found that lucerne would grow on an unlimed sandy loam of pH 4.0, provided enough phosphorus was added to overcome aluminium toxicity, acting by precipitating the aluminium as aluminium phosphate.

Many tropical legumes, however, thrive at pH figures as low as 4.0 without the addition of lime, provided other nutrients are adequately supplied.

Andrew and Norris (1961) grew five tropical forage legumes and four temperate legumes in pots of calcium-deficient soil, with treatment representing the addition of calcium carbonate at rates of 0 to 25 tonnes per hectare. With the nil treatment, the yields of the temperate legumes *Medicago truncatula (M. tribuloides)*, *M. sativa* and *Trifolium repens* were 1.5, 2.0 and 6.5 percent respectively of their maximum yields with adequate lime, while the yields of the tropical species *Desmodium uncinatum, Indigofera spicata, Centrosema pubescens, Stylosanthes fruticosa (S. bojeri)* and *Macroptilium lathyroides* (L.) Urb. were 25, 40, 52, 64 and 68 percent of the maximum yields respectively. The progressive responses to lime are given in Figure 18.

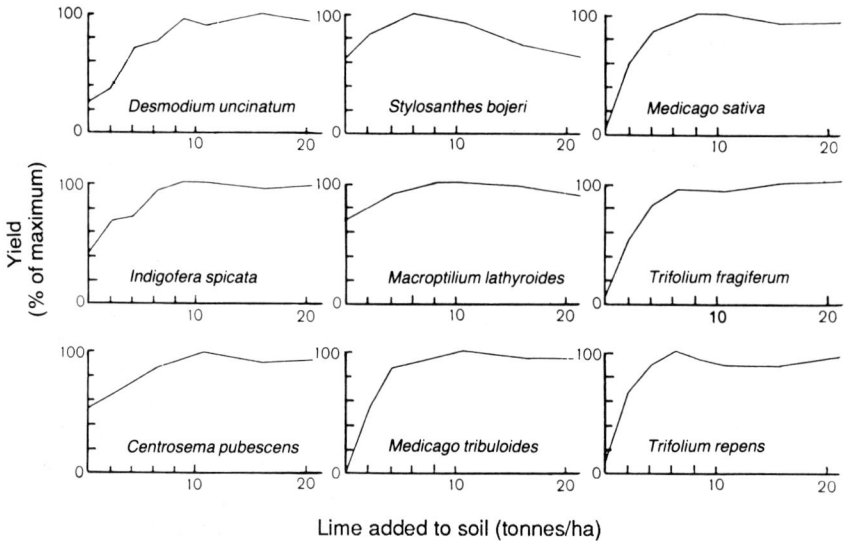

Figure 18.   Progressive responses to lime

76

All the legumes responded to increasing dressings of lime up to a maximum of 750 to 1 000 kg/ha.

Calcium percentages found in plant tops did not differ much between the tropical and temperate legumes. Andrew and Norris decided that the difference between the two groups in relation to calcium as a plant nutrient lay in the efficient extraction of calcium from the soil by the tropical species. It is interesting to note that *Desmodium uncinatum* was the least efficient of the tropical species in this regard and was similar to the temperate white clover. There are, however, some differences among species in their calcium requirements. Döbereiner and Aronovich (1965) found that lime was essential for the growth of *Neonotonia wightii* on high manganese soils, and Mears and Barkus (1970) obtained a response to calcium in the presence of molybdenum.

Having then disclosed the relationship between the two groups in respect to the needs of the plant, Andrew and Norris looked at the needs of the *Rhizobium*. They were able to show that the tropical legumes as a group did not require lime for nodulation and so differed from the temperate group. *Stylosanthes guianensis* gave equal nodulation per gram of root tissue in a soil at 3 percent calcium saturation (nil treatment) as in a soil with 37.2 percent Ca saturation (equivalent to 740 kg/ha of added lime). *Medicago sativa*, on the other hand, gave no nodulation at 3 percent saturation and high nodulation at 37.2 percent saturation. Norris (1967) then found that the rhizobia from tropical legumes were slow growing and produced an alkaline reaction in the culture medium, and that those from the temperate legumes were fast growing and formed acid in the culture medium. Thus the nodules from each group of legumes were able to modify the soil reaction in their vicinity.

The implications of these findings are discussed in Chapter 7.

Phosphorus. Deficiency of phosphorus, more than any other element, limits the realization of the production potential of the tropical legumes as well as the temperate species (Figure 19). The tropical soils, apart from black soils derived from basalts and certain alluvial soils, are usually low in phosphorus.

Alluvial soils gain phosphorus from erosive influences moving the surface soil, where most of the phosphorus is held, down into the valleys. Black soils derived from basalt have a high available phosphorus content. Latosolic soils derived from basalt have high total phosphorus, but little is available. All latosolic soils, whatever their origin, contain high amounts of iron and aluminium oxides and these readily adsorb any added phosphorus and render it unavailable to plants. This process is termed "fixing the phosphorus". Par-

Figure 19. Response to phosphorus by white clover on low-humic gley soils of south-eastern Queensland, Australia. The bare plots (lighter patches) are the nil phosphorus treatments. Adequate growth was obtained with 616 kg/ha of single superphosphate (**Source:** Skerman, 1958b)

bery (1940) found that surface soils of the latosols were capable of fixing into an unavailable form 90 percent of the equivalent of 110 tonnes superphosphate (22 tonnes soluble P_2O_5) per hectare in five days. Hughes and Searle (1964) suggested that all fixation sites on the soil colloids or organic matter have to be blocked by adsorbed phosphorus before a surplus is available to plants. Hence, besides an initial dressing of phosphorus to the latosolic soils, annual additions must be made to supply the current needs of the plant. In sandy soils, little of the phosphorus is fixed at the surface; Ozanne et al. (1961) showed that it moved below the top 10 cm even under the leaching influence of a 550-mm annual rainfall.

Since a large proportion of the soils of the tropics are either latosolic or sandy podzolics and solodics, the demand for phosphorus for legume growth will always be high. As the foreseeable world supplies of phosphorus are limited, this may well be a crucial factor in continuing high production from improved pastures.

The tropical forage legumes vary in their demands for added phosphorus. Andrew and Robins (1969a) grew nine tropical legumes and one temperate species — *Macroptilium lathyroides*, *M. atropurpureum*, *Stylosanthes humilis*, *Centrosema pubescens*, *Neonotonia wightii*, *Lotononis bainesii*, *Medicago sativa*, *Desmodium uncinatum*, *D. intortum* and *Vigna luteola*— in pots of a Samford gley soil with varying levels of phosphorus. The legumes at low rates of phosphorus yielded respectively 82, 64, 55, 55, 53, 49, 43 and 42 percent of their maximum yields when grown at higher rates. *S. humilis*, *Medicago sativa* and *Lotononis bainesii* reached maximum yields at 493 kg/ha, while the more responsive species such as *Neonotonia wightii*, *D. intortum*, *D. uncinatum*, *Macroptilium atropurpureum* and *M. lathyroides* required up to 1 232 kg/ha. The total amount of phosphorus taken up from the soil per unit of soil mass was greatest for *S. humilis* and *L. bainesii*, and Andrew (1966) showed that the uptake of phosphate from dilute solutions per unit of time by excised roots of *S. humilis* was greater than that of *M. lathyroides, D. uncinatum* and *M. sativa*. This ability of *S. humilis* to absorb greater quantities of phosphorus from a soil with low available phosphate is extremely valuable in developing countries where the cost of phosphate fertilizers is high and the soils poor. It enables some early development of improved pastures, especially with *Stylosanthes* spp., at a level of productivity higher than the existing one. It should be noted that maximum yields of *S. humilis* were obtained at 493 kg/ha superphosphate, and it will be shown in Chapter 14 that it is economic to apply higher dressings of superphosphate to increase both live-weight gains and animal fertility levels. On soil of high phosphorus status no fertilizer may be required for some time, but on most of the poorer tropical soils successful initial livestock performance from such species as *Stylosanthes* will lead to greater use of this basic element. In coastal southeastern Queensland, most of the sandy soils require 493 kg/ha single superphosphate as a basic establishment application, and the annual application thereafter is 123 kg/ha or 246 kg/ha.

Andrew and Robins (1969b) showed that the addition of phosphorus to soil-growing legumes increased the nitrogen percentage in the tops of the legumes beyond the maximum increase in dry-matter production. They attributed this increase to better root development, better nodulation and improved plant metabolism.

The addition of phosphorus upset the balance of the other cations, although generally the total cation content remained at the same level. When phosphorus (as sodium dihydrogen phosphate) was added, the sodium concentration of *Vigna luteola*, *Medicago sativa* and *Lotononis bainesii* increased; there was an increase in magnesium and a decrease in potassium in *Macroptilium lathyroides* and *M. atropurpureum*. *Lotononis bainesii* and *D. intortum* increased in potassium, and *Centrosema pubescens* and *S. humilis* in calcium.

These results showed that tropical legumes vary in their cation status, and that there is a tendency toward balance within the cations.

The forms of phosphorus available for pasture fertilization are listed below.

Rock phosphates. Often cheaper than other forms, but their phosphorus content varies and the phosphorus is only slowly available. It must be finely ground to speed up its solubility. It is neutral in reaction.

Single superphosphate. A complex mixture of phosphates with some gypsum, contains about 9.6 percent phosphorus. It also contains 20 to 22 percent calcium and some 13 percent sulphur and so is a useful contributor to general fertility. In normal dressings it often satisfies the plant needs for all three elements. Russell (1966) has shown how difficult it is to interpret the response of a leguminous pasture to superphosphate.

In areas where molybdenum is required by the plant, it is usual to add 0.03 or 0.06 percent molybdenum trioxide to the superphosphate for ease in its application, the product being sold as molybdenized superphosphate.

Basic superphosphate. Single superphosphate is acid in reaction and it affects the viability of the *Rhizobium* inoculant if planted with the seed. Hence, neutralized superphosphate is prepared usually by mixing superphosphate with lime to overcome this problem.

Triple superphosphate. Prepared with a high percentage of phosphorus; in areas where transport costs are high, it is cheaper per unit of phosphorus. In contrast to single superphosphate, it contains little sulphur (1.5 percent) and, if needed, this element must be added separately to a fertilizer mixture.

Complex polyphosphates. At present, these are prepared with high levels of both phosphorus and nitrogen (e.g. the metal ammonium phosphates). These are slow-release fertilizers which will be useful in areas where leaching losses of phosphates are likely to be high.

Potassium. Tropical leguminous pastures are rarely visibly deficient in potassium as yet because of the small area of such pastures and because symptoms do not appear unless there is a severe deficiency. Occasionally, however, the typical chlorotic margins described and illustrated by Andrew and Pieters (1970a) appear, as in *Desmodium intortum* (in Colombia) and in soybeans (see Colour plate III).

As productivity increases and stocking rate intensifies, it is likely that more potassium deficiency will appear because of removal in beef, milk and wool. In the *wallum* country of southeast Queensland, Andrew and Bryan have shown that at least 123 kg/ha/year of potassium chloride are needed to keep the pastures productive.

Andrew and Robins (1969c) studied the effect of potassium on the growth of eight tropical and four temperate pasture legumes in pots with a potassium-deficient soil. *Macroptilium lathyroides, M. atropurpureum, Desmodium intortum, D. uncinatum, Stylosanthes humilis, Lotononis bainesii, Centrosema pubescens, Neonotonia wightii, Medicago sativa, M. truncatula, Trifolium repens* and *T. fragiferum* produced in the nil treatment 28, 25, 30, 47, 23, 52, 51, 56, 34, 45, 38 and 53 percent of their optimum yield. The optimum yield for all except *Lotononis bainesii* was given at an application of 123 kg/ha KCl. *L. bainesii* produced most at 246 kg/ha.

In a further paper, Andrew and Robins (1969d) showed that addition of potassium chloride reduced the Ca and Mg contents of the plant tops but did not increase sodium or nitrogen levels. Chlorine content was increased with increasing application of potassium chloride. Species differed in chloride uptake; *Desmodium uncinatum, D. intortum* and *Trifolium fragiferum* were high.

Addition of potash decreased phosphorus levels. This bears out the experience with siratro reported by Jones (1966) in which a very strong increase in yield accompanied application of both high phosphorus and potassium.

Teitzel (1969a) found that, in a grantie loam soil in coastal north Queensland, siratro responded to potash only when adequate copper was present.

Sulphur. Sulphur is an important element in legume nutrition, particularly so in lucerne in which it almost reaches the status of a major element.

The symptoms of sulphur deficiency are much like those of nitrogen deficiency. Sulphur deficiency has been recorded chiefly in sandy soils derived from granite and from chocolate-coloured and black soils derived from basalt. In southeast Queensland it has been noted where lucerne is irrigated with underground water, but not where surface water from the Lockyer creek is used, which accumulates sulphur from the catchment. Sulphur may also be deficient in soils high in organic matter.

81

Sulphur and phosphorus are often associated as deficiencies because of leaching in sandy soils, inherently low sulphur and fixation of phosphorus in soils derived from basalt. The deficiencies can usually be corrected with superphosphate containing some 13 percent sulphur, sulphate of ammonia, or sulphate of potash. Superphosphate fortified with 30 percent sulphur can be applied where the deficiency is greater than can be corrected with single superphosphate. It must be remembered that triple superphosphate contains little sulphur.

Annual applications of sulphur at 22 to 34 kg/ha are generally required to correct sulphur deficiency.

Copper. Andrew and Bryan (1955) showed that the addition of copper sulphate to a low-humic gley soil improved the growth of *Trifolium repens, T. pratense* and *Paspalum commersonii.* Andrew and Bryan (1958) found that *Desmodium uncinatum* did not give a response to copper on a lateritic podzolic soil, but *T. repens* gave a marked response. Andrew and Thorne (1962) then tested five tropical legumes grown in summer and five temperate legumes grown in winter for their response to copper; lucerne was sown with both groups. *Desmodium uncinatum, Macroptilium lathyroides, Centrosema pubescens, Indigofera spicata, Stylosanthes guianensis, Trifolium repens, T. fragiferum, T. alexandrinum* and *Medicago tribuloides* when grown with copper produced 60, 46, 45, 21, 7, 38, 31, 3 and 14 percent of the maximum yields produced when copper was added to the soil. Lucerne produced 35 percent in summer and 17 percent in winter. With increasing additions of copper up to 8.8 kg/ha, the copper contents of the foliage increased in all but *Desmodium uncinatum*, showing it to be relatively insensitive to copper deficiency, whereas berseem clover and *Stylosanthes guianensis* were most sensitive. Silverleaf desmodium, white and strawberry clover were the most efficient at extracting their needs for copper from soils of low copper status.

Some of the podzolic soils derived from granite in the wet tropical coast of north Queensland are deficient in copper and the application of about 11 kg/ha of copper sulphate is needed to correct the deficiency. Siratro gave a very good response to potash only in the presence of adequate copper and stylo responded to copper in the same soil after the application of 250 kg/ha of superphosphate (Teitzel, 1969a).

Copper is usually applied to the soil at the rate of 7.7 kg/ha copper sulphate. The frequency of application has to be determined. Young and Chippendale (1970) believed that it should be applied at less than five-year intervals to heath (*wallum*) country in coastal southeastern Queensland.

82

Molybdenum. This element is especially necessary for nodulation in legumes (see Chapter 7). Molybdenum deficiency is widespread in acid sandy podzolic and latosolic soils. Liming to raise the pH renders more molybdenum available and many of the recorded responses to lime are really molybdenum responses. Truong, Andrew and Skerman (1967) showed a response to molybdenum in siratro growing on solodic soils in southeast Queensland and the effect of liming on molybdenum release for white clover.

Molybdenum deficiency is usually corrected by applying 70 to 140 g/ha of molybdenum as sodium molybdate or molybdenum trioxide. On latosolic soils where *Neonotonia wightii* is to be grown, application of up to 200 g/ha is advisable (Mears and Barkus, 1970). Luck and Douglas (1966) suggested that on some soils up to 312 g/ha are needed.

Zinc. High calcium soils are often deficient in zinc. Often visible in citrus groves in the tropics, symptoms are interveinal chlorosis and little-leaf. Fertilization of maize with high amounts of phosphorus will cause striated chlorosis of the leaves due to zinc deficiency. It has been found that, as the pH of soil is increased, plant uptake of zinc declines. Russell (1966) found this occurred with *Macroptilium lathyroides* in southeast Queensland. Zinc deficiency is also worse at low temperatures and on wet, short days.

Only relatively small amounts of zinc sulphate (about 3 kg/ha) are needed to correct this deficiency; Young and Chippendale (1970) believe it should be applied at less than five-year intervals in the *wallum* area of southeastern Queensland.

Manganese. In acid soils, especially lateritic and latosolic types, high levels of available manganese may cause toxicity symptoms in legumes. Numerous authors have reported manganese toxicity in leguminous crops and in temperate leguminous pasture, but few have recorded manganese toxicity with tropical species. Andrew and Hegarty (1969) tested eight tropical and four temperate legumes to ascertain their response to excess manganese. Of the tropical species on a high manganese treatment (40 ppm), *Centrosema pubescens*, *Stylosanthes humilis*, *Lotononis bainesii*, *Macroptilium lathyroides*, *Leucaena leucocephala*, *Desmodium uncinatum*, *Neonotonia wightii* and *Macroptilium atropurpureum* yielded 70, 50, 42, 29, 23, 24, 11 and 9 percent of their yield under low manganese additions. The temperate legumes *Trifolium fragiferum*, *T. repens*, *Medicago sativa* and *M. truncatula* yielded 49, 18, 17 and 13 percent respectively. Thus centro (*C. pubescens*), Townsville stylo (*S. humilis*), lotononis (*L. bainesii*) and white clover (*T.*

83

repens) are fairly tolerant of manganese excess. The three tropical legumes have potential for pasture improvement on the lateritic soils in the tropics.

Manganese toxicity in *Medicago sativa* growing on a lateritic red earth from Charleville, southwest Queensland, was reported by Cowie and Skerman (1970).

Manganese toxicity is a problem in the latosolic and grey hydromorphic soils of Brazil. Souto and Döbereiner (1970) have shown that liming to a pH of 6.5 removed the toxicity, but that phosphorus actually enhanced manganese absorption when excessive amounts of manganese were present.

Boron. Boron is not normally deficient, but cases of deficiency in lucerne (alfalfa) occur on coastal sandy soils in southeast Queensland.

The application of ordinary borax at the rate of 8 kg/ha usually corrects any deficiency.

Aluminium. Acidity permits high concentrates of ionic aluminium in the soil. Russell (1966) suspected aluminium toxicity in low calcium soils in Queensland. Munns (1965) working with lucerne found that aluminium toxicity was associated with 10 to 100 μM concentrations of aluminium in the soil solution and 3 to 10 μ atoms/g dry weight in the plant tops. In unlimed soils, the addition of phosphorus overcame aluminium toxicity as well as P deficiency. Both Ca and high levels of P lowered the concentration of aluminium in the soil solutions and in the plants.

Application of lime and fertilizers

These materials are usually applied to pastures by broadcasting or drilling them in with the seed. Several broadcasting methods are available:
— with a hand-operated centrifugal broadcaster for small areas;
— with a spinner-type broadcasting unit, power driven directly from the tractor;
— with a seed and fertilizer drill with the rubber, plastic or steel chutes removed from the planting shoe; and
— from the air.

In each of these situations the particle size of the individual ingredients of the mixture has an effect on distribution, but in aerial broadcasting an even size is important.

More effective placement of fertilizer can be achieved by drilling it in with the seed and especially by placement in bands just under the seed of legumes. Care must be taken in using acid superphosphate with inoculated legume seed as it affects the rhizobia. In such cases, the seed should be pelleted or a neutralized fertilizer used. Where lime is to be added to very acid soils for tropical legumes, it should be applied by itself ahead of other fertilizers. As particle size is important in determining the speed of action of lime, only very finely ground material should be used. Mixing lime with sulphate of ammonia will release ammonia, which will be lost to the atmosphere.

Establishing the need for fertilizers. Fertilizers are usually costly and, there-fore, it is important to find out if fertilizer is necessary, and the kind and amount needed for each soil and legume.

Preliminary trials should be undertaken to ascertain likely fertilizer response. Andrew (personal communication) has drawn up a design (see below) for a trial for arriving quickly at an estimate of likely requirements for nutrients. It consists of (a) a trace or minor element trial, and (b) a rate of phosphorus trial carried out concurrently, and depends on a prior knowledge of the likely requirement of the particular soil.

Design for plant nutrition trials

A. Half factorial experiment based on 2^7 (one replication)

● Treatments	kg/ha
1. KCl	100
2. $CaCO_3$	2 000
3. $CuCl_2$	5
4. H_3BO_3	2.5
5. MoO_3	0.5
6. Na_2SO_4	200
$MgCO_3$	50
7. $MnCl_2$	5
$ZnCl_2$	5

● Basal nutrient added

$CaH_4(PO_4)_2$ (aerophos)   240 kg/ha

B. Phosphate rate experiment (4 to 6 replications)

- Treatments - $CaH_4PO_4(PO_4)_2$ kg/ha

1.	Nil
2.	40
3.	80
4.	160
5.	240
6.	400

- Basal nutrient added

KCl	100
$CuCl_2$	5
H_3BO_3	2.5
MoO_3	0.5
Na_2SO_4	200

In using this design the following points should be observed.

- The use of a basal phosphate application for the factorial experiment is based on a known phosphate deficiency as revealed by soil analyses or field experiments previously conducted.
- The combination of Mg, Mn and Zn as a single treatment is based on general experience for the region in question; should a positive growth response be obtained from such a treatment, the degree of response from the individual components can be judged by plant chemical composition with reference to known standard sufficiency values for Mg, Mn and Zn, or a subsequent experiment should be done. It is important that an initial factorial experiment for a given site should always embody all essential plant nutrients, in suitable combinations commensurate with soil chemical analyses, geology and general experience.
- The chosen basal phosphate application should be one of the treatments used in the phosphate rate experiment; should the particular rate be insufficient for optimum plant growth as judged by the rate experiment, an additional application should be made to the factorial experiment (this implies that the two experiments must be done concurrently).

6. Nitrogen nutrition of tropical pastures

E.F. Henzell

Although most of the livestock in tropical countries at present obtain their feed from unimproved grasslands, there is a growing interest in improved pastures. Pasture improvement usually involves (*a*) the use of new plants, and (*b*) the use of fertilizers and/or legume nitrogen fixation to increase soil fertility. New plants are needed because the amount of plant and animal production that can be supported by unimproved grasslands is usually very much below the level that could be obtained in the same climatic environment with specially selected species and cultivars. A higher level of soil fertility is needed because selected, high-producing pasture plants have higher requirements for mineral nutrients than the less productive plants found in unimproved grasslands, and because most tropical soils release nutrients too slowly to meet the larger needs of high-producing plants. For improved pastures, nitrogen is usually of primary economic importance, because of the costliness of the relatively large quantities needed for good pasture growth. Other nutrients, such as phosphorus, sulphur, and molybdenum, may also be deficient.

The three main sources of available nitrogen for growth of tropical pastures are (*a*) soil organic matter, (*b*) nodulated legumes, and (*c*) synthetic nitrogenous fertilizers. Mineralization of soil nitrogen is the most important source for the growth of unimproved grasslands. In the case of improved pastures containing nodulated legumes, soil nitrogen is supplemented by nitrogen fixed by the *Rhizobium* symbiosis (Chapter 7). This fixed nitrogen is used first for the growth of legume plants and later contributes (via grazing animals and soil) to the growth of other plants in the pasture. In grassland receiving heavy dres-

E.F. Henzell, CSIRO, Cunningham Laboratory, St. Lucia, Australia.

sings of nitrogenous fertilizer, most of the available nitrogen supply comes from recent additions of fertilizer; other sources, including soil organic matter, are comparatively insignificant.

Soil nitrogen and the nitrogen economy of unimproved grasslands

Mineralization of soil nitrogen. The total quantity of nitrogen in soil profiles under tropical grasslands in Queensland, Australia, ranges from approximately 4 500 to 24 000 kg N/ha (G.D. Hubble and A.E. Martin, quoted by Henzell, 1968). Soils in other continents exhibit a similar range of nitrogen content (Nye and Greenland, 1960). This nitrogen is composed largely of relatively old, inert "humified" soil organic matter, with a smaller, more available fraction derived from recent additions of dead plants and animals and animal excreta. A significant proportion of the nitrogen in the deeper horizons is present in the clay as fixed ammonium, which is almost entirely unavailable to plants.

The rate of release of soil nitrogen under old undisturbed grassland is very slow indeed, usually less than 1 percent per year for the whole profile and a little higher for the topsoil. Thus, the quantity of nitrogen available for plant growth is usually less than 100 kg/ha/year and often only 10 to 20 kg (Brockington, 1960; Norman, 1963; Lansbury, Rose-Innes and Mabey, 1965).

Somewhat higher rates of nitrogen mineralization occur for a short period after the soils have been disturbed by cultivation or clearing from forest, but under grass usually they persist for only a few years. Small quantities of mineral nitrogen are provided by rain and by the urine of grazing animals. Smith (1965) calculated that the urine of cattle grazing *Hyparrhenia* veld in Zambia (2 to 4 ha per animal) would return 1 to 3 kg N/ha/year.

In spite of the fact that the grasses commonly found in unimproved tropical grasslands are adapted to growth and reproduction with relatively low levels of mineral nitrogen, it is rare to find an unimproved tropical grassland that does not contain some species capable of responding to an increased nitrogen supply, provided other nutrients are not deficient. For example, in the Transvaal, *Trachypogon spicatus* and other purple veld grasses were killed by nitrogenous fertilizers, whereas *Cynodon dactylon* and *Eragrostis* spp. gave a very good response (Hall, Meredith and Altona, 1950; Roux, 1969). At Katherine (Northern Territory, Australia) *Sorghum plumosum* and *Themeda australis* gave a much better response than *Chrysopogon fallax* (Norman, 1962). In both these examples, superphosphate was required to achieve the maximum response to nitrogenous fertilizers, whereas in Zambia a *Hyparrhenia* grassland responded only to nitrogenous fertilizer (Smith, 1964).

Nitrogen concentration in forage. The nitrogen concentration in the forage produced by unimproved tropical grasslands generally is quite low by temperate area standards (French, 1957), especially in regions with a marked dry season. Apart from a short period at the commencement of the growing season, probably associated with rapid mineralization of soil nitrogen, nitrogen concentration falls rapidly with growth and reaches a low level before flowering (see Figure 20); this fall continues into maturity as nitrogen is translocated from the tops to tissues in stems and roots (Christian and Shaw, 1951; Brockington, 1960; Norman, 1963). The nitrogen concentration in dry-season forage may, therefore, fall below 0.5 percent of the dry weight, and even with very selective grazing the concentration in the diet may fall below 1.1 percent (7 percent crude protein), the level at which nitrogen begins to limit intake of feed (Milford and Minson, 1966a, b).

Gains and losses of nitrogen. There is practical evidence that soils under old, moderately grazed, unimproved tropical grasslands maintain stable contents of nitrogen and organic matter — that is, gains and losses are approximately balanced. Soil nitrogen is more likely to be depleted if the grassland is grossly overgrazed or if dung and urine are regularly removed (see losses below), but the decline in soil fertility is still likely to be less than when the same soils are cultivated and cropped.

Large amounts of nitrogen may be lost for several years after fertile tropical soils are cleared of forest and sown to grass (McGarity, 1959; Bruce, 1965). In this particular situation, rapid mineralization of organic matter can release high levels of available nitrogen, similar to those found with nitrogen-fertilized grassland.

Gains. The amount of nitrogen added to tropical grasslands by dust or rain is likely to average about 5 kg/ha/year and will not exceed 10 kg/ha/year (Henzell and Norris, 1962a, b; Wetselaar and Hutton, 1963; Jones and Bromfield, 1970). The rate of nitrogen fixation by naturally occurring legumes is governed by the total yield of nodulated legume plants in the community. Unimproved tropical grasslands often contain such a small proportion of legume that gains of nitrogen from this source must be negligible.

It is more difficult to assess the role of nonsymbiotic nitrogen-fixing organisms (Moore, 1966). Nitrogen-fixing bacteria in the genera *Azotobacter*

89

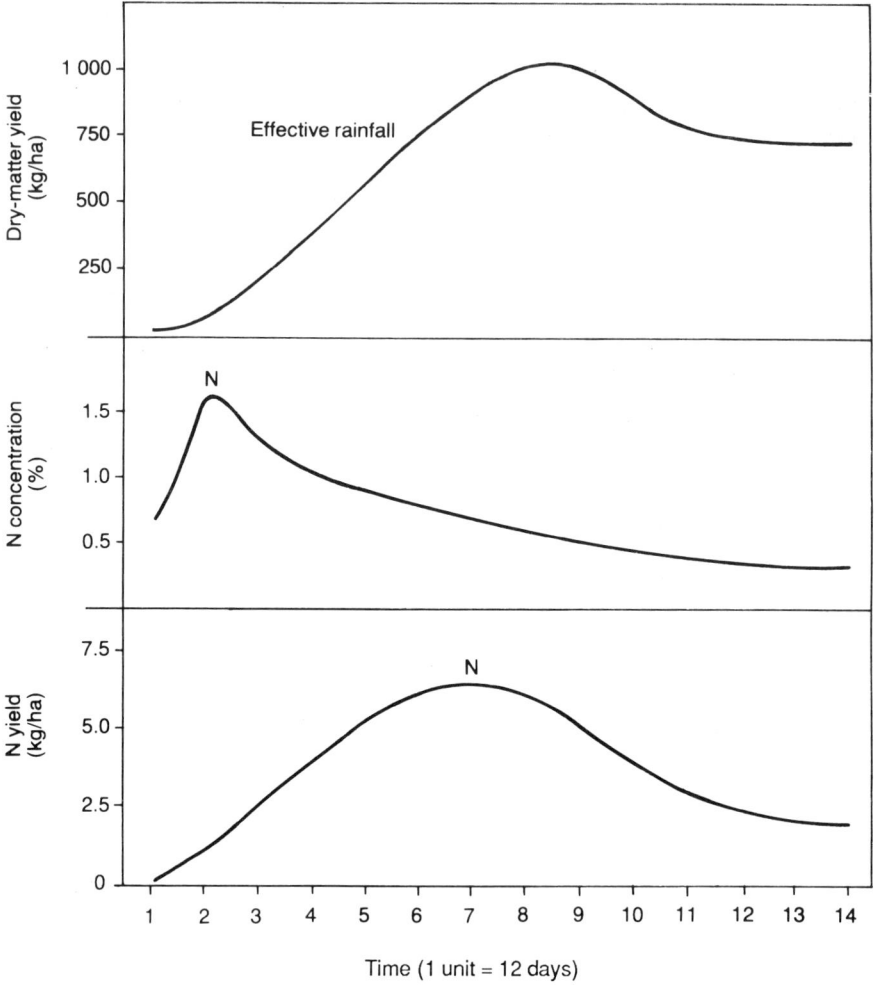

Figure 20. Time trends in dry-matter yield, nitrogen concentration and nitrogen yield of an unimproved tropical grassland in northern Australia (**Source:** Norman, 1963)

and *Beijerinckia* are known to occur in tropical grassland soils (Tschan, 1953; Becking, 1961; Döbereiner, 1969), but it has not been proved that they fix a useful amount of nitrogen under practical conditions. Algae may fix some nitrogen. Nonsymbiotic fixation is more likely to be of importance in very nitrogen-deficient soils, because available soil nitrogen can substitute for fixation (Henzell and Norris, 1962).

Studies carried out in higher latitudes have shown that nonsymbiotic nitrogen fixation may be significant in waterlogged soils (Barrow and Jenkinson, 1962; Mulder, Lie and Woldendorp, 1969; Porter and Grabble, 1969), but in well-aerated soils it is likely to be negligible (Mishutin and Shilnikova, 1969). Algal crusts are capable of adding some nitrogen to grassland soils, even in semi-arid climates (Mayland, McIntosh and Fuller, 1966).

Losses. Removal of nitrogen from grassland in the form of milk or beef is likely to be quite small under extensive conditions. As nitrogen concentration in cattle is about 2.4 percent of their live weight (Agricultural Research Council, 1965), the removal of nitrogen from systems that turn off less than 100 kg/ha/year of live weight is likely to be less than 2 kg/ha/year. The quantity removed in whole milk may be somewhat greater; cows' milk contains 0.5 to 0.6 percent nitrogen by weight.

A large number of measurements have been carried out (e.g. in Uganda by Mills, 1953; in Queensland by Martin and Cox, 1956; in Ghana by Greenland, 1958; and in Nigeria by Moore and Jaiyebo, 1963) which show low levels of exchangeable ammonium (< 10 mg/kg) and very low levels of nitrate in soils under unimproved tropical grasslands. Higher levels of nitrate have been reported occasionally under old swards (Hardy, 1946), although usually in unimproved grassland nitrate-nitrogen concentrations of 10 mg/kg or more are found only during the first year or two after the soil has been disturbed by cultivation (Theron, 1951; Botha, 1963; Warren, 1956). In fact there is evidence that mineralization is suppressed in old grassland that receives little or no input of nitrogen from fertilizers or legumes (Theron, 1951). Because of the very low concentrations of mineral nitrogen normally found in soil under unimproved tropical grasslands, losses of nitrogen by volatilization or denitrification (Hauck and Bremner, 1969) and leaching are likely to be very small indeed.

Nitrogen is lost when grasslands are burned, though the quantity remaining in the top growth toward the end of the dry season (when they are normally burned) may be only about one-third of the amount present at the height of the growing season (Brockington, 1960; Norman, 1963). About 5 kg N/ha are lost from burning the native pasture at Katherine, Northern Territory (Norman and Wetselaar, 1960). The effect of this loss may not be obvious because burning can increase the rate of release of soil nitrogen.

Nitrogen may be partially lost by volatilization of ammonia to the air from dung and urine on the pasture, or totally if the excreta are deposited off the area of grassland (e.g. when cattle are kept in kraals or in bomas at night). Since a much larger proportion of the nitrogen in the diet of ruminants is excreted than is retained in animal tissues, this source of loss is potentially much larger than that caused by removal of milk and beef.

Mineral nitrogen and legume growth. Legumes grow just as well on mineral nitrogen as on nitrogen fixed from the air.

Legume seedlings grown on a nitrogen-deficient soil usually exhibit a period of nitrogen starvation after the seed reserves are exhausted and before the nodules become fully effective. Addition of mineral nitrogen at this time can cause an increase in legume growth that persists after the mineral nitrogen has been exhausted and the plants become dependent on symbiotic fixation. This effect has been demonstrated a number of times in pots, but efforts to obtain the same response in the field have usually been unsuccessful, probably because of competition from non-legumes. Studies using the 15 N isotope (Henzell, 1970a) have shown that pasture legumes generally obtain less than their share of the available nitrogen when grown in competition with grasses and broad-leafed weeds (if one judges their share by the proportion they contribute to the yield of forage). The fact that legumes usually die out from mixed swards if they fail to nodulate is additional evidence of their inability to compete for mineral nitrogen.

With older, effectively nodulated legume plants it is unusual to obtain a large response to additions of mineral nitrogen (e.g. as fertilizer). Indeed, growth with mineral nitrogen is the common yardstick for assessing whether a legume is effectively nodulated or not. A large response to fertilizer nitrogen by established pure stands of pasture legumes indicates a defect in the symbiosis that may be corrected by further research.

Nitrogen and grass-legume competition. The supply of available nitrogen is one of the main factors governing the degree of competition for light, water, and mineral nutrients in grass-legume pastures.

In temperate grass-clover pastures, the taller grasses tend to shade the shorter clovers (Donald, 1963). The relatively low-growing tropical legumes *Stylosanthes humilis* and *Lotononis bainesii* show essentially the same

92

response to grazing and mowing as the clovers (Shaw, 1961; Bryan, Sharpe and Haydock, 1961) but other tropical legumes, including *Desmodium intortum* and *D. uncinatum, Neonotonia wightii, Macroptilium atropurpureum* and *Centrosema pubescens*, climb over or twine upward on taller vegetation as they grow.

When mixtures containing these species are cut or grazed to near ground level and are allowed to grow again, the grass grows more rapidly at first and tends to shade the young legume shoots (as with clover and ryegrass). Later, these tropical legumes can grow above and shade the grass, and the competition is then reversed (Jones, 1967b). Some of the tropical browse legumes (e.g. *Leucaena leucocephala*) grow tall enough to shade their companion grasses all the time (Gray, 1968).

In the case of ryegrass and clover, any increase in the supply of available nitrogen, whether from soil, nitrogenous fertilizers or previous legume fixation, increases the shading (Donald, 1963), and may also intensify competition by the grass for nutrients such as phosphorus, potassium and sulphur (Mouat and Walker, 1959). Similarly, repeated application of nitrogenous fertilizers to tropical pasture mixtures usually reduces the proportion of legume (Henzell, 1963; Horrell and Court, 1965; Jones, 1967b; Stobbs, 1969a; Clatworthy, 1970; Whitney, 1970). Colman (1970) found with *Neonotonia wightii* and Kikuyu (*Pennisetum clandestinum*) that the reduction in legume yield due to nitrogen application was greater as grazing frequency increased. Presumably, this interaction resulted from the ability of *N. wightii* to grow over the Kikuyu under lenient defoliation. The effect of nitrogenous fertilizers on pasture mixtures containing browse legumes apparently has not been studied yet.

The grasses in tropical legume-based pasture mixtures are usually quite nitrogen-deficient. This is shown by the generally low concentration of nitrogen in the grasses and also by the fact that they always respond to nitrogenous fertilizers. It seems to be impossible in practice to combine a high rate of symbiotic nitrogen fixation by a legume with an optimum supply of available nitrogen to the associated grass, presumably because the balance of competition is too much in favour of the grass. This also helps to explain why some of the most spectacular successes with legume-based sown pastures, in cool and warm climates, have been achieved in regions with adequate rainfall and very poor soils, particularly soils that are acutely deficient in nitrogen. Under these conditions the use of fertilizers to correct nutrient deficiencies, other than the acute nitrogen deficiency, gives a large competitive advantage to the legume. On soils that mineralize nitrogen rapidly, it is usually more difficult to maintain a significant proportion of legume in the mixture.

On some tropical soils, grasses and weeds show a larger response to super-phosphate than do their associated legumes (Moore, 1965; Jamieson, 1969), in direct contrast to what commonly occurs with clover-based pastures on nitrogen-deficient soils in cooler climates (Trumble and Shapter, 1937; Sears, 1953). It seems that these particular tropical soils contain sufficient available nitrogen to permit a considerable increase in grass growth when the overriding phosphorus deficiency is corrected.

Transfer of nitrogen from legumes to associated grasses. There are two pathways for transfer of nitrogen from legumes to associated non-legumes in mixed pastures: (*a*) transfer via the excreta of grazing livestock; and (*b*) underground transfer (Figure 21).

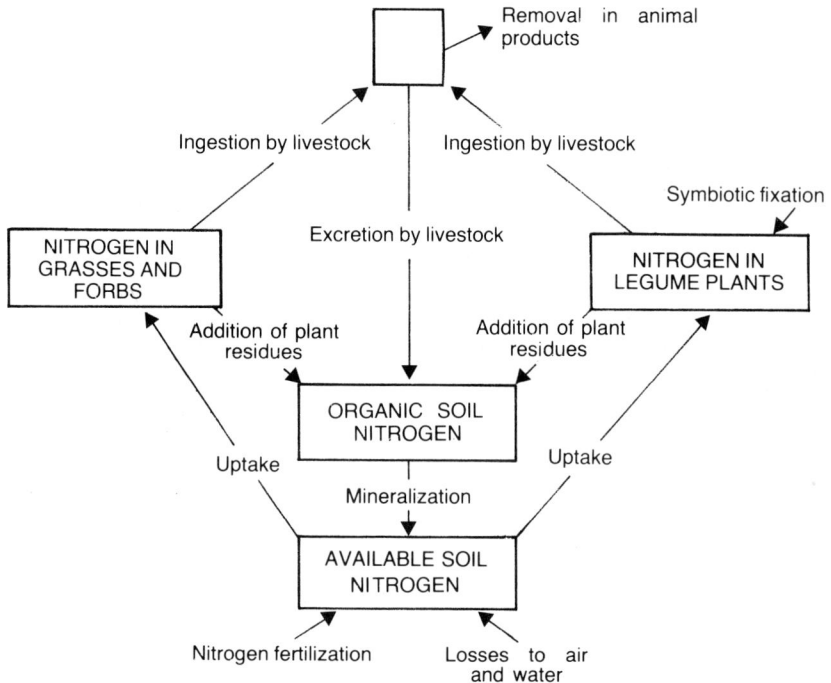

Figure 21. Flow diagram for nitrogen in grazed legume-grass mixtures (**Source:** Henzell, 1970a)

94

Transfer in excreta. Livestock excrete most of the nitrogen they ingest. Apparently the proportion that is digested and excreted in urine rises with the concentration of nitrogen in the diet (Milford and Haydock, 1965; Robinson and Stewart, 1968), and the nitrogen in urine is more readily available for plant uptake than that in dung.

However, the nitrogen in dung and urine is deposited on the pasture very unevenly, particularly by large animals such as cattle. Even with high stocking rates, excreta affect the nitrogen supply of only a very small fraction of the total area of pasture at any point in time (Lotero, Woodhouse and Petersen, 1966).

Under some conditions, substantial proportions of the nitrogen in dung and urine are lost to the atmosphere (Henzell, 1970a).

Underground transfer. This term covers all the other pathways of transfer, including those involving the legume tops (Henzell, 1970a). Recent results with tropical legumes showed negligible transfer from young, actively growing plants (Vallis *et al.*, 1967; Henzell *et al.*, 1968). Quite low rates of transfer over periods of six months to two years were observed by Henzell (1962a) and by Whitney, Kanehiro and Sherman (1967), who showed that *Desmodium canum* could actually depress the yield of nitrogen in associated grasses.

The main transfer is caused by senescence, death and decomposition of plant material — either from old tissues that died naturally, or from tissues of any age killed by grazing, drought, disease, etc. Opinions differ on the relative importance of the different plant parts. In *Centrosema pubescens* and *Desmodium intortum*, Whitney and Kanehiro (1967) found that leaf fall was a more important source of transfer than leaching from live legume tops or release from nodules and roots following defoliation.

Grazing pressure has an important influence on the mechanism of transfer. There will be very little leaf fall under heavy grazing but probably a faster turnover of root and nodule tissues.

No figures have been published for the rate of underground transfer by tropical legumes under practical grazing management. Special techniques are required to separate underground transfer of nitrogen from nitrogen transfer in dung and urine. However, some information is available on the effect of the two forms of transfer on the nitrogen concentration of grasses in mixed pastures.

Effect of legumes on the nitrogen concentration in associated grasses. In addition to the direct contribution that legume forage makes to the nitrogen

95

(crude protein) content of livestock diets, the presence of a legume in a pasture can also result in increases in the nitrogen concentration of the associated grasses, compared with the levels found in unimproved grasslands.

On plots that were grazed intermittently, Jones, Davies and Waite (1967) found that *Macroptilium atropurpureum* and *Lotononis bainesii* increased the nitrogen concentration in the tops of *Paspalum plicatulum* from about 0.8 to 1.0 percent to as much as 1.2 to 1.4 percent. Similarly, the nitrogen percentage in shoots of star grass (*Cynodon plectostachyus*) in Nigeria was raised from 1.8 to 2.4 as a result of association with *Centrosema pubescens* (Moore, 1962), and *Desmodium uncinatum* caused an increase of about this size with three gràsses in Kenya (Birch and Dougall, 1967). Under cutting, that is, with underground transfer only, *Desmodium intortum* gave a higher nitrogen percentage in pangola grass in Hawaii (Whitney, Kanehiro and Sherman, 1967), but five legumes had no effect on the nitrogen concentration of *Cenchrus ciliaris* at the end of a short-term trial in western Tanzania (Walker, 1969a).

Rates of nitrogen fixation. Under the moderately nitrogen-deficient conditions normally found in legume-based pastures, the amount of mineral nitrogen available for uptake by the legume is relatively small, consequently most of the nitrogen used for its growth has to be fixed from the air. The amount fixed is closely correlated with the yield of the legume (or the size of the plant).

Figures have been presented (Henzell, 1968) suggesting that the normal range yield (i.e. of tops cut off at some arbitrary height above the soil) from tropical pasture legumes in northeastern Australia is 1 000 to 5 500 kg/ha/year of dry matter and that the rates of nitrogen fixation associated with these yields would probably be about 20 to 180 kg N/ha/year. Higher values could be expected with faster growing species in more favourable tropical climates.

Failure to distinguish between fixation and net change in soil nitrogen has sometimes led to confusion about the nitrogen-fixing capacity of tropical legumes. Because nitrogen can be added to and lost from pasture systems simultaneously, nitrogen fixation by legumes does not necessarily mean that there will be an increase in soil nitrogen. In general, the losses are likely to be higher, particularly those due to leaching and volatilization, in legume-based pastures than in unimproved grasslands because of the higher nitrogen status of the system.

In Australia, measurements of net nitrogen changes of soil under tropical pastures have shown increases of up to 145 kg N/ha/year with good legume growth, though in other cases the change has been zero (Bruce, 1965) or even negative (Vallis, 1969). A *Centrosema pubescens* pasture in Nigeria added 280 kg N/ha/year to the soil (Moore, 1962).

The role of legumes in pasture improvement. The chief objective of including legumes in pastures is to increase livestock production, which may have no direct connection with nitrogen fixation. It is more likely to be a function of the feeding value of the legume forage. Through selective grazing quite a small quantity of legume in the pasture can markedly improve the nutrition of livestock at critical times of the year (Christian and Shaw, 1952).

The functions of symbiotic nitrogen fixation by pasture legumes can be summarized as follows.

● The ability to fix nitrogen from the air gives the legume a competitive advantage in association with grasses and weeds.
● The presence of the legume in the forage provides a better diet for live-stock, because legumes in general have a relatively high feeding value, and specifically because nodulated legumes can maintain a higher nitrogen concentration than grasses, particularly in mature forage (Norman, 1970). Nodulated legumes do not show the rapid dilution of plant nitrogen concentration that occurs when unnodulated legumes and non-legumes are grown with a limited nitrogen supply (Haydock and Norris, 1967).
● The legume contributes available nitrogen to non-legumes in the sward.
● The legume contributes nitrogen that may be useful during a subsequent period of arable farming.

Nitrogenous fertilizers

Yields in small plots. The high yields obtainable from nitrogen-fertilized tropical grasses in small-plot experiments have attracted a great deal of attention. Among the studies conducted before 1960 (Henzell, 1962b), the work in Puerto Rico is outstanding. With natural rainfall only, napier grass (*Pennisetum purpureum*) yielded 84 800 kg/ha/year of dry matter when fertilized with 897 kg N/ha/year and cut every 90 days (Vicente-Chandler, Silva and Figarella, 1959). Fertilized Para and Guinea grass also gave spectacular yields in the humid (1 430 to 2 610 mm of rain per year), warm (24 ± 6°C) environment of Rio Piedras (lat. 18°N).

Many other trials have confirmed the ability of tropical grasses to respond to relatively heavy (by temperate standards) dressings of nitrogenous fertiliz-

97

ers. They include experiments carried out in Georgia (Burton and Jackson, 1962), Florida (Wallace *et al.*, 1957) and Texas (Ashley *et al.*, 1965) in the United States; the Caribbean (Richards, 1965; Salette, 1970); Colombia (Crowder, Michelin and Bastidas, 1964); Hawaii (Whitney and Green, 1969a, b); northern Australia (Henzell, 1968); East Africa (Stobbs, 1969a); and Zimbabwe (Clatworthy, 1967). There is now ample proof of the efficacy of nitrogenous fertilizers in producing and maintaining high grass yields throughout the wetter subtropical and tropical lands of the world.

In the experiments quoted above, the grass was usually given substantial basal dressings of superphosphate and potash. Although this field of work is not as well documented, there is evidence of response to phosphorus, potassium and sulphur by nitrogen-fertilized grass (McClung and Quinn, 1959; Vicente-Chandler, 1966), and to lime when the soil has been acidified by heavy dressings of nitrogenous fertilizers (Abruna, Vicente-Chandler and Pearson, 1964; Adams *et al.*, 1967). No report has been seen of responses to micronutrients.

The amount of rainfall at which water becomes a more important limiting factor than nitrogen for growth of tropical grasses has never been precisely defined, but useful responses have been obtained down to 600 to 700 mm per year (Visser, 1966).

Effect of fertilization on nitrogen concentration. Nitrogen deficiency is one of the main reasons for the low nitrogen concentrations observed in the forage produced by unimproved tropical grasslands; much higher levels can be achieved in nitrogen-fertilized grassland.

The effect of fertilizer is strongly influenced by the relative timing of application and measurement. The following argument assumes that there is sufficient rainfall to wash the fertilizer into the soil and for grass growth. If there is a very short interval between fertilization and harvest, say about one week, nitrogen concentration may be increased without much change in yield (Barnes, 1960). Over a longer period, both yield and nitrogen concentration may be increased, but if the interval is 8 to 12 weeks or longer only the yield may be increased, except with very heavy fertilizer dressings (Oakes and Skov, 1962a; Henzell, 1963). Nevertheless, when sufficient nitrogen fertilizer is applied to satisfy the demand for increased yield and to meet the needs of a higher internal nitrogen concentration, it is possible to maintain a higher concentration even in mature forage (Henzell and Oxenham, 1964). Late fertilizer dressings have been used successfully to raise the nitrogen percentage of Coastal Bermuda and pangola grass during winter (Blue, Gammon, and Lundy, 1961; Kretschmer, 1964, 1965).

98

With heavy fertilization, the nitrogen concentration in tropical grasses is still likely to fall below the level needed by some classes of livestock (e.g. milking cows) even with selective grazing. The nitrogen percentages in tropical grasses tend to be lower than those in temperate grasses, other factors being equal. Among the tropical species, Kikuyu (Milford and Haydock, 1965), *Entolasia imbricata* (Bogdan, 1963) and some grasses from wet sites in Suriname (Dirven, 1965) have been reported to maintain higher than average nitrogen concentrations (see Colour plate II).

Livestock production. Discovery of the high dry-matter yields obtainable from nitrogen-fertilized tropical grasslands was followed by work on livestock production.

Beef. A trial in Puerto Rico showed live-weight gains (LWG) of 890 kg/ha/year on Guinea grass receiving 269 kg N/ha/year (Rivera-Brenes *et al.*, 1958). Later experiments using higher rates of nitrogen, up to 504 kg N/ha/year, under natural rainfall or with irrigation gave production in excess of 1 000 kg LWG/ha/year (Caro-Costas and Vicente-Chandler, 1961; Caro-Costas, Vicente-Chandler and Burleigh, 1961; Caro-Costas, Vicente-Chandler and Figarella, 1965). Rates of gain exceeding 1 000 kg/ha/year have also been recorded in other parts of the Caribbean (Oakes, 1960; Richards, 1966, 1970), in Hawaii (Plucknett, 1970) and in Queensland (Queensland Department of Primary Industries, 1967; Evans, 1969).

Table 6.1 summarizes results of other experiments in which a large response to nitrogen fertilization was recorded. The reader is referred to the original papers for details about the amounts of other nutrients that were used, the type of cattle and the type of grazing management. Particular attention should be given to the latter point, because there is a real doubt in some of these experiments whether the form of management (for instance, reduction of the stocking rate, or no stocking at all, during the dry season) is compatible with the organization of the local commercial beef cattle industry.

Not all grazing experiments have shown large increases in beef production (Quinn *et al.*, 1966; Henzell, 1968), and some negative results may not have been published. In general, the proportional increases have not been as large for beef production as for forage yield, indicating an inefficient conversion of forage into beef. The feeding value of nitrogen-fertilized grass is considered further in the next section.

Dairying. A number of experiments have indicated that nitrogen fertilization

99

TABLE 6.1 **Results of some experiments with beef cattle on nitrogen-fertilized tropical grasses**

Country	Grass	Rate of nitrogen	Rate of live-weight gain	Reference
		(kg/ha/year)		
Australia (Queensland)	*Chloris gayana* *Setaria anceps*	336	538	R.J. Jones, quoted by Henzell, 1968
Brazil	Colonial guinea (*Panicum maximum*)	0 200	241 595	Quinn *et al.*, 1970
Brazil	Pangola (*Digitaria decumbens*)	0 100	349 531	Aronovich, Serpa and Ribeiro, 1970
South Africa (the Transvaal)	Veldt	0 71	66 226	Hall, Meredith and Altona, 1950
Zimbabwe	Star (*Cynodon plectostachyus*)	270	830	Rodel, 1970
Uganda	Guinea, Rhodes (*Panicum maximum, Hyparrhenia rufa*)	0 157	217 376	Stobbs, 1969c
United States (Georgia)	Coastal Bermuda (*Cynodon dactylon*)	0 224	290 767	Burton, 1954
(Florida)	Pangola (*Digitaria decumbens*)	112 336	336 637	Hodges *et al.*, 1967

results in relatively high per hectare production of butterfat (Colman *et al.*, 1966) or milk (Blydenstein *et al.*, 1969; Caro-Costas and Vicente-Chandler, 1969; Richards, 1970). However, Dirven (1965) drew attention to the fact that energy, not protein, is likely to be the limiting factor for milk production on nitrogen-fertilized tropical grasses. While milk production of 25 kg/cow/day is attainable on grass alone in temperate areas, only 8 to 9 kg might be possible on tropical grasses. Since then other authors have arrived at a similar conclusion (Hardison, 1966; Hamilton *et al.*, 1970) and experimental measurements have confirmed it in most cases (Blydenstein *et al.*, 1969; Hamilton *et al.*, 1970). Caro-Costas and Vicente-Chandler (1969) recorded a slightly higher per-cow milk yield (11.4 kg/day during an eight-month lactation period). The only exception to the generalization has occurred with Coastal Bermuda grass fed as silage or green chop with concentrates at Clemson,

South Carolina (King, 1964; Brannon, King and Cook, 1966). Relatively high milk yields (> 20 kg/cow/day) were obtained.

Although fattening beef animals are less demanding in their nutrition than dairy cows, the relatively low feeding value of tropical grasses probably restricts production of beef too. It is not uncommon to record only about 0.5 kg LWG/day on (apparently) good pasture with cattle that have the genetic potential to gain > 1 kg/day. This inefficient level of production per animal tends to be concealed by high stocking rates, which result in relatively high rates of production per hectare.

Commercial use of nitrogenous fertilizers. Despite a vast amount of research, the commercial use of nitrogenous fertilizers at significant rates on grassland remains confined to a small group of countries in western Europe and to parts of the United States. No truly tropical country uses much nitrogen on grassland at present though some is used in the southeast United States on grasses of tropical origin, such as Coastal Bermuda, pangola and Bahia grass. In Florida and Georgia, about half of the 1.1 million hectares of improved permanent pasture were fertilized with nitrogen in 1964, and the average rate of application per fertilized hectare was 50 kg in Florida and 77 kg in Georgia (Ibach and Adams, 1967).

Why is nitrogen not being used in the tropics? In some areas it is difficult to obtain supplies of nitrogen fertilizers because of the absence of domestic nitrogen-fixing plants or shortage of foreign exchange to import fertilizers, but in most countries nitrogen in not used because it is too expensive in relation to the price of livestock products. Detailed economic studies are regrettably scarce, but the current opinion is that nitrogen fertilization of grass is uneconomic for beef production in such places as Brazil (Aronovich, Serpa and Ribeiro, 1970), Uganda (Stobbs, 1969b), Tanzania (Walker, 1969b) and most of northern Australia (Henzell, 1970b). It is commercially viable in Florida and Georgia and there are indications of profitability in the coastal region of southern Queensland (Henzell, 1970b) and in Jamaica. Jamaica imports increasing quantities of beef, there is pressure for intensified land use (Richards, 1970) and economic studies have proved that fattening, but probably not breeding, should be highly profitable (Nestel and Creek, 1964, 1966). There seems to be general agreement, however, that milk production is more likely to give a profitable return from nitrogen fertilization than beef (Nestel and Creek, 1964, 1966; Caro-Costas and Vicente-Chandler, 1969; Henzell, 1970b).

There is no reason to believe that the price of nitrogenous fertilizers will change dramatically because of advances in the technology of nitrogen fixation. The costs of distribution and marketing already make up a significant part of the sale price. In the long run, research may improve the feeding value of tropical grasses and there is scope for the more efficient use of nitrogenous fertilizers. Losses of nitrogen, presumably by leaching and volatilization, appear to be very large with heavily fertilized grassland. Consequently, nitrogenous fertilizers often, but not always (Mott, Quinn and Bisschoff, 1970), have little residual value.

The present position in regard to the sources of nitrogen for tropical pastures can be summarized as follows: costs of livestock production in the tropics generally increase in this order: (a) extensive production from native vegetation exploiting native soil nitrogen; (b) more intensive production from sown grass (soil nitrogen) or grass-legume pastures (nitrogen from soil plus symbiotic fixation); (c) highly intensive production from nitrogen-fertilized grassland, where the main source of combined nitrogen is a fertilizer factory. While in most tropical countries the economic returns from dairying can probably justify stage (c), returns from beef production cannot.

Where this situation exists, priority must go to exploiting native soil nitrogen (which has no direct cost) through replacement of inedible trees and shrubs by edible plants and by control of grazing. The next step should be the use of phosphatic and other non-nitrogenous fertilizers on legume species that are adapted to growing on acid tropical soils. Despite the problems in obtaining truly reliable and persistent pasture legumes for some tropical environments, they still offer a cheaper method of development than nitrogenous fertilizers.

7. *Rhizobium* relationships of legumes

The Leguminosae have long been known as plants which in nature are associated with a group of symbiotic bacteria, the mutual relationship under favourable conditions producing nodules of bacteria-containing tissue on the living roots of these legumes. However, not all species of the Leguminosae are nodulated, notable exceptions being members of the family Caesalpiniaceae, of which 65 percent are not nodulated (Allen and Baldwin, 1954) including the browse species, *Bauhinia*. Nor are all root-nodulated plants Leguminosae; certain *Casuarina* and other species also have this characteristic, though they associate with different micro-organisms. Thus, the symbiotic bacteria of legumes, or *Rhizobium* as they are called, are not the only beneficial organism associated with roots or with nodules. *Rhizobium* technology has become a specialized field of study in relation to the Leguminosae.

The early study of *Rhizobium* and the technique of adding these bacteria to seed at planting to improve growth were directed to the temperate legume species of the tribes Trifolieae and Vicieae. Attempts to translate the results of this work directly to tropical species have given variable results.

Although Norris (1956, 1967) was not the first to suggest that the legumes originated in the tropics, he has used this hypothesis in his studies with *Rhizobium* and developed the concept that host specificity within the *Rhizobium* genus is a much later development associated mainly with the temperate legume, the original tropical species being unspecific or "promiscuous" in relation to their *Rhizobium* requirements.

Of the three families of legumes — the Caesalpiniaceae, the Mimosaceae and the Papilionaceae — the generally non-nodulated Caesalpiniaceae are regarded as the primitive group. The Mimosaceae, which are mainly trees and shrubs, are more advanced and are nodulated but rather restricted in their distribution. The Papilionaceae, although still containing the most valuable tropical pasture species, contain also the specialized temperate tribes

Loteae, Vicieae and Trifolieae, which encompass most of the commonly used temperate forage legumes.

Norris suggests that the original tropical legume was a tree, and that its evolution has been: tree → shrub → woody climber → perennial herb → annual herb. These annual legumes which have reached great importance in agriculture both as pulse crops and pasture species are thus far removed from the original typical tree and their specialized *Rhizobium* relationships are therefore understandable. They are further specialized in that they are adapted to fertile soils, require a high calcium status and are inefficient in extracting calcium, phosphorus and other nutrients from the soil. However, most tropical legumes are adapted to poor acid soils, where they have originated and on which they nodulate freely. An examination of roots of the naturally occurring legumes in the moister tropical areas will generally reveal abundant and apparently effective nodule development.

The nodulation of a legume

The mode of infection of clover by the nodule bacteria has been recorded by Nutman (1958). Before nodules appear on the clover root, nodule bacteria in the soil nearby are stimulated to multiply by a secretion from the root of nutrients and growth factors. The clover root is then infected through the root hair, which has been deformed by bacteria secretions. Infection occurs at a few well-separated zones on the root and not at random, and the number of hairs infected varies with the species of legume. This variation in infection among legumes is related genetically to the host legume's capacity for lateral root development. The infected root hair can be recognized under the microscope by the formation within it of an infection thread or tube containing the *Rhizobium* bacteria. The infection thread is laid down by the host plant. Some special tetraploid cells of the root cortex are stimulated into meristematic activity and form the nodule, activated by the infection thread.

Nodule formation can itself inhibit further nodule formation and the size of the nodule has a bearing on its inhibitory effect. Large nodules are mostly sparse, and small nodules are usually abundant. The life of a nodule in herbaceous species is usually short, the longevity and size being a function of the *Rhizobium* strain and the host plant. The total nodule volume per plant is constant, irrespective of the virulence of the *Rhizobium*, but differs with different strains or lines of the plant host.

The time taken for the appearance of the first nodule differs with species of legume. Gates (1970) recorded in the laboratory initial nodulation in

104

Stylosanthes humilis on day 17 from planting, active nodules on day 23, and active fixation on day 26. Oke (1967a) reported that *Centrosema pubescens* nodulated two weeks after germination, and *Cajanus cajan* and *S. guianensis* in three weeks.

Nodules occur on lucerne plants in the field about the twenty-third day after seed inoculation, and on white clover about the eleventh day.

The quantity of atmospheric nitrogen fixed by legumes varies with the type of legume host and the *Rhizobium* strain which lives in symbiosis with it. The amount of fixation is closely related to the volume of *Rhizobium*-containing tissue in the nodule and the duration of life of this tissue. Nitrogen fixation is thus closely related to nodule structure. Size of plant will also have an effect.

The presence of haemoglobin in root nodules is necessary for nitrogen fixation. Nodules formed by ineffective strains are white in cross section and do not contain haemoglobin. The red pigment of haemoglobin in effective nodules gradually changes to green as the bacterial tissue ages and nitrogen fixation ceases.

The two main types of *Rhizobium*

"Associated everywhere with tropical legumes is a type of *Rhizobium* which by common usage has come to be called the 'cowpea type'" (Norris, 1966). Basically this type of organism is a rod with a single subpolar flagellum which in culture grows slowly and produces an alkaline reaction in the growth medium. This type of organism normally cross-inoculates from one host species to another, from one genus to another, from one tribe to another and from one family to another. For lack of a better term it is referred to as "promiscuous" (some, however, can be quite specific). By contrast, the organism associated with the highly specialized tribes Vicieae and Trifolieae tends to show peritrichous flagella and very fast growth, and to produce a strong acid reaction in the culture medium. It also exhibits symbiotic specialization; usually there is a marked inability to cross-inoculate except between closely related species within a genus or between closely related genera.

Closely related legume species were formerly grouped according to their *Rhizobium* affinities into cross-inoculation groups, but there are many species which do not fit into this pattern. The promiscuous tropical legumes were assigned to the cowpea-type cross-inoculation group but Norris prefers to drop the grouped concept for these and simply refer to those legumes which do not require a specific *Rhizobium* as being nonspecialized.

Legumes should be treated with the cowpea-type inoculum. Those species

105

which will not nodulate with this type of *Rhizobium* are strainspecific and require a special inoculum.

In the process of evolution and the development of certain leguminous species within a special ecological niche, host specificity with regard to *Rhizobium* has been a natural process. Several instances of such specialization have been recorded, among them the clovers of the East African highlands (Norris, 1959a; Norris and 't Mannetje, 1964) and in *Trifolium ambiguum* (Hely, 1957, 1963), *Centrosema* (Bowen and Kennedy, 1961), and *Desmodium* (Diatloff, 1968). Therefore, each legume must be carefully tested as to its *Rhizobium* partner. Several of the tropical legumes will not require seed inoculation, though inoculation with a known effective strain of the cowpea type is an insurance against nodulation failure.

When a strain of *Rhizobium* is sought for a new legume, there is no substitute for conducting tests with a number of *Rhizobium* strains isolated from the species itself. For this reason, when plant explorers are collecting seeds of legumes for introduction purposes it is important also to collect nodules from the plants. These can be dried or stored in small plastic tubes containing a dehydrating agent such as calcium chloride.

If the strains from the species itself are not available, then strains from its nearest botanical relative should be tried.

In current *Rhizobium* research, a spectrum of strains is tested with each leguminous species and the nodulation performance is recorded and catalogued. Even within a species there can be specificity — a *Rhizobium* strain which effectively nodulates one variety of soybean may be quite ineffective on another.

To ascertain whether a *Rhizobium* strain is acid- or alkali-forming, Norris (1967) suggests that the strain should be cultured for a month on a standard yeast-extract mannitol medium containing bromthymol blue pH indicator. If the strain is an alkali producer, the medium turns blue; if it is an acid producer, the medium turns yellow. This test can be a useful diagnostic test for the plant introduction officer. If the examination shows that the bacteria from a newly introduced legume are acid producers, it is a strong indication that the legume will probably respond to lime additions on acid soils.

The problem of native *Rhizobium* and other biological antagonisms

Native strains of *Rhizobium* of the cowpea type can be very common in the soil and may be quite effective in nodulating the sown legume. The fate of an uninoculated seed planted in a new environment will depend in part on the effectiveness of the subsequent plant in its symbiosis with these native *Rhizobium* species. It may be compatible and nodulate and develop well, it may be incompatible and fail to nodulate, or it may slowly select an effective strain from the native *Rhizobium* and gradually build up a successful population of effective nodules over a period. Thus, some tropical legumes such as *Glycine, Desmodium* and *Centrosema* may grow much better in their second year than in the year of establishment.

The acceptance or rejection of a *Rhizobium* strain is genetically controlled so that as each newly bred legume is produced it must be fitted with its effective strain of *Rhizobium*. Every new variety must be screened against the inoculant currently in use before embarking on field experiments. Even with a promiscuous legume, inoculation with a known effective strain is desirable and can often be beneficial; thus, *Stylosanthes guianensis* at Pitangueiras in Brazil grows much better when inoculated with the Australian cowpea-type inoculum (*Rhizobium* strain CB756) than when uninoculated.

Specificity of a host enables it to be inoculated with an effective strain and gives it a better start (provided inoculation is successful) than in the case of a promiscuous legume which may be nodulated by ineffective strains in the soil. Enlarging the dose of the chosen strain in order to obtain high *Rhizobium* counts on the seed increases chances of nodulation by the applied strain.

In Western Australia an interesting case of biological antagonism was encountered. A lateritic podzolic soil carrying dwarf tamma scrub (*Casuarina humilis* Otto and Dietr.) was cleared and sown to inoculated subterranean clover seed immediately afterwards. Nodulation failed, the plants were severely stunted and a sward could not be established. When the land was allowed to lie fallow for two to three years, establishment was successful. After clearing, the soil is left with 10 percent by weight of the surface 5 centimetres composed of root tissue from the *Casuarina*. On these roots a dense growth of fungi develops successively. *Penicillium* spp. dominate the early phase and these are succeeded by *Ascomycetes* and *Basidiomycetes*, at which stage toxicity disappears and nodulation follows. The antibiotics produced in the early phases by numerous organisms are toxic to both plant and *Rhizobium*. This case is worth remembering when problem areas are encountered.

Types of inoculants. Some early attempts to introduce legumes into a new area involved the spreading of soil from the original habitat over the new land to be planted. Later, the culture of the nodulating organism, *Rhizobium*, in the laboratory in various liquid or on semi-solid media to which the necessary nutrients had been added and the treating of the seed with this culture prior to planting was adopted. Loss of viability of the *Rhizobium* in the field was rapid.

As knowledge of the *Rhizobium* strain requirements of each particular legume increased, it became necessary to hold type or reference cultures of each in the laboratory, releasing starter cultures from these to be multiplied for use in the field as commercial inoculants.

Type cultures are usually stored in the laboratory on agar slopes covered with sterile paraffin, on porcelain beads, or by freeze-drying (lyophilized and held at 5°C storage). In Australia, commercial inoculants are prepared by growing *Rhizobium* species from mother cultures prepared from the type culture in yeast mannitol broth to give a high count of viable bacteria (at least 500×106 per ml and preferably better than $2\,000 \times 106$), injecting this broth into unsterilized peat at 45°C or gamma-irradiated peat at a moisture content of 45 to 60 percent on a dry-weight basis, incubating for a week at 26°C to "mature" (multiply) and then packaging in bags made from 0.038-cm-thick polythene for distribution. These polythene bags are best stored in a domestic refrigerator until used, but the bacteria will normally survive room temperature of 26°C for up to 26 weeks on sterile peat. Nonsterile peat requires cold storage for maximum survival of bacteria.

For commercial sale, each package is labelled with the serial number of the inoculant, the legumes for which the culture is to be used, the date of expiry of viability, and storage methods. Some notes on inoculating seed may also be added.

Quality control of legume inoculants. Date (1969) has described the system of quality control in legume inoculants in Australia. It was initiated in 1955 by the Microbiology Group of the Faculty of Agriculture of Sydney University under the name of U-DALS, and now functions under the name of Australian Inoculants Research and Control Service (AIRCS).

The function of AIRCS can be divided into three main aspects: (*a*) selection, testing and maintenance of suitable rhizobial strains; (*b*) control of quality of legume inoculants; (*c*) advice to, and research for, manufacturers, distributors and users of inoculants on the problem of production handling and application that affect the quality and efficiency of inoculant cultures.

In the inoculation process, the peat mixture is made into a slurry with water or preferably with a dilute suspension of adhesive (1 to 2 percent Mellofos,

15 percent gum arabic) for better adherence. The seed is heaped on a concrete floor and as the inoculum is added it is thoroughly mixed with a shovel; or the operation can be carried out in a concrete mixer or similar machine. The aim is to have the seed covered without excessive moisture.

The next stage is most important — the seed should be spread out to dry in a thin layer in the shade and should be planted as soon as possible thereafter. For most tropical legumes this treatment is all that is required.

Australian recommendations have been based on the use of one or two strains, each with a wide host spectrum, rather than multistrain cultures (up to ten separate strains) as used in other parts of the world. The apparent purpose of multistrain cultures was their wide host range, but this can be dangerous where some strains may be competitive for nodule formation on a particular host or may dominate in the inoculant culture. The simpler approach has the advantage of greater ease of preparation of the inoculant itself and of consumer comprehension.

Broth cultures from manufacturers are tested for quality and provisional results are given to the manufacturer within 24 hours. The broth is then impregnated into peat and held as single strain material. This peat material is then tested for *Rhizobium* counts and also for effective nodulation of the host plant growing aseptically in a nitrogen-free plant nutrient medium before the expiry date is finalized.

The inoculation of legume seed. The aim of inoculation of legume seed is to coat the seed with a sufficiently high number of viable rhizobia of the correct strain to provide early and effective nodulation of that legume in the field.

Vincent (1970) regards 300 organisms per seed as minimal to ensure satisfactory nodulation but in current practice much higher numbers are used. Diatloff (personal communication) suggests 510 to 51 000 cells per seed for commercial cultures.

Table 7.1 indicates the quantity of peat culture to use in legume inoculation. It is based on figures supplied by the CSIRO Division of Tropical Agronomy, Cunningham Laboratory, Brisbane, Australia.

Effect of physical factors on legume symbiosis (van Schreven, 1958)

Among physical factors affecting symbiosis are air, moisture, light, temperature and soil reaction.

109

TABLE 7.1 **Guide to the quantity of peat culture to use in legume inoculation**

Seed size group	Species	No. seeds/kg	70-g culture treatments (kg)	1-g culture treatments (g)
2.2 million/kg	*Lotononis bainesii*	3 828 000	0.45	7
	Lotus uliginosus	2 310 000	0.45	7
660 000 to 2.2 million/kg	*Indigofera spicata*	660 000	2.3	35
	Lotus corniculatus	811 800		
	Medicago polymorpha	666 600		
	Trifolium fragiferum	792 000		
	Trifolium repens	1 936 000		
132 000 to 660 000/kg	*Desmodium intortum*	440 000	7	100
	Desmodium uncinatum	268 000		
	Macrotyloma axillare	138 600		
	Medicago sativa	446 600		
	Medicago truncatula	268 400		
	Neonotonia wightii	151 800		
	Stylosanthes guianensis	338 800		
	Stylosanthes humilis	380 600		
	Trifolium alexandrinum	440 000		
	Trifolium pratense	554 400		
	Trifolium subterraneum	88 000-154 000		
22 000 to 132 000/kg	*Calopogonium mucunoides*	66 000	13.5	200
	Centrosema pubescens	41 800		
	Leucaena leucocephala	26 400		
	Macrotyloma uniflorum	39 600		
	Macroptilium atropurpureum	96 800		
	Macroptilium lathyroides	121 000		
	Pueraria phaseoloides	88 000		
	Teramnus uncinatus	33 000		
	Vicia angustifolia	59 840		
22 000/kg	*Cajanus cajan*	16 000	27	400
	Lablab purpureus	4 400		
	Stizolobium deeringianum	1 800		
	Vigna unguiculata	13 200		

NOTE: A guide to the *Rhizobium* culture needed to inoculate the various legumes used in tropical pastures has been prepared by Norris (1967) and Date (1969). This guide and a list of *Rhizobium* cultures is given in Appendix 2.

Air. A well-structured soil allowing free access of air has a beneficial effect on the activity of rhizobia.

Moisture. The viability of rhizobia is affected by moisture content of the soil and rhizobia are sensitive to excessive drying when exposed to the open air. However, small numbers survive for a long time in air-dried soil, probably in the films of hygroscopic moisture which surround the soil particles. Excess water may limit aeration and hence survival of the bacteria. Maximum growth and nodulation of legumes generally occur in soil with a water content between 75 and 85 percent of its water-holding capacity. Survival of rhizobia in arid regions is important. The technique of seed pelleting is designed to increase survival of selected strains on the seed sown.

Light. An optimum light intensity is necessary for maximum nodulation and nitrogen fixation. Under shaded conditions it has been shown that spraying with sugar solution increases nitrogen fixation. Under grazing many nodules are shed or decay and are replaced only when the plant has produced new growth that can provide soluble carbohydrates.

Temperature. Gukova (1945) found that a decrease of 5°C below optimum soil temperature reduced the amount of nitrogen fixed by 4.5 percent, whereas an increase of 4°C above optimum reduced nitrogen fixation by 5 percent.

With tropical legume species sown during high summer temperatures and high humidity, the survival of *Rhizobium* is important. Bowen and Kennedy (1959) found that on agar, 68 *Rhizobium* strains of the cowpea type had a mean upper limit of growth of 35.4°C compared with 33.2°C for nine strains from *Trifolium* and 41°C for eight strains from *Medicago*. Some cowpea strains had a growth threshold as low as 30°C. In moist sandy soil three or four cowpea-type strains failed to survive 12 hours at 40°C. Field measurements at Coolum, Queensland, Australia (lat. 27°S), showed that soil at 2.54-cm depth reached 47.5°C on a February day, remaining above 40°C for six hours. Norris has drawn attention to the fact that rhizobia in peat cultures survive much better than those grown on agar slopes, and the experiments listed above should be repeated with peat culture. Souto (1969) showed that high nitrogen and phosphorus levels in the soil improved tolerance of *Neonotonia wightii* to high soil temperatures. This was a nodulation effect.

Reaction. Strains of *Rhizobium* differ in their tolerance of acidity. In pure culture, lucerne organisms will not grow below about pH 5.0, clover organisms below pH 4.5. Munns (1968) showed that acidity inhibits nodulation of lucerne in the early stages but has no effect on nodulation after nodule infection. The acidity prevents root-hair infection.

Soil acidity causes problems in mineral nutrition of both *Rhizobium* and the legume host such as deficiencies of calcium, magnesium and potassium. Often phosphorus and nitrogen are deficient and toxic amounts of manganese and aluminium and of hydrogen ions may occur. One of the reasons for poor growth of certain legumes on acid soils may be reduced uptake of molybdenum.

Seed tolerance to insecticidal treatment

Jones (1965) and Russell and Coaldrake (1966) showed that seed treatment with normal dressings of the insecticides dieldrin, aldrin, endrin, telodrin, DDT and lindane had no adverse affect on nodulation, but the latter workers found that increasing the dose of endrin by a factor of 3 had a strong suppressive effect on both nodulation and growth.

Nutrition of the nodulated legume

Nitrogen. Small additions of nitrogen to the soil enhance nodulation by legumes. Large additions of nitrogen seem to depress nitrogen fixation probably owing to too low a carbohydrate/nitrogen ratio in the plant. Root-hair infection is inhibited by high nitrate. Fixation of atmospheric nitrogen decreases with increasing rates of fertilizer nitrogen. Gates (1970) working with *Stylosanthes humilis* has shown that, by correctly balancing nitrogen and phosphorus, growth can be improved without impairment of nodulation. Increase in nodule dry weight was proportional to the nitrogen added, up to at least 16 mg nitrogen. In the case of mixed pastures, the addition of fertilizer nitrogen stimulates the grass component which then suppresses the legume by increasing competition.

Phosphorus. Gates (1970) showed that the demands for both phosphorus and nitrogen are as high in the nodule as in the young laminae in a number of tropical legumes including *Stylosanthes humilis, Macroptilium atropurpureum*

112

and *Leucaena leucocephala*. The nodule dry weight increases with increasing levels of phosphorus up to at least 250 kg/ha P. The number and density of nodules are greatly stimulated by phosphorus; also the rate of fixation of nitrogen by the nodule is enhanced by the addition of phosphorus. Consequently, the addition of phosphorus to a pasture usually results in an increase in the percentage of legumes as their competitive ability is improved by phosphorus in relation to the associated grasses.

Potassium. This appears to have a beneficial effect on nodulation only in the presence of adequate levels of phosphorus.

Manganese. High levels of manganese have an adverse effect on symbiosis (Souto and Döbereiner, 1970), much more so than on the plant itself. Liming helps to overcome this toxicity, and high levels of phosphorus tend to enhance manganese absorption.

Magnesium. Norris (1959b) found that magnesium, rather than calcium, was the major requirement of *Rhizobium* and subsequently Vincent (1962) found that the ratio of magnesium to calcium should be about 8:1.

Sulphur. Sulphur is important in nitrogen metabolism because it is a component of proteins. Gates (1970) showed that it is important in initiating nodulation but has little subsequent effect.

Boron. In the absence of boron, nodular tissue is affected. With broad beans, Brenchley and Thornton (1925) found boron deficiency reduced nodulation in inoculated plants to 10 percent of that in healthy plants. Anderson (1952) found a boron response with subterranean clover only in the presence of molybdenum.

Copper. Hallsworth, Greenwood and Yates (1964) showed that, with increasing levels of copper from 0.1 to 5 µmol per litre, the weight of nodules of *Trifolium subterraneum* increased. It appears that the host plant competes with the nodules for copper.

Cobalt. The addition of cobalt to culture solutions stimulates nitrogen fixation and increases the size of nodules. Vitamin B_{12} is present in nodules, and in pink nodules of *Medicago sativa* it is four times as plentiful as in ineffective white nodules (Hallsworth, Wilson and Greenwood, 1960).

113

Molybdenum. It has been shown by numerous workers that molybdenum is essential for nitrogen fixation by legumes. The molybdenum needs differ with legume species and the tropical legume *Neonotonia wightii* has a high requirement. Shaw, Gates and Wilson (1966) showed that higher nitrogen concentration and increased yields of *Stylosanthes humilis* were obtained with phosphorus and molybdenum. Molybdenum deficiency is found mainly on acid soils. Molybdenum is released from the soil by liming to raise the pH level but it is usually cheaper to add a few ounces of molybdenum to the fertilizer mixture, e.g. molybdenized superphosphate.

Calcium. Traditionally calcium has been regarded as essential for the growth of *Rhizobium* and it has been added to culture media as well as to the soil in which legumes are to be grown. Norris (1959b) found that *Rhizobium* was not a calcium-sensitive organism but needed calcium only in trace amounts. It was, however, very sensitive to magnesium. Bergersen (1961) and Vincent (1962) showed that response to calcium by *Rhizobium* ceases at about 1 ppm, which is higher than the level normally present in culture media by contamination. They also showed that for optimum growth in culture the Mg:Ca ratio should be about 8:1. Andrew and Norris (1961) proved that the process of nodule formation is dependent on calcium in both temperate and tropical legumes, but that the host legume would die from calcium deficiency in the field long before the bacteria were affected.

Pelleting to protect the inoculum on the seed

Cass-Smith and Pittman (1939) showed that mixing superphosphate with inoculated seed for even one hour before planting gave a marked reduction in establishment, and contacts for 24 hours completely killed the *Rhizobium*. Contact with some nonacid phosphate such as rock phosphate or basic superphosphate is not harmful.

In semi-arid areas or even in moister areas where a dry period occurs after sowing or where legume seed is sown from the air on to a dry soil surface, especially in the tropics, the inoculum may not survive the adverse effects of sunlight and dryness. Seed may also be removed by seed-harvesting ants.

To protect the seed after inoculation, a process of seed pelleting has been developed. It generally involves coating the inoculated seed with either lime or rock phosphate using a sticker such as 45 percent wt/vol gum arabic solution or 4 percent pure methyl cellulose solution.

Properly pelleted seed may be held at room temperature up to one month before sowing and still produce good nodulation. If the pelleting material is calcium carbonate, the technique enables the pelleted seed to be mixed for a short time with superphosphate at sowing time, which may facilitate operations. However, to obtain survival in pellets, it is essential to use only peat culture. Equivalent pellets made with broth or agar cultures lose viability quite quickly and may be of little use.

Adhesive. Several alternatives are available. The most widely available is powdered commercial gum arabic (= gum acacia). This is used as a 40 percent wt/vol solution. It takes considerable time to dissolve and may require gentle heating. Alternatives are the commercially available methyl cellulose materials Methophas P.M. 125 granular supplied by ICI, and Methocel supplied by Dow Chemical Company. It is essential that the grade supplied should be guaranteed to be free from preservatives. These are used as a 3 to 4 percent water solution and are best allowed to stand overnight to thicken. Large-scale operations allow costs to be cut because 1.5 kg do the work of 15 kg gum arabic. The celluloses are inert and neutral substances, unlike gum arabic which is a nutrient material and has a pH of about 3.9. If seed-harvesting ants are a problem, cellulose pelleting seems to make the seed less attractive than gum-arabic pelleting. Normally no additives such as sugar are used, though they do occur in certain commercial pelleting operations.

Coating material. This must be very finely ground to the consistency of dust (100 percent to pass at least a 300-mesh sieve). For instance, in the case of lime, plasterers' whiting and microfine lime are suitable, but ordinary agricultural grade lime is not. About 25 percent of the seed weight (e.g. 3.5 kg lime) is required for 14 kg clover seed. The smaller the seed, the more material will be required for pelleting (see Table 7.2).

Use either lime or finely ground rock phosphate for pelleting. In New Zealand a mixture of dolomite and rock phosphate is preferred for clover seed.

There are two distinct types of *Rhizobium* to be considered. The type associated with clovers, medics and some others is a fast-growing acid producer for which lime is used as a coating. The type associated with most tropical species (but there are exceptions) is a slow-growing alkali producer, for which rock phosphate should be used.

The following list is intended only to be a broad guide. In exceptional circumstances, some of the species in the rock phosphate list may benefit from a lime pellet.

115

TABLE 7.2 **Recommended coating materials for pelleting legume seeds**

Lime	Rock phosphate
Adesmia	Acacia
Anthyllis	Aeschynomene
Astragalus	Alysicarpus
Cicer	Arachis
Coronilla	Cajanus
Hedysarum	Calopogonium
Lathyrus	Canavalia
Lens	Cassia
Leucaena	Centrosema
Lotus corniculatus	Clitoria
L. maroccanus	Crotalaria
Medicago	Cyamopsis
Melilotus	Desmodium
Onobrychis	Glycine
Phaseolus vulgaris	Indigofera
P. coccineus	Lablab
Pisum	Lespedeza
Psoralea	Lotononis
Sesbania	Lotus major
Trifolium	Lupinus
Trigonella	Macroptilium
Vicia	(all Phaseolus spp. except vulgaris and coccineus)
	Ornithopus
	Pueraria
	Rhynchosia
	Stilozobium
	Stylosanthes
	Tephrosia
	Teramnus
	Vigna
	Zornia

Döbereiner and Aronovich (1965) found a beneficial effect from lime pelleting on nodulation of *Centrosema pubescens* growing in a manganese toxic sand in Brazil.

Neonotonia wightii, which seems to prefer a soil with a pH above 6.5, can give a response to lime pelleting. *Desmodium uncinatum* has also given a response, but *Desmodium intortum* has not.

Quantities required. These vary according to seed size and the purpose for which the pellet is to be used. The smaller the seed, the more material is required. There are two main reasons for pelleting: as a method of preinoculation, in which case only a light coating of pellet materials is required; and to

116

protect bacteria against adverse conditions such as when sowing in direct contact with acid fertilizers like superphosphate, sowing into dry soils or when sowing members of the "lime" group into acid soils, in which case a heavy coating is preferable. The information on quantities (see Tables 7.3 and 7.4), is therefore subdivided for seed size and for light and heavy pellets.

Four basic seed sizes are recognized: small (e.g. white clovers, lotus, *Lotononis*); small to medium (e.g. *Desmodium intortum*, lucerne, *Stylosanthes*); medium (e.g. subterranean clover, glycine, siratro, *Medicago*); large (e.g. vetch, cowpea, *Lablab purpureus, Leucaena*).

TABLE 7.3 **Seed size for light pellets**

Seed size	Large lots			Small lots		
	seed (kg)	adhesive (ml)	coating (kg)	seed (g)	adhesive (ml)	coating (g)
Small	6.8	284	3.4	25	1	12.5
Small to medium	10.2	284	3.4	37.5	1	12.5
Medium	13.6	284	3.4	50	1	12.5
Large	27.2	284	3.4	100	1	12.5

TABLE 7.4 **Seed size for heavy pellets**

Seed size	Large lots			Small lots		
	seed (kg)	adhesive (litres)	coating (kg)	seed (g)	adhesive (ml)	coating (g)
Small	6.8	1.136	6.8	25	4	25
Small to medium	10.2	1.136	9	37.5	4	30
Medium	13.6	1.136	9	50	4	30
Large	27.2	1.136	6.8	100	4	25

Hints on procedure. There is no substitute for experience in pelleting. Do not expect the first batch to be perfect. The peat culture is thoroughly mixed and stirred into the adhesive just before the pelleting is to be done. Then the adhesive is stirred into the seed. The object is to get the seed thoroughly wet and sticky without it becoming a soggy mass and without fouling the mixing vessel.

117

Pelleting is best done by some type of spinning or tumbling action, not by shaking. Small lots may be done in a beaker, can or bucket; large lots in a cement mixer, preferably without vanes. Lacking a cement mixer, larger batches may be done by rolling the material backward and forward on a large sheet of plastic. In small lots, the best pellets result from dumping the sticky seed into the coating material and stirring briskly with a round rod. In larger batches, you must add the coating material to the sticky seed, but this should be dumped in one batch, not sprinkled in a little at a time. If you intend to spin the seed, do not overload the vessel, because seed must have room to tumble.

Pelleted seed may be used immediately if desired, but the coat will be soft. A few hours' drying (in the bag) allows the pellet to harden sufficiently to be sown through the seed box. If larger pellets are used, a drying period is desirable, otherwise the coat will tend to break off.

The pelleting vessel should always be sterilized at the conclusion of operations. This is best done by washing it thoroughly with hot tap water, and then rinsing it out with alcohol. If doing a series of species requiring different *Rhizobium* strains, you may have to sterilize between strains. This depends on the cross-inoculating facility of the legume species being pelleted. For instance, one could pass from *Lotononis* to almost any other species or vice versa without sterilizing, and from almost any member of the "lime" group to any member of the "rock phosphate" group or vice versa, or from lucerne to clover. But it would be unwise to pass from white clover to *Trifolium cherleri* (requiring a different strain), because enough carryover might occur to cause a significant amount of ineffective nodulation.

In heavy pellets there may be a surplus of coating material which can be sieved off if required.

Is pelleting really necessary? For many normal plantings in which seed is sown by hand or machine without contact with fertilizer, a pellet coat is quite unnecessary. Breaking down the strength of the adhesive (1 to 2 percent solution for the celluloses, 15 percent for gum arabic) will firmly glue the bacteria to the seed without the seed bonding together on drying. The bacteria will persist very well and seed inoculated this way may be safely held under cool conditions for two weeks before planting, if necessary.

Identification of effective nodulation in the field

Generally, effective nodulation takes place in the field within four weeks of planting. Effective nodulation is generally indicated by vigorous growth of

118

the legume itself and improved growth and colour of associated grasses; by the presence of large nodules on the taproot and main laterals in the top 2.5 to 5 cm of soil; and on cross-section the cut surface of the nodule shows a pink coloration due to haemoglobin. An old nodule may have a green area at the base of a restricted red zone indicating previous haemoglobin. (Colour plate V shows an effectively nodulated plant.)

In a few instances, distinctive nodule colour provides a reliable and ready means of identification of effectiveness, e.g. a strain of *Rhizobium* on *Lablab purpureus* and *Macrotyloma uniflorum* gives black nodules (Cloonan, 1963).

Major references

Date, R.A. (1969); Norris, D.O. (1967); Norris, D.O. (1970); van Schreven, D.A. (1958); Vincent, J.M. (1970).[1]

---

[1] The book *Tropical pasture research. Principles and methods*, edited by N.H. Shaw and W.W. Bryan, CSIRO Division of Tropical Agronomy, Australia, published (1977) by the Commonwealth Agricultural Bureaux, Hurley, Berkshire, UK, contains the most recent information on legume bacteriology from the late Dr D.O. Norris and Dr R.A. Date as Chapter 7.

8. Management of tropical pasture legumes

The most critical stage in the life of a pasture is the period of establishment. The aim is high-yielding pasture, free of weeds and capable of a high stocking capacity. All the expenditure involved in establishing the pasture is made with one end in view — a high-quality pasture which will give a return on the investment. Should good establishment not be attained, the pasture may take years to reach a highly productive state — and in many instances may never yield to the full capacity of the soil in which it grows, resulting in loss of profit or even loss of capital to the producer.

With adequate knowledge of soil requirements, land preparation, *Rhizobium* and nutrient needs of component species, it is now comparatively easy to establish legume-based pastures, but the maintenance of the botanical composition of the sown species over an economic period of land use requires considerable skill. It involves a knowledge of the growth rhythm of the different species in the sward and their responses to defoliation by the different grazing habits of the animal species using the pasture.

Knowledge of the growth rhythm of individual species is gradually accumulating, but much more is needed to fully understand the dynamic ecosystem of soil, plant and animal. Competition between pasture plants for water, light and nutrients, the effects of temperature and occasional fire and flooding, as well as selective grazing, and deposition of dung and urine by the animals affect the day-to-day and season-to-season dominance of the component species. Management is the intelligent manipulation of these factors in order to increase output in the most economically advantageous way.

"Animal production involves the synthesis of nature's basic materials into plants, the conversion of these plants into animal products, and their interactions. In most cases the management of grazing lands, considered in its broadest sense, is the weakest link in the chain. Not infrequently, however, one finds that disease, poor marketing facilities, use of animals of low potential, and other factors are limiting". (Peterson, 1962).

Early management of the pasture

While pasture growth is rapid during and immediately after the rainy season in the tropics, weed growth is equally vigorous, especially in the better soils. Careful seed-bed preparation to eliminate weeds and adequate fertilizer to give the sown pasture the best possible conditions for rapid establishment assist in early weed control; preplant herbicides also help. However, as weeds will still be a problem, it is important to allow the pasture to become well established before grazing so that it can compete with them.

In tall grass/legume mixtures, the grass is usually the more vigorous, especially in the first year, and the pasture may appear ready for grazing before the legume is well established. Usually the grass will seed in late summer to autumn and if it is shading the legume too much, a light grazing only should be given to top the grass and allow more light into the legume, particularly if the latter is a nonclimbing type. If the stand is thin, however, deferring the first grazing until after the first seed has fallen will allow self-sown seed to thicken up the sward. The light grazing during the first year establishes the ground cover and thereafter the stocking rate can be increased.

It is important to develop a balance of species, and it is especially important to preserve the legume component. In the first year, the legume is often dominant, but as it makes nitrogen available the grass component is strengthened and the need for balance becomes more important. At least 30 percent of legumes should be maintained in a pasture for high production (Bryan and Evans, 1968). Higher seeding rates for the legume in comparison with the grass, and adequate phosphorus fertilization will assist the legume to attain early dominance. Some leguminous species such as common stylo (*Stylosanthes guianensis*) and, to a lesser extent, siratro are unpalatable early. Thus the companion grass is the first to be grazed, allowing the legume to compete successfully and increase.

Grazing management to suppress the grass should be aimed at encouraging the stoloniferous growth of the legume.

As a rule the tropical pasture species should not be grazed too closely. The reaction of the companion grass as well as the legume to defoliation at varying heights must be taken into account. Pastures based on molasses grass (*Melinis minutiflora*) and *Brachiaria mutica* should not be grazed too closely, whereas

121

pastures of pangola grass (*Digitaria decumbens*), Kikuyu grass (*Pennisetum clandestinum*), Townsville stylo (*Stylosanthes humilis*) and Miles lotononis (*Lotononis bainesii*) can be grazed heavily.

Mechanical slashing to control tall growing weeds will help, but slashing should not be lower than 15 to 20 cm, as damage to the legume may result. Spot spraying with herbicides can be used, but only when it is known that the herbicide will not affect the pasture species or the grazing animal. Slashing and annual application of fertilizer are better than mechanical renovation of tropical pasture mixtures. Withdrawing a pasture temporarily from grazing will allow climbing legumes to overtop the pasture and suppress unwanted weed growth and should be done at least once a year (Luck and Douglas, 1966). If labour is cheap, hand grubbing of individual bushy species may be more efficient and economical.

Grazing management

Pasture is of value only if converted into animal products, apart from its soil fertility-building properties. The return from money invested in pastures, therefore, depends upon the operators' skill in converting herbage into saleable products and also upon the length of time the vigour of the sward can be maintained. Grazing management is a compromise between plant and animal needs: there is skill in knowing when to sacrifice pasture for the sake of the animals, and when to accept immediate detrimental effects on the stock for later benefits from the pasture.

Good grazing management should therefore aim at:

— supplying nutritious grazing throughout the year at low cost;
— minimizing physical waste and inefficient utilization of herbage; and
— maintaining the productivity of the sward (Wendt *et al.*, 1970).

Provision of nutritious grazing throughout the year

The provision of nutritious grazing throughout the year is an ideal which is difficult to achieve in the tropics unless rainfall is adequate in amount and uniform in distribution. Much of the tropics has a well-defined wet season when pasture growth is excessive and an equally well-defined, and often prolonged, dry season when there is a scarcity of animal food.

Forage availability varies from season to season during the year, and from year to year in most countries. It is probably the most general and important single problem in livestock feeding. It is the main cause of slow and irregular growth of steers — which often require four to five years to be ready for marketing. It is the direct and indirect cause of high mortality during droughts and is a contributing factor to low reproduction rates.

Stock numbers should be adjusted to ensure that there is a reasonable carryover of forage, however poor its quality, into the unfavourable seasons. This means, in the beef industry, disposal of fat cattle and culling of low-quality animals to provide adequate forage for young cattle and breeders.

Improved pastures can fill this latter role. Choice of species to fill in the gaps in the feed year have been discussed in Chapter 4. Autumn-saved and deferred feed can be provided by such legumes as *Neonotonia wightii*, *Macrotyloma axillare*, *Desmodium intortum* and especially *Stylosanthes guianensis* and *S. humilis*, and management of the grazing operations to allow these to be kept for such strategic feeding is desirable. An area of irrigated pasture is also invaluable for feeding special classes of stock during critical periods.

In extended dry seasons, conserved or purchased fodder and supplements will probably be needed. For shorter dry seasons, it is usually cheaper to conserve standing feed in the paddock if weather conditions permit (i.e. so that the feed will not seriously deteriorate from fungal attack).

This deferred feed need not be leguminous. Deferred grass supplemented with protein and nitrogen-rich sources such as cottonseed pellets can be utilized or molasses-urea made available in drum licks.

Minimizing physical waste and inefficient utilization of herbage

Stocking rate and grazing pressure. Both overgrazing and undergrazing are detrimental to the sward. Bryan (1968) found undergrazing encouraged bush regrowth at Beerwah in southeastern Queensland, Australia. The unknown factor in the management of tropical pastures has always been the optimum grazing rate at which a productive pasture can be maintained and high weight gains in grazing cattle can be obtained. In early grazing experiments in north Queensland grazing rates were very low, mainly because not much information was available on the extent to which pastures could be grazed without causing degradation. It was thought that the tall grasses would not persist if they were heavily grazed or cut short. Rates of up to 1.2 hectares per beast were employed and live-weight gains per hectare were low. In 1958, good Guinea grass/centro pasture at Utchee Creek, north Queensland, grazed at

123

an average rate of 0.68 hectare per beast through the year, gave a live-weight gain of 264 kg/ha. In 1969, the stocking rate was increased to 0.36 hectare per beast and live-weight gain increased to 572 kg/ha. The pasture was still not fully utilized and grazing intensity was increased to 7.5 beasts per hectare (0.13 hectare/beast) during the wet season, and an annual average of 0.32 ha/ beast; the live-weight gain increased to 610 kg/ha/year. These results were achieved on a newly cleared and planted fertile deep red basaltic latosol in an annual rainfall regime of 3 750 mm and would exceed the performance on poorer soils. It does emphasize the importance of stocking rate in maximum utilization of pastures and in economic returns. The most influential factors in management are the number of grazing animals per unit area of land (stocking rate) and the number of animals per unit of available herbage (grazing pressure). The intensity with which a pasture is grazed can affect both animal production and botanical composition. At low stocking rates, animal production per hectare is poor. Understocking results in underutilization of land and herbage, although production per animal is high.

As stocking rate increases, the output per animal declines but production per hectare increases (McMeekan, 1956; Mott, 1960; Bryant et al., 1965; Peterson et al., 1965). Maximum production per hectare is obtained when loss in production per animal by increasing the stocking rate is offset by the increased number of animals carried. The optimum carrying capacity can only be determined after long experience. Data available from stocking-rate studies suggest that the stocking rate at the point of no gain would be approximately twice the number of animals required to produce the maximum gain per hectare (Riewe, 1961).

A too heavy stocking rate results in low animal production. It can cause damage to the sward, allows weeds to invade the pasture, and in extreme cases leads to soil erosion (Figure 22).

Stobbs (1969h) studied the effect of stocking rate on a *Hyparrhenia rufa/ Stylosanthes guianensis* pasture at Serere, Uganda. The grazing treatments were:

— fixed stocking at 1.67 beasts per hectare;
— fixed stocking at 2.5 beasts per hectare;
— fixed stocking at 5 beasts per hectare;
— fixed stocking at 5 beasts per hectare for the wet season (April to December) reduced to 1.6 beasts per hectare for the dry season (January to March); and
— a variable (put-and-take) stocking rate.

Figure 22. High stocking rate with consequent overgrazing (**Photo:** Uganda Department of Agriculture)

The trial lasted for three years. Results are shown in Figure 23.

Total animal production at 1.67 beasts/hectare was only 485 kg/ha compared with 1 244 kg/ha at 5 beasts/ha, despite the fact that the beasts on the heavily stocked treatments lost weight in the dry season.

Analysis of the botanical composition of the pastures at the beginning of the experiment disclosed approximately 70 percent *Hyparrhenia*, 20 percent *Stylosanthes* and 10 percent weedy grass species. The light and medium grazing treatments maintained this botanical composition, but in the heavily grazed plots the percentage of *Hyparrhenia* declined and weedy species increased. The legume content increased slightly. Heavy wet-season grazing and light dry-season grazing preserved a high percentage of the original botanical composition. Heavy stocking with concurrent low availability of forage and less opportunity for selective grazing gave a higher output per hectare and lower individual animal production than the low stocking rate, but raised problems of weed invasion, indicating that it should be undertaken with caution. The first year's production per hectare from the heavily stocked treatment was almost equivalent to the total production from the lightly grazed treatment over the three-year life of the trial. Only at the higher level of stocking was the pasture being fully utilized.

More lenient grazing during the dry season, and more particularly at the start of the rainy season, resulted in good animal performance and maintenance of a high proportion of planted species in the sward. In Uganda, heavy grazing during the wet season (from April to December) and light grazing during the dry season (from January to March) appeared best for the *Hyparrhenia/Stylosanthes* pasture.

The stocking rate used on a pasture depends on the purpose of the pasture, that is, whether it is to be the main or sole source of feed throughout the year, or is to provide feed only for one season of the year. In the former case, the dry season becomes critical and may determine the overall stocking rate; alternatively, conservation of feed for the lean period or use of purchased food is required. In the latter case, the maximum grazing intensity consistent with pasture survival and well-being can be used, but heavy grazing during the growing period decreases root penetration and reduces future vigour and production in most species.

A feature of subtropical legume-based pasture is the incidence of frosts, to which many tropical legumes are susceptible. In frost-prone areas, it is important to make use of the tropical legume before frosts are expected, as the nutritive value decreases sharply after frosting (Jones, 1967a).

Fouling of pastures by urine and dung patches can cause the grazing animal

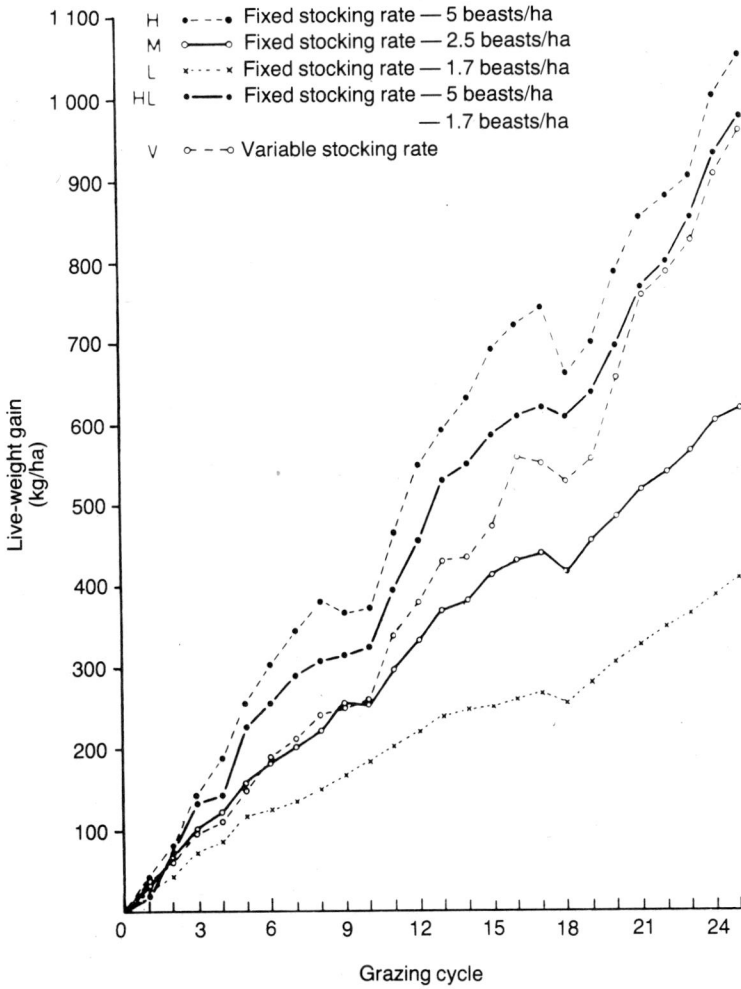

Figure 23. Cumulative live-weight gains from pastures stocked at five different stocking rates: H – high; M – medium; L – low; HL – high, wet season, low, dry season; V – variable (put-and-take) (**Source:** Stobbs, 1969h)

Figure 24. Low stocking rate and poor utilization of available forage (**Photo:** Uganda Department of Agriculture)

to avoid eating the adjacent forage. Usually in high rainfall conditions the high concentration of nutrients is soon leached out and the dung pats disintegrate. In Africa, the dung patches are soon disposed of by dung beetles. Where such uneven grazing occurs, slashing of the vigorous growth around such patches will improve the palatability of the new regrowth and spread some of the dung. Harrowing of tropical pasture species is difficult. Grazing the pastures only in the morning lessens the deposition of dung.

Animals return about 80 percent of their nutrient intake to the pasture in the form of dung and urine, which can be used in improving the general fertility of the property. As animals defecate more frequently at night, by organizing the night paddocks so that they rotate around the less productive portion of the property, a transfer of nutrients takes place from the fertilized or naturally fertile areas to the poorer ones, thus improving the performance of the latter. This system of fertility building has been exploited in Tanzania by Walker (personal communication). At Armidale, New South Wales, Australia, Hilder and Mottershead (1963) have shown that fertility transfer also takes place in moving to and from water, with a build-up of fertility around water points and stock camps, and a consequent depletion of nutrients in the areas removed from water and shade. Such knowledge is useful in determining the location of watering points and the provision of shade.

Grazing systems. There are two basic grazing systems — continuous grazing and rotational grazing.

Continuous grazing. This is the simplest method of handling animals and pasture, and in many areas it is the only system used. It allows the animals free range and so encourages selective grazing. The animals then choose the more nutritious portions of the plant and select a diet of higher food value than an analysis of clippings shows. Weir and Torrell (1959) found that sheep selected herbage containing 4.1 percent more protein and 3.5 percent less crude fibre than found in hand-clipped herbage, and Hardison et al. (1954) found that cattle selected a diet containing 23 percent more crude protein, 37.3 percent more fat, 25.6 percent more ash and 16.8 percent less crude fibre than the hand-collected sample. These figures cannot be taken as absolute, because other factors are involved, but they indicate that selectivity improves the diet.

Continuous grazing also often has advantages in parasite control, and incurs less expenditure on fencing and water.

In extensive beef cattle production, free-range grazing is often the only economic system of management and a species such as Townsville stylo

(*Stylosanthes humilis*), which can stand heavy and continuous grazing, is an excellent legume for this purpose. Small areas should be closed to grazing at flowering for seed production purposes.

Rotational grazing. This requires more control of animals and pasture. Traditionally associated with small paddocks, it involves more fencing and water supplies, although the nomads, by a system of manual herding, achieve broadly the same effect without fences.

It also enables pastures to be used quickly at their highest nutritive value or to be heavily grazed to force animals to eat unpalatable species and weeds. Rotationally grazed pastures can also be closed to grazing at intervals and saved for deferred feeding, fodder conservation, weed control, seeding or rehabilitation. The smaller paddocks permit segregation of sexes and ages in the grazing herd and give more control of mating and supervision at calving or lambing. Where cattle tick is a problem, rotational grazing combined with strategic dipping gives the most effective control available, as pastures can be closed to grazing when larval ticks are most likely to infest animals. On the debit side of rotational grazing are the high cost of fencing and water supplies; the likelihood of heavy infestation with internal parasites; and the danger of disease epidemics with heavy concentration of animals over short grazing periods.

In intensive livestock husbandry, such as finishing off young beef cattle for market, and in dairying, a certain amount of rotational grazing is essential. Young beef animals and milking cows require a high plane of nutrition at all times, and the quality of feed on offer is most important. When animals enter a new pasture, they select the leafy and more nutritious parts of the plant first and, if grazing is prolonged, the quality declines. This has led to a practice of short rotational grazing or "top-grazing" for the milking herd to select the better quality diet, followed by the remainder of the herd as "bottom-grazers" to eat the lower quality remaining pasture at the end of the grazing cycle. This was the principle of the Hohenheim system of grazing in Europe.

Stobbs (1969k) showed that when herbage was plentiful, animals selected mainly the leafy parts of *Stylosanthes guianensis*, but that as the amount of herbage was reduced complete flower clusters containing viable seed were eaten. On the last one or two days of the grazing cycle, a higher proportion of the stemmy parts of the legume were taken, and intake was lower.

Rotational grazing to allow selection of high-quality foliage must be distinguished from rotational grazing to ensure persistence of species. Lucerne or alfalfa (*Medicago sativa*) cannot persist under continuous grazing, but is

130

highly productive under a system of heavy stocking for short periods, sometimes termed "crash-grazing", and then adequate rest for recovery. In southeast Queensland, lucerne and Rhodes grass have been maintained in a mixture for 30 years by grazing the lucerne in the 10 percent bloom stage. Grazing ceases when the lucerne crown are exposed, and the sward is mown immediately after grazing to produce an even regrowth.

Rationed grazing, as distinct from rotational grazing, is an important tool where the area of rain-grown improved pastures or irrigated pastures is limited. One of the problems in an initial venture into improved pasture is that often too small an area is planted, and in dry times, unless rationed, the newly established pasture is likely to be overgrazed and its long-term productivity reduced. It is appropriate to have 0.4-0.6 hectare of improved pasture to every highly productive beast. Willoughby (1959) showed that increasing amounts of dry pasture reduced live-weight loss only slightly, whereas small amounts of green material resulted in marked live-weight gains. Crofts, Geddes and Carter (1963) have shown similar results with dairy cattle at Badgery's Creek, New South Wales, a finding that led Geddes to undertake extensive water harvesting for pasture irrigation. Rationed grazing with the use of an electric or other temporary fence can fill an important role in animal production, particularly in times of pasture stress.

Comparison of continuous and rotational grazing. Stobbs (1969j) undertook a long-term trial to investigate grazing systems at Serere, Uganda, using the tropical pasture mixture *Panicum maximum*/*Macroptilium atropurpureum* over a period of 1 218 days (29 grazing cycles). Three systems of grazing were imposed, using zebu-type steers at a stocking rate of 5 beasts/ha in the following ways: (*a*) continuous or free grazing using only one paddock; (*b*) rotational three-paddock grazing, moving cattle every two weeks; (*c*) rotational six-paddock grazing, moving cattle weekly.

Because of possible interaction between method of grazing and stocking pressure, the stocking rate on one replicate of each treatment was increased to 6 beasts/ha for the last eight grazing cycles (22 to 29). A severe dry period occurred during cycle 29 (see Table 8.1).

Production from the three-paddock grazing system over the whole 1 218 days was 1 544 kg/ha, from the continuously grazed pastures 1 462 kg/ha, and from the six-paddock system 1 310 kg/ha. The cumulative performances are shown in Figure 25.

The continuously grazed pastures were significantly higher yielding than those under the six-paddock system, and there was no significant difference

131

TABLE 8.1 **Summary of live-weight gains under various grazing systems**

Grazing period	Continuous		3-paddock		6-paddock	
	Heavy stocking	Very heavy stocking	Heavy stocking	Very heavy stocking	Heavy stocking	Very heavy stocking
	(kg/ha)					
Mean LWG cycles 1-29	1 462		1 544		1 310	
Mean LWG cycles 1-21	1 157	. . .	1 106	. . .	983	. . .
Mean LWG cycles 22-28	517	508	494	631	458	453
Mean LW loss cycle 29	−154	−242	−99	−150	−92	−166

NOTE: . . . means data not available.

between the final weight gains from the continuously grazed plot and the three-paddock system. During a severe dry season, however, weight losses were heavy for the continuously grazed system.

At the end of the 1 218 days, there was a significantly lower percentage of *Panicum maximum* and the highest percentage of siratro and weed invasion in the continuously grazed plots. At the end of the twenty-eighth cycle, these plots had lower herbage yields and showed signs of deterioration. The continuously grazed plots also suffered most in the dry season, indicating that where feed is limited a rotational system of grazing management is beneficial. Stobbs concluded that over long periods rotational grazing is necessary to maintain a satisfactory sward under Serere (Uganda) conditions, and the three-paddock system with two weeks grazing and four weeks rest appeared to be the most suitable practice.

Frequency of grazing. Length of grazing period on a rotationally grazed pasture and the time to allow for recovery before the next grazing are important aspects of grazing management. The reaction of each pasture species to defoliation determines to a large degree what species are combined in a pasture. A complete pasture mixture at sowing may, after a few years, become a simple mixture of species requiring similar grazing regimes. Thus, many complex tropical pasture mixtures sown on fertile alluvial land in the subtropics finally become simple *Paspalum dilatatum/Trifolium repens* perennial pastures. Stobbs (1969i) subjected pasture mixtures of *Hyparrhenia rufa/ Stylosanthes guianensis* and *Centrosema pubescens* to low frequency rotational grazing (i.e. seven days grazing and 28 days resting) and high frequency rotational grazing (i.e. $3^{1}/_{2}$ days grazing and 14 days resting). Overall differences in live-weight gain were very small, the low frequency having a slight advantage in the wet season, and the high frequency having the advantage in the dry season. Juko and Bredon (1961) have shown that the nitrogen content of the leaves of tropical legumes is higher than that of the stems and with rapid rotational grazing the "top-grazers" would obtain a high plane of nutrition.

Where *Desmodium* spp. are grown with *Setaria* spp., short grazing periods of a few days at a time with rests of a few weeks result in excellent utilization of the *Setaria* shoots, and the legumes are not preferentially grazed. Glycine is preferentially grazed and should be frequently rested to allow regrowth.

Maintaining the productivity of the sward

Under heavy rotational grazing at Serere, the botanical composition of the sward was markedly altered. The manipulation of grazing management can

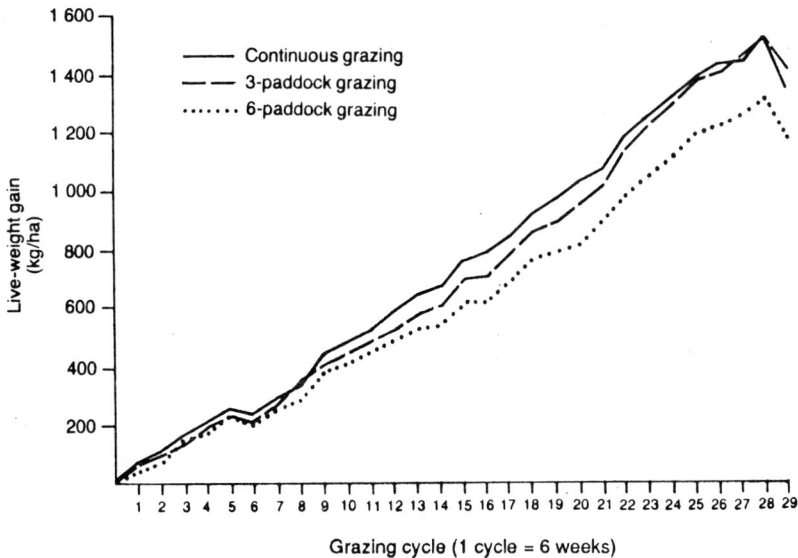

Figure 25. Cumulative live-weight gains from rotationally and continuously grazed pasture

determine the sward composition, and if the original mixture sown is to be maintained, close supervision of grazing management is required. Both heavy grazing and light grazing affect species dominance and are uneconomical in terms of productivity.

In the spring or early growing season, only light grazing should be allowed, to permit the formation of new growing points and leaves and the development and rooting of new runners. This will ensure productivity in summer and ultimate persistence (Ebersohn, 1969).

To maintain composition and high productivity, attention to the fertilization programme is important. Legumes are encouraged by a high phosphorus status and under heavy use potash may become limiting. Annual dressings of superphosphate are required, but potash less regularly. Sulphur may be deficient in some soils and is low in triple superphosphate, but is usually adequate in single superphosphate for most legumes and soils. Molybdenum will be needed for nodulation, but the initial dressing at seeding will usually last for several years. Grasses, on the other hand, are encouraged by nitrogen, and if legumes are taking over the pasture, some strategic nitrogen dressings can be given to suppress the legume, or the rest period between grazings can be increased (especially in the case of irrigated white clover/grass swards).

Guarding against bloat. Where white clover is a constituent of the wetter tropical pastures, it is important to see that it does not become so abundant that bloating is likely. In the case of bloat risk, the pasture should be sprayed ahead of grazing with an "anti-bloat" chemical such as "Bloat-guard" (polyoxpropylene Block polymer No. 18667, Smith, Kline and French Laboratories, Philadelphia, Pennsylvania, United States). Only one case of bloating from a tropical legume has been reported — from *Lablab purpureus* (Hamilton and Ruth, 1968) when it was the sole diet. By allowing a portion of the grazing to include gramineous species no trouble is experienced.

Fencing

In any controlled grazing system apart from nomadic herding, fencing is a prerequisite, but is often more expensive than necessary because more materials are employed than are needed. Single-strand electric fences are useful for rationed grazing of pastures in high rainfall areas. For permanent fences, a new type of suspension fence for cattle has been developed in Australia

(Mawson, 1963). It consists of posts 30 metres apart, to which $12\frac{1}{2}$ gauge high-tension barbed wire is fixed under tension and supported at 6-m intervals with heavy gauge wire spreaders. The top wire is usually 1.25 m from the ground, and if a four-wire fence is desired, the successive distances downward for the wires are 22.5, 20 and 17.5 cm; if a three-wire fence, the distances are 32.5 and 27.5 cm respectively. This fence is adequate for handling cattle and is considerably cheaper than conventional fences to construct.

Grazing trial techniques

For details of techniques for grazing trials, the reader is referred to Bulletin 51, Commonwealth Bureau of Pastures and Field Crops, Hurley, Berkshire, United Kingdom (1976), entitled *Tropical pasture research. Principles and methods*, compiled by N.H. Shaw and W.W. Bryan of the Cunningham Laboratory, CSIRO, Brisbane, Australia.

9. Weed control in tropical pastures

D.R. Bailey

A weed is usually defined as any plant growing where it is not wanted. Thus an edible and nutritious weed which may be a pest in cultivation is not a pasture weed. A food crop plant growing in a field of a different crop would be classed as a weed. Succulent herbs and woody species both occur as weeds.

It was once thought that the presence of weeds in a pasture was a sign of declining fertility, but it is now known that a different suite of weeds is encouraged by the use of fertilizers and many of our worst pasture weeds in fertile soils are those which can thrive on high levels of nitrogen or, as they are termed, nitrophilous weeds. Moore (1970) gives a good account of the ecological succession which leads to the presence of such weeds. Consequently, the presence of some weeds in pastures is rather a reflection of management than of soil fertility.

Weeds account for enormous losses in production of food crops and fibres, as well as of animal products. In Queensland, Australia, it is estimated that losses of the order of $A4 million annually accrue from contamination of merino wool clip with noogoora burr (*Xanthium pungens*). Also in Queensland, about 5 000 tonnes of butter submitted for grading annually is weed-tainted. The weed mainly responsible is lesser swine cress of bitter cress (*Coronopus didymus*). The weed taint does not disappear under ordinary pasteurization and a special ultra-high-temperature process must be used to remove it. Stock deaths from poisonous plants also cause heavy losses and some leguminous species such as *Astragalus* in the United States and *Acacia georginae* and *Gastrolobium grandiflorum* in Queensland (McEwan, 1964; Everist, 1969), which both contain the toxic fluoracetate ion, take heavy toll.

D.R. Bailey, Department of Primary Industries, Queensland, Australia.

Bush regrowth is a major cause of economic loss in the developing countries. Large areas of bush originally cleared by mechanical means or by fire and subsequently regenerated often cause a greater problem than the original forest. The brigalow (*Acacia harpophylla*) scrubs of Queensland constitute a good example (Skerman, 1953). Bush regrowth is a problem throughout Africa. Heady (1960) says: "Bush is the major deterrent to high forage and livestock production from semi-arid areas in East Africa". About half the 4 000 hectares originally cleared for groundnut production at Kongwa in Tanzania have now reverted to bush, which will need to be treated. When tropical rain forests are cleared, there is a well-defined succession back to the original vegetation: weed growth, the first stage, is followed by lianas and then the original forest species. Even if a tropical pasture mixture is established, it is a continuous challenge to management to maintain the pasture in the face of the natural succession of weeds and bush regrowth.

Types of weeds

The weeds of tropical pastures can be classed in two categories: short-lived species and persistent perennials.

Short-lived species. This group comprises annual, biennial and short-lived perennial species. As a rule, because of their prolific germination, they cause most trouble during pasture establishment by hindering the growth of the pasture seedlings. They compete with the sown species for moisture, light, nutrients and space.

Only in a few instances will pasture seedlings establish satisfactorily without the existing ground cover having been destroyed. Cultivation is generally practised to attain this, but each disturbance of soil encourages the germination of further weed seeds because cultivation changes soil aeration and soil moisture and enhances soil nitrate production. Light intensity at the soil surface is also increased, and this may be important for some weed species. Although many of these weeds are short-lived, they are often so numerous that they delay the grazing of a new pasture for many months.

Annual and biennial weeds usually occur in regular cycles in overgrazed and run-down pastures. For example, *Ageratum* spp. frequently grow vigorously in winter in coastal pastures and *Hyptis suaveolens* is a recurring pest of Townsville stylo (*Stylosanthes humilis*) pastures in northern Queensland.

Persistent perennials. These weeds live for a number of years and often reproduce by both sexual and vegetative means. The most troublesome

137

species in pasture are those which form dense thickets (e.g. *Acacia* species) and those which are poisonous to stock. Less spectacular, but still very important, are the tenacious perennial grass weeds. The nutritive value of these is usually very low and often their spread is aided by a rhizomatous or stoloniferous growth habit. Blady grass, lalang or Kunai grass (*Imperata cylindrica*) and mat or narrow-leaved carpet grass (*Axonopus affinis*) are examples of such weeds.

Preplanting practices to help control weeds

There are several ways in which weeds can be minimized prior to sowing a new pasture.

Careful seed-bed preparation. The initial cultivation should destroy the existing ground cover, and if persistent species such as blady grass (*Imperata cylindrica* var. *major*) are present, should commence several months before the planned sowing date. This ensures that rhizomes are brought to the surface during the drier months so that they may wither and die. Each cultivation should be shallower than the previous one and aim at forming a firm weed-free seed bed of reasonably fine tilth. Progressive reduction in cultivation depth helps ensure that weed seeds are not being continually brought up from the lower depths. Rolling to firm the seed bed is often helpful on light loamy soils, but can cause undesirable surface sealing on heavier clays.

In the wet tropical areas, the commencement of the wet season causes weeds to germinate and at the same time prevent further effective cultivation before sowing the pasture. In these circumstances, seedling weeds can be sprayed prior to planting with a nonresidual herbicide such as diquat. Likewise, 2,4-D may be used for the control of broadleaved weeds, but planting should be delayed for at least one week to enable 2,4-D residues to be broken down by microorganisms in the soil. The rate is 2.2 to 4.4 kg/ha when the ground is moist.

Cropping prior to pasture sowing is often effective in controlling weeds. On newly cleared brigalow (*Acacia harpophylla*) land, precropping with summer and winter crops for three or four years greatly reduces brigalow suckering, and also brings a cash return to the farmer. Cropping with maize, oats, cowpeas or *Lablab purpureus* in often practised on dairy farms to help reduce weeds prior to sowing permanent pasture.

138

Suitability of species; use of fertilizer. Not only must the pasture species be adapted to the particular geographical region but they must be further selected on a farm basis according to soil drainage, fertility and topography.

Seed of low germination and purity will give thin, patchy establishment providing favourable conditions for weed invasion. It is the responsibility of the farmer to check that the seed purchased is of good quality, and if it has been stored for more than two or three months, a new germination test should be carried out before planting.

Many tropical soils are severely deficient in phosphorus. On these, phosphatic fertilizer must be used to ensure the successful establishment of pasture plants. Even though pasture seeds may germinate in the absence of fertilizer, subsequent seedling growth is generally poor and frequently many seedlings die. On the wet tropical coast, this is often the case with Guinea grass (*Panicum maximum*) if planted on granitic soils without superphosphate.

In some situations many elements can be lacking, for example the *wallum* lands extending north along the coast from Brisbane are deficient in phosphate, potassium, copper, molybdenum, boron and calcium.

Time of planting; method of planting. Pasture seed sown during warm, moist weather as a rule establishes more satisfactorily than that sown during the drier winter months. Although out-of-season machine planting is sometimes used where large acreages are to be planted, this practice is not advisable if the seed is to be broadcast.

Machine planting usually gives more even pasture establishment than broadcasting, and where possible this method should be used. Patchy establishment will result from broadcasting unless seed is distributed evenly. This is most pronounced where seeds of different species have been mixed together and sown from aircraft; light grass seeds often drift widely, while heavy legume seeds fall in narrow strips.

On land than can be cultivated, broadcast seed should be lightly covered by harrowing, but on newly felled land this is usually not possible unless the timber has been stacked into heaps before burning.

Postplanting weed control practices

Mechanical control. Mechanical control methods fall into four categories: treatment by hand, mowing and slashing, bulldozing and treatment with fire.

Hand treatment. Hand grubbing with a mattock is an effective control measure on some perennial species which are not well controlled by herbicides (e.g. lantana (*Lantana camara*) and guava (*Psidium guava*)). Obviously, this method of control is only practicable when plants are fairly small and scattered. Slashing the tops off perennial weeds with a brush-hook is usually ineffective unless they are slashed repeatedly, because many species resprout. Sprouting may occur from either root or stem buds. For example, *Acacia* species sprout from the roots, whereas stem sprouting is quite common with devil's fig (*Solanum torvum*).

Mowing and slashing. Tractor-powered rotary slashers are generally used in preference to reciprocating side-mowers for mowing weeds in tropical pastures. The main reasons are that rotary slashers are more robust and they allow greater flexibility in height adjustment (see Figure 26).

Figure 26. Rotary slasher for weed control

Slashing has to compromise between effective weed control and minimum damage to the pasture. In newly sown pastures, this means cutting at no lower than approximately 20 cm and frequently enough to allow full light penetration to the pasture plants. Rotary slashing gives satisfactory control of erect annual broadleaved weeds but is of no benefit on sward-forming perennial grasses such as couch (*Cynodon dactylon*) or sour grass (*Paspalum conjugatum*). Moreover, because many perennial broadleaved weeds, such as wild tobacco tree (*Solanum auriculatum*) and knobweed (*Hyptis capitata*) resprout, repeated slashing is essential for their control. The cost of slashing varies with the type of growth. It is less, for example, for light weed growth than for tall brigalow (*Acacia harpophylla*) suckers.

Roller-choppers are sometimes used instead of rotary slashers on woody regrowth, such as that of *Eucalyptus* spp. and *Acacia* spp. These machines comprise a cylindrical base with regular spaced blades protruding from its surface. The blades crush and break the regrowth as they roll over it. Because these machines are not power driven, higher tractor speeds can be used and as breakages are few, treatment cost per hectare is low. Their success depends upon the regrowth being reasonably brittle and not too resilient. Moreover, the pasture must be well established; otherwise pasture plants are damaged. As with rotary slashers, follow-up treatment is invariably required to give control of perennial species.

Bulldozing. Bulldozing is often used for the reclamation of dense lantana (*Lantana camara*) thickets. Generally, the uprooted bushes are stacked into heaps for burning and the land then prepared for replanting.

Fire. Fire is used widely to control weedy species, especially in bush regrowth in semi-arid areas. In the higher rainfall areas, fire is seldom used, and because of its generally adverse effects on tropical legumes it is used as little as possible. In some cases where molasses grass (*Melinis minutiflora*) has been seeded initially on a burn to smother weed growth, it is later removed by fire to establish better pastures (e.g. *Panicum maximum/Centrosema pubescens* mixture).

It is in bush control in semi-arid areas that fire exerts its greatest effect. In the brigalow (*Acacia harpophylla*) areas of Queensland, fire generally encourages seedling and sucker regrowth, but a fire used strategically in the spring burning of grass reserved from the previous summer in a stand of young brigalow suckers will exert some effective control.

Many existing grasslands throughout the tropics are believed to occupy a

141

position as a fire climax in what was originally dense rain or monsoon forest: Kunai grass (*Imperata cylindrica*) in New Guinea and mission grass (*Pennisetum polystachyon*) in Fiji are examples. Some of the new tropical legumes such as siratro (*Macroptilium atropurpureum*) can be established in burnt *Imperata cylindrica* pastures by oversowing and adequate fertilizing.

Lamprey (personal communication) in the Serengeti Game Reserve in Tanzania found that burning more frequently than every two years eliminated regeneration of *Acacia drepanolobium*, and Pratt in Kenya (1966, unpublished) found a better control of this species with a combination of chemical treatment with fenuron combined with subsequent burning.

Biological control. To most people, biological control of weeds implies control by insects; but insects are not the only biological means for controlling weeds. Grazing animals, other plants, and fungi are also biological agents for weed control.

Grazing management. Continuous heavy grazing usually encourages weeds because palatable plants are eaten out and the unpalatable and undesirable weedy species have a better chance to flourish. Nevertheless, stocking is sometimes used to control some palatable perennial weeds; for example, young sucker regrowth of *Acacia harpophylla* can be controlled by heavy stocking with sheep (Skerman, 1953). In Africa, the goat is used a good deal in bush control work (van Rensburg and Hornby, 1948). Bryan (1968) has pointed out that grazing by cattle is necessary to control bush regrowth in the early years of grass/legume establishment in the *wallum* areas of southeastern coastal Queensland.

By regulating the time, duration and intensity of grazing, it is often possible to encourage the desirable species and to crowd out the weedy ones. This requires a fairly detailed knowledge of the growth rhythm of both the pasture and weed species. Restricting the grazing of poorly established pastures often helps the sward to thicken by allowing the sown species to set seed; new seedlings then fill the interplant spaces. Moreover, it gives the pasture plants a better opportunity to crowd out the weedy species.

Control by other plants. Sometimes it is possible to use a plant to suppress other less desirable species. This is particularly so with the tropical leguminous species and has been frequently demonstrated with leguminous cover crops. In some cases, the legume has itself to be weeded in the early stages of growth, but later it dominates the pasture and weeds are suppressed most frequently by competition for light, but also for nutrients and water. *Pueraria*

142

phaseoloides is a most effective legume in this regard, climbing up trees and shrubs and fence-lines and smothering other growth. At Maracay in Venezuela, it is very effective in suppressing nut grass (*Cyperus rotundus*). Siratro (*Macroptilium atropurpureum*), centro (*Centrosema pubescens*), glycine or perennial soybean (*Neonotonia wightii*) and greenleaf desmodium (*Desmodium intortum*) all have a useful smothering effect when needed, and by strategic management can be used to suppress undesirable weeds and grasses such as couch grass (*Cynodon dactylon*).

Control by insects. Biological control of weedy species by insects has played an important part in reclaiming land for development, especially in the case of the prickly pear (*Opuntia* spp.) in Australia, where about 24 million hectares were eliminated by the caterpillar of the moth, *Cactoblastis cactorum*. However, in tropical forage-legume management, biological control by insects is not a major tool in weed control.

Herbicides. Herbicides used in chemical control of weeds include:

2,4-dichlorophenoxyacetic acid	(2,4-D)
4-(2,4-dichlorophenoxy) butyric acid	(2,4-DB)
2,4,5-trichlorophenoxyacetic acid	(2,4,5-T)
2,4,5-T/2,4-D	(a mixture of 2,4,5-T and 2,4-D)
4-amino-3,5,6-trichloropicolinic acid	(picloram)
1,1'-ethylene-2,2'-dipyridylium dibromide	(diquat bromide; "diquat")

Descriptions are given below.

● 2,4-D. This is a crystalline powder which is almost insoluble in water. Commercially it is available in salt and ester forms. The salts are naturally water-soluble, and the esters are formulated into an oily liquid which readily forms a milky emulsion with water. Because broadleaved plants are generally more susceptible to 2,4-D than monocotyledonous species, this herbicide is widely used for the selective control of broadleaved weeds in cereals and pastures.

When describing commercial preparations of 2,4-D or making recommendations for their use, reference is always made to the amount of actual 2,4-dichlorophenoxyacetic acid which makes up the salt or ester, that is, their "acid equivalent". Similar terminology is also applied to other hormone herbicides.

The most commonly used salt is in the amine form, which usually contains 0.5 kg of 2,4-D acid equivalent per litre and incorporates wetting and sequestering agents. Sequestering agents stabilize the spray mixture when hard

143

water is used. Amine salts are nonvolatile and wet leaves well, but several hours of fine weather are required after spraying to ensure complete absorption into the plant.

Ester formulations of 2,4-D are generally considered more phytotoxic than amines, and are prepared by combining 2,4-D acid with an alcohol to form the oil-soluble ester. For example, the most common form, ethyl ester, is derived from ethyl alcohol. Esters are usually more effective than amine salts when plant growth is not very active or when plants are particularly difficult to wet. Moreover, because esters penetrate the leaves fairly rapidly, they are less affected by light rain shortly after spraying. Most are volatile and may damage nearby susceptible crops. The danger is greatest on warm windless days. To reduce this hazard, esters of flow volatility such as the butoxyethanol ester have been developed.

Commercial ester formulations contain from 200 to 800 g of 2,4-D acid per litre, and on an acid equivalent basis 2,4-D ethyl ester is about twice the price of 2,4-D amine.

● 2,4-DB. This is closely related to 2,4-D but is in itself only slightly toxic to plants. Its phytotoxic action relies on an enzyme system within the plant to convert it to lethal amounts of 2,4-D. Many broadleaved plants are capable of bringing about the conversion to 2,4-D fairly rapidly, whereas in some leguminous species it is so slow that the concentration of 2,4-D is never sufficient to cause serious plant injury. These characteristics enable broadleaved weeds to be selectively controlled in lucerne and some temperate pasture legumes. Unfortunately, in many instances, 2,4-DB is no more selective than 2,4-D toward tropical pasture legumes and is usually slow to kill weeds unless they are small and growing vigorously. Moreover, it is expensive, costing about five times as much as 2,4-D. For these reasons, where a choice can be made, the less expensive and quicker acting 2,4-D should be used.

2,4-DB is available in liquid formulations of either the potassium or the sodium salt; there is no discernible difference between the two.

● 2,4,5-T. This is also closely related to 2,4-D and is more effective than it on most woody plants, but is often less active than it on herbaceous species. Nevertheless, tropical pasture legumes are severely damaged by 2,4,5-T and where practicable its use should be confined to spot treatment of individual weeds.

It is commercially available as esters and amine salts containing between 200 and 1 070 g of 2,4,5-T per litre. The butyl ester, containing 400 g of 2,4,5-T per litre, is used most frequently.

● Mixtures. Mixtures of 2,4,5-T and 2,4-D are sometimes used. Usually they contain equal amounts of 2,4,5-T and 2,4-D esters which form a milky emulsion when mixed with water. Amine and salt formulations are also available but they are not widely used.

The merit of straight 2,4,5-T as against the mixture is a matter of debate. For most woody species, 2,4,5-T alone is more economical, but for a few mixed populations of herbaceous and woody plants the mixture may be better. It is also effective on some perennial broadleaved weeds such as devil's fig (*Solanum torvum*) which become difficult to kill with 2,4-D beyond the seedling stage. Aircraft operators often claim better results with the mixture even though the species may be quite susceptible to 2,4-D ground spraying.

● Picloram. Discovered in the early 1960s, picloram introduced a completely new chemical group into the field of growth-regulating herbicides. This highly active systemic herbicide has proved valuable in the control of a wide range of weed and brush species which are tolerant to 2,4,5-T. Grasses are unaffected at normal application rates, but tropical pasture legumes are very easily killed. Consequently, picloram cannot be used for overall spraying in grass/legume pastures. It is, however, very useful for injection of trunk and woody plants such as *Eucalyptus* spp.

The initial response of broadleaved plants is often one of upward or downward cupping of leaves. Picloram can be absorbed by both roots and leaves and is readily translocated within the plant. Moreover, this chemical is the most soil-persistent of all the organic herbicides and must be used with great caution in areas where susceptible crop and pasture plants are to be grown after spraying.

The commercial formulation most frequently used contains 50 g picloram and 200 g 2,4-D per litre. This mixture is usually diluted 1 to 4 with water for injection, and 1 to 100 for high-volume foliage spraying.

● Diquat. This destroys the normal photosynthetic process and causes fairly rapid desiccation of the green parts of the plant. Although not widely used in tropical pastures, some farmers are now using it in preference to slashing for the control of billygoat weed (*Ageratum* spp.). It is not particularly well translocated within the plant and, therefore, is generally not effective against perennial weeds. As a rule, diquat "scorches" pasture grasses but these recover quickly.

Another important property of diquat is that on contact with soil it is immediately inactivated by the clay particles.

The chemical is sold as the dibromide salt in aqueous solution. Both the herbicidal activity and the organic chemical reactions of diquat are dependent solely upon the diquat cation and are not influenced by the nature of the associated anion.

145

To ensure adequate foliage wetting, nonionic or cationic wetting agents must be used with diquat sprays. Household and industrial detergents should not be used, because they are often of the anionic type which can interfere with the phytotoxic action of the herbicide.

The commercial concentrate contains 200 g diquat cation. As a rule, 140 to 280 g diquat cation are required per hectare for selective control of annual weeds in tropical pastures.

Methods of herbicide application. These include various forms of spraying, and stem and base treatment.

● Overall spraying

Boom sprayers, boomless sprayers, tractor-mounted misting machines and aircraft are used for overall pasture spraying. Of these, boom sprayers are the most satisfactory because they give uniform distribution across the spray swath.

The boomless sprayer consists of an assembly of one to five nozzles supported on a single bracket mounted at the rear of the tractor. The spray can be pointed in any desired direction and will cover a swath from about 6 to 13 m wide depending upon nozzle height and spray pressure. Cross-winds seriously affect the pattern of distribution of these sprayers.

Aircraft and tractor-mounted misting machines are used where weed growth is tall or inaccessible. The delivery rate from aircraft is usually below 55 litres/ha and that of tractor-mounted misting machines between 55 and 110 litres/ha. Both methods allow large acreages to be treated quickly, but uneven coverage, overlapping and spray missing are not uncommon.

● Spot spraying

Hand-operated knapsack sprayers and knapsack misting machines are sometimes used for spraying individual weeds, but both are laborious to use for long periods. Knapsack sprayers require large amounts of water, and on leafy weeds 0.7 to 1 metre tall 1 650 litres of solution may be needed to cover 1 hectare effectively.

High-volume sprayers mounted on either tractors or vehicles are also used for spot spraying; these permit large volumes of spray mixture to be transported at one time. A commonly used sprayer consists of a double-acting piston pump powered by a two-stroke petrol engine. A long delivery hose permits the operator to move over a relatively large area and a special hand-piece enables herbicide to be shot into inaccessible locations.

● Stem and base treatment

Where weeds are large and scattered, it is usually more economical to apply a concentrated mixture of herbicide to the base of the plant than to spray the foliage. Moreover, herbicide mixture sufficient for several hours work can be carried by one person.

Plants may be cut about 30.5 cm above ground and the exposed butt painted, or on some species effective control can be achieved by painting a band of herbicide 30 to 40 cm wide around the base close to the ground. Usually an ester formulation is used and mixed with diesel distillate or sump oil to improve bark penetration. Often herbicide is injected through the bark of standing trees without felling. Either the base of the trunk is encircled by overlapping axe cuts or individual injections are made every 8 to 15 cm around its circumference.

Effect of herbicides on some tropical pasture species and common pasture weeds. If herbicides are used in pastures, it is essential that a minimum of damage be done to the sown species and at the same time that satisfactory control of weeds be achieved. Heavy loss of legume in a newly sown pasture would severely reduce its potential productivity, whereas slight loss or temporary suppression in either a new or established pasture may do no harm. Moreover, the correct use of herbicides may often permit earlier grazing of new pastures.

The effect of herbicides on some of the more common tropical pasture species is discussed in Chapter 14 under each individual species. At present, only three chemicals are suitable for overall spraying: 2,4-D; 2,4-DB and diquat. Picloram; 2,4,5-T and 2,4,5-T/2,4-D mixtures are extremely damaging to tropical pasture legumes and should be confined to spot treatment.

10. Irrigation of tropical pasture legumes

F.H. Kleinschmidt and P.J. Skerman

Irrigation of tropical leguminous pasture species is not undertaken extensively, apparently for the following reasons:

— irrigation water is usually expensive and cash crops are given preference;
— if it is possible to grow and irrigate lucerne (alfalfa) or white clover and ryegrasses, the returns will be greater than with tropical species (many alluvial areas in the tropics are suitable for the production of these species); and
— if it is not possible to grow temperate legumes under irrigation for the cool season, greater returns can be achieved by irrigating nitrogen-fertilized grass.

At Mareeba, north Queensland, Australia, the most productive use of irrigation water has proved to be for pangola grass (*Digitaria decumbens*), with added nitrogen. Evans (1969) obtained a live-weight gain of 2 000 kg/ha from such pastures. Irrigated tropical legume-based pastures cannot approach this figure, probably because intake is not high enough. In addition, the legume is not able to supply enough nitrogen in mixtures to satisfy the high demands of a vigorous summer-growing grass and so the grass is always producing below its potential (Chapter 6).

Evans (1970) recorded similar results with rain-grown grass/legume and pure grass pastures at Beerwah (southeast Queensland) under an annual rainfall of 1 650 mm. Live-weight gains ranged from 290 to 545 kg/ha/year for a grass/legume mixture to 1 200 kg/ha/year gain from pangola grass fertilized with nitrogen (State Agricultural College, Lawes, Queensland).

Roberts and Carbon (1969) grew several species of tropical and temperate grasses and legumes under irrigation in a Mediterranean-type climate at Perth, Western Australia. In this environment at 30 to 33°20′ south latitude,

148

although winters are cold, summer temperatures are high enough to promote vigorous growth of tropical species. Solar radiation often exceeds 600 cal/cm^2/day. Yields of dry matter from irrigated elephant grass (*Pennisetum purpureum*) reached 42 931 kg/ha with *Chloris gayana, Brachiaria mutica, Digitaria decumbens, Setaria sphacelata* and *Paspalum dilatatum*, each yielding over 20 000 kg/ha. Such yields compare favourably with the best yields in tropical countries. The important finding was that although the protein content of the high-yielding tropical grasses was generally about half that of the legumes, the highest yield of protein per hectare was still obtained from the grasses because of their high dry-matter yields. The highest yielding grass/legume mixtures were with grass in association with siratro, lotononis, glycine and lucerne. The siratro/pangola grass mixture yielded 16 226 kg DM/ha, followed closely by other legumes in the above order. The authors concluded that irrigation of nitrogen-fertilized tropical grasses was the most economic use of land and water. Jones *et al.* (1968) studied the relative merits of irrigating tropical and temperate legumes in the subtropics at Samford (southeast Queensland). They showed that *Centrosema pubescens, Lotononis bainesii* and *Desmodium uncinatum* were inferior to white clover for grazed irrigated pastures in this environment. Frost damage, competition from grasses, and high grazing pressure all contributed to the lack of persistence in tropical legumes.

The habit of growth of most tropical legumes makes water movement through the sward difficult to manage, and where spray irrigation is used, shifting the pipes through the tangled mass is difficult and tiresome.

Experience with irrigated browse crops

In Hawaii, Takahashi and Ripperton (1949) reported the irrigation of the browse shrub Koa haole (*Leucaena leucocephala*), grown in contoured rows about 1 metre apart. Furrows 10 to 15 cm deep were opened on the contour at intervals of about 1 metre. Seeds were sown on the downward slope near the bottom lip of the furrow. Irrigation water was delivered along the furrow, and as the plants developed the original furrow was gradually filled in and a new broad, shallow furrow some 50 to 70 cm wide was shaped for future watering.

Irrigation water was applied only during dry spells, to bring the total amount of water supplied by rainfall plus irrigation to 1 250 to 1 750 mm. The irrigation water was applied after cutting the plants.

149

Experience with irrigated tropical pasture mixtures

For some years, mixtures of the tropical grasses Para (*Brachiaria mutica*), Rhodes grass (*Chloris gayana*) and guinea grass (*Panicum maximum*) with associated centro (*Centrosema pubescens*) and stylo (*Stylosanthes guianensis*) were grown under irrigation in the Burdekin River delta area (north Queensland). The soil is deep, fertile loam to clay loam overlying alluvial sand and gravel. Water was pumped from an underground aquifer. The pastures were used to finish shorthorn beef steers for slaughter (see Figure 27).

A variable stocking rate was employed, using 3 beasts per hectare in winter and 4.5 beasts per hectare in summer. Water was applied every two weeks at 5.8 to 7.5 cm for each irrigation. The grazing sequence was one week grazing with three weeks' rest for the pasture.

Centro gave a better overall performance than stylo, although it was dormant for a short period in the winter. Rhodes and stylo were the best species for the winter period July to September (see Table 10.1).

TABLE 10.1 **Performance of selected irrigated tropical pasture mixtures**

Mixture	Average stocking rate (beasts/ha)	Average live-weight gain for 6 seasons (kg/day)
Chloris gayana + *Stylosanthes*	3.72	0.60
Brachiaria mutica + *Centrosema*	4.07	0.74
Panicum maximum + *Centrosema*	4.0	0.7
Panicum maximum + *Stylosanthes*	3.65	0.64

Source: Allen and Cowdry, 1961a

The carrying capacity of the native pasture was 1 beast to 5 hectares on the unimproved natural grassland adjacent to the experiment. Further inland, from whence most steers would come for finishing, the carrying capacity is 1 beast to 10 hectares.

From the above figures, it can be anticipated that the live-weight gain per hectare would be 440 to 495 kg over the 300 days. This is not a high figure for irrigated pastures and compares very unfavourably with irrigated grass fertilized with nitrogen. The amount of feed on offer is shown in Figure 28.

The above pastures could probably have carried more stock. In the early years of the experiment, experience was being gained in the use of such pastures. Evans (1969), using a Para grass/centro mixture at Parada, north Queensland, was able to attain an average live-weight gain of 0.96 kg/head/day.

The place for irrigated tropical legumes

In the hotter and frost-free tropics where lucerne and white clover will not thrive, there may be a place for finishing beef cattle for market by dry or cool-season watering of tropical species such as siratro, lotononis and the

Figure 27. Shorthorn steers being finished for market on an irrigated Para grass-centro pasture in the Burdekin Delta, north Queensland, Australia (**Photo:** Queensland Department of Primary Industries)

151

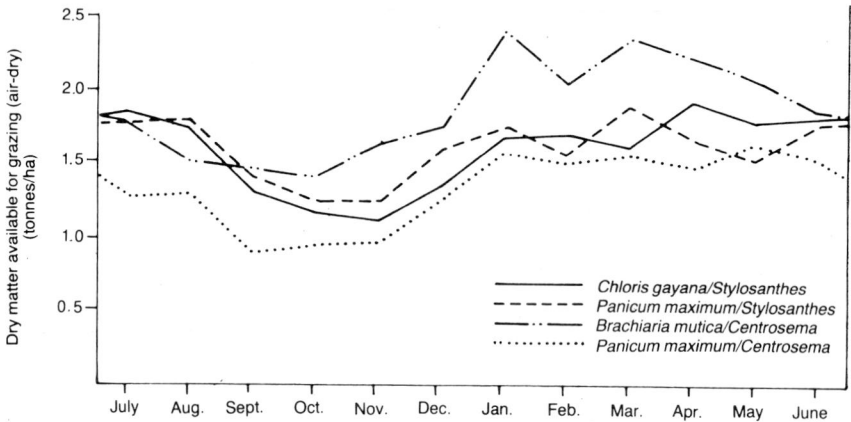

Figure 28. Amount of feed on offer from irrigated pastures in the Burdekin Delta, north Queensland (**Source:** Allen and Cowdry, 1961b)

desmodiums. This would mean dependence on rain for the summer growth with supplemental irrigation in the dry season, which would be possible only when temperatures were high enough to give a growth response in the legume. Selection for cool-season performance would be worthwhile.

Another hazard would be the possibility of extensive *Rhizoctonia* damage in dense moist tropical legume swards. Leaf-eating insects would also descend on an oasis of such irrigated pasture in a dry season.

Irrigation layout

Simple designs for the gravitational irrigation of pastures are given in Figures 29 to 32. Pastures are usually irrigated by a controlled flooding system in which water is reticulated by gravity. Spray systems are used, but the power needed to operate spray under pressure is usually about double that needed to pump through an open pipe.

Border check layout. In this design (Figures 29, 30), water is supplied from a head ditch into irrigation bays defined by low check banks. Any drainage water from the irrigated bay is taken off by the drain at the lower end of the bay and goes to the drainage ditch. Experience with the soil and the pasture will gradually lead to little waste of water into the drain. The bays are usually

152

about 10 m wide and 120 to 160 m long depending on the soil type, with the longer run for the clay soils.

Contour ditch layout. In this system (Figure 31), contour ditch irrigation, which is similar to uncontrolled flooding, is used on slopes of 2 to 10 percent. It consists of a supply channel large enough to take the available flow of water built with a fall of 0.3 to 0.4 percent. Water is fed out of the supply ditch by means of pipes through the bank or by siphons over the bank. The supply ditches are about 30 m apart.

Water can be more evenly distributed between the ditches by opening up small furrows on the true contour with a single furrow plough at 5- to 6-metre intervals. For large flows, a canvas or plastic sheet attached to a piece of timber or pipe can be used in the supply ditch as a stop to make water over-flow the ditch onto the pasture below. Manipulating this canvas stop controls the supply of water to any pasture adjacent to the ditch.

Contour check layout. In this system (see Figure 32), furrows are laid out on the true contours and the water is directed into each contour bay by a check placed across the supply ditch as required.

Cross-section of bay

5 to 10 m

1-2 m   Maximum cross slope 1° per bay,
preferably level

Supply ditch

Underground delivery pipe

Underground delivery pipe   Check bank

Drain

Supply ditch   Pump-house
and well

Supply ditch

Check bank

Bay

Drain

Drain

Drainage outlet

Bay

Drain

Drain

Labour power required:
one worker/4-5 ha/day

Length of watering run:
120-160 m, depending on soil type

Drainage outlet

Species suitable for watering
with this method:

lucerne
pasture mixtures
cereals
sorghums
stone-fruits on heavy soils

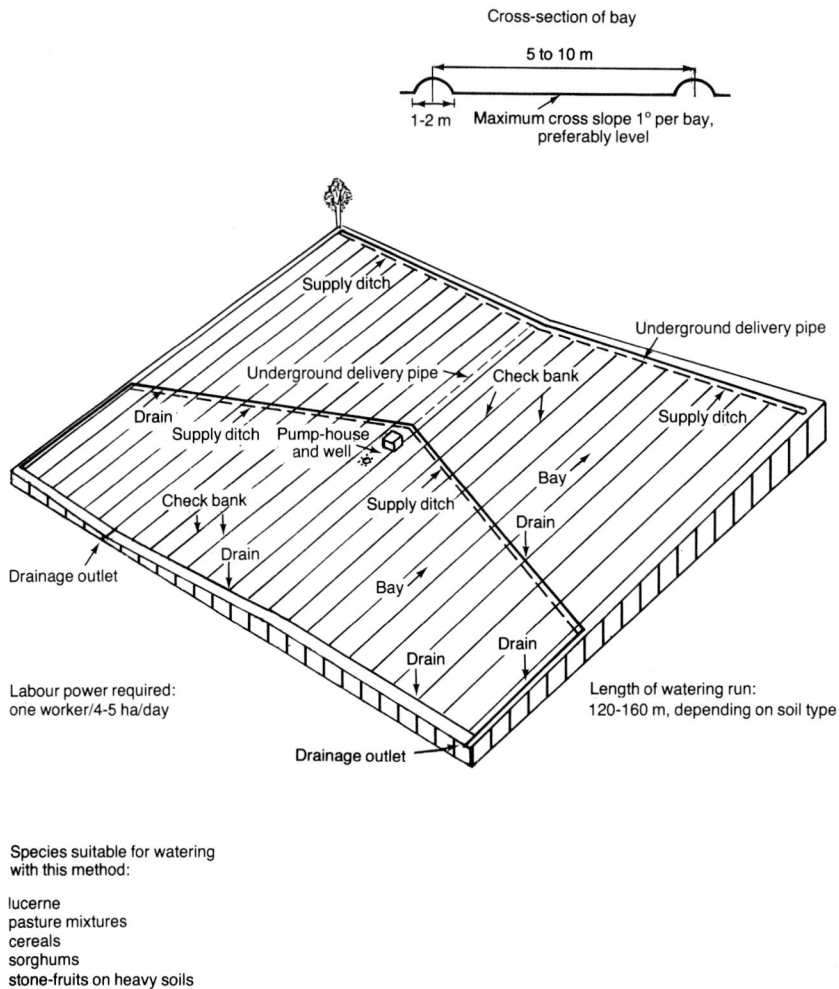

Figure 29. Border check method

154

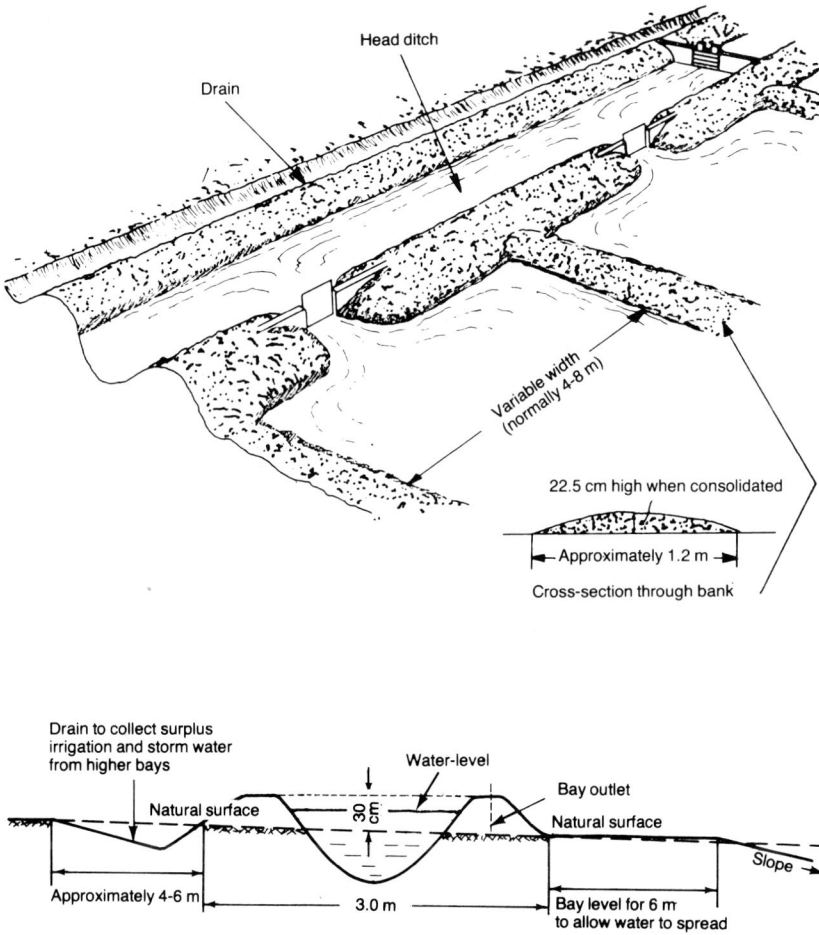

Head ditch

Drain

Variable width
(normally 4-8 m)

22.5 cm high when consolidated

Approximately 1.2 m

Cross-section through bank

Drain to collect surplus
irrigation and storm water
from higher bays

Water-level

Bay outlet

Natural surface

30 cm

Natural surface

Slope

Approximately 4-6 m

3.0 m

Bay level for 6 m
to allow water to spread

Figure 30. Border check irrigation

155

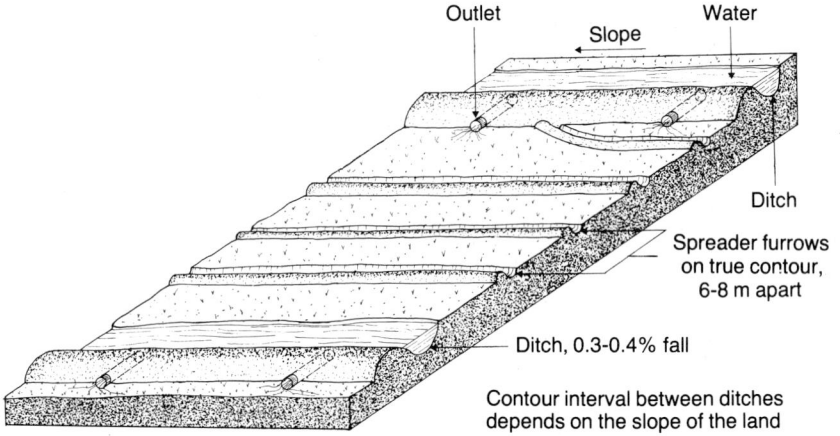

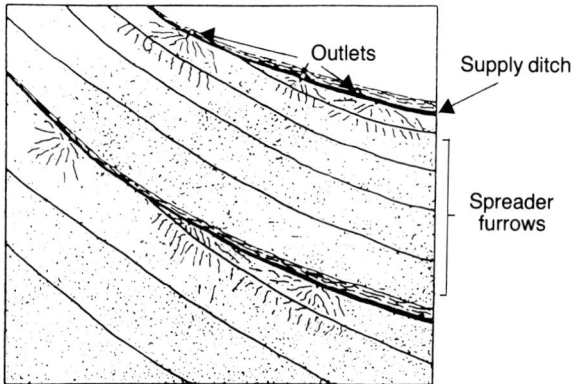

Figure 31. Contour ditch irrigation (above, plan; below, detail)

156

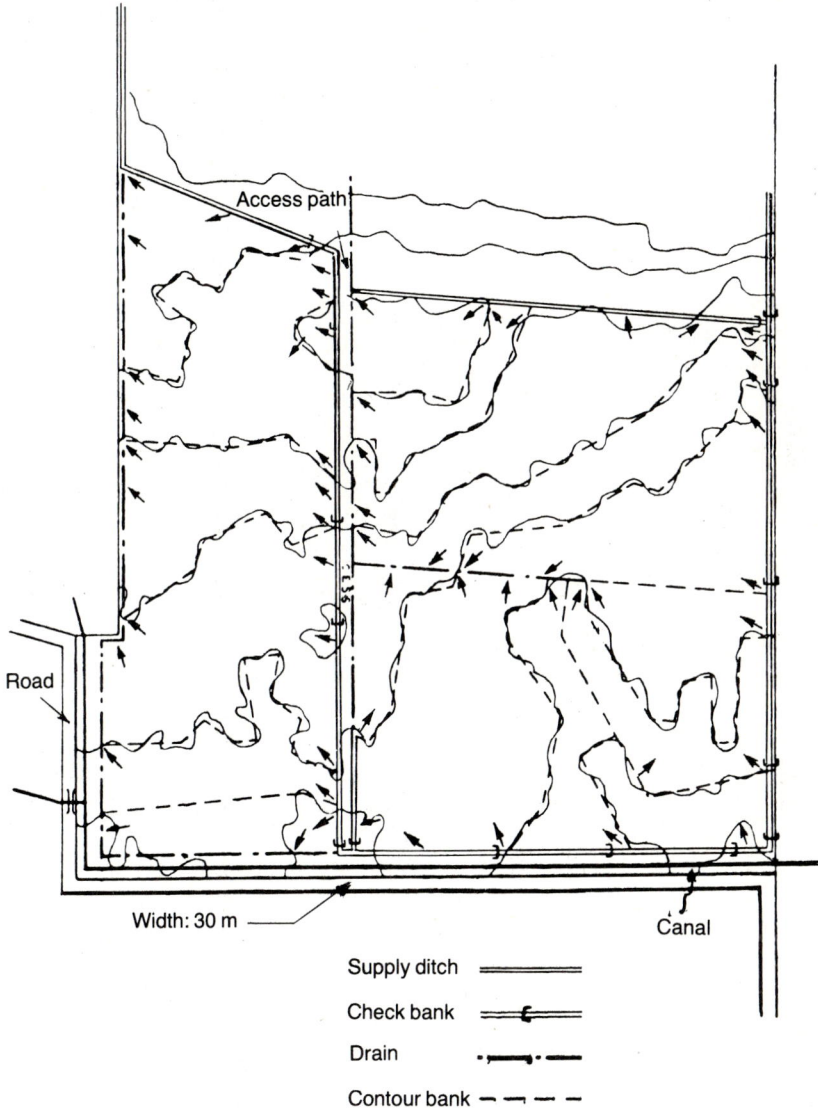

Figure 32. Contour check method

11. The production of forage legume seed

In a successful scheme to extend the use of improved pastures, much depends upon the availability of cheap seed of high purity and high viability. One of the problems of the present "pasture revolution" has been the high cost and often the unavailability of pasture seeds. The position has been aggravated by droughts, incidence of frosts, and the high cost of seed harvesting and cleaning.[1]

It is inevitable that in the early years the small amounts of seed available of the new varieties or cultivars command a higher price. But price should decline as seed multiplication grows, so that farmers and graziers can afford to plant adequate acreage to meet the requirements of the whole herd or flock.

One of the main causes of disappointment in returns from improved pastures has been the lack of adequate area, particularly in drought years, so that young pastures have been overutilized, with loss of the legume and deterioration of stand. Cheaper seed would help farmers to establish larger areas and thus eliminate some of the problems in the new venture. There is still some hesitancy among farmers to plant improved tropical species because of the short-term knowledge of their performance.

A good deal has yet to be learned about seed production by pasture plants and techniques of harvesting, cleaning and storing. In many cases, pasture-seed production is very difficult to achieve, as some cultivars possess characteristics which make preparation of seed extremely laborious, and also because the seed potential of our tropical species is measured in hundreds of kilograms per hectare as compared with thousands of kilograms for temperate species. Thus, small yields of seed difficult to harvest mean high produc-

---

[1] For more information, especially on the physiology of seed production the reader is referred to L.R. Humphreys *Tropical pasture seed production*, Rome, FAO (1974), which deals with both grass and legume seed production.

158

tion costs per hectare and high-priced seed for the buyer. As more efficient techniques are developed, seed supplies and prices should become more in keeping with the demand for an economic but efficient pasture. To obtain a large quantity of seed at a reasonable price in countries where labour is not cheap, mechanization is playing an increasingly important part and the specialist seed-producer is emerging.

Seed production

The multiplication of foundation seed from the plant breeder's laboratory to commercial sale requires careful supervision. Usually the breeder will release seed to government experimental or selected commercial farms where "mother seed" will be produced and then released more widely for commercial seed production.

"Mother seed" production. The "mother seed" is produced from a very small sample of seed or vegetative material and, because of its small quantity, is planted in pots or in nurseries. A description follows of the procedure adopted by one large producer of tropical legume seed (Redrup, 1966).

Normally each accession from the plant breeder is divided into four parts, of which one is placed in storage in case all planted material is lost in the first season. Because of limited knowledge of a new species under consideration, the remaining three parts are divided, first into early, midseason and late season sowings, and then into nursery pots with accompanying direct-seeded field plots. The nursery mother-seed plots are scattered widely as to soil type, elevation and location, both to correct for seasonal variations and to reduce the risk of loss through randomly occurring flood, insect attack, fire or frost. Simple fertilizer treatments are superimposed over the mother-seed and field plots as a guide to likely nutrient requirements. These mother-seed plots are maintained indefinitely free of seedling regrowth, and of off-type individual plants. As this becomes difficult with vine-type or prostrate legumes, or as mother-seed supplies prove inadequate to meet demand, fresh mother-seed plots are established by clones.

Commercial seed production. *Choice of site.* In the choice of location for commercial seed production, the need for isolation of plots (to avoid cross-pollination) must receive attention. Such isolation also helps to reduce risks due to hazards of the season. Most commercial seed production is by natural

ground speed. Ensure an even flow of material from the header front to the cylinder to prevent "plugging" or blockage.

● Cylinders and concaves. For rasp bar cylinders, set the clearance at 9 mm in front and 3 mm at the back. Cover the open concaves on hard-to-thresh crops. The cylinder speed adjustment is made on the basis of peripheral speed. Cylinder peripheral speed (fpm) = 3.14 times cylinder diameter in feet times cylinder revolutions per minute (rpm). For easily threshed crops, use 4 200 fpm, that is, 730 rpm with a 55-cm diameter cylinder. For harder-to-thresh crops, use 6 044 fpm, that is, 1 050 rpm. Use lowest possible cylinder speed for a satisfactory job. High speeds can cause grain damage, overthresh, chop up straw and overload the sieve, with resultant seed loss over the shoe.

● Straw walkers or rack. Under all conditions, check that these run at the correct speed. Check for plugged or blocked straw walkers and see that the curtains are in the correct position and are not torn.

● Cleaning shoe. Most harvesters have adjustable chaffer sieves and do a satisfactory job with correct settings. Under most conditions they can be set with openings of 12.5 to 13.5 mm without plugging or entangling tailings and returns. A small adjustment will materially affect the workings of the machine. A slight increase in the opening will reduce the seed passing out of the machine. Keep the chaffer extension board raised and open to trap unthreshed pods. The wind should be adjusted for maximum seed recovery rather than for a clean sample. Start with excess wind and reduce to obtain maximum seed recovery. There should only be an occasional light seed in the tailings from the wind blast. The wind blast should be directed toward the front to the centre section of the sieves. The volume of air may have to be reduced by either blocking of the fan, or, where possible, removal of opposite fan blades, taking care the fan is not out of balance. The clean grain sieve can either be adjustable or with punched holes. Details of the seed harvesting methods for individual species of tropical legumes are given along with other specific notes in Chapter 14.

Seed drying. It is difficult to obtain all fully matured dry seed at harvest because of the uneven ripening of many of the legumes. Consequently, there will be immature pods and seed in the harvested sample. If left in bulk, the green material will continue to respire and heating will take place. Hence, it is important to spread the freshly harvested seed on tarpaulin or concrete floors to dry in thin layers for several days before cleaning and grading.

Cleaning. Several types of seed cleaners are available commercially. Legume seed is usually winnowed employing an air blast and a series of sieves to separate portions of the plants, seed pods, immature seeds, etc.

Seed grading. Heavy seed always gives the best performance. Grading to discard any small, light or damaged seed is necessary for sound commercial seed products.

Seed grading may depend on air aspiration, on vibration, or on selective indented discs moving through the seed and removing all seeds that do not conform to the desired shape.

Sealing and labelling. Most commercial seed sellers seal the bags and label them before putting them into storage. If a certified seed scheme is in operation, sealing of the bags is a protection ensuring that the seed has passed inspection and is true to the label.

Storage. Seed for storage should contain less than 12 percent moisture to prevent heating. It may need to be dried in a hot-air drier periodically.

Some seeds are treated with insecticidal and fungicidal dusts to protect them in storage, but it is better to treat the storeroom rather than the seed itself in the case of legumes, as treatment may later interfere with the effective functioning of the *Rhizobium* in the inoculum.

For further details on the storage of seed in warm climates, see Arvier (1983). He stresses the role of modern moisture-proof packaging in maintaining low moisture levels, especially for relatively small quantities of seed. The most satisfactory packaging materials are laminated and incorporate tough flexible polymers in sheet or woven form. For large-scale commercial enterprises, the value of the commercial cold room for seed storage is also stressed, as is the role of good hygiene and pest control.

Legume seed quality control and testing

Most governments have regulations governing the quality of agricultural seed offered for sale. This is a wise precaution, as noxious and worthless weeds can be introduced and seed-borne diseases disseminated in uncontrolled seed. The quality of the seed is controlled by seed-testing officers who generally operate under international rules for seed testing.

A seed test is the detailed examination of a representative sample from

163

which it is possible to estimate and report certain properties of the particular line of seed, including:

— percentage by weight of pure seed;
— the nature and respective percentages by weight of various impurities which are present; and
— the percentage of pure seeds which germinate with normal roots and shoots, and the percentages of hard seeds and fresh ungerminated seed present at the final germination count.

Samples are collected at random at an intensity according to the bulk to be tested, mixed, and then reduced by accurate techniques to obtain a representative sample for testing.

The purity test involves separation of:

— pure seed of the legume being tested;
— seeds prohibited by legislation;
— inert matter;
— weed seeds; and
— other crop or pasture seeds.

Each species of prohibited seed, weed seeds and other crop or pasture seed is identified, each of these components is expressed as a percentage by weight and the nature of the inert material is determined.

Regulations prescribe a minimum percentage of pure seed (usually over 90 percent of legume seeds), and maximum percentages of weed seed, other crop seeds and inert matter. A list of prohibited seeds is drawn up for each country and lots found to contain any prohibited seed are withdrawn from sale.

The germination test is conducted to assess the viability of seed for field planting. Samples of 100 seeds in replications of three or four are germinated under specified temperature conditions in a moist atmosphere and examined after a specified number of days for their performance.

At the time of the final count of seedlings after germination, the number of seeds are to be interpreted as falling into five categories:

— number of normal seedlings which have produced normal roots and shoots;
— number of fresh, ungerminated seeds;

— number of hard seeds, that is, seeds which have imbibed water and remain firm to the touch;
— number of abnormal seedlings, that is, seeds which have not produced a root or shoot, or seeds which have malformed or damaged roots or shoots which would prevent establishment of the plant in the field;
— number of nonviable seeds.

Germination methods and standards for legume seeds in Queensland, Australia, are shown in Table 11.1.

TABLE 11.1 Germination methods and standards for tropical legumes

Seed	Percentage germinable seed		Percentage allowable by weight				Dormancy treatment	Germination	First count (day)	Final count (day)
	Minimum of germinable seeds	Maximum of hard seeds	Pure seed minimum	Weed seed maximum	Other crop seed maximum	Inert matter maximum				
Alysicarpus vaginalis (Alyce clover)	50	..	93.0	0.5	5.0	2.0	Pierce ungerminated swollen seeds at 21 days. Continue test until 35 days	35°C, cover	4	21
Arachis hypogaea (groundnuts)	80	..	95.0	0.2	—	5.0	Dry at 40°C, up to 14 days. Test at 30°C	25°C, dust seed	5	10
Cajanus cajan (pigeon pea)	70	10	98.8	0.2	—	1.2		25°C, cover	4	10
Calopogonium mucunoides (calopo)	50	..	93.5	0.2	5.0	1.5	Acid treat 20 minutes	25°C	3	10
Centrosema pubescens (centro)	50	..	93.8	0.2	5.0	1.2	Cut seed coat	25°C	4	10
Clitoria ternatea (butterfly pea)	50	..	93.5	0.2	5.0	1.5	Acid treat 20 minutes	25°C	3	10
Crotalaria sp. (crotalaria)	70	..	98.0	0.2	—	2.0		20-30°C	4	10
Cyamopsis tetragonoloba (guar bean)	70	10	94.5	0.5	0.5	5.0		20-30°C	5	14

TABLE 11.1 Germination methods and standards for tropical legumes (continued)

Seed	Percentage germinable seed		Percentage allowable by weight				Dormancy treatment	Germination	First count (day)	Final count (day)
	Minimum of germinable seeds	Maximum of hard seeds	Pure seed minimum	Weed seed maximum	Other crop seed maximum	Inert matter maximum				
Desmodium spp. (tick clovers)	70	..	94.5	0.5	0.5	5.0		20-30°C	4	10
Glycine max. (soybean)	60	10	98.8	0.2	—	1.2		25°C, cover	5	8
Lablab purpureus (dolichos)	75	10	97.5	0.5	0.5	2.0		25°C, cover	4	10
Lespedeza spp. (lespedezas)	40	10	93.5	0.5	5.0	1.5		20-35°C, cover	5	14
Leucaena leucocephala (leucaena)	60	..	97.5	0.5	0.5	2.0	Cut seed coat	25°C	4	10
Lotononis bainesii (lotononis)	50	45	93.0	0.5	5.0	2.0	Acid treat 20 minutes	20-30°C	7	21
Lotus spp.	75	30	93.5	0.5	5.0	1.5	Pre-chill fresh seed	20°C	5	12
Lupinus angustifolius (lupins)	80	10	98.3	0.2	0.5	1.2	Pre-chill fresh seed	20°C, cover	5	12
Lupinus luteus (lupins)	80	10	98.3	0.2	0.5	1.2	Pre-chill fresh seed	20°C, cover	5	12

167

TABLE 11.1 Germination methods and standards for tropical legumes (continued)

Seed	Percentage germinable seed		Percentage allowable by weight				Dormancy treatment	Germination	First count (day)	Final count (day)
	Minimum of germinable seeds	Maximum of hard seeds	Pure seed minimum	Weed seed maximum	Other crop seed maximum	Inert matter maximum				
Lupinus varius (lupins)	80	60	96.5	0.2	0.5	3.0	Pre-chill fresh seed	20°C, cover	5	12
Macroptilium atropurpureum (siratro)	70	..	97.5	0.5	0.5	2.0	Acid treat 25 minutes Alternative mechanical scarification	25°C	4	10
Macrotyloma axillare	60	10	97.5	0.5	0.5	2.0	Cut seed coat	25°C, cover	4	10
Macrotyloma uniflorum	60	..	97.5	0.5	0.5	2.0		25°C, cover	4	10
Medicago littoralis (harbinger medic)	75	20	96.0	0.5	2.0	2.0		20°C	4	14
Medicago lupulina (black medic)	75	20	98.0	0.5	0.5	1.5	Pre-chill fresh seed	20°C	4	10
Medicago orbicularis (button medic)	50	20	95.0	0.5	0.5	4.5		20°C	4	10
Medicago polymorpha (burr medic)	75	20	95.0	0.5	0.5	4.5		20°C	4	14
Medicago sativa (lucerne)	80	30	98.0	0.5	0.5	1.5	Pre-chill fresh seed	20°C	4	10
Mucuna spp. (velvet bean)	70	..	98.8	0.2	—	1.2	Cut seed coat	32°C, cover	2	7

TABLE 11.1 **Germination methods and standards for tropical legumes** (continued)

Seed	Percentage germinable seed		Percentage allowable by weight				Dormancy treatment	Germination	First count (day)	Final count (day)
	Minimum of germinable seeds	Maximum of hard seeds	Pure seed minimum	Weed seed maximum	Other crop seed maximum	Inert matter maximum				
Neonotonia wightii (glycine)	60	..	97.5	0.5	0.5	2.0	Acid treat 25 minutes	25°C	4	10
Ornithopus compressus (yellow serradella)	75	75	93.5	0.5	5.0	1.5		20°C	7	14
Ornithopus sativus (pink serradella)	75	10	93.5	0.5	5.0	1.5		20°C	7	14
Phaseolus coccineus (scarlet runner bean)								20°C, cover	5	9
Phaseolus limensis (lima bean)	75	..	98.8	0.2	—	1.2		25°C, cover	5	9
Macroptilium lathyroides (phasey bean)	70	..	98.0	0.5	0.5	1.5	Acid treat 20 minutes	25°C	4	10
Phaseolus vulgaris (French bean)	75	..	98.8	0.2	—	1.2		25°C, cover	5	8
Pueraria spp. (puero, kudzu)	50	..	93.5	0.2	5.0	1.5	Acid treat 20 minutes	25°C, cover	5	14
Stylosanthes guianensis (common stylo)	40	20	90.0	0.5	0.5	9.5	Water 55°C for 20 mins	32°C; alternative 20-35°C	4	10

TABLE 11.1 Germination methods and standards for tropical legumes *(continued)*

Seed	Percentage germinable seed		Percentage allowable by weight				Dormancy treatment	Germination	First count (day)	Final count (day)
	Minimum of germinable seeds	Maximum of hard seeds	Pure seed minimum	Weed seed maximum	Other crop seed maximum	Inert matter maximum				
Stylosanthes humilis (Townsville stylo)	40	20	90.0	0.5	0.5	9.5	Cut seed pod	32°C	1	4
Trifolium alexandrinum (berseem clover)	75	10	98.0	0.5	0.5	1.5	Fresh seed, test at 15°C	20°C	3	7
Trifolium campestre (large hop clover)							Fresh seed, test at 15°C	20°C	4	14
Trifolium dubium (yellow suckling clover)	75	20	93.5	0.5	5.0	1.5	Fresh seed, pre-chill, test at 10° or 15°C	20°C	5	14
Trifolium fragiferum (strawberry clover)	75	20	93.5	0.5	5.0	1.5	Fresh seed, test at 15°C	20°C	3	7
Trifolium glomeratum (clustered clover)	75	10	93.5	0.5	5.0	1.5	Fresh seed, test at 15°C	20°C	4	10
Trifolium hirtum (rose clover)	75	10	96.5	0.5	2.0	1.5	Fresh seed, test at 15°C	20°C	4	10
Trifolium hybridum (alsike clover)	75	10	93.5	0.5	5.0	1.5	Fresh seed, pre-chill; test at 15°C	20°C	3	10
Trifolium incarnatum (crimson clover)	75	10	98.0	0.5	0.5	1.5	Fresh seed, pre-chill; test at 15°C	20°C	4	7

TABLE 11.1 **Germination methods and standards for tropical legumes** *(continued)*

Seed	Percentage germinable seed		Percentage allowable by weight				Dormancy treatment	Germination	First count (day)	Final count (day)
	Minimum of germinable seeds	Maximum of hard seeds	Pure seed-minimum	Weed seed-maximum	Other crop seed-maximum	Inert matter-maximum				
Trifolium lappaceum (lappa clover)							Fresh seed, test at 15°C	20°C	3	7
Trifolium pratense (red clover, NZ cow grass)	75	10	96.5	0.5	2.0	1.5	Fresh seed, pre-chill; test at 15°C	20°C	4	10
Trifolium repens (white clover)	75	20	93.5	0.5	5.0	1.5	Fresh seed, pre-chill; test at 15°C	20°C	3	10
Trifolium resupinatum (shaftal clover)	75	30	93.5	0.5	5.0	1.5	Fresh seed, test at 15°C	20°C	3	7
Trifolium subterraneum (subterranean clover)	80	10	97.3	0.2	0.5	2.2	Fresh seed, test at 15°C	20°C	4	14
Vicia angustifolia (narrow leaf vetch)	60	10	98.8	0.2	—	1.2	Fresh seed, diffuse light	20°C, cover	5	14
Vicia articulata (monantha vetch)	60	10	98.8	0.2	—	1.2		20°C, cover	5	10
Vicia faba (broad bean)	70	..	98.8	0.2	—	1.2		20°C, cover	4	14
Vicia dasycarpa (woolypod vetch)	60	10	98.8	0.2	—	1.2	Fresh seed, pre-chill 10°C; test 5 days at 15°C	20°C, cover	5	10

TABLE 11.1 **Germination methods and standards for tropical legumes** *(concluded)*

Seed	Percentage germinable seed		Percentage allowable by weight				Dormancy treatment	Germination	First count (day)	Final count (day)
	Minimum of germinable seeds	Maximum of hard seeds	Pure seed minimum	Weed seed maximum	Other crop seed maximum	Inert matter maximum				
Vicia pannonica (Hungarian vetch)	60	10	98.8	0.2	—	1.2	Fresh seed, pre-chill	20°C, cover	5	10
Vicia sativa (common vetch)	60	10	98.8	0.2	—	1.2	Fresh seed, diffuse light	20°C, cover	5	10
Vicia villosa (hairy vetch)	60	10	98.8	0.2	—	1.2		20°C, cover	5	14
Vigna angularis (adzuki bean)								20-30°C, cover	4	10
Vigna marina	70	..	97.5	0.5	0.5	2.0	Cut seed coat	20-30°C	4	10
Vigna radiata (mung bean)	70	10	98.8	—	0.2	1.2		25°C, cover	3	7
Vigna sesquipedalis (asparagus bean)	75	..						25°C, cover	5	8
Vigna umbellata (rice bean)			98.8	0.2	—	1.2	Cut seed coat	32°C, cover	3	8
Vigna unguiculata (cowpea)	70	..	98.8	0.2	—	1.2	Cut seed coat	32°C, cover	2	8

NOTE: The acid used to break dormancy is commercial sulphuric acid of the specific gravity of 1.8 (Prodonoff, 1968).

12. Selection and breeding of tropical pasture legumes

E.M. Hutton

A high proportion of the pasture legumes currently used in the tropics have been developed and commercialized in northern Australia. The main cultivars are listed in Table 12.1.

For over 30 years there has been an active programme of plant introduction (Hutton, 1970a) into Australia, which was proved to be deficient in native legumes useful in tropical pastures. As seen in the table, South America and Central America have been particularly important sources of tropical pasture legumes followed closely by several countries on the African continent. Southeast Asia and the Philippines have also provided some legumes.

In South and Central America and in Africa, there is still a wealth of naturally occurring ecotypes of different legumes awaiting investigation. At present, Australia is making a special effort in the introduction of *Stylosanthes* species, so that new cultivars, particularly perennials, will eventually be released. There is a distinct need for all tropical countries with promising native legume material to intensify their efforts in its classification and evaluation. For example, at the IRI Research Institute in Matão, Brazil, at the Instituto de Zootecnica in Nova Odessa, central Brazil, and at Kitale, Kenya, work is in progress on evaluation of both commercialized and native legumes. It is hoped that research workers in other tropical countries will develop new and promising cultivars from their own native leguminous flora.

E.M. Hutton, CSIRO, Division of Tropical Pastures, Brisbane, Queensland, Australia.

TABLE 12.1 **Main Australian cultivars and their origin**

Cultivar	Species	Origin
Calopo [1]	*Calopogonium mucunoides*	South America
Centro [1]	*Centrosema pubescens*	South America
Greenleaf desmodium [1]	*Desmodium intortum*	El Salvador, Guatemala
Silverleaf desmodium [1]	*Desmodium uncinatum*	Brazil
Archer axillaris [2]	*Macrotyloma axillare*	Kenya
Rongai lablab [3]	*Lablab purpureus*	Kenya
Leichhardt uniflorus [4]	*Macrotyloma uniflorum*	Africa
Clarence glycine [1]	*Neonotonia wightii*	The Transvaal
Cooper glycine [1]	*Neonotonia wightii*	Tanzania
Tinaroo glycine [1]	*Neonotonia wightii*	Kenya
El Salvador leucaena [1]	*Leucaena leucocephala*	El Salvador
Peru leucaena [1]	*Leucaena leucocephala*	Peru
Miles lotononis [1]	*Lotononis bainesii*	The Transvaal
Siratro [1]	*Macroptilium atropurpureum*	Bred in Australia from Mexican ecotypes
Murray lathyroides [3]	*Macroptilium lathyroides*	Caribbean, Philippines
Puero [1]	*Pueraria phaseoloides*	Malaysia, Indonesia
Fine-stem stylo [1]	*Stylosanthes guianensis*	Paraguay
Townsville stylo [4]	*Stylosanthes humilis*	South America
Gordon stylo [4]	*Stylosanthes humilis*	Selected from naturalized Australian populations
Lawson stylo [4]	*Stylosanthes humilis*	Selected from naturalized Australian populations
Paterson stylo [4]	*Stylosanthes humilis*	Selected from naturalized Australian populations
Dalrymple vigna [2]	*Vigna luteola*	Costa Rica

[1] Perennial. — [2] Short-term perennial. — [3] Annual or biennial. — [4] Annual.

Criteria for selection of introductions and bred lines

The cultivars listed in the table have been developed by selection at research stations in Queensland, Australia, from a range of introductions on the various species. Siratro, the only bred one (Hutton, 1962), was developed from a cross between two Mexican ecotypes of *Macroptilium atropurpureum*. These cultivars have proved to be valuable pasture components and have made extensive planting of tropical pastures possible. Where they have been grown in conjunction with superphosphate, they have markedly increased animal production (Shaw, 1961). New legume cultivars to be successful have to perform better in the soil-pasture-animal system than the current cultivars. Long-term grazing experiments under differing environments are required to prove that they are better.

The characters required in successful tropical pasture legumes are listed below.

1. Nitrogen fixation must be efficient so that sufficient nitrogen and protein are made available for a productive pasture-animal system to be maintained. All the cultivars which have been selected are capable of fixing substantial quantities of nitrogen under favourable growing conditions. Henzell (1962) has shown that under optimum conditions tropical legumes fix almost as much nitrogen as white clover and lucerne.

When selecting legumes, it is essential to supply them with the right *Rhizobium* and to investigate their nodulating ability. For example, Miles lotononis (*L. bainesii*) was a failure until its specific *Rhizobium* was isolated and supplied by Norris (1958a). Heritable variation in nodulating ability with an effective strain of *Rhizobium* occurs among ecotypes of *Neonotonia wightii* (Nicholas, 1970) and within an ecotype of centro (Bowen and Kennedy, 1961). Other examples of poor nodulating ability with a genetic basis will be found in other tropical pasture legumes. As an efficient *Rhizobium* symbiosis is essential for the proper functioning of a legume in a tropical pasture, all aspects concerning the nodulating ability of a promising line should be thoroughly investigated before it is released.

2. The plants must have the ability to increase dry matter and protein with superphosphate application so that they are able to exploit the extensive areas of poor soils in the various tropical countries. Legume-based pasture regularly fertilized with superphosphate is the cheapest intensive system for the production of beef in the tropics.

175

At present, a high proportion of tropical countries do not have relatively cheap supplies of superphosphate available so that it is not possible to fully capitalize on the potential for animal production inherent in legumes. In some countries, there is still the philosophy that tropical legumes should be able to grow without applications of superphosphate and this has resulted in failure except in the limited areas of very fertile soils. The grazing animal needs a constant supply of phosphorus, sulphur and calcium as well as protein in the herbage, and this is attained by regular fertilization of a legume-based pasture with superphosphate.

3. Persistence in a pasture in spite of competition from associated grasses and periods of climatic stress is essential. For a tropical legume to be a significant factor in a pasture, it should comprise a third or more of its dry matter production. The factors implicated in the ability of a legume to maintain its place in a pasture in competition with grasses are not fully understood. In addition to vigour, the amount of seed produced, capacity to regenerate, palatability compared with the grasses, growth rhythm, and ability to compete for water and nutrients are all important factors in persistence. If a legume is too palatable, particularly during the first part of the season, it could be eaten preferentially and so suffer undue competition from the grass. Townsville stylo and siratro are relatively unpalatable until the latter half of the season, which aids their persistence in the pasture. Except in the wet tropics, pasture legumes are subjected to periods of hot dry conditions and in some areas to frosting. It is an advantage for the legume selected to have tolerance to a wide range of climatic conditions.

4. Tolerance to the range of soil conditions found in most tropical countries is an essential if legumes are to be widely adaptable. The *Stylosanthes* cultivars, siratro, calopo, centro and the desmodiums have this adaptability, but legumes like *Neonotonia wightii* and Miles lotononis are adapted to specific soil conditions. The cultivars of glycine grow best on kraznozems, self-mulching black soils and alluviums, while Miles lotononis thrives on moist, sandy and self-mulching soils.

5. High production of seed which can be harvested easily by mechanical means is necessary if it is to be sold at a reasonable price, and so be attractive to the farmer. Most of the legumes listed in the table are capable of giving high commercial seed yields if grown and managed properly for this purpose. Some difficulty has been experienced with legumes like siratro, Miles

lotononis and greenleaf desmodium, but techniques are now available for obtaining reasonable commercial yields of these.

A high seed yield that can be readily harvested must always be a major aim in the selection and breeding of new cultivars. For example, in siratro, new lines have been bred in which pods do not shatter so readily, and this has markedly increased seed yields. There is some evidence that seed yield is a heritable character and can be selected for within a species. Environmental conditions, however, can have a marked effect on seed yield and mask any genetic effects.

6. Resistance to diseases and pests is a necessary attribute if legumes are to thrive in a tropical environment. Most tropical areas are noted for a range of insect pests and fungal and other diseases as well as root knot nematode (*Meloidogyne* species). With increasing areas of tropical legumes, there is usually a gradual build-up of the various diseases and pests adapted to them. Most of the legume cultivars that have been commercialized have good tolerance to a range of diseases and pests, although under wetter tropical conditions legumes like siratro and glycine are often badly affected with fungus diseases.

Legume little leaf, caused by a mycoplasma (Bowyer *et al.*, 1969), can at times markedly reduce the stand of legumes like the desmodiums and Miles lotononis. It can also cause some difficulty in siratro, but the seeding ability and regenerative capacity of this legume usually prevent it from being badly affected by this disease (Hutton, 1970b).

Most of the current cultivars have good field resistance to root knot nematode. Leaf-eating insect pests have been a sporadic problem. Root-eating native weevils, such as *Amnemus* (Braithwaite, 1967), have caused serious damage to legumes like glycine and desmodium in some areas in Australia.

7. A high feeding value throughout the season is usually taken for granted. Tropical legumes are not as digestible as the temperate ones like white clover and lucerne, but their voluntary intake by the animal is usually much higher than that of tropical grasses of similar digestibility (Milford and Minson, 1965). In spite of apparent coarseness, tropical legumes maintain a high intake throughout the season (Milford and Minson, 1966a). The most positive way to increase the feeding value of a tropical pasture over the years is to increase its legume content.

Differences occur between the various tropical legumes in feeding value (Milford and Minson, 1968), and can be detected in small samples with the *in*

177

vitro digestibility technique (McLeod and Minson, 1969). In the evaluation of legume ecotypes and bred lines, *in vitro* digestibility should be used as one of the criteria in the future. For a number of bred lines of three legumes grown in the one trial at the Samford Pasture Research Station near Brisbane in the 1967/68 season, the mean and range of *in vitro* digestibilities were as follows: *Desmodium intortum*, 53.1 percent (range 48.1 to 55.4); *Macroptilium atropurpureum*, 64.9 percent (range 58.2 to 72.0); and *Vigna vexillata*, 68.3 percent (range 62.3 to 75.4). This indicates scope in selection for this character both between and within legume species.

8. Freedom from undesirable compounds, particularly if they are toxic, is important. Oestrogens are at a low level in the main cultivars and so are unlikely to influence fertility.

There has been some concern with the tree legume leucaena, as 0.5 percent of the nitrogen of its herbage is present as mimosine, a depilatory. Hawaiian work and research in Queensland (Hamilton *et al.*, 1970) have shown that a diet rich in leucaena does not affect conception and calving rate in cows. However, continuous grazing of leucaena for long periods causes loss of rump hair and weight in steers. It appears that this effect can be obviated by periodic transfer of the animals to a nonleucaena diet. Selection of lines free from mimosine would be advantageous, as leucaena is a most promising tropical legume.

Indigofera spicata, or creeping indigo, is a vigorous stoloniferous legume, but all ecotypes introduced into Queensland have proved to contain the hepatoxin now characterized and named indospicine (Hegarty and Pound, 1968). Until lines are developed which are free from indospicine, this potentially valuable legume should not be exploited. When fed *I. spicata*, pregnant cows abort and sheep suffer liver damage and die.

Completely new and relatively unknown legumes being considered for release should always be tested for toxicity. Rabbits are satisfactory test animals as they are herbivorous and usually react quickly to the presence of toxins in herbage.

Preliminary considerations in a breeding programme

It must be emphasized that it is inadvisable to embark on breeding and genetic work with a legume until a large range of the ecotypes of the species have been assembled and properly evaluated. Only in this way can breeding objectives be defined which are a necessary prerequisite to meaningful genetic improvement in a species. Also, it avoids the situation in which a breeder's work can be nullified by the introduction of a type superior to the line which he has painstakingly bred over a period of years.

As a basis for a breeding programme with a species, work is needed on the cytology, breeding system and flowering behaviour of the various ecotypes that have been collected. Special attention should be given to the ecotypes possessing the character that it is desired to recombine.

Cytology. The chromosome number of the selected species is necessary information. Even if it is already known, the chromosome number of a range of ecotypes should be determined to detect whether a polyploid series is involved. For example, difficulties in crossing some ecotypes of *Neonotonia wightii* in a breeding programme at the Cunningham Laboratory were shown by Pritchard and Wutoh (1964) to be due to the presence of $2n = 22$ and $2n = 44$ chromosome races in the species. This had not been suspected previously and indicates that investigation of the chromosome numbers within a species is always desirable.

Where ecotypes of a species differ in chromosome number, crosses can sometimes be effected after the numbers are matched by the use of colchicine. Recombination of the characters of two species within a genus would be a distinct advantage at times, but this is difficult to achieve. When interspecific crosses have been obtained, as in the genus *Stylosanthes* (Cameron, 1968), the resultant hybrids are usually sterile, but restoration of fertility is possible by inducing polyploidy with colchicine. Closeness of relationship between species in a genus can be assessed from chromosome morphology at meiosis and from the amount of pairing in any interspecific hybrids obtained.

Breeding systems and crossing techniques. A high proportion of the tropical legumes are closely self-pollinated as in the *Stylosanthes* species, siratro, centro, glycine and Miles lotononis (Hutton, 1960). The desmodiums, which are both self- and cross-pollinated, are exceptions. To obtain crosses between ecotypes of the self-pollinating tropicals, it is necessary to maintain a high relative humidity around the emasculated flowers and then for ten to 12 hours after cross-pollination. The flowers of desmodiums usually require tripping by insects to set seed. This results in self-pollination or exposes the stigmas to cross-pollination. Apparently, self-pollination in desmodiums has no effect on the vigour of the subsequent generation. In crossing desmodiums, flowers are carefully tripped early in the day by hand, before the anthers are fully

179

burst, to prevent deposition of the pollen on the stigmas. Cross-pollination is then effected by dusting the stigmas with pollen of the other parent, but a high relative humidity is not necessary for seed set.

The apparently exposed round heads of small flowers of leucaena are usually self-pollinated owing to the way the pollen falls directly into the stigmatic cups (Hutton and Gray, 1959). Emasculation can be effected soon after anthesis by dipping the heads in a weak aqueous solution of a nontoxic spreader. The heads are then dried and the crosses made.

By comparison with the tropicals, the temperate legumes, white clover (*Trifolium repens*), *T. semipilosum*, and lucerne (*Medicago sativa*), often grown in the cooler elevated tropical areas, are self-incompatible and cross-pollinating. Thus, there are no difficulties in breeding programmes with these species.

Day length. In most tropical legumes, the main period for flowering and seed setting is at the end of the wet season. Of those studied, Townsville stylo ('t Mannetje, 1965) and glycine (Wutoh *et al.*, 1968a) are short-day plants. There is also good evidence available that the desmodiums, centro and siratro have a short-day response. In a breeding programme, it is sometimes desirable to induce flowering in plants of the various legumes by subjecting them to a ten- to 12-hour day at a temperature around 30°C.

In the annual Townsville stylo, there is a range of early- to late-flowering ecotypes. In northern Australia, Cameron (1967) found that collections from areas with an annual rainfall over 1 140 mm were generally late-flowering and those from areas with a rainfall of 580 to 890 mm were early to midseason in maturity. This work has provided a basis for development of cultivars adapted to the various ecological niches in northern Australia and similar tropical areas. High-yielding late types only persist in areas with a long growing season; early types are adapted to areas with a short growing season.

In a comprehensive group of collections of the perennial *Neonotonia wightii*, Edye and Kiers (1966) also found a range of early- to late-maturing ecotypes. This has provided promising lines adapted to the various climatic zones without recourse to breeding.

Owing to their latitudinal zones of origin, a high proportion of tropical legumes will be short-day plants. However, as shown with Townsville stylo and glycine, this does not preclude the occurrence of a range of maturities among the ecotypes of a legume.

180

The breeding programme

In the formulation of breeding objectives for a legume, high dry-matter and protein yields and persistence should be given high priority. The parents for a breeding programme need to be carefully chosen, but this does not necessarily ensure that superior recombinants for quantitative character can be selected from the segregating populations following crossing. Sometimes the best results are achieved from crosses between widely divergent ecological types.

In the self-pollinated legumes, the pedigree method of plant breeding or a variation of this is used (Allard, 1960). With desmodiums, the pedigree method of breeding coupled with mass selection appears to be the best technique. The induction of mutations in legumes with the use of various mutagenic agents has not been a very profitable method of plant breeding. With more knowledge of the mutation process becoming available, it is possible that this method could be of value in the future where specific characters are not available in the natural populations.

Where the growing season is six months or more, concentration should be on perennial legumes, as they form stabilized pastures with associated grasses. The segregating populations of single plants from crosses between ecotypes of a spreading perennial legume often require planting at a density of only 283 plants per hectare for their evaluation. Wherever possible, they should be interplanted with a standard bunch grass such as *Panicum maximum* of medium height, *Nandi setaria*, or buffel (*Cenchrus ciliaris*). If this simulated pasture is grazed at appropriate intervals during the growing season, the competitive ability of the different lines and their contribution to pasture nitrogen can be assessed. This practice also controls weeds which can seriously interfere with meaningful selection in a population. After the legume population has been established for two years, the most persistent lines can usually be selected.

At the F_4 and, preferably, F_5 generation stage, replicated seeded swards as well as single plant populations of the legume should be established from the most promising lines. If the swards can be grown with and without an associate grass, better definition of the potential of the different lines under grazing or cutting is obtained. Replication of single plant populations as well as swards of the selected lines is necessary for their proper evaluation. To select for overall adaptability, populations at the F_3 and later stages should be grown in several areas representative of the ecological conditions under which the legume will eventually be grown.

181

With annual self-regenerating legumes like Townsville stylo, segregated single plant populations need to be kept free from weeds and vigorous grasses as the plants are susceptible to strong competition. When the F_4 stage is reached, the lines of the annual legumes are best assessed through the use of replicated seeded plots. Suitable grasses such as *Urochloa mosambicensis* for use with annual legumes are now being selected from current work with a range of introductions.

Breeding work with a number of the important tropical legumes is in progress in the Division of Tropical Agronomy, CSIRO. Some of the programmes are cited below as examples of the type of approach needed to improve a cultivar. As stated earlier, it is difficult to breed a legume cultivar better adapted to the pasture system than one that is already producing satisfactory results.

Townsville stylo. Cameron (1965) made a number of collections in the naturalized populations of Townsville stylo found across northern Australia. In these there was variation in growth habit, maturity and seed yields, and this enabled him to select three vigorous upright new cultivars, Paterson (early), Lawson (midseason) and Gordon (late). Cameron (1968) has shown in crosses involving all maturity types that late flowering was a simple dominant. He was able to produce hybrids from crosses between Townsville stylo and both *S. guianensis* and *S. hamata*. The interspecific hybrids were sterile, but fertility was restored by treatment with colchicine.

Siratro. This cultivar combines the best characters from two Mexican ecotypes of *Macroptilium atropurpureum* (Hutton, 1962), and has proved to be adaptable over a wide range of conditions. There is a need, however, to replace it eventually in the pasture system with a new type possessing higher dry matter and seed yields and a longer growing season. Since the release of siratro, F_2 to F_5 populations of 50 crosses involving siratro selections and several Central American introductions of *M. atropurpureum* have been raised. The aim has been to develop new lines with the main characters mentioned as well as a number of associated characters including resistance to *Rhizoctonia solani*. Only two of the crosses have produced lines, now in the F_7 generation, that are a significant advance on siratro. As evidenced by their performance in three different environments in northeastern Australia, both their dry matter and seed yields are around 40 percent better than siratro, and their growing season is almost two months longer. It is still necessary to evaluate them in a suitable grazing system to ascertain whether they are a definite advance on siratro.

182

Glycine. Wutoh *et al.* (1968a,b,c) have studied segregation of important agronomic characters in a number of crosses between ecotypes of *Neonotonia wightii*. They showed that time of maturity and seed weight are heritable and that it is possible to combine early flowering, high yield and good stolon development in one variety. Field establishment of glycine is often slow and irregular, which indicates poor nodulating ability in the main cultivars. Studies on the inheritance of nodulating ability are in progress at the Cunningham Laboratory (Nicholas, 1970) and it now appears possible that quicker and better nodulating lines with superior agronomic characters can be bred.

Leucaena. The tree legume leucaena has considerable potential in spite of its mimosine content because it gives high and constant weight gains in cattle while it is producing shoots. The main disadvantage of the current cultivars is their tendency to grow out of reach of the animals under favourable conditions. Most of the genetical and breeding work done by Gray (1967a,b,c) has aimed at producing bushy types of leucaena with high leaf yields and which are easier to keep within reach of the animal. Length of main stem and stem number were highly correlated and appear to be controlled by multiple genes affecting vigour, so these characters and plant vigour are interdependent. Up to the present, the advanced lines from the cross Peru × Hawaii, combining genes for dense branching and high vigour, are the most promising. Now that an agronomic advance has been made, attention will have to be paid to developing lines low in or free from mimosine. This may be difficult, although Gonzales *et al.* (1967) have obtained wide variation in mimosine content in populations from interspecific leucaena crosses.

Desmodium. Main emphasis has been placed on *D. intortum* because of its vigour and adaptability. Progress in improving this species has been slow because the factors influencing its persistence in pastures are not understood. Crosses have been made between *D. intortum* and the closely allied *D. sandwicense* and selections made in the resultant populations. None of the bred lines have been any more persistent or less susceptible to legume little-leaf than greenleaf desmodium. Some of the lines have, however, been earlier than greenleaf and have produced better seed yields. An understanding of the effects of selfing and the extent of cross-pollination in *D. intortum* is still required. Also a greater range of contrasting ecotypes is needed if real progress is to be made in improving greenleaf. In order to introduce new variability in desmodium, intercrosses were made between greeenleaf and silverleaf desmodiums and between greenleaf and *D. sandwicense* (Hutton and Gray, 1967) but the results from these have not been promising.

183

Breeding work with associated temperate legumes

The temperate legumes, white clover, *Trifolium semipilosum* and lucerne, have a place in the cooler parts of the tropics.

No planned work has been done with white clover, but in southeast Queensland it behaves mainly as an annual. This, together with its crosspollinating habit, gives a rapid change in the population, which in a few years results in the production of ecotypes of white clover adapted to the particular environment. *T. semipilosum*, with its deep taproot, is more tolerant to drought than white clover and will find an increasing place in the white clover areas of the tropics when the main problems affecting its establishment have been solved. Its slow nodulation may be due to competition between white clover *Rhizobium* and its own specific *Rhizobium* for the active sites on the roots. Also, various legume viruses severely infect a proportion of the population. Here again, natural selection appears capable of producing stabilized and adapted cultivars in this potentially valuable legume (R.J. Jones, personal communication).

The ordinary Hunter River cultivar of lucerne has proved to be valuable on both granitic soils and clay loams in the Australian subtropics, particularly under a rotational grazing system. However, it does not persist for more than four or five years, owing to root rot, and decimation of the stand is accelerated by continual grazing. It is thought that transfer of the creeping-rooted character from rambler (Heinrichs, 1954) to a vigorous cultivar that grows mainly during the autumn/late spring period would overcome this problem. Edye and Haydock (1967) and Bray (1967, 1969) have shown that the creeping-rooted segregates from appropriate crosses have a significantly better persistence than the noncreeping ones. Also there is a high correlation between creeping-rootedness and yield. Some promising synthetic varieties are now being developed for more extensive trial under different grazing systems.

13. The chemical composition and nutritive value of tropical legumes

D.J. Minson

The value of any feed depends on the quantity eaten and the extent to which the food consumed supplies the animal with energy, protein, minerals and vitamins. Many tropical legumes have been studied but the results are widely dispersed in the scientific literature. Ranking the legumes in order of their nutritional value is not possible, since very few legumes have been compared in the same environment. As the composition of legumes is influenced by the fertility of the soil in which they are grown, the age of the material, method of handling and storage and feeding, one must be cautious about comparisons.

Voluntary intake

The quantity of dry matter voluntarily eaten by an animal is the most important factor controlling the productive value of a feed. It is obvious that if an animal eats only a small quantity of a tropical legume, production of meat or milk will be low no matter how high the protein, digestible energy or mineral content of each unit of feed.

Ideally, the voluntary intake of any pasture plant should be measured in the field where selective grazing is allowed to operate. As there are at present no accurate methods of measuring the voluntary intake of grazing animals,

D.J. Minson, CSIRO, Division of Tropical Crops and Pastures, Cunningham Laboratory, Brisbane, Australia.

185

the data for tropical legumes in Table 13.1 were derived from indoor feeding trials in which sheep or cattle were fed fresh or dried material. These intakes varied from 17.3 of mature *Crotalaria lanceolata* (Milford, 1967) to 76.0 g/kg $W^{0.75}$ (W = body weight) of immature *Vigna unguiculata* (Milford and Minson, 1968). For sheep there is a general trend for the voluntary intake to decrease with decreasing digestibility of the dry matter. The relationship between voluntary intake (V) and dry-matter digestibility (D) for the legumes fed to sheep is shown below, together with the corresponding regressions for *Chloris gayana* (Milford and Minson, 1968) and *Panicum* species (Minson, 1971).

Legumes $V = 1.76D - 44.5$ RSD ± 8.5 r $= 0.86$

Chloris gayana $V = 0.74D + 3.4$ RSD ± 3.7 r $= 0.73$

Panicum spp. $V = 1.42D - 20.6$ RSD ± 6.8 r $= 0.76$

(RSD = regression standard deviation; r = correlation coefficient; + 1 is perfect positive correlation)

Thus the tropical legumes appear to have a higher voluntary intake than *Chloris gayana* but a similar intake to that reported for *Panicum* species.

The voluntary intake of cattle is about twice as high as that of sheep fed *Vigna unguiculata* when intake is expressed as g/kg $W^{0.75}$ (see Table 13.1), but the intakes were similar when the results are quoted as kg per 100 kg body weight (Miller *et al.*, 1964). Care must be therefore taken in comparing the voluntary intakes of different pastures when measured with different classes of stock.

Energy. The most desirable way of expressing the energy value of a feed is as net energy since this is the quantity of energy that the animal can use for maintenance or production. Owing to the elaborate equipment required to determine net energy, only one cut of one tropical legume (*Desmodium uncinatum*) has ever been studied in this way (Graham, 1967). Most energy values have been determined in digestion trials and the results expressed as digestibilities of the dry matter or organic matter, or as total digestible nutrients (TDN) as shown in Table 13.2. Results have also been quoted as starch equivalents (Milford, 1967) but these are not presented in the table. Since there is a high correlation between dry matter (DM) digestibility and TDN (r = 0.96; Milford, 1967) results will be discussed only in terms of DM digestibility.

TABLE 13.1 The voluntary intake of tropical legumes (g/kg W$^{0.75}$)

Species	Condition	Animal	Mean	Range	Reference
Centrosema pubescens	Fresh	Sheep	33.9	28.4-39.4	Miller and Rains, 1963
Crotalaria lanceolata	Fresh	Sheep	17.3		Milford, 1967
Desmodium uncinatum	Fresh	Sheep	50.8	30.8-62.2	Milford, 1967
Desmodium scorpiurus	Fresh	Sheep	33.4		Miller and Rains, 1963
Lablab purpureus	Dried	Sheep	62.7	53.8-73.6	Milford and Minson, 1968
Lotononis bainesii	Fresh	Sheep	57.6		Milford, 1967
Macroptilium atropurpureum	Fresh	Sheep	37.5		Milford, 1967
Macroptilium atropurpureum	Dried	Sheep	59.5		Minson and Milford, 1966
Macroptilium lathyroides	Fresh	Sheep	35.1		Milford, 1967
Stizolobium sp.	Dried	Cattle	83.5	79.6-88.3	Miller et al., 1964
Stylosanthes guianensis	Fresh	Sheep	41.3		Miller and Rains, 1963
Stylosanthes guianensis	Fresh	Sheep	33.9		Milford, 1967
Stylosanthes humilis	Dried	Sheep	60.7		Playne, 1969
Vigna unguiculata	Fresh	Sheep	28.4		Miller et al., 1964
Vigna unguiculata	Fresh	Cattle	64.3	62.2-66.4	Miller et al., 1964
Vigna unguiculata	Dried	Sheep	66.4	57.4-76.0	Milford and Minson, 1968
Vigna vexillata	Fresh	Sheep	56.6	52.8-60.5	Milford, 1967

187

TABLE 13.2 **Dry-matter digestibility and TDN (total digestible nutrient) value of tropical legumes**

Species	Dry-matter digestibility		TDN		Reference
	Mean	Range	Mean	Range	
Crotalaria lanceolata	36.0		36.0		Milford, 1967
Crotalaria medicaginea			58.6		Singh and Talapatra, 1963
Desmodium intortum	54.4	53.0-56.0			Jones, 1969
Desmodium scorpiurus	44.7 (OM)				Miller and Rains, 1963
Desmodium uncinatum	50.2	47.3-54.1	50.2	46.5-59.9	Milford, 1967
Lablab purpureus	57.8	56.0-59.6	55.7	53.7-57.4	Elliot and Fokkema, 1960
Lablab purpureus	55.5	50.9-59.2			Milford and Minson, 1968
Lotononis bainesii	59.7		57.6		Milford, 1967
Macroptilium atropurpureum	51.6	40.3-56.8			Minson and Milford, 1966
Macroptilium atropurpureum	50.4		48.0		Milford, 1967
Macroptilium atropurpureum	65.2	62.2-69.0			Milford, 1967
Macroptilium lathyroides	54.9	41.7-62.0	56.7	43.4-65.0	Milford, 1967
Neonotonia wightii	54.4		53.2		van Wyk et al., 1955
Neonotonia wightii	58.3	55.7-61.7			Holder, 1967
Pueraria phaseoloides	46.0	43.4-48.7			Dijkstra and Dirven, 1962
Pueraria thunbergiana			49.9		van Wyk et al., 1955
Stizolobium deeringianum (Mucuna pruriens)	53.4	50.8-55.7	52.5	50.2-55.8	Elliott and Fokkema, 1960
Stizolobium deeringianum	54.9 (OM)	53.4-57.0 (OM)			Miller et al., 1964
Stylosanthes guianensis	45.9 (OM)				Miller and Rains, 1963
Stylosanthes guianensis	48.4				Milford, 1967
Stylosanthes guianensis	46				Newman, 1968
Stylosanthes humilis	45				Newman, 1968
Stylosanthes humilis	58.7				Playne, 1969
Vigna unguiculata	59.0				Ranjhan et al., 1967
Vigna unguiculata	57.4	55.8-59.1	53.2	51.6-54.7	Elliott and Croft, 1958
Vigna unguiculata	59.5	54.4-65.1 (OM)			Miller et al., 1964
Vigna unguiculata	59.8	56.6-63.8			Milford and Minson, 1968
Vigna vexillata	63.5	57.7-69.3	63.5	57.0-70.0	Milford, 1967

The DM digestibility percentages of the tropical legumes shown in Table 13.2 ranged from 36.0 to 69.3, with a mean of 54.0 percent. These values are similar to those found for 543 tropical grasses (Minson and McLeod, 1970) but it is not possible to ascertain whether the legume and grasses were similar in average maturity.

In the tropical grasses, dry-matter digestibility declines with maturity and a similar though less rapid effect has been found with the tropical legumes *Lablab purpureus* and *Vigna unguiculata* (Milford and Minson, 1968), *Macroptilium atropurpureum* (Minson and Milford, 1966) and *Macroptilium lathyroides* (Milford, 1967), although Milford (1967) found no decrease with *Desmodium uncinatum*. The leaves of tropical legumes might be expected to have a higher digestibility than the stems (Jones, 1969, reported a difference of 3.3 to 8.7 digestibility units beween the leaves and stems of *Macroptilium atropurpureum*).

Protein. The crude protein percentages reported for tropical legumes (see Appendix 1) vary from 5.6 for a cut of *Stylosanthes humilis* (Newman, 1968) to 35.8 for the leafier parts of *Leucaena leucocephala* (Hutton and Bonner, 1960), with a mean crude protein percentage of 17.2 percent for all legumes (Appendix 1). These levels of crude protein are much higher than those found in tropical grasses — for example, the mean content for the sample reviewed by Butterworth (1967) was 7.7 percent. Thus the presence of legumes in a sward raises the crude protein of the mixture and the high protein of the legumes can act as a supplement for the low protein grasses.

As tropical legumes mature there is usually a decrease in the crude protein percentage. This effect is well illustrated by the data for *Desmodium uncinatum* (Milford, 1967), *Lablab purpureus* (Milford and Minson, 1968), *Macroptilium lathyroides* (Milford, 1967) and *Stylosanthes humilis* (Fisher, 1969). The apparent digestibility of the crude protein on 61 cuts of tropical legumes has been measured with sheep and cattle (French, 1937; Work, 1964; Odejar, 1949; Robinson, 1950; Loosli *et al.*, 1954; Reyes, 1955; van Wyk *et al.*, 1955; Elliott and Croft, 1958; Lansbury, 1959; Elliott and Fokkema, 1960; Dijkstra and Dirven, 1962; Miller and Rains, 1963; Miller, Rains and Thorpe, 1964; Peixoto *et al.*, 1966; Holder, 1967; Milford, 1967; Newman, 1968). The values vary considerably but this variation is mainly associated with the level of crude protein in the legume; low crude protein digestibilities are associated with low crude protein percentages in the plant and vice versa. More important than digestibility percentage is the actual quantity of crude protein apparently digested per 100 units of feed. Using the published results

189

of digestion trials with tropical legumes, a useful prediction equation has been calculated relating digestible crude protein (DCP) per 100 units of feed to the crude protein (CP) percentage of the dry matter, as follows:

Legumes $\quad$ DCP = 0.93 CP − 3.99 RSD ± 1.17 r = 0.96

Grasses $\quad$ DCP = 0.90 CP − 3.25 RSD ± 0.84 r = 0.98

This equation is very similar to one based mainly on tropical grasses (Milford and Minson, 1965) and illustrates that the protein in tropical legumes is digested with the same efficiency as protein in tropical grasses of *similar* crude protein content. However, since the level of crude protein in the tropical legumes studied is much higher than that in the grasses, the mean quantity of digestible crude protein in the tropical legumes is 12.0 g/100 g feed compared with a mean of 3.7 g/100 g feed in the tropical grasses cited by Butterworth (1967).

Crude fibre. Crude fibre percentages have been determined in many tropical legumes (Appendix 1). These varied from 12.4 for *Leucaena leucocephala* (Farinas, 1951) to 43.4 for *Macroptilium lathyroides* (Milford, 1967), with a mean of 30.6 percent.

This value for tropical legumes is slightly lower than the mean value of 33.4 percent for tropical grasses reported by Butterworth (1967).

The crude fibre content of legumes tends to increase with increasing maturity and decreasing dry-matter digestibility, and where no facilities are available for the determination of *in vitro* or *in vivo* digestibility then the level of crude fibre (CF) can provide a rough indication of the dry-matter digestibility. Using the data in Table 13.2 and the corresponding crude fibre percentages in Appendix 1 a regression has been calculated:

DM digestibility = 84.5 − 0.94 CF $\quad$ r = −0.79 $\quad$ RSD = ± 4.1

Mineral composition. *Phosphorus.* The phosphorus contents of tropical legumes are shown in Appendix 1. The maximum level recorded was 1.0 in *Centrosema pubescens* (Guayadeen, 1951) but this is rather atypical (0.65 was recorded in *Pueraria phaseoloides*, Blasco and Bohorquez, 1968); and the lowest, 0.06 in *Stylosanthes humilis* (Shaw *et al.*, 1966; Fisher, 1969). Increasing maturity leads to a decrease in phosphorus content (Fisher, 1969), while increasing the level of fertilization with superphosphate increased the phosphorus percentage of *Stylosanthes humilis* from 0.06 to 0.21 percent (Shaw *et* *al.*, 1966). Thus large differences within species can occur owing to stage of growth and level of phosphorus fertilization and this makes comparison between species presented in Appendix 1 of doubtful value. However Andrew and Robins (1969b,c) in two experiments compared a range of legumes under the same conditions, and valid comparisons between species can be made within each study. No major species differences in phosphorus content are apparent.

Calcium. The level of calcium found in tropical legumes is generally much higher than that found in the grasses (see Appendix 1). In nearly all cases it is well above the normal requirements of grazing cattle.

Sodium. The sodium levels in tropical legumes shown in Appendix 1 were determined in pot experiments with different species being studied at the same time (Andrew and Robins, 1969b,c). Sodium levels below 0.05 were generally found with *Centrosema pubescens*, *Desmodium intortum*, *Desmodium uncinatum*, *Neonotonia wightii*, *Leucaena leucocephala* and *Macroptilium atropurpureum*. Higher values were reported for *Lotononis bainesii* and *Macroptilium lathyroides*, while *Vigna luteola* had a mean sodium level of 0.27 percent. Level of phosphorus and potassium fertilizer influenced the sodium content (Andrew and Robins, 1969b,c).

Magnesium. Magnesium is a very important element in the incidence of hypomagnesaemia. Fortunately the level of magnesium in the tropical legumes studied is generally high compared with the levels found in the grasses. These data suggest that these species contain adequate magnesium for ruminants.

Copper. Copper levels determined in five tropical legumes fertilized at different levels by Andrew and Thorne (1962) were: *Centrosema pubescens*, 2.0 to 5.9 ppm; *Desmodium uncinatum*, 3.0 to 5.1 ppm; *Indigofera spicata*, 1.8 to 4.8 ppm; *Macroptilium lathyroides*, 1.7 to 3.9 ppm; and *Stylosanthes guianensis*, 2.4 to 4.4 ppm. All these values are close to or below the normal requirements of beef cattle.

Undesirable factors. *Toxic factors.* Toxic factors have been identified in several tropical legumes. Best known is probably mimosine, which occurs in *Leucaena leucocephala* in concentrations ranging from 2 to 9.4 percent of dry weight (Oakes, 1968). Mimosine causes a loss of hair (Oakes, 1968), and

191

Figure 33. Two-year-old stand of Townsville stylo at Mountain Valley Station, Northern Territory, Australia. It was aerially seeded at a rate of 6.6 kg seed pods and 137 kg single superphosphate per hectare into natural woodland, Australia (**Photo:** N. Tiver)

Hamilton *et al.* (1970) reported enlarged thyroid glands in calves born to heifers fed *Leucaena leucocephala*. Despite these problems this plant is used widely in Hawaii, as it does not affect reproduction and gives rapid liveweight gains in steers.

Indigofera spicata contains an amino acid, indospicine (Hegarty and Pound, 1968), which damages the liver of grazing animals to such an extent that this legume cannot be used in pastures.

Tannins have been found in both the leaves and stems of different species of *Desmodium* (Rotar, 1965). The leaves of *Desmodium intortum* and *Desmodium uncinatum* grown in Hawaii contained 3.2 to 8.8 percent and 6.1 to 6.9 percent tannin respectively and it was suggested that levels in excess of 7 percent might inhibit rumen cellulase activity (Rotar, 1965). Hutton and Coote (1966) compared five species of *Desmodium*, finding leaf tannin contents varying from 3.6 percent for *Desmodium uncinatum* to 9.3 percent for *Desmodium canum*.

Oestrogens. Aglibut and Castillo (1963) measured the diethylstilbestrol equivalent of 14 legume species, using mice. They reported positive values in *Calopogonium mucunoides, Phaseolus aureus, Pueraria phaseoloides, Vigna unguiculata* and *V. sesquipedalis* but none of the values exceeded 0.018 micrograms diethylstilbestrol per gram of dried feed. Also using mice, Bindon and Lamond (1966) found that feeding 24 different tropical legume samples had no adverse effects on reproduction.

Conclusions

The tropical legumes appear to be a rich source of protein and most minerals and have a range of digestibilities similar to the tropical grasses, while the limited data on voluntary intake indicate that they are equal to or better than the tropical grasses. Some tropical legumes contain toxic factors, but oestrogenic activity is generally low.

193

14. The tropical pasture legumes

In the catalogue following this chapter each of the pasture legumes is characterized in varying detail, depending upon the information available. Botanical descriptions have been gleaned from technical publications dealing with the flora of various countries and from popular descriptions recorded by several pasture workers. The characterizations include the following information (if available).

Scientific name. The valid botanical name has been recorded as far as is known at the time of writing. New taxonomic studies have recently led to changes in some of the well-known names, but for valid reasons.

Synonyms. Some older scientific names have been listed. There is some confusion in identifying some species.

Common names. The most usual common names for the plants are listed. In Africa there are numerous local or tribal names which cannot be listed in a publication of this size.

Description. The botanical description has been made as simple as possible.

Distribution. The place of origin of the species has been recorded whenever possible, and a general world distribution listed.

Season of growth. The plant is listed as an annual, biennial, short-term perennial or perennial. A normally perennial plant grown out of its most favoured environment may behave as an annual.

Optimum temperature for growth. Only a few of the tropical pasture legumes have been actually grown in a controlled environment, notably by Whiteman (1969), and there is room for more information in relation to this factor.

Minimum temperature for growth. In the subtropics, a knowledge of the reaction of pasture legumes to low temperatures is particularly important, as frosts vitally affect their growth pattern. There are three critical temperatures — that at which growth ceases, that at which defoliation takes place, and that at which the species is killed. A knowledge of these temperatures is needed in formulating pasture mixtures to provide for year-long grazing. Ludlow and Wilson (1970) have provided a useful lead in evaluating many of the tropical legumes.

194

Frost tolerance and regrowth after frosting. This factor is closely related to (and is usually included with) the other temperature data. The onset of frost subsequently precipitates a shortage of forage and a deterioration in quality (Jones, 1967a). Consequently, cold-tolerant species such as *Lotononis bainesii* are important in providing green feed following frosts, but more productive and reliable legumes are needed. Recovery after frosting is also important and those legumes which, although frost-susceptible, can grow quickly in the spring, fill a valuable niche in the feed-year programme.

Latitudinal limits. This factor is also related to temperature and day length, and is important in relation to plant exploration. There is need for collection and testing of pasture legumes from the subtropical temperate border regions for use as cool climate legumes for autumn-winter production. The variation with latitude in flowering times provides an important breeding tool.

Altitude range. Altitude is also related to temperature and photoperiod. Atkinson (1970) has suggested that collection of native legumes from high-altitude-low-latitude environments might also be useful in finding species suitable to autumn and winter production. Many of the tropical countries such as Colombia, Guatemala, Kenya, Peru and Tanzania have mountainous areas for development of improved pastures as well as low coastal and subcoastal regions.

Rainfall requirements. The approximate figures for rainfall requirements have been given, though little detail of incidence has been recorded. In areas where bimodal rainfall is experienced, the annual rainfall is usually not as effective as in areas where the season is more defined. The length and intensity of the dry season should be known, to assess the likely persistence of plants in a new environment.

Drought tolerance. Persistence under drought conditions is particularly important in all but the wet tropics, and more controlled environment studies with tropical legumes are required to record their response to stress imposed at varying times in their growth cycle. The severity of droughts may be such that self-seeding annuals such as *Stylosanthes humilis* may perform better than perennials in that they survive the drought as seed. Deep-rootedness is another important factor in drought survival.

Tolerance of flooding. Little accurate information on this important character is available. In the wet tropics, waterlogging is a feature of a number of swampy areas (e.g. the Llanos *inundables* of Venezuela). Where monsoonal downpours cause widespread short-term flooding, it is important to know the response of forage legume plants to flooding and their survival as seed. Nodulation under waterlogged conditions should also be known.

195

Soil requirements. The adaptability of the forage legumes to varying soil textures and pH has been listed. This gives an idea of the extent to which soil modification may be necessary to ensure adequate growth. Andrew and Norris (1961) have shown that extensive liming is not necessary for the tropical forage legumes as a group and is of more importance in modifying pH upwards to release molybdenum and reduce manganese and aluminium toxicity.

The response of the legumes to saline soil conditions is as yet little known; but, as development encroaches on coastal saline flats or inland salt-affected areas, there will be increasing demand for salinity tolerance in forage plants.

Rhizobium *relationships*. This important factor has been dealt with in Chapter 5 and also individually in this chapter at the species level. Information from Norris (1967), Date (1969) and Batthyany (1970) has been freely used. More information is needed relating to the date of initiation of effective nodulation in individual species and the nodulation sequences as reported by Whiteman (1969).

Ability to spread naturally. An important characteristic in any forage legume is its ability to spread naturally, especially into native pasture. The extent of environmental modification required to achieve this objective should be known.

Land preparation for establishment. It is known that most forage legumes will establish successfully if land preparation is well done. However, ploughing and frequent cultivation to provide a seed bed are expensive, and more information is needed on methods of adequate establishment at the lowest cost. Species with large seeds usually establish well and breeding for increased seed size may well prove worth while.

Sowing methods. Various methods of sowing have been listed. Simple methods for use in small areas have been outlined as well as large-scale aerial seeding procedures.

Oversowing into natural pastures. If oversowing is successful, the cost of pasture establishment can be reduced. The response to oversowing of the tropical species has been listed.

Sowing depth and cover. Depth of planting is important, especially for small seeded legumes such as *Lotononis bainesii* that can easily be sown too deeply. Some cover is desirable to prevent too rapid moisture loss and removal by harvesting ants.

Sowing time and rate. Choice of sowing time is often dictated by incidence of seasonal rainfall. It is important to obtain good establishment with sown pastures and moist conditions are necessary for good germination. Larger seeded legumes can be planted after rain has fallen but, by the time planting machinery can negotiate the wet land after rain, the surface is often too dry to ensure germination of small-seeded species. For this reason sowing dry, ahead of the likely rainy season, is often practised. As excessive heat and evaporation during midsummer are often hazards in pasture establishment, late planting is often adopted to coincide with cooler weather (as in the case of *Neonotonia wightii* and *Lotononis bainesii*). Care, however, must be taken not to sow so late that subsequent cool weather will interfere with establishment.

Sowing rate determines the rapidity of establishment of a sward, the density of many species closely following the planting rate (Middleton, 1970). The amount of seed planted, however, is also determined by cost and therefore a compromise must be reached.

Number of seeds. It is important to know the number of seeds per kilogram for calculation of effective stand and for inoculation purposes, and to assess seed size for planting techniques.

Percentage of hard seed. Hard-seededness protects a species against adverse conditions but, unless some presowing treatment is given, initial stands of many tropical legumes will be low.

Seed treatment before planting. Notes are given of methods for breaking dormancy, the need for inoculation, pelleting and treatment for disease, and insect prevention.

Nutrient requirements. Although some fertile soils may not need added fertilizer for good pasture establishment, the better soils are usually used for cash-cropping and pastures are relegated to the poorer areas. In most cases these will need a complete fertilizer dressing. Although a blanket dressing is often required, some species require a more specific nutrient than other species (e.g. *Neonotonia wightii* needs more molybdenum than other legumes).

Specific reactions to deficiencies of calcium, phosphorus, potash and copper and to toxicity by excess manganese have been given for some of the main forage legumes, based mainly on the research of Dr C.S. Andrew of CSIRO, Australia, and his co-workers.

Response to photoperiod. Species have been listed as short-day, long-day, or day-neutral plants and, where information is available, the optimum number of hours of sunlight for flowering and seed production are listed. More information is required for this characteristic for many legumes.

Response to light. This characteristic has been interpreted generally as tolerance to shading, especially by taller grasses.

Compatibility with grasses and other legumes. Companion grasses and legumes have been listed for the tropical pasture legumes. The competitive effects of tall grass and sward-forming grasses limit the performance of some species.

Ability to compete with weeds. Many of the tropical forage legumes were originally developed as cover crops and hence have been selected over the years for their ability to smother weeds. Quite often they were hand weeded in the early stages until they could successfully compete. Hence, some are initially slow in forming a complete cover. Management to encourage the smothering ability of such legumes as siratro and *Macrotyloma axillare* is important in some situations.

Tolerance of herbicides. Most of the work on the response of forage legumes to both pre- and post-emergent herbicides has been carried out by D.R. Bailey in north Queensland, Australia. There is still a good deal of this kind of work to be done with the newer legumes.

Seedling vigour. This characteristic is important in permitting a legume to make an early impact in mixed pastures before the associated grasses either shade or starve it out. Most large-seeded legumes have an early advantage in this characteristic.

Vigour of growth and growth rhythm. A vigorously growing legume will successfully compete with companion species and also give an early bite for the grazing animal. The growth rhythm is most important in the choice of species and more legumes should be studied and catalogued for this feature. Two aspects require study — growth throughout the season ungrazed or uncut, and the response to cutting or grazing at different times or at different physiological growth stages. The growth curves for the species which have so far been recorded are presented.

Nitrogen-fixing ability. The main role of legumes in a mixed pasture is to provide nitrogen for the associated grasses. The usual method for recording this ability is to grow grass with or without a known addition of nitrogen and to compare the yields with those of the same grass plus a legume. In other cases, estimates may be made from the yield and nitrogen contents of the individual components. Figures derived from such methods are recorded for each species. The accuracy of such determinations has been discussed by E.F. Henzell in Chapter 6. Leaf fall and subsequent mineralization are also shown to be important.

Response to defoliation. The ability of a forage legume to withstand defoliation is an important feature under grazing and determines persistence. The growth habit of a creeping stoloniferous or rhizomatous legume gives it an advantage as the growing points are near the ground and are less subject to damage by grazing. Most of the recorded data refer to hand cutting trials in

198

small plots. On a field scale, persistence under different grazing pressures, as revealed by periodic checks on botanical composition, gives a more realistic appreciation.

Grazing management. It is comparatively easy to establish a legume if satisfactory climatic soil and rhizobium requirements are met. The persistence and contribution of the legume to a pasture under grazing require skilled management. Generalizations such as light grazing in the year of establishment and withdrawing areas from grazing to allow for seeding apply to most legumes, but there is still a good deal to be learned about the management of mixed pastures under grazing to ensure that each species makes its best contribution at the required period of the year.

Response to fire. In many tropical and subtropical pastoral areas, fires are common in the winter or dry season. Burning may be a planned annual or biennial event to control pasture or brush growth, but more often depends upon some haphazard system of burning to provide a new shoot, or results accidentally. In areas subject to fire, a legume must be able to resist the heat of the fire by reason of a well-developed crown or fire-resistant stems or by annual seeding ahead of the passage of fire. Where information is available with regard to the individual species, it has been recorded. More knowledge is needed about fire tolerance.

Breeding systems. A knowledge of the breeding systems of pasture species is essential for the plant breeder and the pure seed producer. The chromosome numbers for the various species have been listed and E.M. Hutton has dealt with selection and breeding of tropical forage legumes in Chapter 12.

Green- and dry-matter yields. The yield, intake and digestibility of the dry matter of fodders are important in animal nutrition. The yield of dry matter indicates the approximate carrying capacity of a pasture based on an intake of 2.5 to 3 percent of the body weight of the grazing animal per day.

Green- and dry-matter yields have been recorded for the various species. The varying conditions of production, such as irrigated or rain-grown, fertilized or unfertilized, annual or perennial mixture, hand cutting or mechanical harvesting, nursery or field plots, make the figures presented only an approximate guide to the performance of each species in a particular situation. Where average yields have been recorded over three successive seasons, a better assessment can be made.

Suitability for hay and silage. Where information is available about these features it has been recorded. A good deal of information on hay making has come from Otero (1952) in Brazil relating to species found in that country, but generally information is scanty and recorded more as an afterthought in

199

most of the literature. Otero (1952) has also recorded his experience with silage manufacture from some tropical species in Brazil and the analyses of the final product. Risopoulos (1966) had success with *Stylosanthes guianensis* in Zaire, and Catchpoole (1970) has more recently recorded problems in ensiling tropical forage legumes on their own. Although silage for short-term feeding of beef cattle is losing popularity, it is still a potent source of feed for dairy cattle in winter and the dry season. More research is needed on the ensilage of mixtures of the gramineous fodder crops and grasses with tropical legume species and the feeding value of the material.

Value as standover or deferred feed. This feature is important, as the dry season is often rainless and well defined in tropical areas. Any species which will hold its leaf and maintain a high nutritive value in the "standing hay" stage, such as *Stylosanthes humilis*, is invaluable in filling the need for animal nutrition at this time. Where information is available, it has been recorded for each species.

Feeding value. Where the information was available, chemical analyses and digestibilities of the component groups have been recorded for each species and in Appendix 1. Again, the varying conditions of soil, climate, time, stage of growth and component parts of the plants listed make comparisons difficult, but some generalizations can be made. Young plants have higher protein values than older ones, and the leaf a much higher figure than the stems. Leguminous browse has high protein but also high crude fibre figures.

Much more information is needed about the digestibility and intake figures for the various species.

Palatability is a relative term. Any specific information has been listed in connection with individual species.

Toxicity. As a few of the tropical legumes contain material toxic to livestock, a brief outline of these occurrences has been given. The tropical legumes generally do not contain as much oestrogenic substances as the temperate species. *Leucaena leucocephala* and *Acacia georginae* can be quite toxic. The reader is referred to the original papers for detailed information relating to toxicity.

Seed harvesting methods. Information about harvesting tropical legume seeds is scanty, although improved techniques are being developed. Since this aspect is important in keeping seed prices at a reasonable figure, as much information as possible about harvesting methods has been given in the text in Chapter 11 and for individual species in this chapter.

Seed yields. Actual seed production by tropical legumes is usually much

200

higher than the figures quoted for recovery in the harvesting process. Seed shattering, uneven ripening and inaccessible seed all reduce the yield.

Minimum germination and quality of seed for commercial sale. This section has been based mainly on the requirements of the Queensland Department of Primary Industries, Australia, which has now had several years' experience in testing and regulating seed supplies for the tropical legumes.

Cultivars. The Australian cultivars have been officially described and registered for release, for identification of mother seed and as a protection for the farmer. Commercial lines of seed from other countries have been listed where information is available. A systematic approach to cultivar registration is recommended. Information about the Australian system can be obtained from the Division of Plant Industry, CSIRO, Box 1600 Canberra, ACT, Australia.

Diseases. Diseases known to attack the various tropical legumes have been recorded. Notes relating to the etiology of the diseases and their control have not been recorded, and the reader is referred to the original references or to local plant pathologists for this information.

Pests. Insect pests recorded as attacking forage legumes have been listed. More detailed information is contained in the original papers referred to and is available from local entomological sources.

Main attributes. These have been listed to give a quick appreciation of the merits of each legume species.

Main deficiencies. These notes refer to certain disadvantages possessed by the legume in question.

Performance. This section is devoted to a record of the performance of the forage legume, mainly under grazing, as contrasted with yields under cutting treatments recorded earlier.

Much more information is needed for the tropical legumes on their performance in terms of animal production. Inevitably most of this will come from grass-legume association. There is need in most countries for a rapid transfer from hand cutting of nursery plots to a measurement of performance by livestock grazing the species in the field.

Sources of seed supplies. This section has been included to enable workers to obtain small samples of seed of various species for trial. The author is indebted to Mr J. León of FAO who has collected much of the data (Appendix 3).

Main references. Although a bibliography is provided, the multiplicity of published papers makes it difficult to determine the major references. A selection of such works has been listed for the important species.

201

the
pasture species

CATALOGUE

Aeschynomene spp.

Aeschynomene americana L.

Synonyms. *A. javanica* Miq.; *A. glandulosa* Poir.; *A. guayaquilensis* G. Don; *Hippocrepis mimosula* Norona; *A. tricholoma* Standl. and Steyerm.

Common names. American joint vetch (United States and Australia), thornless mimosa (Sri Lanka), bastard sensitive plant (Jamaica), pega pega, pega ropa, antejuela, ronte, cujicillo, dormilonga and others (Latin America) (Rudd, 1955).

Description. An erect-ascending, annual or short-lived perennial, shrub-like legume, 1 to 2 m tall. Stems hairless, often hard but pithy in the lower regions; upper sections hairy, usually glandular. Leaves 3 to 7 cm long. Leaflets 8 to 38 pairs, 5 to 15 mm long, 1 to 2 mm wide, linear or linear to oblong, sensitive, folding together when touched. Stipules persistent, linear-lanceolate, 1 to 2 cm long. Inflorescence a loose, few-flowered raceme, often branched. Calyx about 4 mm long, deeply two-lipped. Corolla varying in colour from white tinged with mauve or light violet to purple to flesh coloured, sometimes pinkish orange, 6 to 10 mm long. Pod usually slightly curved, four to eight jointed, up to 4 cm long, the upper margin entire, the lower deeply indented at each segment joint; each segment one-seeded; semi-circular, about 4 mm long. Seeds grey-green to dark and light brown.

Distribution. Native to Central America and tropical South America, extending as far south as Argentina and north to Florida, United States, and the West Indies. Var. *flabellata* is found mainly in wet or moist places at elevations up to 2 800 m. Used in Indonesia for green manure and in Florida and Queensland, Australia, for forage.

Rudd (1955) identified three varieties: var. *americana,* found predominantly in the Caribbean and adjacent areas; var. *glandulosa* (Poir.) Rudd, native to Central America and tropical and subtropical South America; and var. *flabellata* Rudd, from Central Mexico.

205

Figure 34. A well-grown seed production stand of *Aeschynomene americana* cv. Glenn

Season of growth. Germinates in the first spring storms and grows through summer. Flowers and matures from autumn till early winter.

Frost tolerance and regrowth after frosting. None.

Altitude range. Sea level to 1 000 m in Jamaica (Adams, 1972).

Rainfall requirements. Essentially a wetland plant, requiring a minimum of 1 000 mm mean annual rainfall.

Drought tolerance. Slight.

Tolerance of flooding. Excellent. Its main natural habitats are low-lying wet areas and drainage ditches. *A. indica* is probably better adapted to very wet sites and longer periods of flooding.

Soil requirements. Quite widely adapted, but particularly suited to wet, low-lying, sandy soils and to similarly located solodics of coastal lowland country.

Rhizobium *relationships*. Promiscuous and nodulates readily with broad-spectrum cowpea-type inoculants.

Ability to spread naturally. Seed is widely distributed by cattle, both on their coats and in dung. Spread is rapid and effective, generally with no rhizobial limitations, particularly in suitable wet habitats.

Figure 35. *Aeschynomene americana.* **A**-Leafy branch **B**-Flowering and fruiting branch

Land preparation for establishment. The better the seed bed the better the establishment, but it has been satisfactorily established into existing pastures in the Mackay region of central Queensland, with a simple renovation and subsequent broadcasting of seed. In subsequent years, regeneration occurs without further soil disturbance.

Sowing methods. Seed in pod or dehulled seed can be drilled into fully prepared seed beds or introduced into existing sown or native pastures using strip cultivation and planting methods. Herbicidal treatments and sod-seeding can also be used.

Sowing depth and cover. Hodges *et al.* (1982) recommend planting no deeper than 2.5 cm since seedling emergence is reduced at greater depths.

Sowing time and rate. Planting is best done at the onset of the wet season (from June 1 to July 1 in central Florida) (Hodges *et al.*, 1982). Initial season production is reduced from later plantings. The recommended planting rate in Florida is 5 kg/ha of dehulled seed or 12 kg/ha of seed in pods (Hodges *et al.*, 1982).

Number of seeds per kg. For cv. Glenn, 368 000.

Percentage hard seed. Seed in pod can be greater than 90 percent hard or as little as 55 percent, while the mere act of dehulling samples can reduce the hard-seed level to 20 percent or less. If necessary, further scarification can be applied.

Seed treatment before planting. Generally, dehulling of the seed is sufficient, but if the sample already has an adequately low hard-seed content and seed is readily available, an increase in the sowing rate to that for seed in pods is all that is required.

Nutrient requirements. On phosphate deficient soils, American joint vetch responds strongly to applied phosphate. At Mackay, on soils with only 3 to 6 ppm available P, applications of 20 kg/ha of P have given two- to fourfold increases in dry-matter yield (H.G. Bishop, personal communication).

Response to lime at 2.24 tonnes/ha ground limestone was clear on very acid soils at Ona, Florida, but responses to P and K have been variable on deficient soils (Hodges *et al.*, 1982). American joint vetch at least will grow on low-phosphorus soils. Liming should be aimed at adjusting the pH to the range 5.5 to 6.0 (Hodges *et al.*, 1982).

Compatibility with grasses and other legumes. In the seedling regeneration phase at the beginning of each growing season, joint vetch can be susceptible to competition from more vigorous grasses. Once established, the tall and erect growing plants can stand competition. There are few legumes adapted to the wet locations, so these rarely worry American joint vetch. The same remark applies to weeds.

Tolerance of herbicides. In seed-production stands, American joint vetch is tolerant to both trifluralin and 2,4-D (J.M. Hopkinson, personal communication).

Seedling vigour. In Florida, initial growth is slow. Plants require 60-70 days to reach a height of 60 cm and need special care at this stage. In existing stands the area should be heavily grazed over winter, after seed set is complete, but grazing should cease when seedlings reach a height of 5 cm, and not recommence until they are 45 cm tall (Hodges *et al.*, 1982).

Vigour of growth and growth rhythm. Once seedlings are established, growth is vigorous and difficult to control. It continues until full flowering is achieved. Once maturity commences, growth ceases. After frosting, residues rapidly break down.

Nitrogen-fixing ability. It is estimated in Florida that the equivalent of 112 kg/ha N are added each year to the pasture system by American joint vetch (Hodges *et al.*, 1982).

Response to defoliation. American joint vetch withstands heavy grazing well; under these conditions it branches readily close to the ground and develops a low, dense, leafy canopy.

Grazing management. The only stage when special attention to grazing management is required is during seedling establishment. In Australia, continuous grazing is then practised on most pastures and American joint vetch accepts this well. In Florida, with rotational grazing, a 3- to 5-week regrowth period between grazings is recommended for maximum production (Hodges *et al.*, 1982).

Response to fire. Fire during the growing season is unlikely and a fire during the late dry season is most likely to increase the germination of the hard seed, at the same time reducing competition for the establishing seedlings.

Breeding system. Uncertain, but believed to be essentially cleistogamic, pollination taking place before the flowers open.

Dry- and green-matter yields. In Malawi, a Rhodes grass/*A. americana* mixture yielded more than a Rhodes grass/*Neonotonia wightii* mixture. *A. americana*, however, was less palatable than glycine: only 65 percent of the herbage was consumed by animals, compared with 74 percent for the glycine (Papua New Guinea, 1961). Parbery (1967a) obtained a maximum dry-matter yield of 13 325 kg/ha on Cunnunurra clay in 220 days. Applied nitrogen depressed the yield but increased the protein percentage from 9.4 to 13.4 percent. The plant was 1 m high at maturity and woody at the base. In Sri Lanka, five cuttings, each two months apart, yielded 26.4 tonnes/ha of green material (Paul, 1951).

Suitability for hay and silage. It has been used for green chop, silage and haymaking in Florida, but the high moisture and mucilaginus nature of the green material make it difficult to handle, while the hay dries brittle and twiggy. As a result, it is not ideal for these uses (Hodges *et al.*, 1982).

Value as standover or deferred feed. Slight, as it becomes dry and brittle once mature, especially after frosting, and disintegrates readily.

Feeding value. The mature pods contained 19.6 percent protein at Cunnunurra, northwest Australia (Parbery, 1967a), while, when grown in dense stands at Peradeniya, Sri Lanka, hay had an analysis of 12.72 percent moisture, 21.27 percent crude protein, 7.7 percent ash and 22.5 percent crude fibre (Paul, 1951). In Florida, stands of American joint vetch cut to 13 cm stubbles when 30 cm tall, over two years averaged 17.5 percent crude protein and 70 percent *in vitro* organic matter digestion. Crops allowed to grow to 90 cm before cutting gave poor-quality forage (Hodges *et al.*, 1982).

Toxicity. None recorded.

Seed harvesting methods. Seed is readily harvested with a conventional wheat header. Crops need to be sprayed with insecticide at flowering to control podboring insects. Seed crops in northern Queensland are invariably attacked by powdery mildew late in the season, but this does not appear to cause any trouble (J.M. Hopkinson, personal communication).

Seed yield. Yields of seed in pod as high as 2 tonnes/ha have been obtained from small plots in Queensland (J.M. Hopkinson, personal communication).

Cultivars. An unnamed cultivar under the common name American joint vetch has come into use in Florida in recent years. This comes from material tested in the early 1950s. Unnamed material has also been used as a green manure crop in Indonesia (Whyte, Nilsson-Leissner and Trumble, 1953) while it was studied as a forage and soil cover in Sri Lanka (Paul, 1951).

In 1983, cv. Glenn was released in Queensland for use on the wet coastal lowlands around Mackay. 'Glenn' derives from seed collected in 1971 some 45 km south of Tampico, Mexico, at an altitude of 15 m and a similar latitude (22° N) and rainfall (1 000 mm/year) to Mackay. It is very similar to the Florida common material, maturing by mid-season, with mature seed present by mid-May. One earlier-flowering and numerous later-flowering accessions were present in the Mackay collection (H.G. Bishop, personal communication).

Diseases. The only disease noted to date has been powdery mildew on ungrazed seed crops.

Main attributes. Ability to thrive on low-lying, wet, rather infertile soils is its outstanding feature.

Main deficiencies. Lack of carryover forage for dry season use.

Performance. Cattle readily graze *A. americana*, and weight gains of 0.54 kg/head/day have been recorded over an 85-day period (Hodges *et al.*, 1982), while gains of 0.51 kg/head/day for 177 days (December 5 to June 20) were recorded at Mackay (H.G. Bishop, personal communication). In the latter case, the area was stocked at 2 steers per 1.25 ha.

Main reference. Hodges *et al.* (1982).

Aeschynomene falcata **(Poir) DC** (see Colour plate IV)

Common name. Joint vetch (Australia).

Description. A prostrate herbaceous perennial with a short, tough taproot. Slender stems branch at soil level from a bulbous woody crown. Stem decumbent, to about 6 cm long, pubescent and sometimes also hispidulous; stipules lanceolate, acuminate, 5 to 8 mm long; five to seven (to eight) foliolate the petiole and rachis pubescent like the stem; leaflets obovate-elliptic, about 6 to 10 mm long, 3 to 4 mm wide, obtuse, mucronate, pubescent on both surfaces, the base often oblique, entire; inflorescences usually with one or two flowers developing, longer than the subtending leaves, the peduncles and pedicels hispidulous: the bracts and bracteoles subovate, acuminate, 1 to 2 mm long, about 1 mm wide, pubescent; ciliolate; yellow flowers 7 to 9 mm long, the claw about 1 mm long; the blade orbiculate, 6 to 7 mm in diameter, entire, the outer face puberulent; wings about 7 mm long, the claw 1 mm long, the blade about 6 mm long, 1.5 to 2 mm wide at maximum; keel 7 to 8 mm long, the claws 1 mm long, the blades 6 to 7 mm long, about 2 mm wide; stamens about 8 mm long; legume usually falcate six to eight articulate, the stipe 6 to 14 commonly 8 to 10 mm long, with spreading, glandular hairs, 1.5 to 2 mm, the articles 3 to 4 mm long, 2.5 to 3.5 mm wide, puberulent with crispate or appressed hairs, sometimes also sparsely hispidulous, the body of the articles tending to break away from the margins; seeds about 2 mm long and 1.5 mm wide, dark brown (Rudd, 1955).

Distribution. The Australian cultivar was collected in Paraguay. It is also recorded in Venezuela (Pittier, 1944).

Characteristics. The only commercially available cultivar (cv. Bargoo) is distinguished from the species by seeds in the sample varying in colour from light yellowish-brown to almost black. There are approximately 375 000 seeds per kilogram. Cv. Bargoo is predominantly summer growing, but possesses a degree of tolerance to low temperatures and mild frosts. It is adapted to a variety of poor to moderately fertile soils, derived from sandstone, shale and granite of the Clarence Basin in northern New South Wales, Australia. It responds to superphosphate. It tolerates temporary waterlogging but it does not persist in permanently damp situations. It is easy to establish but requires inoculation with a specific strain of *Rhizobium* (CB 2312) for best results. However, effective indigenous rhizobia are present in some soils.

Flowering begins in the spring and continues into autumn. Seed is usually ready to harvest in mid-summer and autumn. The seed pods shatter readily.

Possesses a high degree of resistance to *Amnemus* weevil and is resistant to root knot nematodes.

Very palatable and is grazed selectively by cattle, sheep and marsupials. It withstands heavy grazing and will also persist in rank ungrazed pastures saved for autumn grazing. It combines well with *Axonopus affinis, Paspalum dilatatum, P. notatum, Cynodon dactylon* and a wide range of native grasses (Mackay, 1973).

Aeschynomene indica L.

Synonyms. *A. cachemiriana* Camb.; *A. glaberrima* Poir.; *A. pumila* L.; many others.

Common name. Budda pea (Australia).

Description. An erect, sub-shrubby, generally annual but sometimes perennial herb, 0.3 to 2.5 m tall. Stems mostly slender, about 5 mm wide at the base, but sometimes thick and spongy and up to 2.5 cm wide, pubescent with rather sparse, tubercular-based, sometimes glandular hairs. Leaves sometimes sensitive, 16- to 50- (sometimes 70) foliate. Leaflets linear-oblong, 0.3 to 1.3 cm long, 1 to 3 mm wide, rounded and mucronate at the apex, obliquely rounded at the base, entire or very finely serrulate, glabrous. Stipules elliptic-lanceolate, straight, spurred, 0.3 to 1.5 cm long, 1 to 3.5 mm wide, deciduous. Inflorescences leaf opposed or axillary, one- to six-flowered. Rachis 1.5 to 6 cm long. Peduncle 0.8 to 2.1 cm long. Pedicel 1 to 2 and finally up to 8 cm long. Calyx glabrous, two lipped, the lips oblong, 4 to 6 mm long, 2 to 3 mm wide. Standard yellow or whitish, mostly lined and suffused with red outside, or purplish, elliptic, 0.7 to 1 cm long, 4 to 7 mm wide. Wings and keel greenish white or pale yellow, the petals of the keel not laciniate. Pod linear, straight or slightly curved, 2.4 to 4.8 cm long excluding the 6 to 9 mm stipe, 5 to 13

212

jointed, one suture more or less straight, the other slightly constructed between the articles. Articles oblong, 3 to 5 mm long and as wide, compressed with sparse, short, tubercular-based hairs, central part raised. Seed dark olive-black or brownish, oblong, slightly beaked near the eccentric small hilum, 2.8 × 2.55 mm, 1.3 to 1.8 mm thick (Gillett, Polhill and Verdcourt, 1971).

Distribution. Widespread in tropical Africa from Senegal to Ethiopia and from northern Somalia southward to the Transvaal and Namibia, and also on Sao Tome and Madagascar. Widespread also in tropical and subtropical Asia, Australia and North America. In Australia, it occurs as far south as northern New South Wales and the northern parts of the Flinders Ranges of South Australia. The North American material is often identified as *A. evenia*. However, Rudd (1959) states that the two are distinct species, both of which occur in North America.

Habitat. Mostly found in wet and muddy places on flood plains, in seasonal swamps and around the margins of more permanent swamps into quite dry country. In Fukuoka Prefecture, Japan, it is among the pioneering annuals that colonize paddy fields when cultivation stops. It occurs from sea level to 1500 m altitude.

General features. Of low to moderate palatability, it is not readily eaten but is generally considered a useful legume where it occurs. There is some suspicion of occasional toxicity. Boyland (1974) indicated strong field evidence of toxicity in western Queensland, Australia, and Bogdan (1977) reported the death of animals in Malawi. Stock deaths implicating *A. indica* have been reported also from Papua New Guinea (Papua New Guinea, 1961).

Nodulation has been reported from Queensland (Bowen, 1956) and Mali (Sanogho, 1977), the latter estimating nitrogen-fixing abilities of the different strains of nodule bacteria isolated.

Chromosome number. 20n = 40.

Main reference. Gillett, Polhill and Verdcourt (1971).

Alysicarpus spp.

Alysicarpus monilifer (L.) DC

Description. Diffuse herb. Leaves ovate or oblong or lanceolate, cordate at the base, 2.5 to 7.5 cm long, firm, both surfaces pale green, prominently nerved, glabrous or sparsely pubescent beneath. Racemes axillary and terminal, early four- to six-flowered. Calyx not longer than first joint of pod. Pod 1 to 2 cm long, 3 to 4 mm thick, three- to five-jointed, distinctly moniliform, glabrous or sparsely pubescent (Andrews, 1952).
Distribution. Widespread in the central Sudan (lat. 15° N) in the 300- to 400-mm rainfall zone on deep sand and heavy clay.
Characteristics. A free seeding annual. At Redland Bay, Queensland, Australia (lat. 27°S), it produces an attractive leafy plant and seeds heavily. No feeding trials yet attempted.

Alysicarpus rugosus DC

Synonym. *Hedysarum rugosum* Willd.
Description. Upright legume growing from 30 to 60 cm, subglabrous. Leaves 2.5 to 7.5 cm long, oblong, two to three times as long as broad. Racemes dense, 2.5 to 10 cm long, pedicels 0.2 to 0.3 cm; flowers adpressed to subglabrous rachis. Pod three- to five-jointed, glabrous, 0.2 cm broad; joints rather broader than long, marked with close prominent transverse ribs (Hooker, 1879).
Distribution. In India, up to 1 400 m in the foothills of the Himalayas; Sri Lanka, Burma, West Indies and elsewhere in the tropics.

214

Figure 36. *Alysicarpus monilifer,* showing plant in seed

Characteristics. Leafy, palatable, fairly persistent and moderately productive in Zambia. Grown satisfactorily on cultivated land but not yet grown in association with grass pastures (van Rensburg, 1967). Produced 5 775 kg DM/ha/year [1] over two years' trials in Zambia from two cuttings per year. Average protein content of dry matter 15.86 percent, giving an average protein yield per hectare of 780 kg; 42.26 percent of the total dry matter was consumed by grazing. Stem comprised 61 percent of the dry matter, containing 8.50 percent crude protein; leaf contributed 1 367 kg DM/ha of dry matter of 17 percent protein content.

Analysis of plant: 16.4 percent crude protein; 33.8 percent crude fibre; 1.4 percent fat; 6.7 percent ash; 41.7 percent nitrogen-free extract with 0.82 percent calcium, 0.19 percent phosphorus, 2.04 percent potash, 0.13 percent chlorine and 0.03 percent sodium in the dry matter (Verboom, 1965).

---

[1] DM = dry matter.

215

Alysicarpus vaginalis (L.) DC

Common name. Alyce clover (United States).

Description. Low spreading summer annual, but in thick stands the stems are more nearly erect and attain a length of up to 1 m. Stems moderately branched and moderately leafy. Leaves, simple and broadly ovate in shape, up to 2.5 cm long. Flowers, reddish yellow or pale purple, borne in racemes of six to 12 flowers each. The seeds, dark red, oval or oblong, 1 to 1.5 mm long.

Distribution. Native to India, southern Asia, Pacific Islands (Fiji, Tahiti), East Africa; naturalized in South America and the United States.

Characteristics. Self-seeding. Prefers sandy loams, but is very susceptible to nematodes and so is often grown on clay soils. Does not tolerate wet lands; good drainage is a prerequisite. Responded well to nitrogen on Cunnunurra clay in northern Australia, but nitrogen depressed its growth in Cockatoo sand (Parbery, 1967b). Responds to phosphorus but not to potash. Grows commonly on coral sand in Tanzania and Tahiti and in clay loams on basaltic slopes in Fiji (Payne, 1955).

Rhizobium *requirements*. Non-specific; does not require inoculation.

Sown in spring to midsummer at the rate of 11 to 16 kg/ha of scarified or machine-hulled seed on a fine seed bed at a depth no greater than 1 cm and lightly covered or rolled. Used primarily as a hay crop in Mississippi and Florida, United States, and is a useful component of natural grazing with *Desmodium heterophyllum* and *Atylosia scarabaeoides* on basaltic slopes at Sigatoka, Fiji.

Provides green grazing in the hot dry months of August and September in the southern United States. When harvested for hay is cut in early bloom with an ordinary mower and raked as soon as drying permits. Dries readily and can be baled the following day. Hay yields average from 3 750 to 6 500 kg/ha according to fertility. Parbery recorded 4 050 kg DM/ha on unfertilized Cunnunurra clay and 5 925 kg with 40 kg N. In Holly Springs, Mississippi, United States, unfertilized *A. vaginalis* yielded 4 340 kg hay per hectare and a yield of 6 120 kg was obtained from a dressing of 20 kg P_2O_5 plus 10.25 kg K_2O per hectare. Highest yields are obtained from the first crop in any field; subsequent crops decline owing to nematode injury.

Seed crop can be harvested by combine or mown and threshed. Seed yields range from 100 to 300 kg/ha.

Main reference. Wheeler (1950).

216

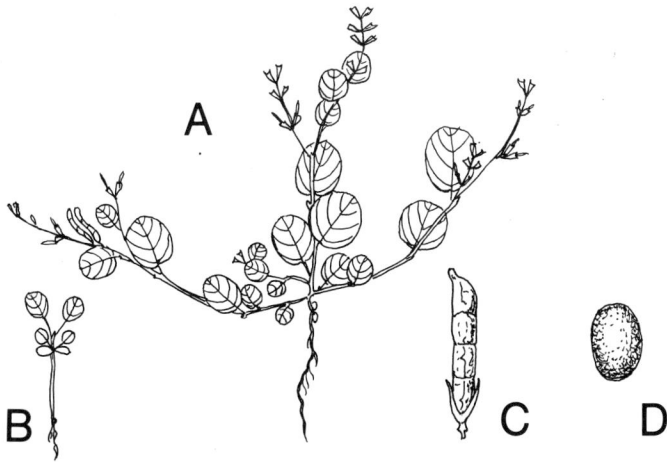

Figure 37. *Alysicarpus vaginalis.* **A**-Habit **B**-Seedling **C**-Pod **D**-Seed

217

Arachis spp.

Arachis glabrata Benth.

Common name. Arb (United States).

Description. Low perennial plant arising from central branching root, primary branches prostrate, creeping, often attached to ground here and there by adventitious roots, forming stools 15 to 30 cm in diameter and producing short sub-erect flowering stems at the crown. Leaves small, glabrous, four-foliolate, leaflets broadly ovate to elliptic, 6 to 30 mm long and 5 to 14 mm wide, papery. Flowers yellow, axillary. Fruit small (10 × 5 to 6 mm), ovoid, acute; pericarp thin, brittle (Hermann, 1954).

Distribution. Collected from a black soil at Campo Grande, Mato Grosso, Brazil, and listed as P.I. 118457. Southern limit about 35°S, northern Georgia, United States; occurs in Uruguay, northern Argentina, Paraguay and Brazil.

Main attributes. Highly valued companion for pasture grasses; competes with all pasture grasses, though slow in developing a sod. Withstands drought, grazing, trampling. Tops killed by frost, but rhizomes survive. Propagated by cuttings. Sometimes attacked by *Phyllosticta* leaf spot.

   A long-lived summer perennial. Adapted to deep sandy soils; rhizomatous. Persists under close mowing and grazing, gives good yields under low fertility.

Main deficiencies. Slow to establish in first season; gives relatively low yields under high fertility; fairly low palatability (Higgins, 1951).

Arachis monticola Krapovickas and Rigoni

Description. Annual with quadrangular root. Central plant erect, growing to 30 cm. Leaves oval to oblong. Five sessile axillary flowers (Krapovickas and

218

Rigoni, 1959). Chromosome number 2n = 40.

Distribution. Argentina.

Characteristics. Offers possibilities as a reseeding annual in perennial grasses in the Americas. Grows early in the spring and persists until frost. Pods small, single-seeded, ripen over a long period; not concentrated under any particular part of the plant, making seed harvest a problem.

Grows well in grass sods, especially with *Pensacola bahia*. Quite palatable, apparently produces top-quality forage (Higgins, 1951).

Atylosia spp.

Atylosia scarabaeoides (L.) Benth.

Synonym. *Dolichos scarabaeoides* L.

Common name. Wild kulthi (India).

Description. A perennial climber or trailer, forming thick mats. Root-stocks woody, slender. Stems often reddish, covered with short, ferruginous pubescence. Leaflets elliptic, the laterals slightly oblique, 0.8 to 7.3 cm long, 0.5 to 3 cm wide, rounded to subacute at both ends, sparsely to densely greyish-velvety pubescence, and gland dotted on both surfaces. Petioles 0.7 to 2.7 cm long, ferruginous pubescence. Petiolules 1 to 1.5 mm long. Inflorescence axillary, few-flowered. Peduncles 1 to 3 mm long. Pedicels about 4 mm long. Calyx about 6 mm long, the lobes longer than the tube. Standard yellow, flushed with crimson outside, obovate, 9 to 10 mm long, 5 mm wide, glabrous. Pods oblong, compressed, 1.5 to 2.5 cm long, 5 to 7.5 mm wide, three- to six-seeded. Valves conspicuously grooved between the seeds, densely, shortly pubescent and with much longer ferruginous, more or less adpressed hairs

219

Figure 38. A sparse stand of *Atylosia scarabaeoides,* planted on poor, red earth in central Queensland, Australia

and small glands. Seeds reddish, mottled brown or black, oblong, 4 to 5 × 2.5 to 3 mm, 1.5 to 1.8 mm thick (Gillett, Polhill and Verdcourt, 1971). *Distribution.* Widespread throughout Asia, especially in southern India (north to 24°N lat.), China, southeast Asia, northern Australia and New Guinea. Undoubtedly introduced to Africa, where it occurs in West Africa and Zanzibar.

General features. In northern Australia two varieties have been described, var. *scarabaeoides* (Syn. var. *queenslandica* Domin.), which is common and usually grows in open or cleared woodlands on heavy soils in north Queensland, and var. *pedunculata* Rey. and Pedley, which has a more open habit, leaves less rugose, and more membranous. It is rarer and found from northeastern Queensland to Western Australia, usually on sandy lateritic soils in open forest (Reynolds and Pedley, 1981).

In India, where *A. scarabaeoides* has been studied as a potential pasture legume, it occurs with *Sehima nervosum,* but not with *Heteropogon contortus* in arid areas (Shankar, Velayudhan and Toivedi, 1975), but has been dibbled into *Heteropogon* grassland at Jhansi before the monsoon, giving 40 percent

220

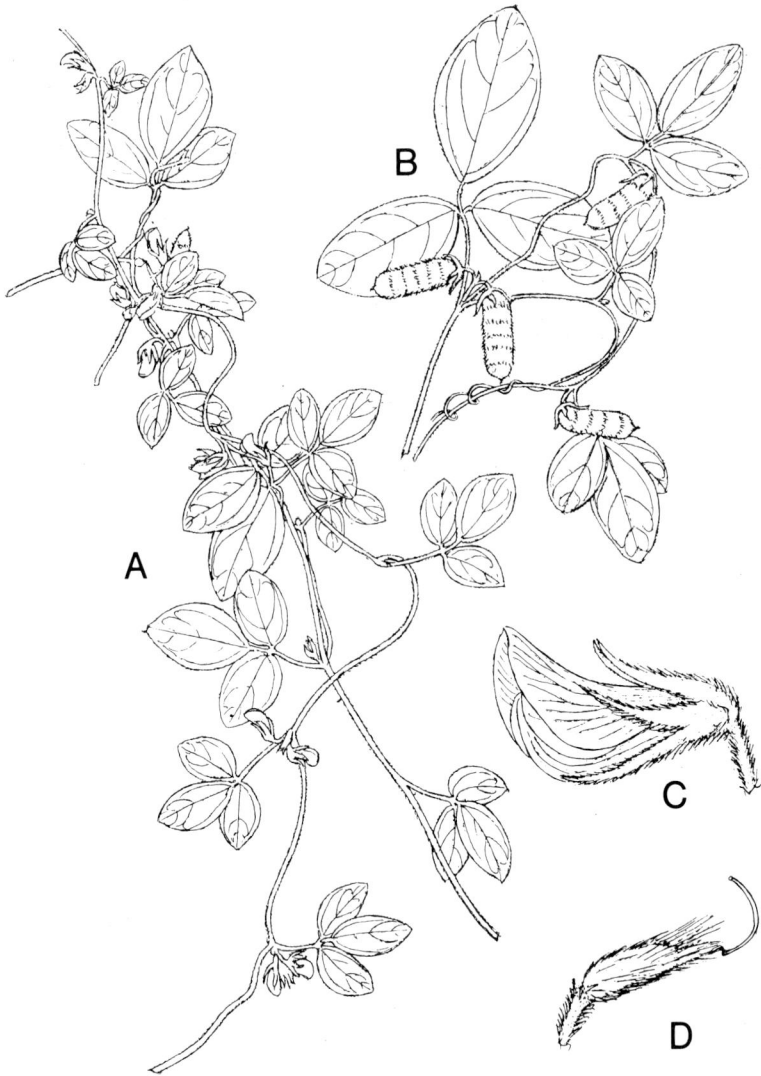

Figure 39. *Atylosia scarabaeoides.* **A**-Flowering branch **B**-Fruiting branch **C**-Flower **D**-Gynoecium

221

establishment. When planted in rows during the monsoon with 9 kg/ha of P_2O_5, 61 percent establishment was achieved, while siratro gave only 20 and 36 percent establishment under these conditions (Shankarnarayan *et al.*, 1975). At Jodhpur, when sown with the first showers but without inoculation, it was deep rooting and nodulated, but nodules were few (Satyanarayan and Gaur, 1965). It was also considered suitable for sowing into forest pastures in Uttar Pradesh, where it remained green in summer.

Plants grown at ICAR, Agra, had a protein content of 8.6 to 12.9 percent throughout the year and averaged 1.6 percent Ca and 0.15 percent P (Singh, 1962).

A. scarabaeoides was considered to be among the most suitable legumes for permanent pastures in the Guinea zone of Northern Nigeria by Foster and Mundy (1961).

Chromosome number. 2n = 22.

Main references. Reynolds and Pedley (1981); Gillett, Polhill and Verdcourt (1971).

Calopogonium spp.

Calopogonium caeruleum (Benth.) Sauv.

Synonym. *Stenolobium caeruleum* Benth.

Common names. Namie napirang, Klein Kau (Suriname).

Description. A sturdy, somewhat woody, climbing, perennial legume. Stems twining, pubescent to glabrescent. Leaves alternate, trifoliate. Petioles to 12 cm long, pubescent. Leaflets rhomboid-acute to ovate, the lateral ones oblique, pubescent above, velvety pubescent below, 5 to 20 cm long, 5 to 15 cm wide. Inflorescence elongated. Many-flowered, axillary, spike-like raceme, 25 to 50 cm long. Peduncle very short. Rachis sulcate, tomentose. Calyx campanulate, villous, 4 mm long, the teeth triangular-lanceolate, shorter or nearly as long as the tube. Corolla blue to violaceous. Standard 10 mm long, 8 mm wide, emarginate. Wings about as long as the standard. Keel shorter. Pods linear-oblong, pubescent, 4 to 8 cm long, 8 mm wide, impressed between the seeds. Four to eight seeds, orbiculate, compressed, shining (Graham, 1933; Pulle, 1976).

Distribution. Native to Central America, Mexico and the West Indies, and eastern tropical South America to southern Brazil.

General features. *C. caeruleum* reached the grazing assessment stage of evaluation at South Johnstone, on the wet tropical coast of north Queensland, Australia. It had grown better in winter than common centro (*Centrosema pubescens*) and had been high yielding. It also showed strong stolon development and, under grazing, maintained a better legume proportion in guinea grass pastures than centro or puero (*Pueraria phaseoloides*). By 1976, it had dominated an 0.8 ha area planted in 1973. By 1979, this dominance was so complete that animal performance was declining drastically. The *C. caeruleum* proved sufficiently unpalatable not to support animals on almost pure stands (Middleton and Mellor, 1982).

Main reference. Middleton and Mellor (1982).

223

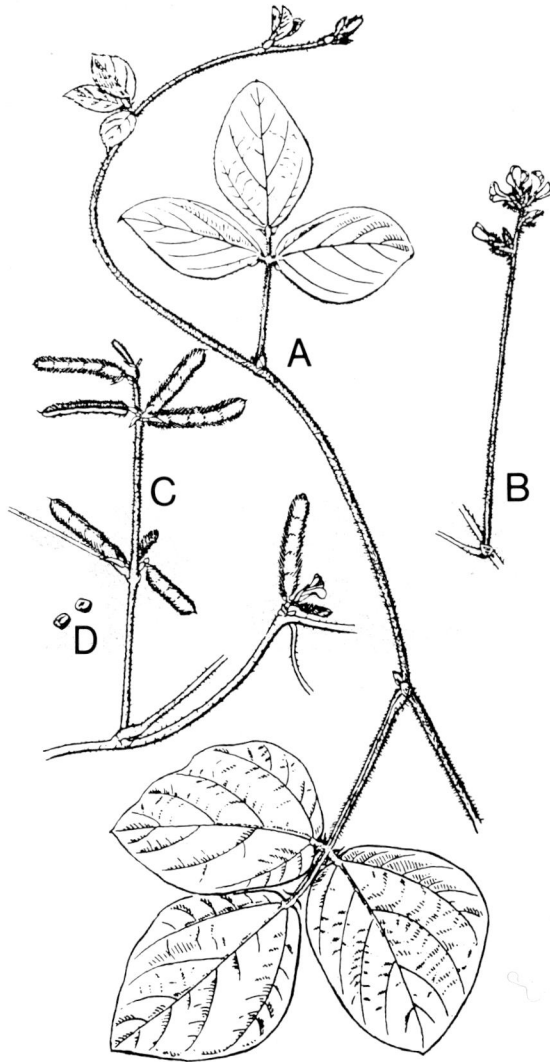

Figure 40. *Calopogonium mucunoides.* **A**-Leafy branch **B**-Flowers **C**-Pods **D**-Seeds (**Source:** Malaya Planters Bulletin)

Calopogonium mucunoides **Desv.** (see Colour plate VI)

Synonyms. *Calopogonium orthocarpum* Urb.; *Stenolobium branchycarpum* Benth.

Common names. Calopo (Australia), rabo de iguana (Colombia), falso oro (Brazil).

Description. Vigorous, creeping and twining, hairy herb forming a tangled mass of foliage 30 to 40 cm deep. Stems succulent, covered with long brown hairs, creeping in lower part; root at nodes which come in contact with the soil; upper part of stem becomes twining. Leaves trifoliate; leaflets hairy on both surfaces, smaller than those of *Pueraria phaseoloides*; terminal leaflet broadly ovate to ovate-rhomboid, lateral ones are obliquely broadly ovate, about to 4 to 5 cm long and a little less in width. Stipules small and triangular; small flowers borne in short axillary racemes of four to eight to 12 on hairy peduncles. Flowers blue with greenish-yellow blotch. Pods linear, compressed, 2.5 to 4 cm long, yellowish brown, densely covered with long erect hairs, four- to eight-seeded. Seeds broadly oblong to squarish, with rounded corners, compressed, light or dark brown, not mottled, about 3.5 × 2.5 mm (Barnard, 1969) (see Figure 40).

Distribution. Native to tropical South America, widely distributed as a cover crop or weed throughout the tropics.

Season of growth. Summer-growing perennial under wet conditions, annual seed-regenerating legume otherwise, e.g. Sri Lanka, Tanzania (Rijkebusch, 1967).

Optimum temperature for growth. Adapted to the hotter, wetter tropics.

Minimum temperature for growth. Not as cold-hardy as centro or puero. Ludlow and Wilson (1970) obtained only 2 percent of the dry matter, 4.8 percent of the growth rate and 14 percent of the leaf area when grown at 20°C compared with the growth at 30°C. This was the poorest performance of the tropical legumes tested.

Frost tolerance. None.

Latitudinal limits. Occurs from Mexico to northern Argentina (Burkart, 1952) — southern limit at about lat. 29 to 30°S. Some native *Calopogonium* species are listed by Burkart in Argentina.

Altitude range. It will grow up to 2 000 m in Colombia (Crowder, 1960) but prefers low elevations.

Rainfall requirements. Generally at least 1 125 mm and preferably more than this.

Drought tolerance. May die out under dry conditions and regenerate in the wet season as an annual form seed, or simply drop its leaves, depending on the intensity of the dry season.

Tolerance of flooding. Excellent. Grows in very wet conditions in Fiji, Panama and Venezuela, and nodulates well.

Soil requirements. Adapts to a wide range of soil textures and pH, doing quite well at pH 4.5 to 5.0.

Rhizobium *relationships*. Unspecialized in its *Rhizobium* requirement. If inoculating, use cowpea-type, e.g. CB 756 (Norris, 1967).

Ability to spread naturally. Excellent in moist fertile situations, where it has a tendency to become a weed. Seed is also spread by birds (Horrell, personal communication).

Land peparation for establishment. Usually sown on well-prepared seed bed for use as cover crop. For pasture purposes, would establish easily on a roughly prepared seed bed.

Sowing methods. Usually drilled or broadcast; can be broadcast on ground or from air into the ashes after a burn. Oversown into natural pastures under wet moist conditions, it should establish well. Could also be sod-seeded.

Sowing depth and cover. Sown at 1 to 2.5 cm and lightly covered with harrow, or sown in ashes of a burn.

Sowing time and rate. Midsummer, at 1 to 2 kg/ha.

Number of seeds per kg. 65 000 to 70 000.

Percentage of hard seed. Otero (1952) gives 75 percent hard seed in a fresh sample, of which 13 percent germinated and 12 percent was rejected seed.

Seed treatment before planting. To break dormancy: (*a*) treat with concentrated sulphuric acid (sp. gr. 1.8) for 20 min. (Prodonoff, 1968) or with 24 or 36 N sulphuric acid for seven minutes, wash and dry (Black, 1968); (*b*) scarify with sand (Otero, 1952, obtained 99 percent germination); (*c*) infra-red irradiation — Philips Infraphil lamp 13373/479 (150 watts) for eight hours or Osram I.R.R. 4892 (250 watts) for 16 hours (Wycherley, 1960). Inoculation not necessary, but preferable. Pelleting not necessary unless to protect rhizobia when rock phosphate should be used (Norris, 1967). Insect and disease control usually not necessary.

Nutrient requirement. Usually gives marked response in leaf size to application of superphosphate.

Response to light. Can grow in reduced light conditions with tall grasses and is used as a cover in plantation crops; but Schofield (1941) states that it is shade-intolerant.

Compatibility with grasses and other legumes. Excellent. Grows with all the tall tropical grasses such as *Panicum, Hyparrhenia, Setaria, Brachiaria* and

Melinis, and persists with pangola grass if carefully grazed. Often sown with centro and puero in green manure mixtures to give early cover, after which puero and then centro persist.

Ability to compete with weeds. Excellent; quickly suppresses weeds and in some cases becomes a weed itself.

Tolerance of herbicides. No data available.

Seedling vigour. Vigorous seedling enables rapid early growth.

Vigour of growth and growth rhythm. A most vigorous plant, giving a full cover in 4 to 5 months (Crowder, 1960). At Sigatoka, Fiji, it was the only legume making a contribution to the fodder six weeks from sowing (Payne *et al.*, 1955).

Nitrogen-fixing ability. Oke (1967b) rated it as less than that of puero. It fixed 3.8 mg N/day per plant and 87 percent of the fixed N was transferred to the tops. In wet conditions in Panama, it nodulates well and provides a good deal of nitrogen to associated pangola grass early in the life of the mixture, but does not persist with pangola grass in Colombia. Schofield (1945) found that the nitrogen content of soil previously under bare fallow at Innisfail in north Queensland, Australia, for seven months was 34.4 ppm. In soil into which calopo had been ploughed, after 18 months' growth it was 66.7 ppm compared with 171.8 ppm under puero, which was the highest of four legumes.

Response to defoliation. Recovers slowly. Crowder (1960) states that it can be cut at eight-week intervals in Colombia.

Grazing management. Grazing should not begin until the plants begin to grow erect; then rotationally at intervals of 8 to 12 weeks.

Response to fire. Will not tolerate fire, but regenerates from seed.

Breeding system. Probably self-fertile. Chromosome number 2n = 36.

Dry- and green-matter yields. Crowder (1960) recorded 4 tonnes DM/ha in Colombia; van Rensburg (1967) 3 067 kg/ha in Zambia.

Suitability for hay and silage. Not usually conserved for hay. No record of its use as silage.

Value as standover or deferred feed. Will persist in frost-free moist localities. It remained green and continued to produce some growth in the dry season in Zambia (van Rensburg, 1967). At Parada in north Queensland (Downes, 1966) and at Serere, Uganda, it dropped its leaves and gave poor standover feed.

Feeding value.

• Chemical analysis and digestibility. Bermudez *et al.* (1968) recorded 16.7 percent crude protein in the dry matter. Vergara (1967) recorded an average phosphorus content of 0.25 percent and 1 percent calcium in dry matter.

• <u>Palatability</u>. Although calopo is generally recorded as unpalatable, the author has seen a paddock of *P. maximum*, centro and calopo pasture well grazed by dairy cattle near Mombasa, Kenya. Lychatchynsky and Steenmeyer (1968, unpublished) tested several legumes at Matão, São Paulo, Brazil, for palatability, including glycine, siratro, the desmodiums, lotononis and teramnus. Calopo was low in palatability early, but after flowering it became most palatable and, at the rating given it, it proved the most palatable overall. It would appear that some strains are more palatable than others. Bermudez *et al.* (1968) says it is not palatable because of its hairiness. It is eaten well in pastures at Palmira and Turipana, Colombia, and Serere, Uganda (Horrell, 1958).

Toxicity. None observed.

Seed harvesting methods. Much by hand; some directly with an all-crop harvester.

Seed yield. Seeds well in Brazil; van Rensburg (1967) stated that it was a shy seeder in Zambia. Average yield 200 to 300 kg/ha (Davies and Hutton, 1970).

Minimum germination and quality required for commercial sale. Fifty percent germination and 93.5 percent purity, with maximum hard seed content of 10 percent in Queensland. Germinated at 25°C (Prodonoff, 1968).

Cultivars. None recognized commercially.

Diseases. Attacked by virus in Guatemala, Costa Rica and Panama.

Pests. Subject to attacks by leaf-eating caterpillars and beetles, but generally not affected in growth.

Main attributes. Vigorous early growth covers the ground well as a pioneer legume in scrub burns; easy establishment; effective nodulation and tolerance of wet conditions; builds up a lot of leaf mulch.

Main deficiencies. Relative unpalatability, short life and intolerance of dry conditions.

Performance. Was combined with Para grass in pastures in north Queensland but proved unpalatable and was discarded. Provides useful pasture with guinea grass and centro on Kilifi plantation near Mombasa, Kenya, and with a number of species in Colombia, but here *Neonotonia wightii* is the preferred species.

Main reference. Schofield (1941).

Canavalia spp.

Canavalia ensiformis (L.) DC

Common names. Jack bean (southern Africa and Zimbabwe), sword bean (Australia), one-eye bean (West Indies), feijao de porco (Brazil), horse gram, chickasaro lima bean, overlock.

Description. Vigorous herbaceous annual climber or woody shrub; three leaflets, ovate or ovate-elliptic, acutely acuminate at the apex, broadly cuneate at the base, up to 20 cm long, 10 cm broad, smooth, with six to seven pairs of lateral nerves. Flowers rose, mauve or white with red base, about 2.5 cm long; few, on stout axis. Pod variable, sword-shaped, elongate, 30 cm or more long, with two longitudinal ribs near the upper suture; seeds narrowly ellipsoid, white, smooth (Andrews, 1952). Each seed has a brown hilum extending around one quarter of it.

Distribution. Occurs in the tropics of both hemispheres, usually in cultivation.

Characteristics. Hardy, drought-resistant annual, immune to most pests; cultivated extensively for forage and green manuring. Young pods and immature seeds used as a vegetable for human consumption; mature seeds ground for livestock feeding. Planted in prepared seed bed in rows about 0.6 to 1 m apart, with two to three seeds in holes 0.4 m apart (54 kg/ha). Sown with 550 kg superphosphate per hectare. Addison (1957) found early planting gave higher yields in Zimbabwe. Delaying planting one or two months reduced seed yields by 410 and 2 134 kg/ha respectively. 22 kg N/ha reduced yields, but some potash deficiency occurred. Grows slowly at first, flowers three months after planting, at which stage it is ploughed in for green manure for sugar cane in Mauritius, where broken seeds are also boiled and fed to cattle. In Hawaii, half-ripe seeds and sorghum are fed. Forage has low palatability (Otero, 1952), but Krauss (1911) said that cattle acquire a taste for it. Forage yields of 18 to 23 tonnes/ha have been reached in Hawaii (Takahashi and Ripperton, 1949). In Brazil and Cuba, silage has been made successfully from the plant.

Addison (1957, 1958) advocated feeding the whole pod and seed to cattle. In a trial, cottonseed cake fed at 0.7 kg/head/day for 115 days gave a live-weight gain of 51 kg; *C. ensiformis* meal at 1.1 kg/head/day a gain of almost 51 kg and urea at 84 g/head/day a gain of 25 kg. This was additional to a basic ration of 7 kg grass silage and 3.6 kg maize stover per head per day. Addison (1958) found that the meal was unpalatable but cattle would eat it if 18 litres of molasses were added to each tonne of *C. ensiformis* meal.

Affleck (1961) and Shone (1961) have reported toxicity in cattle grazing *C. ensiformis* aftermath and consuming too much seed meal. The beans contain a basic amino acid, canavanine, which can be hydrolyzed to urea and catalyzed by an enzyme contained in an extract from pigs' liver. The seed is an important source of urease. Animals affected by eating too much of the plant or meal reach a temperature of 30°C, have a clear nasal discharge, and exhibit lameness and prostration. Mucus membranes become muddy in appearance and clear urine is passed more frequently than usual. It has been shown that 28 g of seed per 0.73 kg body weight are lethal to cattle and either the meal should not comprise more than 30 percent of the ration or it should be heat-treated to destroy the enzyme before feeding. The husk of the seed is always removed before milling.

Chemical analysis of the seed revealed 91.1 percent dry matter, of which 33.9 percent is crude protein, 2.2 percent fat, 11.2 percent crude fibre, 49.6 percent nitrogen-free extract and 3.1 percent ash. Otero (1952) lists several analyses (see Appendix 1). Crude protein content of the plant about 11.0 percent, of the silage 10.67 percent, of the seeds 31.96 percent and of the ensiled seeds 21.25 percent.

Cassia spp.

Cassia mimosoides **L.**

Common names. Artillery plant; five-leaf cassia (Australia).

Description. An exceedingly variable, prostrate to erect legume up to 1.5 m high, usually annual, sometimes with stems becoming woody above ground level and enabling the plant to perenniate. Stems variable, usually puberulent with short curved hairs, sometimes more or less densely clothed with longer spreading hairs. Leaves linear to linear-oblong, more or less parallel-sided, 0.6 to 10 cm long, 0.4 to 1.5 cm. Gland usually at or near the top of the petiole, sessile, normally orbicular or nearly so, disk shaped when dry, 0.4 to 1 mm in diameter. Rachis glandular, serrate or crenate-crested along the upper side. Leaflets sessile, in 16 to 76 pairs, obliquely oblong to oblong-elliptic or linear-oblong, 2.5 to 8 (2 to 9) mm long, 0.5 to 1.25 (1.9) mm wide, acute or subacute and shortly mucronate, glabrous or nearly so. Midrib somewhat eccentric, lateral nerves obscure to prominent beneath. Inflorescence supra-axillary or sometimes axillary, one- to three-flowered. Pedicels 0.3 to 2.5 (3.0) cm long, usually shortly puberulent, sometimes spreading hairy. Petals yellow, obovate 4 to 13 mm long, 2 to 9 mm wide. Pods linear to linear-oblong, (sometimes 1.5 but usually 3.5 to 8 cm long); 3.5 mm wide, usually adpressed hairy. Seeds brown, more or less rhombic, 2 to 3 mm long, 1 to 2 mm wide (Brennan, 1967).

Distribution. The species as a whole is widespread in the tropics of the Old World and has been recorded from the Americas, but this needs confirmation. The range of variation is wide but cannot be clearly linked to either geographic origins or the effect of a hybrid swarm. At present it is simply divided into seven unnamed groups.

Group A = *C. mimosoides* L. var. *telfairiana* Hook. is from Mauritius and the Seychelles, with closely related plants in the Sudan and the Congo. Grows from 0 to 1 370 m in altitude.

Group B is from the Congo, the Sudan, Mozambique, Malawi, Zambia, Angola and southern Africa. Grows at altitudes from 900 to 1 500 m.

Group C is recorded only from Zanzibar, between sea level and 550 m.

Group D only from northwestern Kenya, between 1 680 and 1 740 m.

Group E is from the Sudan, the Congo, Mozambique, Zimbabwe and the Transvaal, and is closely related to plants in Nigeria, Côte d'Ivoire, Mali and Madagascar between 470 and 1 550 m.

Group F = var. *glabriuscula* Ghesq., and is widespread in tropical Africa from the Gambia to Nigeria and the Sudan and southwards to Angola and Natal; it is also found in Asia from India to Australia.

Group G occurs in Sierra Leone, Liberia, the Congo, Eritrea(?), the Sudan, Mozambique, Zambia and Angola, and also in India between 0 and 2 110 m. It resembles *C. capensis* Thunb. var. *humifusa* Ghesq. (Brennan, 1967).

Habitats. Usually found in clearings in forests, forest margins, wooded grasslands, grasslands, cultivated and waste places, sandy river beds, lake and sea-shores from 0 to 2 740 m. It is particularly suited to moist places. In the Sudan, it is common on sandy soils (Andrews, 1952) but in Queensland, Australia, it can be one of the most abundant legumes on the black earths of the Darling Downs (Diatloff, 1967b).

General features. Generally only a short-lived annual. There are variations, in Japan, in flowering times of collections from different areas, those from cold areas being less bulky and flowering earlier (Sekizuka, 1960). In Japan, it has been tested as a companion crop for sowing early in the season in rice paddies for forage (Kawatake *et al.*, 1959).

In Australia and Japan it is not particularly bulky, yields of only 230 to 510 kg/ha being recorded (Sekizuka, 1960), but it has some pasture value where it occurs naturally. It is not very palatable, however.

It nodulates with the cowpea cross-inoculation group (Bowen, 1956). It is also nodulated in the field under good conditions but not under exceptionally wet or dry conditions (Diatloff, 1967b).

Main reference. Brennan (1967).

Cassia rotundifolia **Pers.** (see Colour plate VII)

Synonyms. *Cassia bifoliolata* D.C.; *C. fabiginifolia* H.B.K.; *C. monophylla* Vell.; *Chamaecrista rotundifolia* (Pers.) Greene.

Common name. Roundleaf cassia (Australia).

232

Figure 41. Seedlings of *Cassia rotundifolia* cv. Wynn approaching first flowering in a new planting

Description. A herbaceous, subwoody, short-lived perennial or self-regenerating annual legume. Stems prostrate to semi-prostrate, 30 to 110 cm long, radiating from the root-stock, pubescent to subglabrous. Leaves bifoliate, small. Stipules lanceolate-cordate, 4 to 11 mm long, ciliate or glabrous, up to 1 cm long. Petiole short, 3 to 8 mm long, not exceeding the stipule, not eglandular, pubescent like the stems. Leaflets asymetrically subrotund to broadly obovate, rounded apically, 0.5 to 3 cm long, sometimes ciliate, without epetiolulates. One to two flowers, axillary, small, yellow. Pedicels more or less filiform, 1.5 to 3.5 cm long, longer than the leaves. Sepals lanceolate, usually ciliate, up to 5 mm long. Petals obovate, about 6 mm long, glabrous, sessile. Fertile stamens five, somewhat unequal, filaments very short. Anthers linear-oblong, up to 2 mm long, essentially glabrous and erostrate, dehiscent by paired terminal pores. Ovary pubescent. Pods linear, flat, 1.5 to 4 cm long and 3 to 5 mm wide, elastically dehiscent, blackish brown when ripe. Seeds obliquely transverse in pod, rectangular, flattened (Woodson and Schery, 1951; Adams, 1972; R.W. Strickland, personal communication).

233

Distribution. Native to Florida, United States and Mexico, through Central America south into Brazil as far as Uruguay, being widespread in northern South America. Also found in Cuba, Puerto Rico and Jamaica. Naturalized in parts of West Africa. Normally occurs in savanna habitats, especially with sandy soils (Woodson and Schery, 1951; Adams, 1972).

Season of growth. The growth habit is indeterminate and growth will continue after flowering has begun for as long as temperatures and soil moisture conditions are suitable.

Rainfall requirements. Early-flowering lines will grow with as little as 500 mm annual rainfall, although more than 600 mm is desirable in Queensland, Australia (R.W. Strickland, personal communication).

Tolerance of flooding. In studies by Whiteman *et al.* (1984), cv. Wynn was in the least tolerant of four groups into which the 17 commercial legumes tested under laboratory conditions were divided. This group died within seven to ten days, did not develop any adventitious roots and lost all nodules when flooded.

Soil requirements. Suited to a wide range of soils, but best adapted to light-textured surface soils and not well adapted to clay soils, especially if inclined to become waterlogged. In general, soil requirements are similar to those of siratro (R.W. Strickland, personal communicaiton).

Rhizobium *requirements*. It does not require specific rhizobia, nodulating readily with native rhizobia in Queensland soils.

Ability to spread naturally. Natural spread is quite rapid on suitable soils, even under heavy grazing pressures.

Land preparation for establishment. Under normal conditions, fully prepared seed beds should be considered, but cv. Wynn has some ability to establish and spread into native spear grass (*Heteropogon contortus*) pastures from minimal seed beds. Seedling survival under harsh conditions at Narayen, in southeastern Queensland, was less than for *Stylosanthes* spp., though it flowered and seeded in the first year. Townsville stylo was the only other legume to do so. For seed production, a fully prepared seed bed and a planting rate of 4 to 5 kg/ha are recommended (D.S. Loch, personal communication).

Number of seeds per kg. 200 000 to 470 000 (cv. Wynn 253 000).

Percentage of hard seed. Usually greater than 90 percent in freshly harvested samples.

Seed treatment before planting. Seed should be mechanically scarified to reduce the hard-seed level.

Nutrient requirements. The only recommendations available at this stage are for seed production stands, to which 250 kg/ha Mo superphosphate and

100 kg/ha of muriate of potash are applied at planting in coastal southeastern Queensland, with 125 kg/ha superphosphate and possibly 50 kg/ha of muriate of potash as annual maintenance dressings. Application of Mo at c. 100 g/ha every three years is also tentatively recommended (D.S. Loch, personal communication).

Tolerance of herbicides. For seed production stands, trifluralin can be used before planting, while bentazone (3 litres/ha of 48 percent product) and dinoseb (4 litres/ha of 20 percent product) appear safe for post-emergence use. Acifluorfen, 2,4-D and 2,4-DB have caused crop damage on cv. Wynn and should not be used. Fluazifop (1 to 2 litres/ha on 21.2 percent product) appears likely to control grass weeds in seed crops (D.S. Loch, personal communication).

Response to defoliation. Cv. Wynn is essentially prostrate; it therefore seems likely that it will withstand fairly heavy grazing, certainly heavier grazing than siratro.

Grazing management. C. rotundifolia is fairly tolerant of a wide range of management regimes and styles. However, care will need to be taken that vigorous, tall-growing grasses and weeds are not allowed to dominate the lower-growing cassia.

Dry- and green-matter yields. Annual dry-matter yields of more than 7 000 kg/ha have been recorded at Beerwah and Gatton, those at Beerwah with cv. Wynn being double the yields of greenleaf desmodium (*Desmodium intortum*) and siratro in spring and early summer. Later in the season, yields were equal to those of the standard cultivars (R.W. Strickland, personal communication).

Toxicity. Six accessions of *C. rotundifolia* have been tested on a rat colony at the Commonwealth Scientific and Industrial Research Organization (CSIRO), Samford, southeastern Queensland, and showed no signs of toxicity. Live-weight gains and digestible dry-matter intakes up to 14 percent higher than with the lucerne (*Medicago sativa*) controls have been recorded. Seeds fed gave approximately 80 percent of the weight gains of the control, autoclaved soybean meal (R.W. Strickland, personal communication).

Seed harvesting methods. C. rotundifolia is suitable for normal, direct-header harvesting. Using this method, two harvests per year should be possible. Unless flowering ceases owing to moisture stress, timing of each harvest does not seem critical, as crops flower continuously and standing seed yield will remain near the peak for prolonged periods, the loss of seed from shattering pods being offset by new pods entering the system (D.S. Loch, personal communication).

Seed yield. At present there is no commercial experience, but yields in excess of 800 kg/ha from two harvests in one season have been obtained at Beerwah from small plots. Lower yields of only 200 kg/ha, from single harvests at Grafton, New South Wales, and Narayen, have also been recorded (R.W. Strickland, personal communication).

Cultivars. The only cultivar released to date is Wynn, derived from introduction CPI 34721, from Valinhos in Brazil. Released by the Queensland Herbage Plant Liaison Committee in August 1983, it is an early-flowering type, reaching peak flowering in January. The CSIRO collection can be divided into four maturity groups, the latest not flowering until April/May in southern Queensland and unlikely to complete seed set before the onset of frosts in that environment (R.W. Strickland, personal communication). It is likely that a mid-season type will also be released.

Diseases. To date, the only known disease attack was a minimal leaf spotting caused by *Pleospora* sp. at Gympie, Queensland. This is unlikely to cause any concern under grazing conditions.

Main attributes. With widespread adaptation to lighter-textured, more acid soils and a fair tolerance to heavy grazing, cv. Wynn appears likely to complement siratro in southern Queensland and possibly to extend into somewhat drier country.

Main deficiencies. Restriction to lighter-textured soils.

Centrosema spp.

Centrosema brasilianum (L.) Benth.

Synonyms. *Clitoria brasiliana* L.; *Centrosema angustifolium* Benth., non H.B.K.

Description. A perennial, herbaceous legume. Stems slender and twining with long slender branches. Stipules lanceolate or ovate, 1 to 4 mm long. Petioles 2 to 5 cm long. Three leaflets, linear to oblong, narrowed, obtuse and mucronate at the apex, rounded at the base, glabrous, 3 to 7 cm long and 1 to 3 cm wide. Inflorescence a few-flowered raceme with a 1-cm-long peduncle. Bractlets ovate, acute, obliquous, 1 to 1.5 cm long and 8 mm wide. Calyx glabrous, the tube 4 mm long, the inferior teeth 4 mm long, the other teeth 2 mm long, all ciliate at the margin. Petals blue or violaceous. Standard 3.5 cm long and 3 to 4 cm broad. Wings about 2 cm long and 7 mm wide. Pods linear, 10 to 15 cm long and 5 mm wide, the valves tordate, coriaceous, ribbed near the margins, the beak 2 cm long. Seeds 3 to 4 mm long with a small hilum (Pulle, 1976).

Distribution. A wide natural distribution in the South American tropics, in subhumid and semi-arid regions. It is particularly common in semi-arid northeast Brazil but is also found in the humid Amazonian regions (Clements *et al.*, 1983). It also occurs in Suriname (Pulle, 1976), Panama, Colombia, Venezuela and the West Indies (Hemsley, 1888).

General features. The main feature of agronomic interest in the plant is the drought resistance of the semi-arid ecotypes. Along with *C. pascuorum*, it has shown promise for the Katherine region of Northern Territory, Australia. The best accessions out of eight, mainly northeastern Brazilian lines, showed a noticeable ability to remain green during the dry season (Clements, Winter and Reid, 1984). They were generally lower yielding than *C. pascuorum* accessions but were more leafy and more persistent. Some promise has also

237

been shown in northern Queensland but not in southern Queensland (Clements, Winter and Reid, 1984).

Accessions in the Australian collection do not nodulate effectively with commercial centro rhizobium cultures, but they have nodulated with native rhizobium at Katherine and an effective strain has now been isolated from that area (Clements *et al.*, 1983).

Rhizoctonia foliar blight has limited the promise of *C. brasilianum* in trials in Colombia, Venezuela and Brazil (Clements, Winter and Reid, 1984).

Main references. Clements, Winter and Reid (1984); Clements *et al.* (1983).

Centrosema pascuorum Benth.

Synonym. *C. virginianum* Benth. var. *B. angustifolium* Grisb.

Common name. Centurion (Australia).

Description. A prostrate annual legume with weak ability to perennate in favourable conditions and the ability to twine and scramble up associated species. Stems are cylindrical, scarcely pilose, up to 2 m long, rooting at the nodes under favourable conditions. Stipules subulate to narrowly triangular, 4 to 9 mm long. Petioles glabrous to scarcely pilose, 25 to 50 mm long. Leaves trifoliate, often held erect. Leaflets linear-lanceolate, 5 to 10 cm long, 5 to 10 mm wide, some ecotypes broadly lanceolate, glabrous to scarcely pilose, apex acute to acuminate. Inflorescence racemose, one- to four-flowered. Flowers resupinate, borne singly or in pairs at the end of a short (7- to 12-mm) peduncle, often two peduncles in one leaf axil. Pedicels 4 to 10 mm long. Bracts of two types, proximal stipuliform, triangular, 1 mm long or less, singly or in pairs at bases of peduncle and pedicels, distal ovate, singly at base of pedicel, 2 to 4 mm long. Calyx minutely hooked hairy, tube 3 to 4 mm long, five linear teeth, lowest 4 to 7 mm long. Corolla crimson, 1.5 to 2.5 cm long and wide. Standard spurred on the back toward the base. Pods when mature rather torulose, slightly curved, laterally compressed, 3.5 to 7.0 cm long, including an awn of up to 1 cm and a 3- to 4-mm-wide dark longitudinal stripe near each suture. Up to 15 seeds, monochromic, greenish-yellow to brown (R.J. Clements, personal communication).

Distribution. Natural distribution is restricted to the arid tropical areas of northeastern Brazil, Colombia, Ecuador, Guyana, Venezuela and isolated pockets in the Central American countries (Clements and Williams, 1980), usually in areas receiving less than 1 000 mm mean annual rainfall. Where it

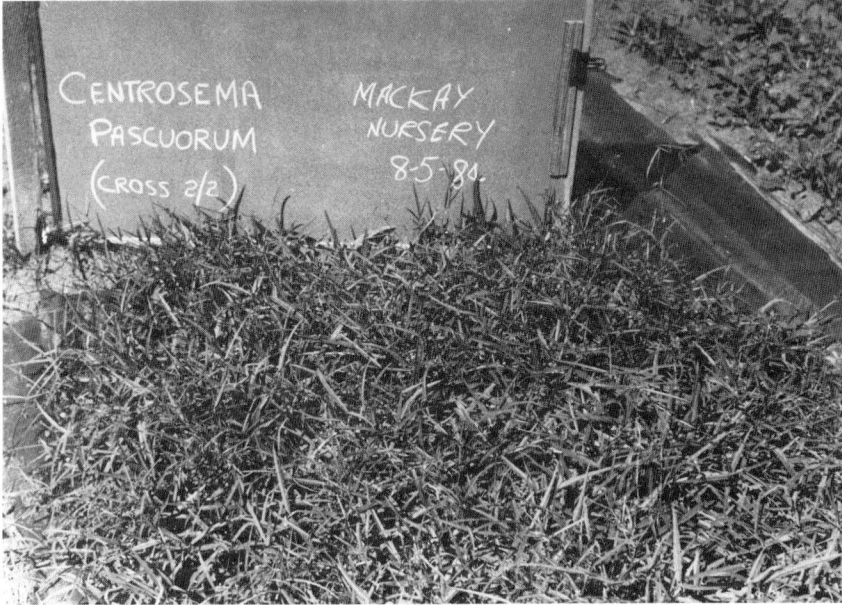

Figure 42. A nursery planting of a breeding line of *Centrosema pascuorum* showing the dense foliage

occurs in Venezuela in areas with an annual rainfall of 1 000 to 1 500 mm, there is an extended and reliable dry season of four to six months with no more than 50 mm of rain. Its natural latitudinal range is from 16°N in Mexico to 10 to 12°S in Brazil (Clements *et al.*, 1983).

Season of growth. 'Centurion' grows only in summer under high temperatures and growth is terminated by any frosts.

Drought tolerance. 'Centurion' is drought escaping and extremely drought tolerant. Under moisture stress, the newly produced leaves become steadily smaller, narrower and more hairy, while the ability to point the tip of the leaf toward the sun, in order to keep it cool and use less water becomes even more pronounced (Clements *et al.*, 1983).

Tolerance of flooding. Cultivar Cavalcade is able to survive prolonged periods of waterlogging and even partial submersion on the seasonally flooded coastal plains of northern Australia (R.J. Clements, personal communication).

Soil requirements. Found naturally on a wide range of soils with a pH range

239

from 5 to 8.5 and textures from sands to heavy clays. There appears to be a good deal of edaphic adaptability (Clements *et al.*, 1983).

Rhizobium relationships. Unknown, but all Australian sowings have nodulated readily with standard *Centrosema* inoculum (CB 1923) (Clements *et al.*, 1983).

Ability to spread naturally. In areas in the Northern Territory, Australia, where it is adapted, it is persisting and spreading away from the original planting sites.

Cultural requirements. The agronomy of the plant has yet to be investigated in detail. To date, all plantings have been in fully prepared seed beds and it has established readily planted at 2 to 4 kg/ha. Seed scarification is necessary, as all freshly harvested samples examined to date have had 100 percent hard seed. In the field, this seed, lying on the soil surface, softens steadily over the dry season, reaching 10 to 30 percent hard seed by the end of the dry season in December, after which the species regenerates aggressively on the break of season rains, competing strongly with associated grasses. It is able to scramble and twine up associated grasses, and in areas where it is adapted, it combines well with them. In the subtropics, spring growth of seedlings is inclined to be slow, probably owing to low soil temperatures. As a result, regenerating stands can be smothered by the more vigorous grasses (Clements *et al.*, 1983).

Dry- and green-matter yields. At Katherine, Northern Territory, herbage yields equalled or exceeded those of *Stylosanthes hamata* cv. Verano and by the end of the third dry season, soil seed reserves exceeded 1 tonne/ha (Clements, Winter and Reid, 1984).

Feeding value. Mineral contents of material grown at Narayen Research Station, in southeastern Queensland, were within the values required for tropical pasture legumes. Concentrations of N and P were lower than those normally expected from *C. pubescens*. Concentrations of Ca, Mg and Na were similar and those of Cu higher than those commonly found in *C. pubescens*. Sodium levels tended to be lower than those required for animals according to Australian Research Council standards (Clements, Winter and Reid, 1984).

Toxicity. None known.

Seed yield. At Katherine, where cv. Cavalcade begins flowering in mid-March, seed yields of 600 to 1000 kg/ha can be expected, usually toward the middle range (Clements *et al.*, 1983).

Number of seeds per kg. 48 000.

Cultivars. One cultivar, Cavalcade, is being released from Katherine for use

Figure 43. *Centrosema plumieri* (**Source:** Pittier, 1944)

in northern Australia. This derives from a breeding line 2/2, which resulted from a cross between CPI 40060 from Ceara, Brazil, and CPI 55697 from Pernambuco, Brazil. Consideration is also being given to the release of a later-flowering line for use on the higher rainfall and wetter, coastal plain areas of the Northern Territory.

Pests and diseases. Most ecotypes show some susceptibility to root knot nematode (*Meloidogyne* spp.); this appears to differ between lines. No completely resistant line is known. Individual plants are attacked by legume littleleaf, but as with root knot nematode the effects are only noted in spaced plants, so that the loss of individual plants is more conspicuous. No other diseases are currently known in Australia (Clements *et al.*, 1983).

Main attributes. Good drought tolerance and strong adaptation to areas with a reliable wet season, high growing-season temperatures, and an extended and completely dry season. It is particularly adapted to the so-called Top End of the Northern Territory of Australia, where it is expected to provide useful, high-quality dry season feed.

Main deficiencies. A narrow tropical adaptation.

Chromosome number. 2n = 22.

Main references. Clements *et al.* (1983); Clements, Winter and Reid (1984); Clements and Williams (1983).

Centrosema plumieri Benth.

Description. Robust, strong-stemmed, glabrous twining plant with a wide ovate leaf, slightly pubescent on the undersurface. Lateral leaves are oblique and terminal leaf larger, on short peduncles, glabrous. Flowers few and large, terminal. Pod 10 to 15 cm long and 1.5 cm broad. Seeds oblong and round. Plant becomes dark coloured when dry. Calyx truncate (Pittier, 1944).

Distribution. Indigenous to tropical America; now widespread in the tropics.

Characteristics. Found at elevations of 200 to 1 100 m in Venezuela. Was used as a cover crop or green manure in rubber, coconut and oil palm plantations until replaced by *Centrosema pubescens* (Burkill *et al.*, 1935). Grof and Harding (1968) found that *C. plumieri* recovered quickly after cutting in north Queensland, Australia, but its yield over five months was less than with other *Centrosema* spp.

Centrosema pubescens **Benth.** (See Colour plate VIII)

Common names. Centro (Australia), jetirana (Argentina, Brazil), bejuco de chivo (Colombia), campanilla (Colombia), butterfly pea.

Description. Vigorous, trailing, twining and climbing perennial herb; in pure stands forms a compact dense cover 40 to 45 cm high in four to eight months from sowing. Very leafy; the slightly hairy stems do not become woody for at least 18 months. Leaves trifoliate; leaflets dark green elliptic or ovate-elliptic, obtuse or shortly obtusely acuminate, about 4 × 3.5 cm, slightly hairy, especially on the lower surface. Stipules long, persistent. Flowers large and showy, borne in axillary racemes. Each flower has two striate bracteoles. Flowers bright or pale lilac on either side of a median greenish-yellow band with numerous dark violet stripes or blotches. Pod linear with prominent margins 7.5 to 15 cm, long, flat, thick, straight or slightly twisted, acuminate, dark brown when ripe, containing up to 20 seeds: septa between seeds. Seeds shortly oblong to squarish with rounded corners, 4 to 5 × 3 to 4 mm, brownish-black, mottled darker blotches with lighter coloured halo (see Figure 44). Wilson and Lansbury (1958) stated it had a shallow root system averaging 30 cm in depth with two taproots to each 900 cm^2, growing in a granitic sandy loam in Ghana. Monteiro and Aronovich (1966) gave some anatomical details of its vegetative organs.

Distribution. Native to tropical South America. Introduced to the Malay Peninsula and Indonesia as a cover crop, probably during the nineteenth century. Now widely grown in the tropics, 50 species occur naturally in South America.

Season of growth; temperature requirements. Summer; prefers moist conditions with a maximum of 25.6°C. Dormant during the winter. Ludlow and Wilson (1970) recorded only 5.3 percent dry-matter production when grown at 20°C compared with the growth at 30°C. Bowen (1959a,b) showed growth ceased at 12.8°C.

Frost tolerance. Low, and is severely damaged. Needs the protection of its crown and lower mature stems. Once established has better survival value.

Latitudinal limits. Probably about 22°N and S latitude.

Altitude range. Widespread below 600 m (Crowder, 1960). Atkinson (1970) records *Centrosema* occurring from sea level to 915 m.

Rainfall requirements. Prefers the wet tropics with a rainfall in excess of 1 750 mm or irrigation, but grows in areas receiving 750 mm or more. It does well at Serere, Uganda, which receives 1 325 mm a year with a five-month dry period (Horrell, 1958). Wilson and Lansbury (1958) state that it requires a

243

minimum of 1 000 mm/year of "twin peak" rainfall in Ghana and gives luxuriant growth when rainfall exceeds 1 750 mm.

Drought tolerance. Deep-rooted and so is fairly drought-tolerant. Dry-season growth slow (Parbery, 1967a), drops its leaves in a prolonged drought (Stobbs, personal communication). Payne *et al.* (1955) state that it provides some green feed in the dry season at Sigatoka, Fiji.

Tolerance of flooding. Fairly good, better than siratro (*Macroptilium atropurpureum*) but not as good as puero (*Pueraria phaseoloides*). Farinas (1966) reported that it survives stagnant water for at least two months in the Philippines.

Soil requirements. Will grow on a wide range of soils, from sandy loams to clays. Grows vigorously on alluvial soils and hill soils of Fiji. In Sri Lanka it prefers the heavier clay loams to clays (Santhirasegaram, personal communication). Will nodulate in soils with a pH as low as 4.0 (Andrew, personal communication), but optimum pH lies between 4.9 and 5.5. Rijkebusch (1967) says it needs a lime-rich soil. Döbereiner and Aronovich (1966) used lime to eliminate manganese toxicity in Brazil, resulting in a 65 percent increase in the amount of nitrogen fixed. Lime pelleting increased the nodule numbers as much as liming, but did not eliminate the serious effects of Mn toxicity on nitrogen fixation.

Rhizobium *relationships*. Native *Centrosema* species are well nodulated throughout Central America, but Bowen and Kennedy (1961) showed that *Centrosema pubescens* has some specificity toward *Rhizobium* with heritable variation in nodulating ability. With culture QA522 they obtained 2 096 g dry matter per pot with 16.7 percent crude protein, whereas uninoculated plots yielded only 1 334 g with 10.5 percent protein. The current *Rhizobium* culture used in Australia is CB1103 (1970). Van Rensburg (1967) showed that centro produced abundant nodules of medium size, sometimes in clusters, most frequently on the secondary roots. Oke (1967a) found that centro developed its first nodules two weeks after germination and the young nodules were the most active in fixing nitrogen.

Ability to spread naturally. Fair to good in a fertile environment.

Land preparation for establishment. Will establish quite well in roughly prepared seed beds provided fertility requirements are met. Establishes well in ashes after burning forest. Gives its best performance on a well-prepared seed bed.

Sowing methods. In small areas, centro can be established with a small "Planet Junior"-type planter; in larger areas it can be drilled or broadcast. When planting with a "Planet Junior", it is sown in rows 1 m apart and 50 cm

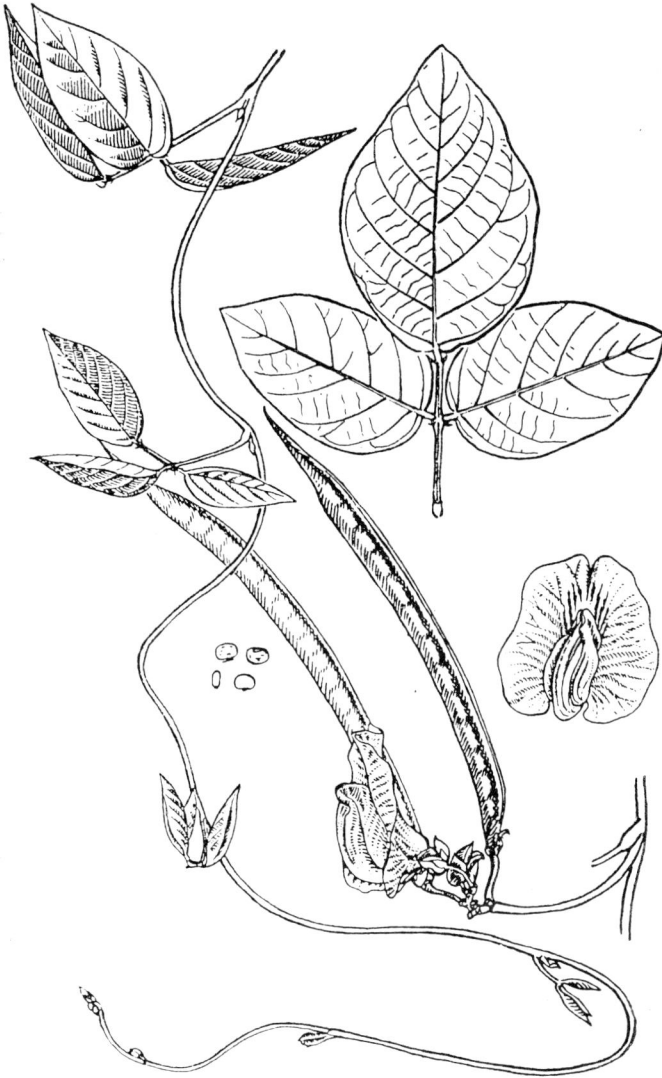

Figure 44. *Centrosema pubescens* (**Source:** Malaya Planters Bulletin)

apart in the row. Should give a complete cover in four to six months. Under certain circumstances can be oversown into existing pastures. Risopoulos (1966) introduced it into *Imperata cylindrica* pastures in Zaire after passage with a "Rome" heavy tandem disc plough. Roberts (personal communication) introduced it into a *Brachiaria humidicola* pasture in Fiji which had been weakened by army-worm attack, and mown. Seed was scarified, inoculated and sod-seeded into the pasture with 550 kg/ha superphosphate. In northern Queensland, it is sometimes introduced into molasses grass pastures after the grass has subdued the weeds, being sown with guinea grass.

Sowing depth and cover. The quite large seed can be sown to depths of 2.5 to 5 cm without affecting germination. Should be rolled or lightly covered by harrow after seeding.

Sowing time and rate. Sow 3.3 to 4.4 kg/ha, drilled in prior to the rainy season. For green manure it can be sown up to 8 kg/ha (Rijkebusch, 1967). For broadcasting, increase the seeding rate.

Number of seeds per kg. 39 600. Percentage of hard seed not more than 75.

Seed treatment before planting. Storage under constant damp conditions depresses total viability and increases hard seed content (Wycherley, 1960). Serpa (1966) showed that germination is hindered by impermeability of the seed coat, which is genetically controlled. The following methods can be used to break dormancy: (*a*) scarify mechanically; (*b*) immerse in concentrated sulphuric acid 24 or 36N for seven minutes, then thoroughly wash with water (Black, 1968); (*c*) immerse in hot water at 77°C for 15 min. (Stobbs, 1969b) or in boiling water, adding 1/4 cold water, and soak seed overnight (Grundy, 1959); (*d*) immerse in warm glycerine at 30°C for two hours (Wycherley, 1960); (*e*) Osram irradiation for 16 hours or more (Wycherley, 1960); (*f*) warm to 50°C for up to eight hours (Wycherley, 1960). These methods increase germination from 9 to 16 percent.

Inoculation is necessary. Pelleting is not usually necessary, except in high manganese soils, or to protect *Rhizobium*. For insect control, use an ant and bean-fly repellent — dust with lindane at 85 g/kg dry seed. If using ant repellent, use lime-pelleted seed to protect the *Rhizobium*.

Nutrient requirements. Centro grows well in fertile soils without fertilizer. In poor soils it responds to phosphorus and molybdenum and sometimes to magnesium. It is not demanding of potassium. In Brazil, dolomite containing 10 percent magnesium is applied three to four months before planting. Parbery (1967a) doubled the yield of centro on Cunnunurra clay and nearly trebled the yield on Cockatoo sand with 100 kg N/ha.

● <u>Calcium</u>. The plant responds to liming. Calcium content of the leaf, stem and nodules was increased by liming to raise the pH from 5.0 to 6.0. The treatment also increased the content of molybdenum and reduced the manganese content (Watson, 1960). Centro gave 52 percent of its maximum yield in the absence of calcium, the maximum yield being at 1 100 kg/ha (Andrew and Norris, 1961). At maximum growth the Ca content is 1.4 percent (Wilson and Lansbury, 1958), or 1.5 percent according to Andrew and Norris (1961). The Ca uptake is 1.7 to 2.0 percent of the dry matter (Andrew and Hegarty, 1969). Kannegieter (1966) recorded 1.66 percent of CaO in the dry matter of centro in Ghana.

● Phosphorus. The critical level for P in the dry matter of the leaves at the immediate preflowering stage is 0.16 percent (Andrew and Robins, 1969a). Grof (1966) showed a straight-line increase in the dry weight of *Centrosema* seedlings up to at least 110 kg P_2O_5 per hectare.

● Potassium. Hamilton and Pillay (1941) found that there were no visible symptoms of K deficiency in centro with a content of 1.35 to 1.88 percent potash in the dry matter. Andrew and Pieters (1970a) showed no deficiency at 1.18 percent. Andrew and Robins (1969) showed that deficiency symptoms occur at a K content below 0.85 percent of the dry matter, and Andrew and Pieters (1970a) have reproduced deficiency symptoms, in colour, of plants containing 0.35 percent K in the dry matter. Deficiency symptoms commenced as a general chlorosis on the mid- to lower positioned leaves of the plant. Chlorosis was interveinal in the early stages, but later full chlorosis occurred with no differentiation of veins except that the base portion of the leaf remained dark green in colour. At this stage there was an associated necrotic spotting and in extreme cases marginal necrosis of the leaflet tips and edges.

Associated with the severe form there was a downward curling of the leaflets from the tips and the surfaces became puckered; the interveinal areas became raised above the veins, which resulted in prominent venation of the underside of the leaflets.

In this species, very little leaf abscission occurred and, furthermore, symptoms were spread over much more of the plant. The parts of the plant remaining visually unaffected were the growing points and a few of the older leaves at its base (Andrew and Pieters, 1970a).

● Copper. Centro gave 45 percent of its maximum yield in the absence of copper when grown in soil culture. The copper content of the seed is 16.5 ppm (Andrew and Thorne, 1962). The initial effects of copper deficiency were reduced plant growth and a general paleness in colour of the younger leaves and slight interveinal chlorosis. Following this the newly expanded leaves

247

showed marginal necrosis, with necrotic tissue curling upward to give a dished or rolled effect. There was also dieback of the growing tip, usually including the newest expanding leaf, but no subsequent axillary growth occurred. Older stem growth and root growth appeared normal.

● Molybdenum. Application of molybdate to centro had a marked effect on the molybdenum content of leaf, stem and nodules, and the nitrogen content was also increased. Liming to pH 7.0 released high levels of molybdenum from the soil (Watson, 1960). Plants with molybdenum deficiency have small white nodules (Andrew, personal communication).

Toxicity levels and symptoms. Centro tolerates fairly high levels of manganese in the soil. Döbereiner and Aronovich (1966) believe that 300 ppm Mn in the foliage is harmful, but Andrew and Hegarty (1969) assessed the "toxicity threshold value" (the Mn concentration in the dry matter when yield is 5 percent lower than the maximum) for *Centrosema pubescens* at 1 600 ppm.

In young plants of this species, the dominant effect of manganese toxicity was a general interveinal chlorosis of the younger leaves and shoots, very like the effect of iron deficiency. However, the young expanding leaves, while being severely chlorotic, also had a bronze colour, the result of numerous minute, rusty coloured spots over the entire leaf surface. The young, fully expanded leaves showed intense interveinal chlorosis, and the margins of the leaflets, particularly toward the tip, became necrotic and curved outward away from the midrib, giving each leaflet a convex dish effect. There was also slight epinastic curvature of the petiolules (Andrew and Pieters, 1970b) — the upper surface growing faster than the lower.

Seedling vigour. Rather slow to establish (but faster than *Neonotonia wightii*). Oke (1967a) found that centro showed a retardation of growth during the third week from emergence due to exhaustion of seed reserves before effective nodulation took place.

Vigour of growth and growth rhythm. When established, quite a vigorous grower, better in the second year. Produces abundant foliage without woody stems up to about 18 months of age with a cover from 50 to 70 cm thick (Crowder, 1960). Bowen (1959) studied the growth rate of *Centrosema pubescens* at Ormiston in south Queensland and at Ayr in north Queensland. Growth rate (see Figure 45) increased to a maximum at the time of runner formation (90 days at both sites), after which this level was maintained until early midwinter, when a decline to zero occurred until regrowth commenced in spring. Decline and cessation of growth were associated at both sites with minimum screen temperatures of approximately 12.8°C. Growth continued at Ayr for approximately five weeks longer than at Ormiston.

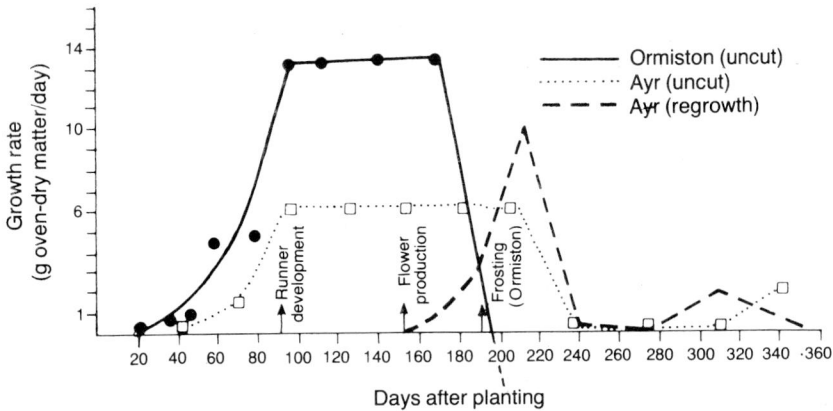

Figure 45. Growth rate curves for *C. pubescens* at Ormiston and Ayr, Queensland, Australia, 1956-57. The curve for Ormiston reached −15 at 223 days

Response to light. Will grow well with grasses of medium height. Seedling growth slow under shady conditions, but mature plants tolerate shade.

Compatibility with grasses and other legumes. Grows well with *Panicum maximum, Hyparrhenia rufa, Melinis minutiflora, Chloris gayana, Pennisetum purpureum, Paspalum dilatatum*; less successfully with *Brachiaria mutica* and *Digitaria decumbens*. Sometimes planted with *Calopogonium mucunoides* and *Pueraria phaseoloides* to give a quick cover, but *Calopogonium* may dominate (Wilson and Lansbury, 1958) in the first year. *Panicum maximum/Centrosema pubescens* pastures were the most successful in Fiji (Payne, 1955).

Ability to compete with weeds. Reasonably competitive, but slashing helps.

Tolerance of herbicides. Bailey (1970) states that from five weeks of age to vining, 2,4-D can be used at 0.825 kg/ha acid equivalent, but some check to growth and loss of plants must be expected. Surviving plants recover and growth on sprayed areas is much the same as on those unsprayed after 12 months. This treatment should only be used when alternative methods of control are unsatisfactory. 2,4-DB is not recommended.

Established centro is checked only slightly by 2,4-D at 1.65 kg of acid equivalent per hectare during the winter months, when growth rate is slow. If the plant is flowering at the time of spraying, germination of the seed set may be lowered by about 15 percent, but this would be of little consequence in a pasture. During the warm spring and summer months, established centro in active growth may be more susceptible to 2,4-D than during the cooler months. Diquat can be used at 285 g/hectare of cation from five weeks of age. Some leaf fall and slight loss of centro numbers can be expected but surviving plants recover rapidly.

Nitrogen-fixing ability. At South Johnstone, north Queensland, *Centrosema pubescens* has fixed 100 kg N/ha in the surface 15 cm of the soil (Bruce, 1967). Bruce (1965) showed that the total N percentages in the top 7.5 cm of soil under 11- and 16-year-old *Panicum maximum/Centrosema pubescens* pastures were 0.406 and 0.432 percent respectively, while that under the adjacent rain forest was 0.423 percent, indicating the ability of a legume-based mixed pasture to maintain fertility after clearing the rain forest (see Figure 46).

Bruce (1967) recorded that land previously under rain forest which had subsequently been supporting pure guinea grass pasture for 16 years contained 0.272 percent N in the surface 15 cm of soil, while that under a guinea

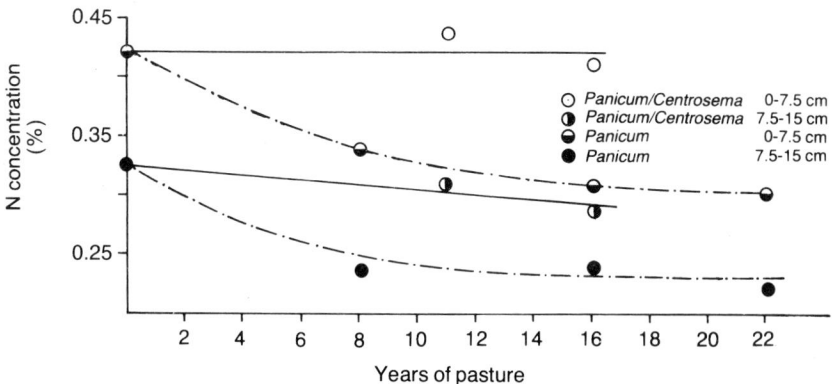

Figure 46. Changes in total soil nitrogen with years under pasture (**Source:** Bruce, 1965)

grass/*Centrosema pubescens* pasture contained 0.346 percent. He also found that an elephant grass (*Pennisetum purpureum*) / *Centrosema pubescens* pasture raised the soil nitrogen in the surface 15 cm of soil by 365 kg N/ha over a three-year period, or at an average rate of 135 kg N/ha/year. Schofield (1945), after ploughing in an 18-month growth of *Centrosema pubescens*, found that the soil nitrogen level rose to 71.7 ppm, compared with 34.4 ppm under a bare fallow for that length of time.

Watson (1957b) recorded the excess nitrogen due to fixation by centro to be 235 kg/ha in five months. He also assessed that 1 hectare of *Centrosema* plants contained the equivalent of 1 303 kg sulphur of ammonia after four months' growth (Watson 1957a). Horrell and Newhouse (1965) at Serere, Uganda, showed that unfertilized grass-legume (*Centrosema pubescens* + *Stylosanthes guianensis*) pastures yielded an equivalent amount of dry matter to a grass pasture fertilized with 165 kg nitrogen per ha. In Uganda, Moore (1962) showed that the nitrogen content of the soil under a *Cynodon plectos-tachyus/Centrosema pubescens* pasture was 275 kg per hectare per year higher than under a pure grass pasture, and the legume raised the nitrogen content of the grass from 1.8 to 2.4 percent. Whitney (1966), in Hawaii, showed that in a volcanic soil *Centrosema pubescens* fixed 264 kg nitrogen per hectare in pure stand, and 121 in a mixed grass sward, and transferred 6 to 11 percent of this fixed nitrogen to the associated grass. Nitrogen transfer to the grass occurred only after six months.

Response to defoliation. Persists well under grazing and has remained in association with *Panicum maximum* pastures at Innisfail, Queensland, for more than 20 years. In Fiji, pastures of guinea grass and centro have been grazed successfully at bimonthly intervals (Payne *et al.*, 1955). Crowder (1960) reported that in Colombia it persists under rotational grazing but not under continuous heavy grazing.

Grazing management. It is advisable to allow *Centrosema* pastures to become well established in the first year before grazing. Slashing or roller-chopping and spot treatment with herbicides may be required to subdue weed growth. Thereafter, the aim should be to keep the legume growing vigorously in association with the grass, keeping the grass in check to avoid dominance. Grass is best kept at a height of 37.5 to 45 cm.

Response to fire. Responds well if established; regenerates from seed after a burn.

Breeding system. Self-pollinated, selections breed completely true. Chromosome number 2n = 20.

Dry- and green-matter yields. Payne *et al.* (1955) recorded an average yield of 4 950 kg DM/ha/year over the three years 1950-52 at Sigatoka, Fiji. Wilson and Lansbury (1958) recorded an annual mean yield of above-ground green matter at 13.5 tonnes/ha. The aerial portion contributed 47 percent of the total plant weight (leaflets 25.8 percent, vines, petioles etc. 21.1 percent), the roots and stolons 53 percent (primary roots 26.1 percent, adventitious roots 10.5 percent, stolons 12.8 percent). In Brazil, an average yield of 40 tonnes of green matter/ha/year has been recorded.

TABLE 14.1 **Analyses and digestibility (percent of dry matter):** *Centrosema pubescens*

Country	Crude protein	Ether extract	N-free extract	Crude fibre	Ash	P₂O₅	K₂O	CaO	Reference
Ghana	20.0	2.3	40.9	30.0	6.2	0.47	1.47	1.76	Wilson and Lansbury, 1958
Australia	18.7	3.0	38.4	30.3	9.6	0.8	—	2.5	Schofield, 1941
Philippines	20.0	4.8	31.2	33.0	8.0	—	—	—	Reyes, 1955
Trinidad	25.5	—		26.5		1.7	—	1.0	Guayadeen, 1951
Malaysia	21.6	2.4	34.0	30.9	9.2	0.44	—	0.8	Gunn, 1951

Source: Wilson and Lansbury, 1958

Suitability for hay and silage. Makes good hay if cut after flowering and before seeding. Hay has been made successfully in Costa Rica (Horrell, personal communication), in Ghana (Wilson and Lansbury, 1958) and in Colombia (Crowder, 1960). The mown plant is allowed to cure for a day or two and is then baled. Leaf shattering is a problem. Satisfactory silage has been made with a mixture of guinea grass and centro in north Queensland (Teitzel, 1969b).

Value as standover or deferred feed. Has proved valuable in Uganda for standover feed in association with *Hyparrhenia rufa* when the dry season starts abruptly (Horrell, 1958).

Feeding value. Valuable for grazing and as a hay crop.

Digestibility of dry matter (Ghana) 53.5 percent; organic matter 53.2 percent; crude protein 62.4 percent; ether extract 44.3 percent; N-free extract 61.3 percent; crude fibre 39.5 percent and ash 56.0 percent. Otero (1952) listed the digestibility of the protein at 65.7 percent and of the fibre at 51.9 percent. Reyes (1955) recorded 35 percent digestibility of the crude fibre in feeding trials in the Philippines. Starch equivalent in Ghana was 10 percent, digestible crude protein 3.2 percent and nutritive ratio 1:3; for the Philippines, Reyes assessed figures of 8.8, 3.2 and 1:2.14 respectively. Miller and Rains (1963) showed that the intake of centro dry matter was low and that the digestibility of the organic matter was low, particularly when immature, relative to that of protein.

Palatability. Fairly palatable.

252

Toxicity. None recorded. Seed fed to mice had no ill effects (Bindon and Lamond, 1966).

Seed harvesting methods. Seed is harvested by hand in many tropical countries. It is difficult to harvest mechanically because of uneven ripening (Crowder, 1960). Best grown on trellises or fence-lines to facilitate hand-picking (Wilson and Lansbury, 1958). In north Queensland (Gude, 1959), centro seed has been harvested mechanically from late winter to early spring from a molasses grass/centro pasture by two methods: (*a*) the pasture is mown and windrowed soon after the first seed pods open. The hay is then cured for a few days in the field and when sufficiently dry for threshing it is collected by a pick-up machine harvester; (*b*) direct heading at a height of 20 to 40 cm. In this case the harvested seed has to be dried on a tarpaulin or concrete floor before cleaning and storage.

Seed yields. At Serere, Uganda, 220 to 275 kg/ha are usual. Wilson and Lansbury (1958) recorded about 1 250 seeds/m^2 collected on a fence-line. Average yield from direct mechanical heading in north Queensland is 140 kg/ha.

Minimum germination percentage of seed for commercial sale. Germination 60 percent, less than 10 percent hard seed, purity 97.5 percent. The seed is germinated at 25°C (Queensland).

Cultivars. There is only one registered cultivar of *Centrosema pubescens*, cv. Belalto, although there are many native ecotypes which include both early- and late-flowering types. 'Belalto' is readily distinguishable from the "common centro" used in Queensland for a number of years by its purple or brown young leaflets and deep mauve to white flowers. It is a vigorous grower in cool conditions and can resist attacks of leaf spot disease and red spider. It also resists weed invasion. 'Belalto' is expected to replace common centro.

Rijkebusch (1967) reports two varieties (M20 and M301) at the Sisal Research Station, Mlingano, Tanzania.

Diseases. Attacked by *Cercospora* leaf spot at Koronovira, Fiji, at Gualaca, Panama, and at Innisfail, Queensland — all areas receiving over 2 500 mm rain a year. The attack can be serious. *Aschochyta* blight affects it in Guatemala. A fungus sometimes affects the pods in Colombia.

Pests. *Meloidae* beetles and thrips in Uganda (Horrell, 1958) and red spider (*Tetranychus* sp.) at Innisfail in north Queensland attack the leaves (Grof, personal communication).

Main attributes. Ability to persist under high rainfall and into the dry season as carryover feed; mixes well with grasses; gives good live-weight gains in the dry season as carryover feed; mixes well with grasses; gives good live-weight gains in the dry season; good seed production, easily obtained; good green manure crop in rubber, coconut and oil-palm plantations.

Main deficiencies. Only moderately stoloniferous cool season dormancy; requires a higher fertility level than stylo; low consumption in the wet season; intolerant of low grazing height.

Performance. Wilson and Lansbury (1958) reported that West African Dwarf Forest sheep grazing *Centrosema* gained 6.7 kg per head in 11 months. Daily dry-matter intake averaged 3 percent of the body weight. As a sole diet, it was thought to be too rich in protein, and a supplement of cassava root was suggested for better performance in Ghana.

An irrigated guinea grass/centro pasture at Ayr, north Queensland, under 41 weeks' grazing with shorthorn beef cattle gave a daily live-weight gain of 0.68 kg/head, and a Para grass/centro pasture 0.74 kg per head. From January to March, weight gains fell owing to high day temperatures and high humidity (Allen and Cowdry, 1961a) (see Figure 47).

In Uganda, the daily live-weight gain by White Fulani cattle grazing *Cynodon plectostachyus/Centrosema* pastures was 0.37 kg compared with 0.31 kg from pure grass (Moore, 1962).

The inclusion of *Centrosema* in a guinea grass pasture at South Johnstone,

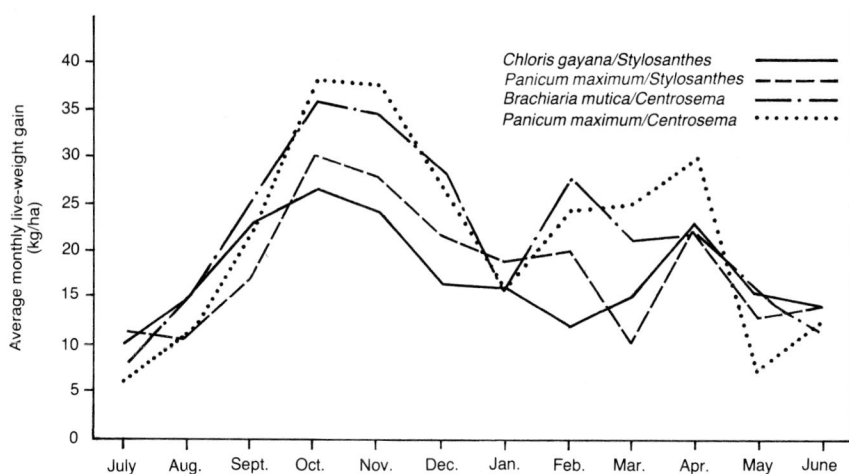

Figure 47. Average monthly live-weight gains per hectare (**Source:** Allen and Cowdry, 1961a)

north Queensland, increased live-weight gain from 440 to 610 kg/ha on pure grass swards (Grof, 1966).

Stobbs (1966) obtained a mean of 495 kg live-weight gain per hectare per year over a two-year period with cattle grazing a *Hyparrhenia rufa/Centrosema pubescens* sward at Serere, Uganda (lat. 1°32′N, rainfall 1 360 mm/year). The effect of introducing the legumes *Stylosanthes guianensis* and *Centrosema pubescens* together with a *Hyparrhenia rufa, Panicum maximum* and *Chloris gayana* mixed pasture with and without added nitrogen at Serere is shown in Figure 48.

Stobbs and Joblin (1966b) found that *Centrosema pubescens* increased the production of *Hyparrhenia rufa* by 18 percent at Serere, Uganda, with an advantage of 11 percent in the second year of grazing and 26 percent in the third year. *Hyparrhenia*/centro pastures produced a live-weight gain of 815 kg per ha over 21 months (465 kg/ha/year) and *Panicum maximum*/centro pastures produced 392 kg/ha/year. The contribution of centro to the dry-season weight gain was 28 percent or 10 kg per month, and during the wet season 15 percent or 6.4 kg per month. Centro produced palatable green material during the three-month dry season.

Main reference. Allen and Cowdry (1961a, b).

***Centrosema virginianum* (L.) Benth.** (see Colour plate IX)

Synonyms. Clitoria virginiana L.; *Bradburya virginiana* (L.) Kuntze; *Glycine pugiunculus* Desv.

Common names. Virginian centro (Australia), wild blue vine, bluebell and wild pea (Barbados).

Description. A climbing, herbaceous, perennial vine. Stems slender, 29 to 160 cm long, trailing and usually vigorously twining. Leaves alternate, pinnately trifoliate, 3 to 10 cm long. Stipules lanceolate or ovate, 1 to 4 mm long, setaceous and often deciduous. Petiole 2 to 5 cm long. Leaflets quite variable, linear to ovate to oblong or lanceolate-oblong, acute or acuminate at the apex, rounded at the base, usually 2 to 6 and sometimes 1 to 8 cm long, 1 to 2.5 cm wide, glabrous or nearly so.

Inflorescence a short raceme of one to four flowers on axillary peduncles, usually shorter than the foliage. Calyx deeply five-lobed, the acute lobes longer than the tube. Corolla purplish or lavender-blue to nearly white. Standard 2.5 cm long and 3 cm wide. Pods linear, 8 to 12 cm long, 3 to 4 cm wide; valves coriaceous, tordate, ribbed near the margin; beak up to 1 cm long.

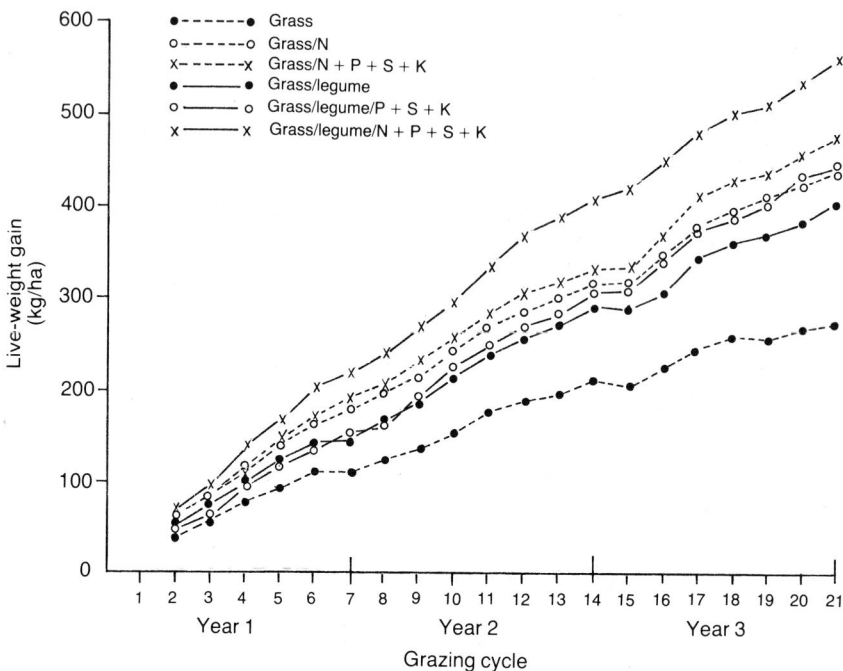

Figure 48. Cumulative live-weight gains per hectare from six pasture treatments (**Source:** Stobbs, 1969d)

Four to ten seeds, dark brown or black, 2 mm long, with a small hilum (Good-ing, Loveless and Proctor, 1965; Correll and Johnston, 1970; Pulle, 1976). Variable in leaflet size and shape and corolla colour. Often confused with *C. pubescens* but less robust.

Distribution. One of the most widely naturally distributed species of *Centrosema, C. virginianum* occurs more or less continuously from Uruguay and northern Argentina to the eastern United States and Bermuda in tropical and subtropical areas. It is found throughout the West Indies and has become naturalized in tropical West Africa.

Latitudinal limits. Occurs naturally from 35°S to 40°N latitude.

Rainfall requirements. Most widely collected from subhumid (500 to 1 000 mm rainfall) and subtropical areas. Occasionally collected from semi-arid, tropical areas in Brazil, but rarely found in the wet tropics.

256

General features. *C. virginianum* is genetically very variable, especially in agronomic features such as vigour, flowering behaviour, frost resistance and drought resistance. It may be useful in subtropical pastures, though to date no natural accessions suitable for use in commercial pastures have been identified in Australia. Overall, it is more drought resistant than *C. pubescens* but less tolerant of waterlogging.

In Australia, it has been grown in both moderately acid sandy soils (pH 5.3 to 5.5) and alkaline clays (pH 6.5 increasing to 9 at depth). Inoculation with commercial centro rhizobium has always ensured nodulation.

In vitro digestibility of the leaves has been assessed at 54 to 59 percent, similar to that of siratro. Digestibility of the stem is somewhat lower (Clements *et al.*, 1983).

Main references. Clements (1983); Clements *et al.* (1983).

Clitoria spp.

Clitoria ternatea **L.** (see Colour plate X)

Common names. Butterfly pea (Australia), Kordofan pea (the Sudan), campanilla (Panama), zapatillo de la reina (El Salvador), papito, bejuco de conchitas (Puerto Rico), pokindang (the Philippines).

Description. Climber, shrubby at base, five to seven leaflets, elliptic to narrowly lanceolate, 3 to 5 cm long, shortly pubescent underneath. Flowers solitary, deep blue, occasionally pure white, very shortly pedicellate, 4 to 5 cm long. Pod flat, linear, beaked, about 10 cm long, slightly pubescent (Andrews, 1952).

Distribution. Native of tropical America, widely grown as an ornamental in the warmer parts of the world (Bermudez, Ceballos, and Chaverra, 1968).

Temperature and moisture requirements. Summer-growing perennial, low frost tolerance, can climb tall grasses and crops. Extends from about lat. 20°N to the Salta district in Argentina at about lat. 24°S (Burkart, 1952), and grows from sea level to 1 800 m (Crowder, 1960). Rainfall requirement from 400 mm, with best performance in the area of 1 500 mm. Fairly drought-tolerant in Zambia (van Rensburg, 1967) and grows in the drier areas of Kordofan, the Sudan, but also performs well under irrigation (Barrau, 1953; Parbery, 1967a). Will not tolerate flooding (Farinas, 1966).

Soil requirements. Adapted to a wide range of soil conditions from sandy to deep alluvial loams and heavy black cracking clays (Lee, 1954) in western Queensland and the Northern Territory, Australia, in the Nile Delta and in India. May have some tolerance to salinity, as it grows on high pH soils of the Nile Delta under irrigation near Khartoum, the Sudan.

Rhizobium *relationships*. Slightly specific in its *Rhizobium* requirement. Bowen (personal communication) obtained 2 497 grams per pot with seed inoculated with strain QA553 containing 19.5 percent crude protein, whereas uninoculated control plants yielded only 140 g/pot with 13.5 percent protein.

258

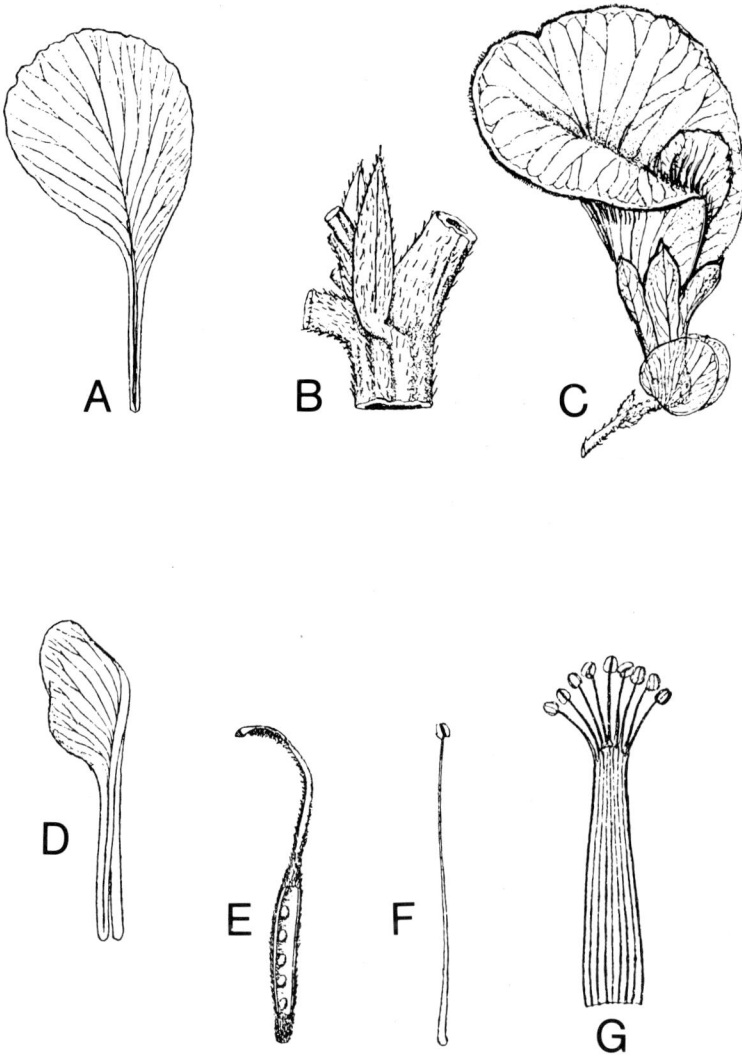

Figure 49. *Clitoria ternatea.* **A**-Wing **B**-Portion of branch, showing stipules **C**-Flower **D**-Petal **E**-Gynoecium (section) **F**-Stamen **G**-Bundle of stamens (**Source:** Wilczek, 1954)

259

Sowing methods. Seed is sown in spring to midsummer at 1 to 3 kg/ha on a well-prepared seed bed at a depth of 1.5 to 4 cm and lightly covered. Can be planted on stakes for seed production. Contains up to 20 percent hard seed depending on the season in which it is produced. Dormancy can be broken by treatment with concentrated sulphuric acid of sp. gr. 1.8 for 20 minutes (Prodonoff, 1968); or by soaking for 12 hours in water, followed by 12 hours in a freezer at —15°C, then defrosted for planting (Lambert, personal communication). Parbery (1967a) obtained no response to a nitrogen application of 100 kg/ha in the Kimberleys (northern Australia).

Seedling vigour and growth habits. Has good seedling vigour and grows rapidly in warm moist weather, producing a dense cover four to six months after seeding; suppresses weeds very well. Grows well with tall grasses such as guinea and elephant grass, with *Andropogon pertusus* in Barbados, and with crops of Sudan grass, sorghum and sunn hemp in the Sudan (Whyte *et al.*, 1969). Should be grazed lightly and in rotation to preserve the pasture. Self-fertile; chromosome number 2n = 16. Flowered in 57 days on Cunnunurra clay, and in 62 days in Cockatoo sand in the Kimberleys, northern Australia (Parbery, 1967a).

Dry-matter yield and feeding value. In Zambia, van Rensburg (1967) obtained 3 330 kg/ha of dry matter from growth from March to June in the first year of establishment. Thereafter, yields declined. Parbery (1967a) obtained an average yield from three varieties in the Kimberleys of 13 350 kg DM/ha/year on Cunnunurra clay under irrigation but only 1 109 kg under dryland conditions on Cockatoo sand. Nitrogen application depressed yields. Crude protein contents ranged from 10.5 to 25.5 percent of the dry matter. In a grazing trial under irrigation in the Burdekin Delta in north Queensland (lat. 20°S), Australia, cattle averaged a daily live-weight gain of 0.68 kg/day grazing on a Para grass (*Brachiaria mutica*)/*Clitoria* pasture, a higher gain than from stylo and centro mixtures (Barrau, 1953).

Seed harvesting methods. Generally by hand.

Diseases and pests. *Clitoria ternatea* is attacked by nematodes at Serere, Uganda (Horrell, 1958), and by grasshoppers and leaf-eating caterpillars (*Prodenia* spp.) in northern Australia (Parbery, 1967a). Under wet conditions it is attacked by virus at Turrialba, Costa Rica, and by *Rhizoctonia microsclerotia* and *Corticium solani* in Zambia (van Rensburg, 1967).

Palatability. Very palatable; its lack of persistence is often due to selective grazing of the legume by cattle. Its drought resistance is also an asset.

Cultivars. Parbery (1967a) used three cultivars in the Kimberleys and all performed well. A Sri Lanka cultivar, CPI 13844, was found to have the important character of basal branching, unusual in other strains. The other two cultivars were the local Australian material and an introduction from south India, CPI 30196. For commercial sale in Queensland it is required to have a minimum germination of 50 percent and purity of at least 93.5 percent with a maximum of 10 percent hard seed. It is germinated at 25°C.

Desmodium spp.

***Desmodium barbatum* (L.) Benth. and Oerst.** (see Colour plates XII, XIII)

Synonym. *Meibomia barbatum* DC.

Common name. Barbadinho (Brazil).

Description. Short-lived perennial, erect, more or less woody herb growing to 1 m; trifoliate leaves with a grey silky pubescence. Basal leaves fall readily and the stem turns quite woody before flowering. Racemes are short, 2 to 4 cm long, borne on the ends of the branches; flowers in clusters, usually red, occasionally white. Pods hairy, each joint 3 mm long, with two to four joints (see Figure 50). There are many ecotypes.

Distribution. Widely distributed from Florida, United States, to northern Argentina (Burkart, 1952). In Guanacaste, Costa Rica, Kretschmer (personal communication) found it growing at 200 to 600 m. In tropical Africa it is found from Zimbabwe to Madagascar.

Characteristics. Prefers well-prepared, deep, fertile soil; is planted with drills in rows 0.6 m apart or broadcast. Contains up to 96 percent hard seed, which should be scarified mechanically or treated with concentrated acid for at least 10, preferably 30, but not more than 45 minutes (Otero, 1952), after which it is washed and dried. Seed is broadcast (18 to 20 kg/ha) or drilled in rows

261

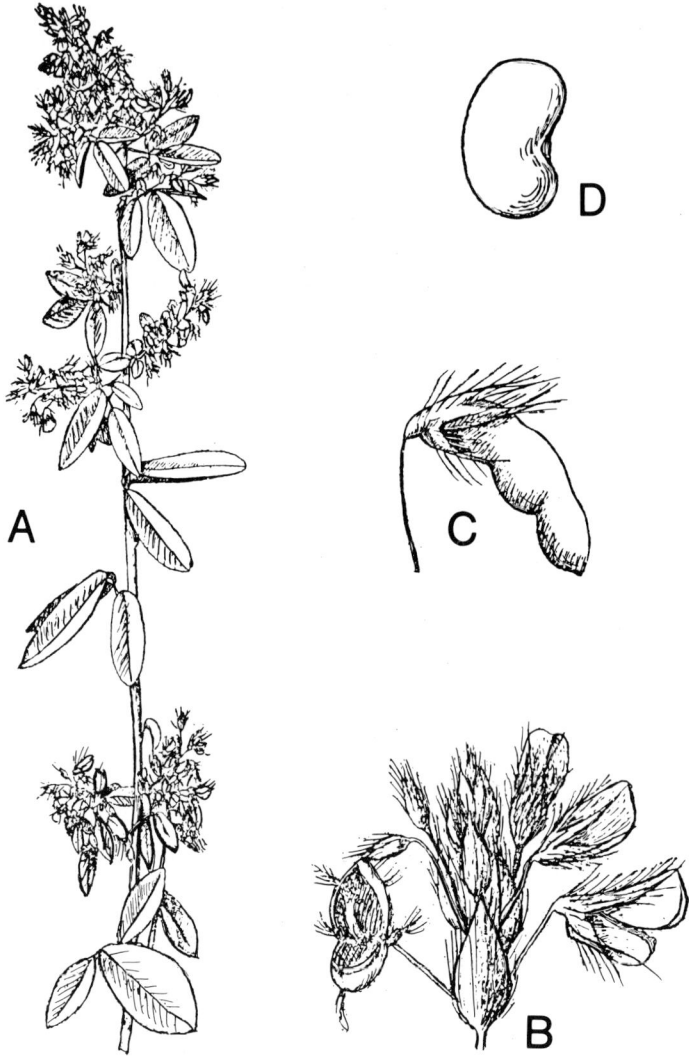

Figure 50. *Desmodium barbatum.* **A**-Habit **B**-Flowering stem **C**-Pod **D**-Seed (**Source:** Otero, 1952)

(14 kg). Cover seed with 3 to 4 cm of soil. Early growth is slow, row planting enabling interrow cultivation to suppress weeds in the first year. It is tolerant of cold and stands grazing, fire and trampling. Should be cut for hay before flowering, taking care to preserve the leaf. If cut later, the material is too coarse. In Brazil, three cuts per year can be taken at 50 to 60 cm. Kretschmer (1964) found it heavily grazed along with jaragua grass (*Hyparrhenia* sp.) in the Guanacaste region of Costa Rica. Bermudez *et al.* (1968) stated that it is highly nutritious and is well accepted by cattle. It makes good hay and green chop cut at a height of 20 to 30 cm; stands cutting and recovers well, though in some areas there is little second growth. Horrell (1963) found it slow in establishment, lacking in vigour and unproductive at Serere, Uganda.

Otero (1952) recorded three cuts per year with a yield of 7 000 kg/ha of green material per cut at a plant height of 50 to 60 cm. Analyses of dry matter revealed 15.16 percent crude protein and 31.1 percent crude fibre at flowering, and 17.27 percent crude protein and 33.98 percent crude fibre before flowering.

Desmodium biarticulatum Benth.

Synonym. *Hedysarum biarticulatum* L.
Common name. Darwin engordo caballo (the Philippines).
Description. Herb, 15 to 30 cm high, slender stems, petiole 0.3 to 0.6 cm long, leaflets rigidly coriaceous, pale green, oblong-ovate 1 to 1.25 cm long, nearly digitate. Racemes 7.5 to 15 cm, lower flowers two to four together. Pod with two joints (rarely one), round-oblong, pubescent, 0.4 cm long, both sutures deeply indented (Hooker, 1879).
Distribution. India, Sri Lanka, Burma, Malaysia, northern Australia.
Characteristics. An erect strong annual; low productivity, high palatability. High resistance to trampling and recovery from grazing. Adaptable to poor soils; and grows from seed. Has no resistance to waterlogging. Produces a large quantity of seed of moderate viability (Farinas, 1966).

Desmodium canum (Gmel.) Schinz and Thellung

Synonyms. *D. incanum* (Sw) DC; *D. supinum* (Sw) DC; *Hedysarum canum* (Gmel); *D. sparsifolium* G. Don.

Common names. Kaimi clover (Hawaii), pega-pega (Brazil, Venezuela); creeping beggar weed.

Description. Perennial; woody upright stems 30 to 60 cm high and fibrous to woody trailing and creeping stems; deep, profusely branched root system. Leaves produced on upright stems are lanceolate and usually have a white mark along the midrib. Trailing stems begin vigorous growth after the first six months of seedling development; leaves are oval or round and normally have no markings. Decumbent stems creep through sward of low-growing grasses, rooting freely at the nodes when in prolonged contact with moist soil or damp, decomposing plant material. Produces reddish or lavender self-fertile flowers on upright stems which bear four to seven seed pods 2.5 to 4 cm long and 0.3 cm wide. Upper edge of pod straight, lower margin indented; entire pod covered with short brown hairs which make pods adherent to clothing and livestock, facilitating rapid and distant spreading of seeds. Seeds are kidney-shaped, light brown, and about 3 mm long and 1 mm wide.

Distribution. Widespread in the wet tropics, common in Fiji, Hawaii, North and Central America and tropical Africa.

Season of growth. Perennial summer-growing legume.

Temperature for growth. Optimum about 30/25°C day/night temperatures. About 5°C night temperature minimum; fairly tolerant of frost, recovering after the cold period.

Latitudinal limits. It extends as far south as Tucuman, Argentina (27°S).

Altitude range. Sea level to 900 m in Hawaii, up to 1 700 m in Colombia (Bermudez *et al.*, 1968).

Rainfall requirements. 1 500 mm to more than 3 075 mm.

Tolerance of drought and flooding. Will tolerate temporary flooding.

Soil requirements. Wide range, from sands to light clays. In Hawaii, it grows well in humic ferruginous latosol, humic latosol and hydrol humic latosols. Occurs on granitic sands in central Paraguay (Fretes, Samudio and Gay, 1970). Grows best in medium acid soils but persists and spreads in very acid (pH 4.5 and lower) soils of low fertility. Range pH 4.0 to 8.0. Tolerance to salinity unknown.

Rhizobium *relationships*. It is "Desmodium" specific and should be inoculated with the special *Desmodium* strain. The current Australian inoculum is CB 627 (Norris, 1967). In Hawaii, the cowpea-type inoculum is successful (Rotar, personal communication).

Ability to spread naturally. Spreads very well by creeping stems which root at the nodes, and from seed spread by adhering to the coats of animals and dropping or being rubbed off.

264

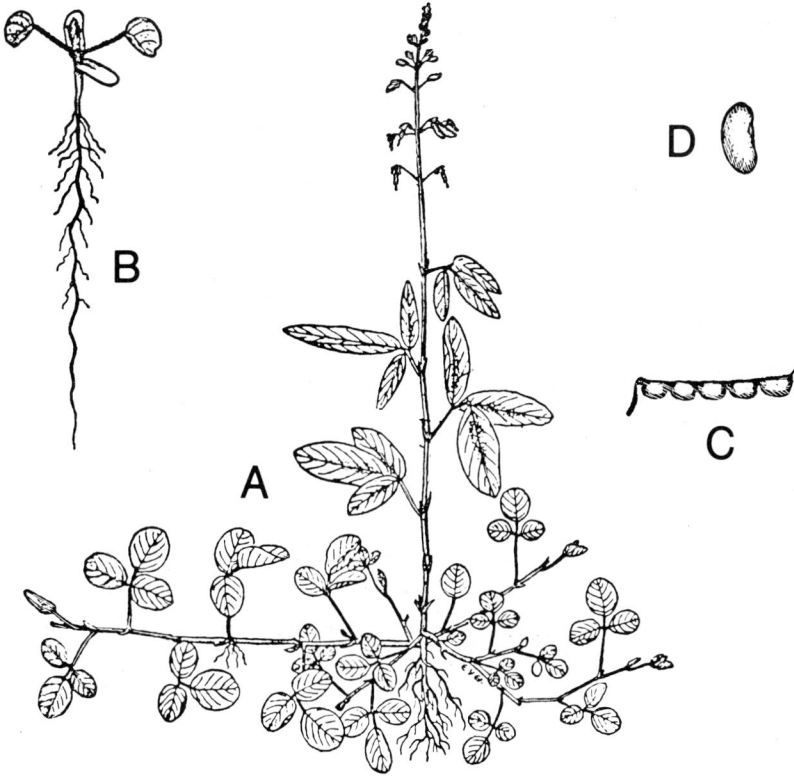

Figure 51. *Desmodium canum.* **A**-Habit **B**-Seedling **C**-Pod **D**-Seed

Land preparation for establishment. Establishes best in a well-prepared seed bed.

Sowing methods. Drill or broadcast onto a prepared seed bed. Can be over-sown into existing pasture by scattering seed (preferably during rainy weather) two to three days before animals are removed from the pasture. Sown 0.6 to 1 cm deep and covered lightly with harrow or roller. Sown in summer at the rate of 5 kg of pure live seed per hectare.

Seed treatment before planting. To break dormancy: treat with concentrated sulphuric acid for ten minutes. Inoculation is necessary. Pelleting is necessary to protect the rhizobia; use rock phosphate (Norris, 1967). Dust with captan, DDT or dieldrin to counteract cutworm damage.

Nutrient requirements. Generally it responds to lime and phosphorus, but they are not necessary. In Hawaii, ammonium phosphate is applied in a band below the seeds when drill planted.

Toxicity levels and symptoms. No toxicity, though tannins reduce its palatability.

Compatibility with grasses and other legumes. Compatible with sward-forming grasses. In Hawaii, it is commonly grown with Kikuyu grass (*Pennisetum clandestinum*), pangola grass (*Digitaria decumbens*), couch or Bermuda grass (*Cynodon dactylon*) and paspalum or dallis grass (*Paspalum dilatatum*). In Argentina it grows with *Paspalum notatum* and *Axonopus compressus* (Burkart, 1952).

Ability to compete with weeds. Excellent; it was first grown as a ground cover in Hawaii for soil conservation. At Topaz, north Queensland, Australia, Kelly (1964) found that it smothered the weedy grasses *Paspalum conjugatum* and mat grass (*Axonopus affinis*).

Seedling vigour. Slow to start growth, but thereafter growth is rapid.

Vigour of growth and growth rhythm. Growth good in wet season, reduced in the dry season.

Nitrogen-fixing ability. Ungrazed, it develops a thick canopy; lower leaves shed and mineralize. Whitney, Kanehiro and Sherman (1967) found it fixed 90 kg N/ha, but transferred none to the associated grass.

Response to defoliation. Will stand heavy defoliation as it is stoloniferous, and carries new growing points near to the ground. It becomes like a mat over the ground with heavy grazing.

Grazing management. Graze heavily for short periods to control weeds and tall grasses. It stands heavy grazing, so management should be designed to control associated grass and other legumes. A rest period (30 to 40 days) between grazings is recommended. Allow grass and legumes to grow to

25 to 30 cm, then heavily graze to a height of 5 to 10 cm during four weeks or less for best results. If grazing of associated grasses is uneven, slash to control growth of taller species.

Response to fire. Little tolerance.

Response to photoperiod. A short-day plant, it flowers in 90 days in north Queensland, in 51 to 81 days in Hawaii. Self-fertilized, but some accessions appear to outcross readily. Chromosome number 2n = 22 (Rotar, personal communication). It is quite shade tolerant.

Dry- and green-matter yields. It yields about 6 500 kg/ha of dry matter in north Queensland (Grof, personal communication). In grass mixtures in Hawaii the yield is 5 000 kg/ha/year (Rotar, personal communication).

Suitability for hay and silage. Unknown.

Value as standover or deferred feed. Fairly good; persists when grass production declines.

Feeding value. Has a protein content of 15.60 percent. Its palatability is affected by a high tannin content.

Seed yield. 20 to 80 kg/ha (Grof, Rotar, personal communications); in small plots, up to 130 to 230 kg/ha.

Cultivars. No commercial lines have yet been developed.

Diseases and pests. Is subject to little-leaf and *Xanthomonas* attack. Is remarkably free from pests, but the rose beetle may attack it in Hawaii.

Main attributes. Its ability to combine with sward-forming grasses and stand heavy grazing. It is stoloniferous and rhizomatous. It is persistent; lasted for 12 years at Topaz in north Queensland (Kelly, 1964).

Main deficiencies. Inclined to become woody, does not produce high yields and has low seedling vigour. Does not fix large amounts of nitrogen.

Performance. In Hawaii, Whitney, Kanehiro and Sherman (1967) grew this legume alone and in combination with *Pennisetum purpureum* and *Digitaria decumbens* on a volcanic soil. *D. canum* gave only low yields and reduced the nitrogen of the grass growing with it.

Main references. Rotar, Brown and Lyman (1967); Younge, Plucknett and Rotar (1964).

Desmodium heterocarpon (L.) DC. var. *heterocarpon* van Meeuwen

1. Carpon desmodium

Synonyms. *Hedysarum heterocarpon* L.; *D. trichocaulon* DC.; *D. polycarpon* (Poir.) DC.; *D. capitatum* Miq.; *D. buergeri* Miq.; *D. ovalifolium* (Prain.) Wall. ex Ridley.

Description. A perennial subshrub, sometimes a shrub or herb, ascending or often creeping with a woody root-stock. Stems and branches up to 1 m long, diffuse and ascending or erect, although prostrate under heavy grazing. Adventitious rooting sometimes occurs from stems lying on the soil surface. Stems range from nearly glabrous to densely covered with white or yellowish hairs. Leaves are generally trifoliate, though unifoliate leaves are common on seedlings and not unusual toward the base of mature stems.

Leaflets smooth on the upper surface and hairy underneath, often with a light green to almost yellow watermark on the upper surface. Terminal leaflet normally broadly elliptic, ovate or obovate, with tips generally retuse. Inflorescence composed of densely flowered, terminal and axillary racemes; flowers pink. Pods erect or ascending, narrowly oblong, compressed and generally four- to eight-jointed, turning from green to dark brown on maturity. Articles quadrangular to semi-elliptic, straight along the upper suture, somewhat rounded below, separating and then dehiscing. Seeds almost quadrate, cream to orange in colour, 2 × 1.5 mm (Kretschmer *et al.*, 1979; Gillett, Polhill and Verdcourt, 1971).

Distribution. Native to southeast Asia; especially found in Sri Lanka, India, the Himalayas, Burma, Thailand, Malaysia, Indochina, Taiwan, Ryukyu and Japan, as well as the Pacific islands and Australia.

Season of growth. Carpon grows steadily in spring and rapidly during summer and autumn until flowering begins. From then until the last frost it grows very slowly. Thirty percent of its growth takes place in spring, 65 percent from summer to early autumn and 5 percent during the rest of the year (Kretschmer *et al.*, 1979).

Rainfall requirements. Requires fairly high rainfall. In Ft Pierce, Florida, United States, carpon grows steadily with 1 250 to 1 500 mm of rain a year.

Drought tolerance. Has sufficient drought tolerance to withstand the regular severe spring drought in Florida (Kretschmer *et al.*, 1979).

Tolerance of flooding. Carpon will not survive extended flooding. Experience in Florida suggests that it can survive up to a week of intermittent shallow flooding (Kretschmer *et al.*, 1979).

268

Soil requirements. Carpon should not be planted in seasonally wet areas, but is adapted to the better-drained mineral soils of southern Florida (Kretschmer *et al.*, 1979).

Rhizobium *relationships*. Cowpea inoculum is recommended in Florida.

Land preparation for establishment. A clean, firm seed bed is desirable. In Florida, a rolling, broadcasting of the seed and rerolling are recommended, especially when a vegetatively established companion grass is being used. Where there is already an adequate grass sod, light disc harrowing or chopping should precede the initial rolling, followed again by broadcasting and rerolling (Kretschmer *et al.*, 1979).

Sowing time and rate. In southern Florida, seed can be planted any time from after the last frost until August, but little or no seed will set the first year unless germination occurs before July. On clean seed beds, the suggested rate is from 3 to 5 kg/ha, but in established grass areas 5 to 10 kg/ha should be used (Kretschmer *et al.*, 1979).

Number of seeds per kg. 770 000.

Percentage of hard seed. One commercial sample in Florida contained 50 percent hard seed, while figures of up to 65 percent have been obtained in experimental harvests (Kretschmer *et al.*, 1976), suggesting hard-seed levels are not particularly high.

Nutrient requirements. In southern Florida, recommendations are about 2.5 tonnes of lime and 90 to 100 kg/ha of P_2O_5 and K_2O applied as an 0-12-6 or similar fertilizer to virgin flatwood soils, with about 300 kg/ha of 0-10-20 annually thereafter (Kretschmer *et al.*, 1979).

Compatability with grasses and other legumes. Carpon is well able to compete with companion grasses in Florida, even when early spring nitrogen applications have been made to stimulate grass growth. In one stand, neither nitrogen application nor the subsequent heavy grazing had any adverse effect on carpon stands over a four-year period. Carpon also combines well with white clover (Kretschmer *et al.*, 1979).

Seedling vigour. Seedling growth is slow.

Response to defoliation. Under heavy close defoliation, carpon develops a prostrate growth habit, and a very low protected crown from which the new regrowth occurs. It is not easily killed by grazing.

Feeding value. The foliage contains 12 to 20 percent crude protein, depending on growth stage, while grass/legume mixtures average 9 to 10 percent crude protein. Some 130 kg/ha N/year was removed from harvested forage. This was equivalent to 190 to 260 kg/ha N applied to pure pangola and bahia grass pastures. *In vitro* organic matter digestibility of the legume grass mixture

269

ranged from 45 to 60 percent, depending on the grass and season of the year. Tannin percentages from two cuts were 2.3 and 3.1 percent (Kretschmer *et al.*, 1979).

Seed harvesting. In Florida, flowering begins in early September and seed maturation is complete in November. Mature pods do not shatter readily, but grass/legume mixtures should be mown when 85 to 90 percent of the pods are mature. They then should be allowed to dry for at least one day, and preferably two to three days, before pick-up thrashing. Pick-up thrashing is preferable to direct heading because the vegetative material still has a high moisture content at this stage, and is likely to gum the thrashing bars of the drum if direct headed (Kretschmer *et al.*, 1979).

Seed yields. Initial commercial yields of 100 kg/ha have been obtained.

Cultivars. One cultivar, Florida, was released in 1979 by the Agricultural Research Center, Ft Pierce, Florida. It derives from USDA Plant Introduction No. 217910 from the Forest Research Institute, Dehra Dun, Uttar Pradesh, India, in 1954. It reached Ft Pierce in 1964 via Tifton, Georgia, United States.

Pests and diseases. Web-worms and other insects can attack the foliage and seed pods in long ungrazed stands during flowering and weed maturation. They are best controlled by grazing until mid-August and only then closing for a short seed-production period.

At least two species of root knot nematodes (*Meloidogyne incognita* and *M. arenaria*) have been found in carpon roots. Old vegetable fields should therefore not be planted. Cv. Florida is particularly susceptible, but two accessions from a collection of eight have apparently proved resistant (Kretschmer *et al.*, 1979).

No diseases have been observed to date (Kretschmer *et al.*, 1979).

Main references. Kretschmer *et al.* (1976), (1979).

2. "Desmodium ovalifolium" (see Colour plates XIV, XVI)

Description. A perennial subshrub, rarely taller than 75 cm. Stems and branches nearly glabrous, rooting freely at the lower nodes, young apical portions clothed with fine silky hairs. Leaves normally trifoliate, sometimes unifoliate. Leaflets round, oval or obovate, the terminal larger than the two laterals, margins entire. Leaflets are densely pubescent on the dorsal surface, with whitish hairs; the ventral surface is glossy, glabrous. CIAT 350 has no leaf markings. Flowers borne in short, crowded racemes, purple or deep pink, turning bluish after anthesis. Lomentum erect or ascending, pods with quadrate articles, 2.5 to 3 mm long (Grof, 1982).

General features. While included in *D. heterocarpon* (L.) DC. var. *heterocarpon* by Ohashi (1973), *"D. ovalifolium"* apparently, is, agronomically, a different plant from the material being developed as a pasture legume in Florida (Imrie, Jones and Kerridge, 1983). Long used as a cover crop in Sri Lanka and Malaysia, especially in rubber plantings, as it is able to thrive in dense shade, *"D. ovalifolium"* is being considered also for pasture. It can be slow to establish but the prostrate, stoloniferous growth habit gives good soil cover. In addition, it withstands frequent defoliation well (Imrie, Jones and Kerridge, 1983).

"D. ovalifolium" is considered of low potential in Zimbabwe (Clatworthy, 1975), on the wet tropical coast of north Queensland (Teitzel, Abbot and Mellor, 1974), and in Uganda (Horrell, 1963), but there is currently interest in it at CIAT because of its tolerance to acid soils and high aluminium saturation. It has yielded well (23.4 and 18.6 tonnes DM per ha per year under six- and eight-week cutting regimes respectively) with *Brachiaria decumbens, B. brizantha, Panicum maximum* and *Andropogon gayanus* on an acid ultisol in the Cauca Valley of Colombia. It has also performed well under grazing on an infertile oxisol in the eastern plains region of Colombia. This material is an international trade line identified as CIAT 350. Widespread testing in lowland tropical regions with an annual rainfall of 2 000 mm or more is recommended (Grof, 1982).

"D. ovalifolium" has specific rhizobium requirements (Halliday, 1979) and apparently no particular problems with diseases or insect pests, although susceptibility to root knot nematodes (*Meloidogyne javanica*) has been reported from Malaysia and Colombia (Grof, 1982). Drought tolerance is also good (Schultze-Kraft and Giacometti, 1979; Rijkebusch, 1967).

It remains to be seen if good animal production can be achieved from *"D. ovalifolium"* pastures on highly weathered acid soils (Jones and Kerridge, 1983). One disadvantage is a rather high level of tannin in the forage, which affects intake and digestibility (Rotar, 1965). Cattle will reject *"D. ovalifolium"* at first, especially early in the growing season. *In vivo* digestibility of Colombian material averaged 57 percent (Grof, 1982).

Main references. Imrie, Jones and Kerridge (1983); Grof (1982).

Desmodium heterophyllum (**Willd.**) **DC.** (see Colour plate XV)

Synonym. Hedysarum heterophyllum (Willd.).
Common names. Desmodium, senivakacegu, wakutu (Fiji), hetero (north Queensland).

271

Description. Perennial, prostrate creeper; stems are reddish brown, hairy, freely branching and root at the nodes. Leaves are trifoliate, the terminal leaflet slightly larger and on a longer stalk than the two lateral leaflets. Flowers are reddish-pink and small, about 3 mm long, two to four flowers in subterminal racemes. Indented pod contains three to six light-brown speckled, shiny, kidney-shaped seeds, 2 mm long.

Distribution. A native legume in the tropics of the Eastern Hemisphere. Occurs in Mauritius, Southeast Asia, Malaysia, Indonesia and is adventive to the South Pacific. It is recorded as a component of lowland pastures in Sri Lanka and Fiji and used as a leguminous cover in the pepper gardens of Sarawak.

Temperature requirements. Perennial summer-growing legume; optimum temperature for growth about 25°C (Grof, personal communication); minimum temperature about 12.5°C. It is easily killed by frost.

Latitudinal limits; altitude range. It occurs from the equator to about 20°S. Recorded from sea level to 200 m', but occurs up to 770 m at Topaz in north Queensland and at 900 m in Fiji (Roberts, personal communication).

Rainfall requirement. At least 1 500 mm, and does well at 3 475 mm at Topaz (Kelly, 1964) and at 4 000 mm in Fiji.

Tolerance of drought and flooding. Roberts (personal communication) states it is quite drought-tolerant in Fiji. It is also tolerant to short-term flooding, and Roberts is interested in it for rice fallows in Fiji.

Soil requirements. It is adapted to a wide range of soils from sands to clays. Its pH tolerance is also wide but it seems to perform best at pH 5.0. It has no tolerance of salinity.

Rhizobium *relationships*. It is extremely specific in its *Rhizobium* requirement. Norris (personal communication) has prepared strain CB 2085 for use with this legume.

Ability to spread naturally. Excellent; has invaded pastures of *Brachiaria* and pangola grass pastures in north Queensland (Grof, personal communication). The segmented pods break up when mature and scatter the seed.

Land preparation for establishment. It prefers a well-prepared seed bed where sown by seed, less preparation for cuttings.

Sowing methods. Seed can be sown by drilling or broadcasting and the cuttings (Payne *et al.*, 1955) dibbled or disced in. Oversown into pastures, it invades short grass pastures of *Brachiaria* and pangola grass. Should be sown no deeper than 0.5 cm, preferably in midsummer.

Seed treatment before planting. To break dormancy: fresh seed gave 50 percent germination. Treatment with concentrated sulphuric acid for ten minutes increased germination to 91 percent (Grof, personal communication). Roberts (personal communication) scarified the seed mechanically for pot trials. Inoculation is absolutely necessary.

Nutrient requirements. Little is yet known of its nutrient requirements but it responds to superphosphate where the soil is deficient in phosphorus.

Vigour of seedlings and growth. It grows vigorously over the summer months.

Response to light. Shade-tolerant. Flowers in 150 days at South Johnstone, Queensland.

Compatibility with grasses and other legumes. It is particularly successful with *Brachiaria decumbens* and pangola grass in north Queensland and with pangola grass and *Brachiaria humidicola (dictyoneura)* in Fiji, where it coexists with *Centrosema pubescens, Ischaemum indicum* and *Dichanthium caricosum.*

Ability to compete with weeds. Excellent, provided the weeds are low growing.

Nitrogen-fixing ability. Grof has recorded 60 kg/ha in north Queensland. The inclusion of *D. heterophyllum* in a pangola grass sward increased the nitrogen content of the grass from an annual mean of 0.86 percent to 1.12 percent. The nitrogen content of the legume averaged 2.8 percent for the season (Grof, personal communication).

Response to defoliation. Highly tolerant of defoliation and will stand heavy grazing.

Grazing management. Keep companion grass short for best performance.

Response to fire. Will not tolerate fire.

Breeding system. Self-compatible but normally cross-pollinated. Chromosome number 2n = 22.

Dry- and green-matter yields. Dry-matter and nitrogen yields obtained by Grof (personal communication) for grass alone, for grass plus legume and for grass plus legume with added nitrogen at South Johnstone, north Queensland, are given in Table 14.2.

Feeding value. Its slight unpalatability enables it to spread early. It is ultimately heavily grazed.

Toxicity. None observed.

Seed harvesting methods. The short stoloniferous habit of growth and uneven ripening of the pods make seed harvesting difficult. Material is harvested by forage harvester and seed reclaimed from the dried material (Grof, personal communication). Some form of vacuum pick-up would probably perform well (Roberts, personal communication).

Seed yields. Grof harvested just over 5 kg of seed from 0.17 hectare at South Johnstone.

TABLE 14.2 Dry-matter and nitrogen yields for *Desmodium heterophyllum* in combination with three grasses, with and without added nitrogen, in South Johnstone, north Queensland, in kg/ha

Grass	Grass alone		Grass plus legume		Grass plus legume plus N					
					110 kg N/ha		220 kg N/ha		330 kg N/ha	
	DM	N	DM	N	DM	N	DM	N	DM	N
Digitaria decumbens (pangola)	10058	97	11866	142	14881	139	17090	188	20822	236
Brachiaria spp. (signal)	14023	132	15146	162	15558	153	19730	199	20767	239
Panicum maximum (guinea)	11439	110	13632	166	12314	144	12836	140	19303	212

Diseases. Susceptible to legume little-leaf, but this is of little consequence under grazing.

Main attributes. Adaptability to a wide range of soils; compatibility and long-term persistence in closely grazed swards of pangola grass and *Brachiaria decumbens*; free-seeding habit (it spreads readily even under heavy stocking). It is accepted by the grazing animal.

Main deficiencies. Difficulty of seed collection; susceptibility to frost.

Performance. No grazing trials.

Desmodium intortum (Mill.) Urb. (see Colour plate XI)

Synonym. *Desmodium aparines* (Link.) DC.

Common names. Greenleaf desmodium (Australia), pega-pega (Philippines, Costa Rica, Venezuela), kuru vine (Zimbabwe), amor seco (Colombia).

Description. Large trailing and climbing perennial; roots at the nodes and has a deep taproot; long, pubescent stems branch freely and are often reddish brown. Has shorter internodes than *D. uncinatum* and is leafier. Leaves usually have reddish-brown to purple flecking on the upper surface. Leaflets, 2 to 7 cm long and 1.5 to 5.5 cm broad, with a length-width ratio of 1.4 to 1, are shorter and more rounded than in *D. uncinatum*. Terminal raceme compact, flower deep lilac to deep pink. Seed pod narrow, bears 8 to 12 seeds, recurves to the main rachis; seed adheres to animals and to clothing, but not as tenaciously as that of *D. uncinatum* (Barnard, 1967).

Distribution. Common in northern parts of South America, on the eastern slopes of the Andes and in a restricted area of Brazil between 18 and 25°S. Native in Panama, Colombia, Guatemala, Venezuela, Ecuador and Peru. Has spread throughout the tropics and is being widely tested for fodder value.

Season of growth. A summer-growing perennial, with main growth from mid-October to mid-May (Queensland).

Optimum temperatures of growth. Requires a long, warm growing season. Whiteman (1968) assessed the optimum temperature for growth at 30/25°C + 3°C. Withstands hot weather better than *D. uncinatum*.

Frost tolerance. Grows well into the autumn, but is susceptible to heavy frosts; retains leaf fairly well after frosting.

Latitudinal limits and altitude response. Extends to 25°S (Bryan, 1969). Is known to grow at 2 400 m near the equator in Kenya. In tropical latitudes in Latin America it occurs from 600 to 2 500 m. In the subtropics it occurs at a lower elevation but is rarely found in warm temperate zones, even at sea level.

Rainfall requirements. Favours a well-distributed rainfall in excess of 875 mm a year. Kelly (1964) found it outstanding in a rainfall up to 3 475 mm at Topaz on the Atherton Tableland, Queensland, but in similar high rainfall in Panama and Guatemala it was badly affected by disease and leaf-eating insects.

Tolerance of drought and flooding. Is susceptible to extended dry spells, but persists well where soils are fertile. Wilts less readily than *D. uncinatum* (Ostrowski, 1966). Stobbs (1969e) showed it to be less drought-resistant than *Stylosanthes guianensis* at Serere, Uganda. It carries little foliage in the dry season, when most of the leaves drop and form a mulch (Horrell, 1958). Will survive temporary flooding and some waterlogging (Boultwood, 1964) but is susceptible to extended waterlogging. Performs better on slopes.

Soil requirements. Grows on a wider range of soils than *D. uncinatum* and does not do quite as well on sandy soils as *S. guianensis* (Stobbs, 1969f). Will grow in a range of soils from light to clay loams. Requires a soil with a pH in excess of 5.0 (Andrew and Bryan, 1958; Moomaw and Takahashi, 1962). Has no tolerance to salinity, and is depressed by high chloride levels (Andrew and Robins, 1969b).

Rhizobium *relationships*. Requires the specific "Desmodium" culture (Date, 1969). The current Australian recommendation is CB 627 (1970). Boultwood (1964) found it unnecessary to inoculate it in Zimbabwe. Whiteman (1970) showed that peak nodule formation occurred three months before flowering in *D. intortum*. Does not spread well from natural seed sources; better by natural vegetative means with its stoloniferous habit.

Land preparation for establishment. Because of its small seed, *Desmodium intortum* requires a well-prepared seed bed (Younge, Plucknett and Rotar, 1964). Will establish from broadcasting into ashes from the air.

Sowing methods. Can be sown by drilling, broadcasting from ground machines or from the air. Has been established by cuttings in Zaire (Risopoulos, 1966) and on steep slopes in Guatemala on contour ridges (Johnston, personal communication). Risopoulos found that planting by cuttings gave only 30 to 40 percent strike but, if the cuttings were rooted in banana fibre baskets under light shade and later transplanted, results were good. Boultwood (1964) established it by transplanted cuttings spaced at 1×1 m and also by undersowing in maize early in the season. Calma, Valera and Santos (1959) obtained best yields with 20×25 cm spacings. It generally does not establish when oversown into existing pastures because of low seedling vigour. Extensive pasture renovation would be required to give any success. Should be sown at no greater depth than 1 cm (Suttie and Ogada, 1967) and rolled or very lightly harrowed.

Sowing time and rate. Can be sown from spring to midsummer or later in frost-free environments at a rate of 1 to 2 kg/ha. Middleton (1970) found no seedling competition at rates of 1.1, 3.3 and 9.9 kg/ha, and under high rainfall conditions seedling density was proportional to sowing rate.

Number of seeds per kg. 594 000. Percentage of hard seed is fairly high. Risopoulos (1966) found that only 25 percent of fresh seed germinated; this figure did not improve up to 18 months after collection.

Seed treatment before planting. To break dormancy: if seed is harvested mechanically, there is no need to treat. Boultwood (1964) advised treatment in strong sulphuric acid for five minutes, then thoroughly washing in water and drying. Inoculation is necessary. Pelleting with lime depresses nodulation (Norris, 1967). Luck (personal communication) obtained nil, 35 and 77 percent nodulation, however, from (a) uninoculated, (b) inoculated and band-sown, (c) inoculated, pelleted and band-sown seed respectively. If pelleting is required, Norris (1967) recommends rock phosphate.

Nutrient requirements. D. intortum usually requires adequate levels of phosphorus, sulphur, potash and molybdenum for growth. Heavy dressings of potassium chloride, however, can cause chlorine toxicity (Andrew and Robins, 1969b). Younge, Plucknett and Rotar (1964) proved that when D. intortum is adequately fertilized with P, K, Mo, and Zn, it is able to compete with Kikuyu and pangola grass. On acid ferruginous and aluminous latosols in Hawaii, Younge and Plucknett (1966b) found that heavy P fertilization was needed. Treatment with 1 650 kg P/ha allowed grazing of

6 beasts/ha compared with 3 beasts/ha pasture treated with 275 kg P/ha.

● Calcium. In Hawaii, Younge, Plucknett and Rotar (1964) found that *D. intortum* grew satisfactorily on a soil of pH 5.5 and showed little response to application of lime.

● Nitrogen. Nitrogen fertilization at 14 kg/ha reduced the *D. intortum* component in mixtures with pangola grass and Kikuyu in Hawaii to less than 10 percent in the ten-week cutting treatment, and to less than 1 percent in the five-week cutting treatment (Whitney, 1970).

● Phosphorus. The critical level for P in the dry matter of the leaves at the immediately preflowering stage is 0.23 percent. On a Samford gley soil fertilized with the equivalent of 60 kg superphosphate per ha, *D. intortum* yielded 43 percent of its maximum yield at 1 230 kg/ha (Andrew and Robins, 1969b). Plucknett and Fox (1966) found that the phosphorus content of unfertilized *D. intortum* remained at 0.11 percent for two years and regarded this as the minimum value for survival of this species.

● Potash. In Swaziland, l'Ons (1968) obtained a linear response to K at rates up to 55 kg/ha, the increase being about 11 kg/ha K. Andrew and Robins (1969c) found that *D. intortum* contained relatively high concentrations of potash. They found that deficiency occurs when the dry matter in the tops at the immediate preflowering period is less than 0.80 percent. Andrew and Robins (1969d) found no deficiency symptoms in plants which had 1.39 g of K per 100 in the dry matter of the tops, but deficiency symptoms were apparent in a plant with 0.47 g/100 g. The first sign of potassium deficiency in *D. intortum* was interveinal necrotic spotting on the midpositioned leaves of the plants. The necrotic spotting was not preceded by any chlorotic effects; however, in severe cases of deficiency some interveinal chlorosis occurred, usually in those areas with intense necrosis. The necrosis and chlorosis occurred in a symmetrical pattern on the tips and margins of the leaflets. Necrotic spots were irregular in shape, equally visible on both leaf surfaces, and mid-brown in colour. In severe form, the small spots coalesced to give large areas of necrotic tissue which usually encompassed the leaf margins. Associated with this there was an inward curling of the leaf margins. In extreme deficiency, even the younger, fully expanded leaves became pale and chlorotic, with the margins of the leaflets curled or rolled inward. Very little leaf abscission occurred in this species (Andrew and Pieters, 1970a).

Vigour of seedling; growth and growth rhythm. Once established, *D. intortum* makes vigorous growth in the wet season; commences growth later in the spring than *D. uncinatum*.

Compatibility with grasses and other legumes. Grows well with *Setaria* spp.,

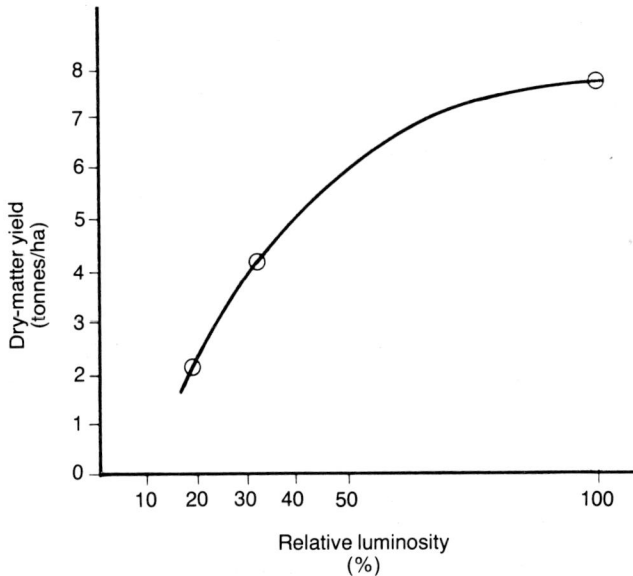

Figure 52. A sensitive short-day species, *D. intortum* flowers late in the subtropics, rendering it likely to be damaged by frost. Flowers mid-May to early June in Queensland, Australia; June to July in Madagascar (Delhaye, 1964). Not as tolerant of low light intensity as *D. uncinatum.* The graph shows the relationship between dry-matter production and light intensity; it is fairly tolerant of shade (**Source:** Compère, 1961)

Paspalum commersonii, Panicum maximum, Pennisetum purpureum, Melinis minutiflora and, if adequately fertilized, with Kikuyu and pangola (Younge, Plucknett and Rotar, 1964). Also grows well with siratro and glycine. Middleton (1970) found siratro more competitive with *Setaria anceps* than with *D. intortum.* Bryan (1966) recorded that *D. intortum* had invaded swards of at least 12 species of *Paspalum.*

Ability to compete with weeds. In the early stages poor, but when well established will suppress weeds (Boultwood, 1964).

Tolerance of herbicides. Exhibits extremely useful tolerance to 2,4-D. At $3^{1}/_{2}$ weeks of age, 1.65 kg of acid equivalent per hectare can be used, thus there is no need to use 2,4-DB. Diquat should be used only at 140 g of cation per hectare from six weeks of age. Diquat at 280 g of cation per hectare is safe on established greenleaf desmodium (Bailey, 1970).

Nitrogen-fixing ability. Directly related to yield. *D. intortum* fixed over 300 kg N/ha/year in Hawaii (Whitney, 1970). Row width had an effect. Whitney and Green (1969b) found that it fixed 213 kg/ha/year in 90-cm rows, and 264 kg/ha/year in 45-cm rows. Whitney, Kanehiro and Sherman (1967) found that it fixed 375 kg N/ha, of which it transferred only 5 percent to the associated grass. Leaf fall could add an additional 1.3 kg N/ha/week.

Response to defoliation. Jones (1967) inposed three cutting heights and three cutting frequencies on a sward of *D. intortum*. It disappeared when cut at a height of 3.75 cm every four weeks, but survived cutting at 3.75 cm at frequencies of 8 and 12 weeks and cutting at 7.5 or 15 cm at 4, 8 and 12 weeks respectively. Risopoulos (1966) found that cutting at 12-week intervals gave the highest yield of cut fodder. Whitney (1970) found a cutting height of 13 cm was better than at 5 cm, N fertilization reduced the legume component to less than 10 percent in the ten-week cutting treatment and to less than 1 percent in the five-week cutting treatment. Under heavy stocking (1 beast to 0.4 ha) the legume decreased; under light stocking (1 beast to 0.8 ha) it increased. Riveros and Wilson (1970) found no significant difference in yield with cutting heights of 7.5 and 15 cm at the first season but a cutting interval of three weeks gave the highest yield; in the second season a cutting height of 15 cm at five-week intervals gave the highest yield. Whiteman (1970) found that the severity of initial defoliation affects the proportion of the original nodule population on the plant and subsequent defoliations affect the number of new nodules developed and their rate of increase in size.

Grazing management. Bryan (1966) illustrates the way cattle graze *Desmodium* pastures: stock normally remove the last part of the shoot and then browse the leaves, leaving large numbers of axillary buds which ensure rapid regrowth. If grazing is intermittent and intense, a greater proportion of the stem and buds may be removed or damaged, with consequent reduction in bud sites and residual leaf material. Recovery from grazing would then be much slower and would ultimately affect persistence. Grazing management must, therefore, first allow the legume to become established and then adjust grazing pressure to allow for retention of bud sites and leaf material. This also involves the companion grass and a compromise must be established to protect the sward.

Response to fire. It is not advisable to allow fire to pass through a sward of *D. intortum*, but it will not be entirely destroyed by it. In an established stand the top is killed but the taproot will shoot again (Boultwood, 1964).

Breeding system. Generally self-fertile, but pollination can be improved by tripping the flowers. Chromosome number 2n = 22 (Rotar and Ukio, 1967).

279

Crosses have been made between *D. intortum* and *D. sandwicense* (Hutton and Gray, 1967; McWhirter, 1969).

Dry- and green-matter yields. Boultwood (1964) recorded seasonal yields of 19 tonnes/ha of green material with a crude protein percentage of 18.8 percent. Younge and Plucknett (1966b) recorded a five-year average of 19 000 kg/ha/year from a pangola/*D. intortum* pasture fertilized with 1 320 kg P/ha in Hawaii. Whitney, Kanehiro and Sherman (1967) recorded a yield of 19 000 kg dry matter/ha/year. Roe and Jones (1966) recorded a dry-matter yield of 12 500 kg/ha at Gympie, Queensland. Riveros (1969) recorded over 17 000 kg DM/ha during a eight-month growing season at Redland Bay, Queensland. Calma, Valera and Santos (1958) obtained 5 875 kg DM/ha in four cuttings spaced at 60, 59, 86 and 101 days respectively.

Suitability for hay and silage. Risopoulos (1966) made good hay at Mulungu, Zaire. It has also been made successfully in Brazil, Queensland and Guatemala (Calma, Valera and Santos, 1958). In Guatemala the hay is ground into meal for stock-feeding. Without the addition of molasses it made reasonable silage with 12.2 percent dry-matter loss and 0.03 percent N loss in Brazil. The pH was 5.0 and was better with the addition of 8 percent molasses on a green-weight basis. Boultwood (1964) made good silage by flail harvesting, adding 2 percent molasses by green weight and compacting well, because the material is light and fluffy. Catchpoole (1970) made stable lactic acid silage from *D. intortum* to which molasses had been added up to 8 percent of the green weight of the material.

Value as standover or deferred feed. Its late flowering allows it to provide good standover feed in frost-free areas.

Feeding value. The meal is an excellent source of protein, riboflavin and vitamin A for chickens (Squibb *et al.*, 1950, 1953; Huang, 1967). In Puerto Rico, Warmke and Freyre (1952) found high intake and good palatability when grazed by cattle.

Chemical analysis and digestibility. Younge and Plucknett (1966b) recorded the chemical composition of a pangola grass/*D. intortum* pasture (mainly *D. intortum*) as 13.1 percent crude protein and 0.34 percent P, 1.44 percent Ca and 0.34 percent Mg in the dry matter. Dry-matter yields for *D. intortum* leaves and stems obtained by Compère (1961) are presented in Table 14.3.

Digestibility of the leaf protein was 54.08 percent, of the stem protein 61.88 percent, of the organic matter, 72.44 and 48.65 percent respectively. Risopoulos (1966) recorded 5.5 percent crude protein of 69 percent digestibility for the green material of *D. intortum* in Zaire.

TABLE 14.3 **Chemical and compositional analysis of _D. intortum_ leaves and stems on a dry-matter basis**

Plant part	Ash	Crude protein	Fat	Fibre	N-free extract	P	K	Ca	Mg
	(%)					(mg/kg)			
Leaves	9.06	23.63	6.12	24.52	36.67	4.7	33.0	9.92	2.88
Stems	7.51	8.92	1.64	43.60	38.33	2.3	34.0	4.62	2.345

Toxicity. None recorded. Boultwood (1964) found no bloat even when comprised the whole ration. Rotar (1965) found 3.2 to 8.8 percent tannin in the leaves and 1.5 to 3.7 percent in the stems. Hutton and Coote (1966) found 7 percent tannin in the dry matter of leaves. Bindon and Lamond (1966) found no toxicity in mice fed with leaves and seed.

Seed harvesting methods. A difficult cultivar to harvest. In Queensland the crop is mown when 40 to 50 percent of the seed is ripe, as seed matures unevenly and progressively from the bottom. Unripe seed matures in the swath, and the ripe seed should not fall in the mowing process. Crops can be either windrowed or left in a swath. Heavy crops are best left in the swath. Leave the crop to dry for 10 to 14 days after cutting, then use a pick-up threshing harvester with the highest possible drum speed and close concave settings. All the material must be threshed in its first passage through the drum or the seed will ball up and be lost over the sieves. Use either adjustable sieves on a fairly close setting or a 25-mm punched-hole bottom sieve. Suttie and Ogada (1967) used this method in Kenya.

Seed yield. In areas where early frosts occur, seed production may be severely curtailed. Greenleaf desmodium yields about 80 to 100 kg/ha clean seed.

Minimum germination percentage required for sale. Seventy percent germination with 94.5 percent pure seed (Queensland).

Cultivars. There is only one Australian cultivar — cv. Greenleaf. It is derived from three introductions, CPI 17916, CPI 18009 and CPI 23189 from El Salvador, Hawaii (ex Guatemala) and the Philippines respectively, each with similar characteristics. In Tanzania there is a cultivar named Tengeru. Rotar (1970) has drawn attention to the great amount of variability within the species and suggests that there is opportunity to develop a fairly large gene pool for _Desmodium_ utilization in many different environments in the tropics and subtropics.

Diseases. *D. intortum* is more resistant to legume little-leaf than *D. uncinatum* (Hutton and Grylls, 1956), but under very wet conditions it, too, is likely to be infected.

Pests. Meloid beetles eat the flowers (Horrell, 1958). *Amnemus* weevil has attacked plants at Gympie, Queensland (Roe and Jones, 1966).

Main attributes. A well-grazed legume with high yield potential in frost-free areas of good rainfall. Has a long growing season and makes vigorous growth in association with grasses; gives early spring growth and is a good fertility builder.

Main deficiencies. Low seedling vigour, poor drought tolerance, poor salt tolerance and relatively low digestibility. It is grazed out unless heavily fertilized. Leaves, flowers and roots subject to attack by various pests.

Performance. Stobbs (1969c) at Serere, Uganda, obtained 500 kg/ha/year live-weight gain from a *Panicum maximum*/*D. intortum* sward. Moomaw and Takahashi (1962) obtained a beef yield of 660 kg/ha per year over a period of $2^1/_2$ years from a fertilized pangola grass/*D. intortum* pasture in Hawaii — a tenfold increase over normal ranch performance. Villareal (1967), at Beerwah, Queensland, obtained a live-weight gain of more than 500 kg/ha from a pangola grass/*D. intortum* pasture containing some *Lotononis bainesii*.

Main references. Bryan (1969); Compère (1961); Younge, Plucknett and Rotar (1964).

Desmodium sandwicense E. Meyer

Synonym. *D. limense* Hook.

Common names. Spanish clover (Hawaii), Sandwitch (the Philippines).

Description. Erect or decumbent perennial herb, 20 to 100 cm tall, occasionally woody at the base; sometimes climbing by means of viscid pubescent recurved hairs on the stems and pods. Internodes of stem as much as 4 cm long. Stems red or green in colour, leaves trifoliate, glabrous above and with a conspicuous lighter irregular area around the midrib (occasionally absent), pubescent underneath, especially on veins, petioles and petiolules; ovate-lanceolate leaflets, 2 to 6 cm long. Inflorescence a terminal axillary viscid-pubescent raceme, 10 to 15 cm long. Flowers white to purple, becoming faded with age; standard with two conspicuous green areas near the base. Pod sub-sessile in the persistent calyx, commonly with eight reticulated joints about 4 mm long and slightly curved on the dorsal suture. Seed yellow olive-green to reddish brown, flat, about 2.5 mm long and 1.75 mm wide (Rotar, personal communication).

282

Distribution. Throughout Hawaiian islands to about 1 075 mm, occasionally to 1 800 m where rainfall is 750 mm or more.

Characteristics. Summer-growing perennial where moisture is available. Whiteman (1970) found the optimum temperature for growth of *Desmodium* spp. to be 30/25°C ± 3°C for day/night temperatures. *D. sandwicense* is less affected by cold weather than *D. uncinatum* and *D. intortum (aparines)*. Its altitude range is from sea level to a maximum of 1 800 m.

D. sandwicense needs an annual rainfall in excess of 750 mm. It is more drought-tolerant than other *Desmodium* spp., and Rotar (personal communication) states that it is valuable for germ-plasm because of this feature. It is not tolerant of flooding. It tolerates a wide range of soil conditions, growing on sands and clay loams with a pH range from acid to neutral.

It is somewhat specific in its *Rhizobium* requirements, and Norris (1967) recommends inoculation with the "Desmodium" strain, CB 627. In Hawaii the cowpea strain is satisfactory (Rotar, personal communication). Whiteman (1970) found that changes in nodulation of *D. sandwicense* during growth do not follow the pattern of steady increase and decline shown by *D. intortum* and *D. uncinatum*. The dry-matter production curve was very similar to the nodule weight curve.

Seed is broadcast or drilled into a well-prepared soil at a sowing depth of 1 to 1.5 cm and lightly covered. Sow just before the rains in summer at 1 to 2 kg seed per hectare, with a dressing of fertilizer containing phosphorus, potash and micronutrients.

D. sandwicense is indeterminate in its flowering response and flowers on three occasions during the growing season in southeast Queensland (lat. 27°22'S): in early December, early February and late April (Whiteman, 1970). It flowers in 45 to 50 days in Hawaii (Rotar, personal communication).

It is capable of growing with tall grasses (which it climbs) and is not affected by normal light conditions. Combines well with *Setaria*, *Paspalum* and *Panicum* spp.

It has fairly good seedling vigour and fixes nitrogen well. It does not respond well to defoliation.

It is self-fertile, usually with some cross-fertilization. Chromosome number 2n = 22. Seed set is improved by tripping the flowers.

Jones, Davies and Waite (1967) grew *D. sandwicense* (CPI 18227 and CPI 18225) with *Paspalum plicatulum* at Samford in southeast Queensland. The legume only persisted for two years, yielding merely 390 kg/ha dry matter in the first year, and contributing 10 percent of the dry matter and 15.2 percent of the nitrogen to the mixture, increasing to 1 300 kg DM/ha in the second

283

Figure 53.   Harvesting seed with a mower in southeast Queensland, Australia

year, which corresponded to 13 percent of the total dry matter and 27.8 percent of the nitrogen. It performed very poorly in comparison with siratro and lotononis. Horrell (1958) found it inferior in performance to *D. uncinatum* at Serere, Uganda, as did Anderson and Naveh (1968) in northern Tanzania. They harvested an average of 3 749 kg DM/ha from three sites.

It is subject to little-leaf and to *Xanthomonas* sp. The rose beetle and cotton cushiony scale attack it in Hawaii (Rotar, personal communication). The main attribute of *D. sandwicense* is its drought resistance. It seeds too heavily, however, and has poor tolerance to grazing.

Desmodium scorpiurus (Sw.) Desv.

Synonym. Hedysarum scorpiurus Desv.
Description. A straggling, climbing or procumbent herb with small blue flowers. Terminal leaflet about 2 cm long and 1 cm broad, rounded at each end,

284

thinly pubescent. Racemes up to 10 cm long, few-flowered. Flowers 4 mm long. Pod five- to eight-seeded, segments twice as long as broad, not deeply indented.

Distribution. Originated in tropical America, now widespread in the tropics. It grows naturally at Townsville in north Queensland.

Characteristics. A vigorous, widely adaptable legume, it spreads quickly because its pods adhere to animals. Forms a good mixture with short grasses such as *Axonopus* and *Paspalum* and with other legumes. It stands treading and is very palatable. Its protein content is recorded by Bermudez *et al.* (1968) as 19.3 percent. In Nigeria, Miller and Rains (1963) found the dry-matter intake per head per day by cattle was 0.42 kg. The plant contained 90.3 percent organic matter, of which 44.7 percent was digestible, 15.5 percent crude protein of which 7.7 percent was digestible, and 29.3 percent crude fibre.

Desmodium triflorum (L.) DC

Synonyms. *Meibomia triflora; Desmodium bullamense* G. Don.

Common names. Amor do campo, trevinho do campo, amorsinho secco, carrapicho (Brazil); hindu pujali (Sri Lanka); pacpaclanhão (the Philippines); hierba cuartillo (El Salvador).

Description. Herb; stems branched, prostrate, creeping and forming dense mats. Leaflets broadly obovate, broad and emarginate at the apex, about 1 cm long, glabrous to finely appressed-pilose beneath. Flowers pink or purplish, one to three in the leaf axils. Pod about one-third indented on one side, approximately five-seeded, pubescent (Andrews, 1952).

Distribution. Semitropical. Common in Queensland, Australia, the southern Sudan, Côte d'Ivoire, Uganda, southern Brazil, Venezuela and Colombia.

Characteristics. A prostrate legume commonly found growing with *Cynodon dactylon* and *Paspalum notatum* in lawns and native pastures in the tropics and subtropics. Does not produce much bulk except where the nitrogen supply makes it possible. Resists drought (Farinas, 1966) and grazing very well. Is extremely palatable to poultry, especially baby chicks (Horrell, 1958). Grows well in Côte d'Ivoire with short grasses (Botton, 1958). In Hawaii, Hosaka and Ripperton (1944) report that it grows well with *Cynodon dactylon*, resists grazing and drought and is palatable to livestock. Payne *et al.* (1955) included it as a suitable component of a pasture ley in Fiji.

285

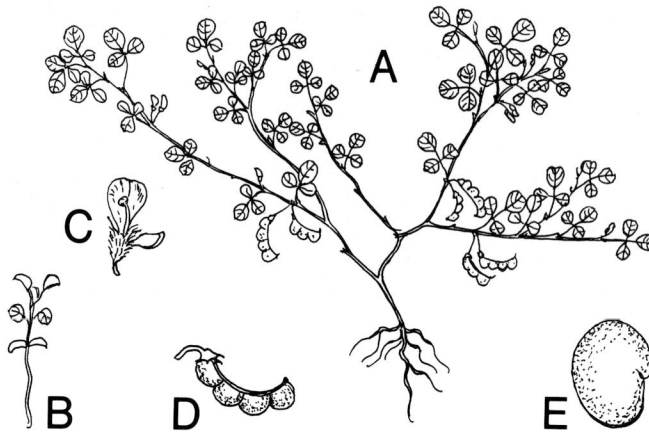

Figure 54. *Desmodium triflorum.* **A**-Habit **B**-Seedling **C**-Flower **D**-Pod **E**-Seed

Otero, 1952, gave the following analysis of the plant on a dry-matter basis:

	Percent
Protein	13.57
Fat	2.58
N-free extract	41.58
Fibre	35.00
Ash	7.27

Bermudez *et al.* (1968) stated that it contained 18 percent protein.

Desmodium uncinatum **(Jacq.) DC** (see Colour plate XVII)

Common names. Silverleaf desmodium (Australia), silverleaf Spanish clover (Kenya).

Description. Large rambling perennial, cylindrical or angular stems densely covered with short, hooked hairs which make the stems adhere to hands, clo-

286

thing, etc. Leaves trifoliate, with short brown stipules that fall off early, and with a petiole 2 to 7 cm long. Leaflets ovate, 3 to 6 cm long and 1.5 to 3 cm wide, the terminal one on a petiole 5 to 15 mm long. Leaflets are dark green on the upper side with an area of white shiny surface near the central vein, often surrounded by a dark shiny area; lower side is lighter green and uniform in colour, both sides hairy. Flowers in long paired racemes on stalks up to 1 cm long. Flowers 1 cm long, widely open when flowering, pink becoming bluish after flowering. Sickle-shaped pods easily break transversely at maturity into four to eight segments 4 to 5 mm long and 3 mm wide, light brown at maturity, densely covered with minute hooked hairs so that the segments adhere to clothing. Seed olive-green, triangular-ovate, 3 mm long, 2 mm wide and over 1 mm thick (Bogdan, 1965). It originally came from Deodora, Brazil, and was introduced to Kenya via CSIRO, Australia, as CPI 8990.

Distribution. There are some 350 species of *Desmodium* in the tropics. *D. uncinatum* is indigenous to northern Argentina, Brazil and Venezuela and is now widespread in pastures and nursery plots throughout the tropical and subtropical world. It is not common in tropical Brazil, being limited to an area of 1 000 m elevation at about latitude 10°S (Bryan, 1969).

Temperature requirements. Summer-growing perennial, starting growth early in the spring. Whiteman (1968) found the optimum temperature for growth to be 30/25°C ± 3°C. It wilts during the high temperatures of summer (Douglas and Luck, 1964). It is fairly cold-tolerant (Huang, 1967). Whiteman and Lulham (1970) put the growth minimum at about 15°C. Ludlow and Wilson (1970) found that *D. uncinatum* at 20°C yielded 23 percent of the dry matter it yielded at 30°C, 42 percent of the relative growth rate and had 14 percent of the leaf area. It is susceptible to frosts but more tolerant than siratro. Heavy frosts affect the tops, but in warm weather the plant recovers quickly (Huang, 1967). Spring and autumn growth is good.

Latitudinal limits. Southern boundary about 30°S (Tucuman, Argentina — Burkart, 1952) extending to about 19°N at Vera Cruz, Mexico (Bryan, 1969).

Altitude response. Grows from sea level to 2 400 m in Kenya. In Zaire it occurs from 1 500 to 1 800 m. The order of altitude response is silverleaf desmodium, *Neonotonia wightii*, greenleaf desmodium and siratro. *D. uncinatum* is much more tolerant of high altitude than other tropical twining legumes (Middleton, personal communication).

Rainfall requirements. Usually in excess of 900 mm (Bryan, 1969). It did not persist in alluvial soils under a rainfall of 3 000 mm in Fiji.

Drought tolerance. Horrell (1958) recorded its dry-season production as poor at Serere, Uganda, and Australian experience is similar.

Tolerance of flooding. Fairly tolerant of flooding and poor drainage. Performs better than *D. intortum* in low-lying land with a water-table close to the surface.

Soil requirements. It is adapted to a wide range of soils, from sands to clay loams; is productive on red basaltic loams and on gleyed podzolics (Mears *et al.*, 1964). It is not as successful on sands as *D. intortum*. Does well on soils with an open texture and not so well on compact heavy clays. Will grow at pH 5.0 (Andrew and Bryan, 1958), do well at pH 5.5 to 6.5 and grow up to pH 7.0. Anderson and Naveh (1968) stated that it was fairly tolerant of soil acidity in Tanzania. Does not tolerate salinity (Andrew and Robins, 1969).

Rhizobium *relationships*. Use a specialized "Desmodium" group *Rhizobium*. The current Australian inoculum is CB 627. It will cross-inoculate with some other species, e.g. *D. intortum* and *Neonotonia wightii* (Diatloff, 1968), but not with others. Peak nodulation occurs 1 month before flowering (Whiteman, 1969).

Ability to spread naturally. Seed pods attach to clothing and the coat of animals and spread widely. In suitable soils (e.g. deep latosolic clay loams) it can spread vigorously.

Land preparation for establishment. Will perform best in a well-prepared cultivated seed bed, but will establish on roughly cultivated land. Luck (personal communication) has established it at Cooroy, Queensland (lat. 26°S, rainfall 1 400 mm), by treating the existing pasture of *Axonopus* and *Paspalum* with Dalapon at 6 kg/ha and sod-seeding *D. uncinatum* into them.

Sowing methods. Sow using a drill, ground broadcasting, aerial seeding or sod-seeding. Some seed cover after sowing is desirable. Very little success has been achieved by oversowing into existing pastures. Sow in early summer at the rate of 2.2 kg/ha, no deeper than 1 cm (Suttie and Ogada, 1967). Whiteman (1969) found that December sowings gave higher nodule weights than later sowings in February and April.

Number of seeds per kg. 198 000 to 220 000. Percentage of hard seed is low.

Seed treatment before planting. Machine-harvested seed usually does not need treatment to break dormancy. For inoculation, use special "Desmodium" culture; the present Australian culture (1970) is CB 627. Lime pelleting stimulates the growth of this species (Norris, 1958).

Nutrient requirements. *D. uncinatum* responds readily to fertilizers; it performs very poorly without phosphorus, potash, sulphur and molybdenum.

- Calcium. A calcium concentration of 1.3 to 1.8 percent of the dry matter of the tops was measured by Andrew and Hegarty (1969). Andrew and Norris (1961) found that *D. uncinatum* gave 25 percent of its maximum yield in the absence of calcium, maximum yield being at 1 000 kg/ha. *D. uncinatum* is one of the least efficient at calcium extraction of the tropical legumes and is only slightly better than white clover.

● Copper. *D. uncinatum* is very efficient in extracting copper from the soil, probably because of its deep root system. In the absence of added Cu it gave 60 percent of its maximum yield — the best performance of ten legumes tested. The copper content of the seed is 6.3 ppm (Andrew and Thorne, 1962). Symptoms of copper deficiency are outlined by Andrew (1963): at an early stage of growth (approximately four weeks), deficient plants, particularly the young growth, become pale- to greyish-green in colour, but this is not associated with interveinal chlorosis. With increasing severity of copper deficiency, the youngest fully expanded leaves show marginal necrosis, especially toward the tip of the leaflets. This is not preceded by interveinal or general chlorosis or by necrotic spotting. Necrotic material tends to curve inwards bilaterally and also from the tip to give a dished effect. Progressively, younger expanding leaflets are pinched and necrotic at their tips and become completely necrotic on expanding fully. With increasing severity of copper deficiency, the young shoots are damaged, malformed, fail to expand, and finally die. Following the death of the young shoots, secondary growth starts from axillary buds, but this also succumbs. Affected leaves abscise early, leaving stems almost devoid of leaves. No visual symptoms occur on the roots or on the mature stems. Growth habit is changed from a prostrate to a semierect form.

● Phosphorus. The critical level for P in the dry matter of the leaves at the immediately preflowering stage is 0.23 percent (Andrew and Robins, 1969a, b). *D. uncinatum* yielded 53 percent of its maximum yield at the equivalent of 250 kg/ha, the maximum yield being achieved at the equivalent of 1 250 kg/ha of single superphosphate.

● Potash. Andrew and Bryan (1958) found that in the field at Beerwah (southeastern Queensland) visual symptoms of potash deficiency occurred when the potash content was below 0.50 percent of dry matter. Plants fertilized with potash contained 0.91 to 1.14 percent K and showed no deficiency. Andrew and Robins (1969c) found deficiency symptoms when the K content of the dry matter fell below 0.72 percent. Andrew and Pieters (1970a) have presented colour plates showing healthy leaves with 1.64 percent, and leaves from plants showing potash deficiency containing only 0.29 percent K in the dry matter. They record that deficiency commences as necrotic spotting on the mid- to lower leaves of the plant. The spots are initially of pinhead size, irregular in shape, and situated interveinally toward the leaflet margins. They are equally visible on both surfaces of the leaflets, and in the early stages create a halo effect around the entire leaflet margin except in the vicinity of the base; in more advanced stages, necrotic spotting occurs secondarily in the full interveinal tissue. Areas of necrotic spotting are preceded by a light form of chlorosis while the remainder of the leaflet is normal green in colour. Very little spotting occurs on the silver-coloured portion of the leaflet; however, in leaves which are severely affected by deficiency, the silver-coloured portions are not discernible. As the deficiency increases in severity, the areas of necrosis enlarge and coalesce, particularly in the distal marginal areas of the leaflets. At this stage, there is a general chlorosis of affected leaves. Associated with the severe form, the lamina of the leaflet curls inwards and there is some degree of surface puckering.

Toxicity levels and symptoms. Manganese: Andrew and Hegarty (1969) determined the toxicity threshold value for manganese of *D. uncinatum* as 1 160 ppm; hence, the plant is fairly tolerant of high manganese levels in the soil. Andrew and Pieters (1970b) recorded a colour photograph of *D. uncinatum* showing symptoms of toxicity and described the symptoms. The dominant effect of manganese toxicity in this species is chlorosis, which commences on the young shoots and newly expanded leaflets as a light interveinal chlorosis. With increasing severity of toxicity, the degree of chlorosis increases; in the extreme case, young leaves are devoid of chlorophyll upon expanding. Older leaves, which show slight interveinal effects initially, continue to exhibit these through to maturity but the thickness of the leaves increases with age (the young chlorotic leaves are very thin). Associated with the decrease in thickness of the latter are a development of a few irregularly shaped brown spots adjacent to the veins and also a narrow band of leaf marginal necrosis with slight inward curling of the leaflet margins. Restricted growth of the primary shoot gave rise to axillary growth which was affected by toxicity (in the same manner as above), resulting in a bushy habit of growth.

Vigour of growth and growth rhythm. It is not very vigorous in the seedling stage. It comes away quickly in the spring, a few weeks earlier than *D. intortum* and two months earlier than *Macroptilium atropurpureum* (Whiteman and Lulham, 1970); also gives an earlier response in spring than glycine. Gives vigorous growth in early summer, again in late summer and autumn, then declines (see Figure 55).

Response to photoperiod. Short days induce flowering in April to May

(Queensland); seed usually sets in June, before winter frosts (Davies and Hutton, 1970). Whiteman and Lulham (1970) found that *D. uncinatum* flowered in 181 days when planted in October and in 132 days when planted in September in southeast Queensland. It is more shade-tolerant than *D. intortum* and *Macroptilium atropurpureum*.

Compatibility with grasses. Combines well with *Setaria* spp., *Panicum* spp., *Paspalum* spp., *Pennisetum clandestinum*, *Chloris gayana* and *Pennisetum purpureum*. Bryan (1968a, b) showed that at Beerwah in southeast Queensland it invaded pastures of *Digitaria decumbens*, *Chloris gayana* and four species of *Paspalum*.

Ability to compete with weeds. When established, it competes fairly strongly.

Tolerance of herbicides. Fairly tolerant to 2,4-D and, even from $3^1/_2$ weeks of age, 1.65 kg acid equivalent per hectare can be used. Early resistance to diquat is not as good as with greenleaf desmodium, but at 0.275 kg of cation per hectare it is safe to spray established swards (Bailey, personal communication).

Nitrogen-fixing ability. Whiteman (1969) showed strong nodule development and nitrogen fixation during summer; although a few nodules remained during the winter, they were not active in nitrogen fixation. Hence, the amount of nitrogen fixed is governed by the length of the winter dormant period. Suttie and Moore (1966) showed that when *Pennisetum purpureum* and *Tripsacum laxum* were grown with *D. uncinatum*, their crude protein contents rose by 18 and 51 percent and that of the mixture of grass and legume by 63 and 112 percent respectively. Henzell *et al.* (1966) found that a stand of *D. uncinatum* in Queensland provided approximately 110 kg N/ha/year in the available forage, and that a similar amount was added to the soil. Suttie (1968) estimated that *D. uncinatum* contributed 160 kg fertilizer N in association with grasses in Kenya; in Malawi the estimate was 90 kg/ha where the legume comprised 30 percent of the sward. Over a four-year period on a nitrogen-deficient soil, Bryan (1962) obtained a mean yield of N in the plant tops of 176 kg/ha/year in grass/*D. uncinatum* swards in southeast Queensland. Henzell (1962) in sand cultures found that about 80 percent of the N fixed by *D. uncinatum* was in the plant tops. Richards and Bevege (1967) found that *D. uncinatum* had an adverse effect on exotic pines but with native *Agathis robusta* and *Araucaria cunninghamii* it trebled the amount of surface litter, doubled tree height and increased dry matter tenfold. Whiteman (1969) has published figures for nitrogen fixed per plant of *D. uncinatum* during the season.

Response to defoliation. Whiteman (1969) showed that cutting or grazing to

291

a height of 5 cm caused *D. uncinatum* to disappear from a Rhodes grass/ *Desmodium* sward.

Grazing management. *D. uncinatum* pastures should be well established before grazing is begun. Cattle eat the young terminal shoots, the leaves and finally the stolons if grazing is continued. As new growth arises, the axillary buds should be protected to allow for rapid regrowth. Weed competition should be reduced in the early stages by slashing, and grazing should commence after about 16 weeks. Top growth should be eaten down rapidly and the stock removed for four to eight weeks between grazings.

Whiteman (1969) showed that close grazing by sheep to a 5-cm height eliminated *D. uncinatum* from Rhodes grass/*D. uncinatum* pasture in four years. In New South Wales, Australia, *D. uncinatum* pastures have survived grazing for at least five years (Bryan, 1969) and at Beerwah, Australia, for eight years (Bryan, 1968).

Response to fire. Will recover from moderate fires if it is well established.

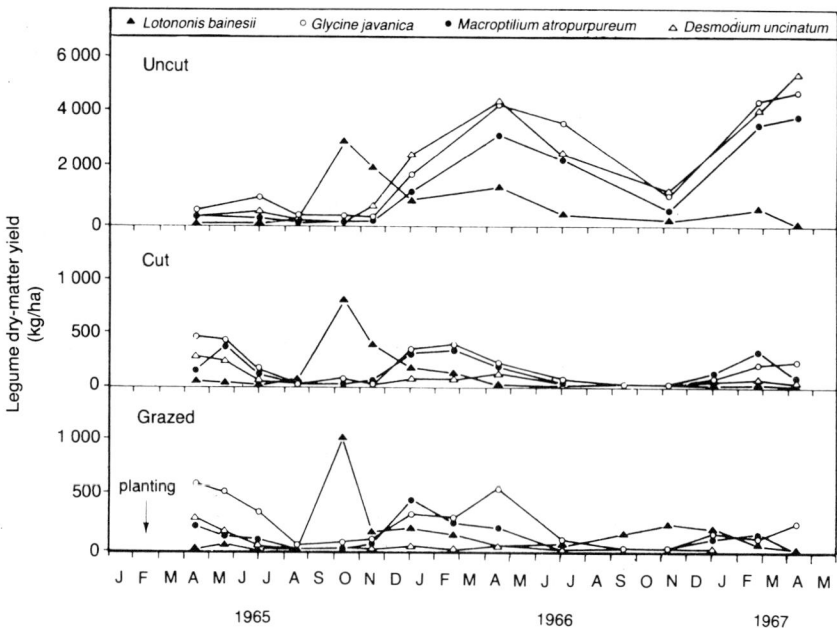

Figure 55. Seasonal changes in dry-matter production of the sown legume components in the grazed, cut and nil-treatment swards

292

Breeding system. Self-fertile; pollination is improved by tripping the flowers. Chromosome number 2n = 22 (Pritchard, personal communication). It has been successfully crossed with *D. intortum* (Hutton and Gray, 1967).

Dry- and green-matter yields. Whiteman (1969) recorded a dry-matter yield of 4 670 kg/ha/year at Samford, southeast Queensland. Bryan and Shaw (1964) measured 730 kg/ha/year of *D. uncinatum* in association with grass which yielded 1 650 kg/ha/year over a period of four years in southeast Queensland. In Swaziland, l'Ons (1968) obtained 3 700 kg/ha of dry matter from *D. uncinatum*; Risopoulos (1966, Zaire) 15 tonnes of dry matter per hectare in a total of five cuttings. Anderson and Naveh (1968) obtained an average yield from five sites in northern Tanzania of 7 020 kg/ha.

Suitability for hay and silage. Luck (personal communication) recorded some useful hay at Cooroy in southeast Queensland. There does not appear to be any record in the literature of the use of *D. uncinatum* for silage.

Value as standover or deferred feed. In frost-free environments, some stand-over feed is retained, but where frosts are severe, losses can be heavy. Whiteman (1969) recorded a 71 percent loss of dry matter over the winter compared with the amount present at the end of summer. The lower leaves tend to drop when the sward becomes dense.

Feeding value. Cattle graze the plant readily and intake is good, even after frosting (Milford, 1967). Leaves and shoots are mainly eaten (Bryan, 1966).

Chemical analysis and digestibility. Luck (personal communication) recorded a crude protein content of 15.88 percent for hay from a first cutting (12 September 1967) and 18.75 percent from the second cutting (27 September

TABLE 14.4 **Chemical and compositional analysis of *Desmodium uncinatum* growing at Lawes and Beerwah**

Date of last cut	Date of harvest	Location	Dry matter	Crude protein	Crude fibre	Ether extract	N-free extract	Ash
			(%)					
December 1957	26 April-6 May 1958	Lawes	22	18.2	32.5	4.6	41.7	7.6
	14-24 June 1958	Lawes	26	11.8	41.3	2.6	37.8	6.8
	6-16 July 1958	Lawes	43	10.9	43.4	3.2	36.6	7.9
Unknown	11 April 1962	Beerwah	29	13.1	36.2	2.0	43.3	5.4
November 1962	22 January 1963	Beerwah	43	11.2	36.4	2.0	45.7	4.7
	19 February 1963	Beerwah	19	13.2	38.2	2.5	41.7	4.4
	17 March 1963	Beerwah	35	12.7	43.3	2.0	38.4	3.6

TABLE 14.5 **Compositional analysis of *Desmodium uncinatum***

Stage of growth	Dry matter	Organic matter	Crude protein	Crude fibre	Ether extract	Nitrogen extract	Starch equivalent	Total digestible nutrient	Intake of		Nitrogen retention
									Dry matter	Digestibility of dry matter	
			(%)						(g/kg W$^{0.75}$/day)		(g/day)
(1)	54.1±0.3	55.0	67.5	33.5	71.2	70.4	39.2	59.9	56.3±[1]1.2	30.4±0.6	+5.90
(2)	50.8±0.5	51.8	53.7	45.2	57.7	60.2	26.5	51.0	57.4±0.2	29.2±0.2	+0.92
(3)	53.0±0.2	53.9	51.5	48.5	23.3	62.8	25.9	47.1	52.8±0.2	28.0±0.1	+0.60
(4)	48.5±0.9	49.2	51.5	41.4	38.0	55.5	25.8	47.5	50.8±3.7	24.6±1.9	+3.24
(5)	47.5±0.6	47.8	50.0	41.4	27.8	53.4	25.2	46.5	30.8±0.4	14.6±0.1	+0.32
(6)	47.3±0.4	48.0	54.9	45.1	51.3	51.4	25.7	48.7	62.2±3.8	28.4±1.8	+6.79
(7)	50.2±0.8	51.2	50.2	52.6	45.8	50.1	25.1	50.9	43.3±4.3	22.7+1.8	

[1] Standard error of mean.

(1) Growing actively, leafy, flowering — Lawes, Australia. (2) Mature, seeding, leafy — Lawes, Australia. (3) Heavily frosted, mature, seeding, dry leafy — Lawes, Australia. (4) Mature, green, leafy, few flowers dried before feeding — Beerwah, Australia. (5) Mature, leafy, but 30 to 40 percent leaf dry, fed dry — Beerwah, Australia. (6) Vigorous growth after rain, leafy, fed dry — Beerwah, Australia. (7) Leafy, leaves starting to dry, fed dry — Beerwah, Australia.

1967). Crude fibre percentage was 42 percent. Milford (1967) gives detailed analyses and digestibility figures for *D. uncinatum* grown on a heavy black clay at Lawes and a sandy soil at Beerwah, at different stages of growth (see Tables 14.4 and 14.5). Mears, Murtagh and Wilson (1964) quote Holder's figure of 54.5 percent digestibility of the crude protein, which was 15.5 percent of the dry matter.

Palatability. Not high; stock take some time to get used to it.

Toxicity. None recorded from grazing. Hutton and Coote (1966) found 3.6 percent tannin in the leaves. Bindon and Lamond (1966) found no toxic symptoms in mice fed on leaves and seeds. Minute amounts of oestrogens have been found in *Desmodium* species but not enough to cause adverse effects in the grazing animal (Bryan, 1969).

Seed harvesting methods. The greatest amount of seed is obtained by mowing when 50 percent of the seed is ripe, allowing it to dry in a swath for 10 to 14 days and then threshing. Wind is a danger to both *Desmodium uncinatum* and *D. intortum* with the plants vulnerable to pod shatter while seed heads are standing; they are protected when laid down in a swath and any seed drop will be caught in the crop. Satisfactory yields can be obtained from direct heading in the field. Use a slower drum speed for silverleaf desmodium — just sufficient to thresh the pods without undue seed cracking (about 200 rpm; less than for greenleaf desmodium). Use an adjustable sieve on a fairly close setting or a 2.5 mm punched-hole sieve.

Seed yields. Suttie and Ogada (1967) obtained 330 kg/ha cleaned seed in Kenya. In Queensland, 220-275 kg/ha are harvested.

Minimum percentage required for commercial sale. Seventy percent germination and 94.5 percent purity are the standards required in Queensland.

Cultivars. There is only one cultivar in commercial use — cv. Silverleaf.

Diseases. It is susceptible to little-leaf (Hutton and Grylls, 1956). In Georgia, United States, Wells and Forbes (1963) found it susceptible to anthracnose caused by *Colletotrichum dematium* f. *truncata*.

Pests. It can be severely damaged by the Amnemus weevil (*Amnemus quadrituberculatus*), the adults of which feed on the foliage in summer and the larvae on the root tissue in winter. In two experiments at Wollongbar, New South Wales, Mears (personal communication) found that the annual dry-matter production of *D. uncinatum* and *D. intortum* declined from 4 400 to 6 600 kg/ha to 165 kg/ha over three years. The foliage can be sprayed with dieldrin at 1.1 kg/ha acid equivalent, but residues may be a problem. White-fringed weevil (*Graphognathus leucoloma*) causes similar (but less) damage. Leaf-eating beetles, caterpillars and grasshoppers are common in moist climates. DDT at 0.55 kg/ha active ingredient can be sprayed onto the foliage to control leaf-eating caterpillars, but after spraying an interval of 30 days should elapse before the pasture is grazed. A pod-borer attacks the pods in Brazil. Colbran (1963) recorded that the nematodes *Meloidogyne javanica* and *Radopholus similis* attack the roots in Queensland, though damage is not severe.

Main attributes. A robust grower which spreads readily and persists well in pastures, it commences growth early in the spring and has a long growing season. It is adapted to a wide range of soils and fixes nitrogen effectively.

Main deficiencies. It has lower digestibility and drought tolerance than *D. intortum*.

Figure 56. *Desmodium uncinatum,* with distinctive foliage, flowers and pods

296

Dolichos spp.

Dolichos sericeus E. Mey ssp. *formosus*

Synonym. *D. formosus* A. Rich.

Description. Robust, climbing herb, wide-twining stems, finely downy when young. Three ovate leaflets, acuminate at the apex, smooth on both surfaces when mature; central leaflet 5 to 8 cm long. Flowers bluish pink, on 3, 4 or more flowered racemes with spreading peduncles 5 to 7.5 cm long. Pod much reflexed, blunt at the apex, 5 to 7.5 cm long, smooth or slightly hairy, eight- to ten-seeded (Andrews, 1952).

Distribution. Widespread in Africa, extending from the Republic of Cameroon at an altitude of 1 525 m through the Sudan, Ethiopia, Uganda, Kenya, Tanzania to Swaziland and Zimbabwe (Hutchinson and Dalziel, 1954).

Characteristics. Van Rensburg (1967) found it very productive in the first year of establishment at Mount Makulu, Zambia, declining in vigour thereafter. From two cuts in January and July 1965 he harvested a total of 6 600 kg DM/ha, and with two further cuts in 1966 the gross yield was 2 285 kg/ha. Crude protein percentage varied from 11.25 at the second cut to 18.95 percent in the first harvest. Crude protein yields were 858 kg/ha in 1965 and 354 kg/ha in 1966.

In a grazing trial, cattle ate 44.74 percent of the dry matter. The stem constituted 55 percent of the dry matter and contained 11.75 percent crude protein, and the leaf yield was 2 470 kg DM/ha with 14.25 percent crude protein.

Naveh and Anderson (1966) obtained more than 15 400 kg DM/ha from a mixed sward of *Panicum maximum/D. sericeus* pasture. Anderson and Naveh (1968) found that *D. sericeus* yielded best of the legumes tried in northern Tanzania. Seed number T 225 was planted in rows 0.6 m apart at about 8 kg/ha. It averaged 20 683 kg DM/ha/year at five sites over periods ranging from 13 to 23 months. It also showed good tolerance of dry conditions. *D. sericeus* is adapted to the heavy black cracking clays (*mbuga*) of

297

northern Tanzania, where *Panicum coloratum* var. *makarikariensis* grows well. At Marandellas in Zimbabwe it is the most vigorous and cold-tolerant legume grown.

Galactia spp.

Galactia filiformis Benth.

Description. Twining herb, leaves oval, oblong, hairy on the undersurface, the rachis of the inflorescence large, with flowers solitary and scattered (Pittier, 1944).
Distribution. Native to Central America.
Characteristics. Is showing some promise in north Queensland (Gartner, 1968).

Galactia striata (Jacq.) Urb. (see Colour plates XVIII, XIX)

Synonyms. *Glycine striata* Jacq.; *Glycine velutina* Bert.; *Galactia cubensis* H.B.K.; *Galactia berteriana* DC.; *Galactia brevistylla* Schlecht.; *Adonia refusa* Rose.
Common names. Galactia (Brazil), Frijolillo (Panama).
Description. A slender, much branched, very variable, twining, climbing perennial legume with stems to 1.5 m long or more. Three leaflets, rounded to obtuse, acute at the tip, 2 to 8 cm long, central leaflet larger than the laterals, slightly more densely puberulent below than above. Stipules lance-deltoid, caudex about 1 to 1.5 mm long. Petioles 1.5 to 3 cm long. Racemes slender, 1 to 24 cm long, flowers from bright pink fading bluish or mauve to purple, few, subsessile to pedicels 1 to 2 mm long. Calyx 4 to 5 mm long,

298

densely puberulent. Petals 8 to 10 mm long. Pods 4 to 7 cm long, 4 to 9 mm wide, narrowly marginate when mature, minutely and moderately puberulent, elastically two valved. Kidney-shaped seeds, 3 to 4.5 mm long and about 2 mm wide, dull black to mottled pinkish brown (Shreve and Wiggins, 1964; Adams, 1972).

Distribution. Native from Mexico south of Sonora State and most of Central America to the West Indies, south through Venezuela to Minas Gerais, Brazil.

General features. A promising legume, native to and being widely tested in South America, especially in Brazil. It has performed well on deep, latosolic soils at Nova Odessa in São Paulo State where it has yielded better than siratro, stylo and centro. Also at Nova Odessa, a guinea grass/galactia mixture yielded as much as guinea grass by itself fertilized with 180 kg/ha of N. Siratro, stylo and centro produced yields ranging only from 40 to 90 kg/ha N when mixed with guinea grass in the same trial. Crude protein, phosphorus and calcium yields were higher for guinea grass combined with any of the four

Figure 57. Galactia striata (**Source:** Pittier, 1944)

299

legumes than for guinea grass by itself fertilized with 225 kg/ha N (de Mattos and Werner, 1979).

Galactia has also equalled siratro and outyielded centro and glycine over four years in Herculandia and Sao Manuel in Brazil (Tosi *et al.*, 1979), while in five locations in Ceara State, galactia, siratro and *Clitoria ternatea* were able to survive the dry season, had high nutritive values and resistance to pests and diseases and produced a large amount of dry matter and seed when sown into native pasture (Gurgel and Fernandes, 1980). At Felixandria, Minas Gerais galactia was among the highest yielding of cight legumes tested, together with siratro and common stylo, giving a fourfold increase in wet-season pasture yields and higher dry-matter yields and P concentrations over the dry season than the native pastures (Andrade, 1981). Galactia is being studied as a native legume also in Costa Rica (Perez-Guerreroz, 1980). It is adapted to the well-drained savanna ecosystem in Colombia (CIAT, 1979).

Rhizobium *relationships*. Galactia nodulates well, with large spherical nodules scattered along the lateral roots. On *cerrado* soils in Brazil, the addition of 75 kg/ha N at planting as well as inoculation with commercial inoculum produced the highest galactia yields (Vargas and Suhet, 1981).

Suitability for hay and silage. In haymaking trials in Brazil, galactia and centro were the fastest drying of the nine legumes tested. While galactia green matter had a high leaf to stem ratio, it lost the most leaves during haymaking (de Moura *et al.*, 1975). At San Miguel, Brazil, it was studied for silage-making together with centro, siratro and glycine. It was concluded that none were really suitable for silage making, ensilage could only be made by wilting at an early stage of growth and that the addition of a preservative, with a high sugar content, would be required to produce a good quality silage (Tosi *et al.*, 1975).

Value as standover or deferred feed. Galactia has been tested as a standover dry season forage on a red-yellow latosol in the Brazilian *cerrados*. It was considered to have considerable potential for this purpose. It has better dry-season growth than most alternative species and apparently retains its quality better as a result. Yields available in the critical August-September period over two years ranged from 1 000 to 5 000 kg/ha dry matter depending on the date grazing was deferred and the year. Leaf contributions to the dry matter decreased from an initial 75 percent to about 35 percent 168 days after deferment of grazing began (Vera *et al.*, 1983).

Feeding value. In the Brazilian deferment study, crude protein over the dry season averaged 15 percent in the standing herbage. The crude protein content of the leaves dropped from 32 percent at 28 days of age to about 20 per-

cent at 112 days and then remained constant. In the stems, it was 15 percent initially, stabilizing at 10 percent at 84 days. The *in vitro* dry-matter digestibility of the available forage fell slowly over the dry season, being in the order of 55 to 60 percent over the most critical period. *In vitro* dry-matter digestibility of the leaves stabilized around 65 percent after 56 days (Vera *et al.*, 1983).
Main references. De Mattos and Alcantara (1976); Vera *et al.* (1983); Perez-Guerreroz (1980).

Glycine spp.

Glycine falcata **Benth.**

Description. A hardy perennial legume with short, erect, ascending or decumbent, non-twining stems arising from a fibrous, woody root-stock. Leaves digitately trifoliate. Petioles 1.5 to 7 cm long, stoutish. Stipules lanceolate, coriaceous, strongly several nerved. Leaflets oblong to long-lanceolate or occasionally oval, grey-green, thick and coriaceous, 1 to 6 cm long, 5 to 20 mm wide, obtuse, usually apiculate, tapering at the base, strigose beneath, sparingly so above, conspicuously net veined beneath. Petiolules very short (1 mm or less), stout and hispid. Racemes axillary, much more so than the leaves, rather loosely two- to ten-flowered, 1 to 6 cm long, on stout peduncles 2 to 12 cm long. Flowers 5.5 to 6.5 mm long on stoutish pedicels (1 to 1.5 mm long) that are shorter than the calyx, inserted singly on the rachis. Calyx 3.5 to 4 mm long, the teeth lanceolate to subulate, equalling or slightly exceeding the tube, the upper pair free above the middle. Corolla light blue to pale mauve, the standard narrowly obovate, slightly exceeding the wings. Keel much shorter than the wings. Pods reflexed, falcate, 15 to 20 mm long, 4 mm wide, biconvex, two- to three-seeded, the valves rigid, with a rigid apiculation, 2 mm long. Seeds oblong to ovoid-oblong, sometimes truncate at

301

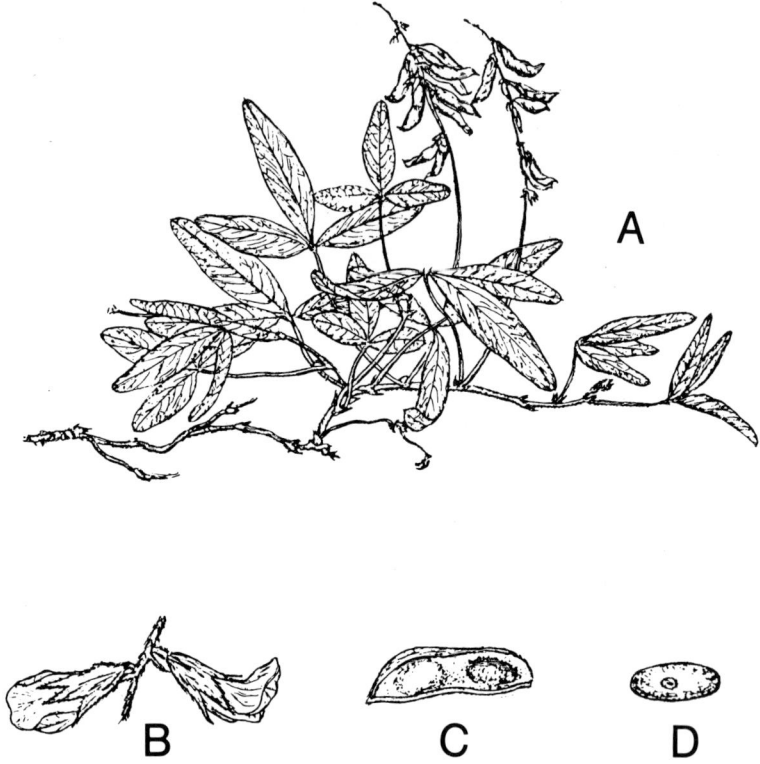

Figure 58. *Glycine falcata.* **A**-Habit **B**-Flowers **C**-Pod **D**-Seed

one end, 4 to 4.5 by 2 mm, olive-green to brown mottled with purple, with a smooth, glossy or dull surface (Hermann, 1962).

Distribution. Native to northern Australia, Queensland and South Australia and the northern interior of New South Wales.

Main features. Grows only from spring till autumn, whenever rain occurs. Quite a drought-hardy, persistent plant, occurring in the arid interior and especially in the semi-arid grasslands on heavier-textured, grey clay soils of the Mitchell grass (*Astrebla* spp.) downs. The native legumes on these soils, of which *G. falcata* is a major one, may contribute only 100 to 200 kg/ha of dry matter a year, but this is of particular importance because of its high quality.

G. falcata was one of the least affected of the native *Glycine* spp. and *Neonotonia* cultivars when artificial salinity was imposed at the five-trionfoliate leaf stage for ten days. It was comparable with *N. neonotonia* cv. Cooper except for lower N content due to salinity but had higher P contents at control levels than the *Neonotonia* cultivars (Wilson, Haydock and Robins, 1970).

Both aerial and subterannean seed pods are set. The aerial pods shatter very easily.

G. falcata is one of the more promising of the native Australian glycines for possible cultivation, though this has not as yet been attempted.

Glycine tabacina (Labill.) Benth.

Synonyms. *Kennedya tabacina* Labill.; *Leptolobium elongatum* Benth.; *L. tabacinum* (Labill.) Benth.; *Leptocyamus elongatus* Benth.; *Desmodium novo-hollandicum* F. Muell.

Common names. Pea glycine; variable glycine; slender sweetroot (Australia).

Description. A prostrate perennial legume. Stems slender and elongated, usually creeping or trailing, occasionally twining, retrorsely strigose to glabrous, the vestiture usually white. Stems arising from a woody, often thickened root-stock. Leaves pinnately trifoliate. Petioles 5 to 45 mm long, the rachis 1 to 4 mm long. Stipules deltoid to oblong-lanceolate, 1.5 to 3 mm long, obtuse to acuminate, coarsely several ribbed, sparsely strigose to glabrous. Leaflets thin, generally strigose beneath, sparingly strigose to glabrous above. Terminal leaflet often larger and longer than the laterals. Leaflets of the lower leaves usually broadly obovate to oval, 10 to 15 × 8 to 14 mm, obtuse to trun-

303

cate, occasionally emarginate, apiculate or acute, tapering at the base, often prominently net veined beneath. Leaflets of the upper leaves usually elliptic-lanceolate to narrowly oblong-lanceolate or linear, 7 to 50 mm long, 3 to 7 mm wide, acute to blunt and apiculate, abruptly tapering to rounded at the base. Petiolules 1 to 2 mm long, hirsute. Occasionally all the leaves oval to broadly oblong, 20 to 40 mm long, 10 to 20 mm wide. Racemes in the upper axils, slender, elongate, 2 to 8 cm long, loosely 4- to 12-flowered on sparsely strigose to glabrous peduncles, 2 to 12 cm long. Flowers in the lower axils often solitary or in few-flowered fascicles. Flowers 6 to 8 mm long on pedicels 0.5 to 2 mm long and frequently subsessile. Calyx 3 to 4.5 mm long, usually sparsely strigose to glabrous, occasionally densely strigose or short-hirsute, the teeth from broadly to narrowly lanceolate, the same length or shorter than the tube. Corolla blue to violet or purplish. Standard obovate reflexed to ascending, slightly longer than the obovate-oblong wings which conspicuously exceed the keel. Pods linear, 14 to 30 mm long, 3 to 3.5 mm wide, compressed, the persistent style variously hooked or curved, three- to six- seeded. Seeds 1.75 to 3 × 1.75 mm, oblong to ovoid, often truncate at the ends, smooth and dull to glossy, muriculate, purplish black in colour (Hermann, 1962).

Distribution. Pea glycine is native from southern Australia through the tropics of northern Australia, South Pacific islands such as Tonga, Fiji and New Caledonia to as far north as Fujian Province, China. It is found in rocky pastures, clearings, dry hillsides and sandy slopes (Hermann, 1962).

General features. Pea glycine is probably the commonest and most widely distributed of the native legumes that are of forage value in eastern Australia, being almost cosmopolitan in its occurrence. It does particularly well on sandy soils but is also found on the black basaltic clays. Intermediates between *G. tabacina* and *G. clandestina* are frequent where the two species occur together.

No toxic features have been recorded for pea glycine and it is readily eaten. It is not however highly productive, but is probably selectively grazed in pastures when stock are present. It responded to applied phosphate at Armidale, New South Wales, Australia, but not as strongly as naturalized legumes or *Lespedeza cuneata*. It did not respond to applied sulphur (Begg, 1963).

Pea glycine was intermediate between *G. falcata* and *G. tomentella* in reaction to artificial salinity applied for ten days in the five-trifoliate leaf stage (Wilson, Haydock and Robins, 1970). It is known to nodulate (Bowen, 1956).

304

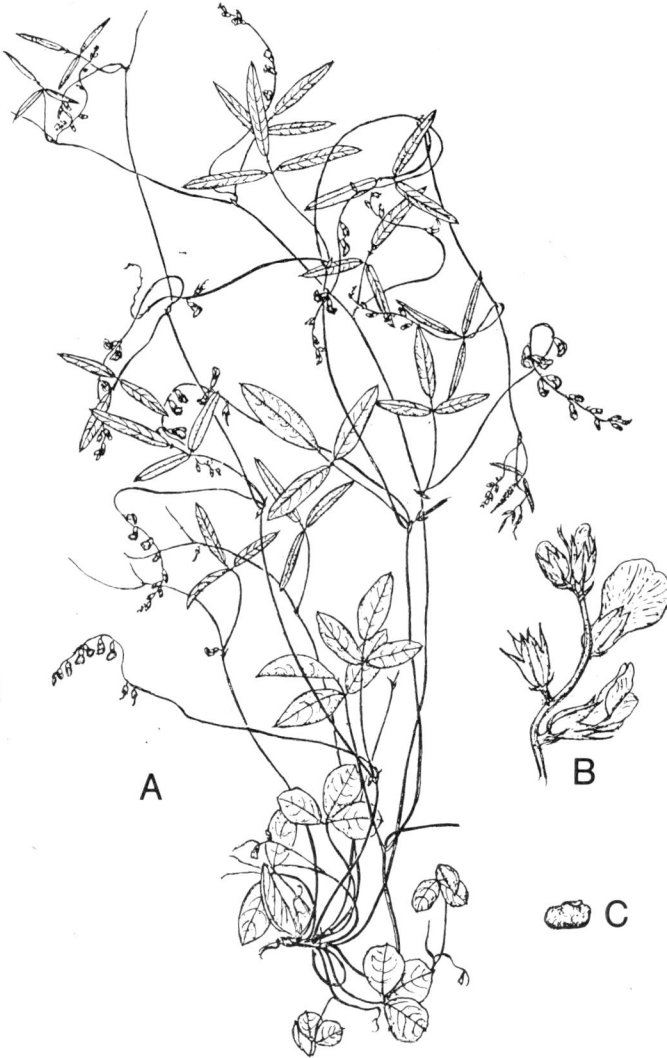

Figure 59. *Glycine tabacina.* **A**-Habit **B**-Portion of inflorescence **C**-Seed

Glycine tomentella **Hayata**

Synonyms. G. *tomentosa* (Benth.) Benth., non L.; *Leptolobium tomentosum* Benth.; *Leptocyamus tomentosus* Benth.

Common names. Woolly glycine, woolly sweetroot vine (Australia).

Description. A perennial legume with prostrate stems that trail or climb over shrubs, usually stouter than those of *G. tabacina* (2 mm in diameter), tomentose-villous, the vestiture usually tawny. Leaves pinnately trifoliate, their villous-hirsute petioles 1 to 9 cm long. Rachis 2 to 9 mm long. Stipules ovate to lanceolate, 2 to 3 mm long, coarsely several ribbed, densely pubescent to glabrous. Leaflets oblong, elliptic to oval or ovate, thin, 14 to 65 mm long, 8 to 35 mm wide, the terminal generally longer than the lateral, obtuse and usually mucronate, sometimes emarginate, abruptly tapering or rounded at the base, more or less velvety tomentose to strigose on both surfaces. Petiolules 0.5 to 1.5 mm long, hirsute. Racemes in the upper axils, 0.75 to 3 cm long, compactly to loosely 7- to 15-flowered, on hirsute peduncles 3.5 to 10 cm long. Flowers frequently solitary, geminate or in few-flowered fascicles in the lower axils, 5 to 7 mm long, on hirsute pedicels 0.5 to 1.5 mm long, inserted singly on the rachis. Calyx 3.5 to 6 mm long, hirsute to densely hirsute or rarely strigose, the teeth lanceolate and usually narrowly attenuate to setaceous at the apex, longer than the tube (2.5 to 3.5 mm long), the upper pair joined to about the middle. Corolla purple to mauve or reddish. Standard obovate, ascending, longer than the obovate, oblong wings, which exceed the keel. Pod linear, 12 to 23 mm long, 2.75 to 3.5 mm wide, hirsute to glabrous, compressed, three- to seven-seeded. Seeds 1.75 to 2.25 × 1.25 to 2 mm, short-oblong to quadrate, usually truncate at both ends, subcylindrical at maturity; smooth, muriculate or papillose; purplish black in colour (Hermann, 1962).

Distribution. Woolly glycine is native from northern Australia, Queensland and northwestern New South Wales, through the islands of New Guinea, New Caledonia and the Philippines to southern China and Fujian and Taiwan provinces. It is found on river banks, dry slopes and in open woods (Hermann, 1962).

General features. Woolly glycine is widely distributed and highly valued as a native legume throughout the subcoastal areas and interior of Queensland. It is particularly conspicuous in good seasons on the heavy, black basaltic clay soils, to which it is well suited. It also occurs among the strand vegetation and along the rocky shores of the Capricorn Coast near Rockhampton, Queensland. When the basaltic downs are cultivated, woolly glycine can grow to weed proportions in the fallows.

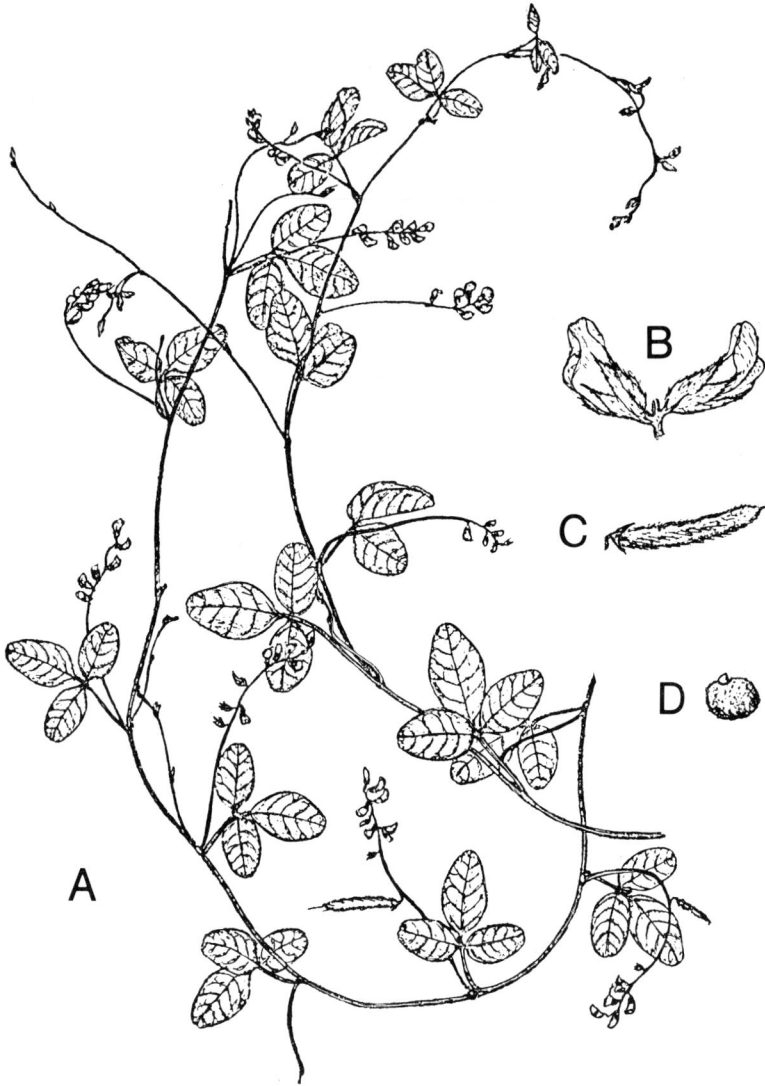

Figure 60. *Glycine tomentella.* **A**-Habit **B**-Flower **C**-Pod **D**-Seed

Woolly glycine nodulates abundantly when growing conditions are good, but the nodules are absent in exceptionally wet or dry times as well as after winter frosts. It shows some ability to retain nodules under adverse conditions, but possibly only because the nodules are more resistant to decay than those of other legumes (Diatloff, 1967b).

Woolly glycine seeds quite profusely, but over a very long period each year. Since the pods shatter readily, conventional harvesting would not give high yields. Vacuum harvesting may be possible, although the natural stands grow on unsuitable soils.

When subjected to artificial salinity for ten days in the five-triofliate-leaf stage, *G. tomentella* showed the least resistance of the three native glycines tested (Wilson, Haydock and Robins, 1970).

Indigofera spp.

Indigofera hirsuta L. (see Colour plate XX)

Common name. Hairy indigo.

Description. An erect or spreading annual, up to 1.5 m tall but generally shorter. Stems cylindrical or slightly ridged, densely clothed with long, fine, spreading, grey or reddish-brown pubescence. Stipules linear, setaceous, up to 1 cm long. Five to seven leaflets, occasionally nine, elliptical-oblong, up to 40 mm long and 25 mm wide, the terminal one rather longer than the lateral, pilose on both surfaces. Inflorescence a dense, many-flowered raceme, hirsute, 20 to 30 cm long including a peduncle more than 25 mm long. Bracts linear-lanceolate, up to 25 mm long. Pedicels around 1 mm long, reflexed in fruit. Calyx stiff, brown and hirsute, about 4 mm long, divided almost to the base into linear, setaceous lobes. Corolla white pubescent outside, brick-red or rose inside. Pods straight, rather tetragonal, with well-developed sutures, 12 to 20 mm long about 2 mm wide, thickly hirsute; many of the hairs, especially the dorsal ones, usually brown. Six to nine seeds, cuboid, angular, strongly pitted (Gillett, Polhill and Verdcourt, 1971).

Distribution. Hairy indigo occurs naturally from Senegal to the Sudan and the Congo, Zambia, Mozambique, Angola and Madagascar in Africa; also natural in southern Asia, northern Australia and Queensland, Australia. It has been naturalized in parts of tropical America.

Season of growth. A summer-growing annual with strains of varying maturity available.

Latitudinal limits. Grown as far south as 30°S in Argentina and as far north as southern Georgia (c. 32°N) in the United States.

Altitude range. Grown from sea level to 1 350 m in Africa.

Rainfall requirements. Occurs naturally in areas with 900 to 1 700 mm annual rainfall, rarely less.

Soil requirements. Tolerates soils of low pH and fertility, and thus can be useful for soil improvement. It is, however, best adapted to moderately acidic soils and fertile sandy loams, thriving with little lime.

Rhizobium *relationships.* The native strains in Queensland nodulate with the cowpea cross-inoculation group (Bowen, 1956), while in Florida, United States, no inoculation is necessary (Whyte, Nilsson-Leissner and Trumble, 1953).

Ability to spread naturally. A free-seeding annual without specific *Rhizobium* requirements, hairy indigo can naturalize readily in suitable habitats.

Land preparation and establishment. A fine, firm, weed-free, well-prepared seed bed is ideal. Like most free-seeding annuals, however, hairy indigo should be susceptible to more simple methods of introduction into pasture situations.

Sowing time and rates. McKee (1952) recommends planting from March to late May, but preferably earlier, at 3 to 5 kg of seed per hectare when drilled in close rows and 6 to 10 kg/ha when broadcast on a well-firmed seed bed, the smaller amount for seed crops, the higher figures for hay or green manure.

Number of seeds per kg. 440 000.

Nutrient requirements. Being well-adapted to low-fertility soils, *I. hirsuta* has not shown major responses to applied fertilizer. In Panama, Silvey and Carlisle (1972) obtained no dry-matter-yield response to the application of up to 30 kg/ha Zn or up to 100 kg/ha P, although forage content of both elements was increased. Liming at up to 2 000 kg/ha Ca depressed extractable Zn and P in the soil and increased forage yield but not crown or root yields.

Response to photoperiod. In Taiwan, *I. hirsuta* was one of the plants that Wang (1961) grew under natural day lengths of from 10 hr 19 min in mid-January to 13 hr 42 min in mid June, photoperiods approximately three hours longer and shorter respectively than normal. It responded to short photoperiods by shortening the period to flowering. During flowering, vegetative growth increased.

Nitrogen-fixing ability. In Florida, it has been estimated that the presence of hairy indigo in mixtures is equivalent to the application of 126 kg/ha N/year on pure grass pastures (Kalmbacher, Hodge and Martin, 1980).

Response to defoliation. Kalmbacher, Hodge and Martin (1980) found that plant height at initial harvest was important: regrowth was inhibited at taller heights because regenerative axillary buds were removed during harvesting. Recovery after cuttting was also slow at Carimagua, Colombia, (CIAT, 1978).

Dry- and green-matter yields. In Florida, a mixture of *I. hirsuta* and pangola grass yielded 4.88 t/ha dry matter, while alone the pangola grass yielded only 3.31 t/ha. *Setaria sphacelata* yield increased from 2.24 t/ha DM alone to 4.52 t/ha when mixed with *I. hirsuta* but there was no increase in yield when *I. hirsuta* was added to a *Paspalum notatum* pasture. The *I. hirsuta* mixtures however yielded less than the same grasses with other legumes (Kretschmer, 1970).

Suitability for hay and silage. One of the first uses for *I. hirsuta* in Florida was hay and silage production, but the hay must be made early in the season as the stems become coarse and woody with age. The aftermath from such an early cut can then be grazed. Silage made from it satisfies only maintenance requirements and was not readily eaten (Catchpoole and Henzell, 1971).

Feeding value. Dougall and Bogdan (1966) analysed *I. hirsuta* forage as containing 23.8 percent crude protein, 2.0 percent èther extract, 15.2 percent crude fibre, 46.8 percent nitrogen-free extract, 1.88 percent calcium and 0.37 percent phosphorus, but Kalmbacher *et al.* (1980) showed that crude protein and *in vitro* organic matter digestibility are greatly influenced by time of sampling, declining rapidly as the plant material ages. Early cut material contained 24 percent crude protein.

Toxicity. Early references such as Bailey (1906) refer to suspected poisoning of stock, but these suspicions apparently have not been sustained.

Seed harvesting methods. *I. hirsuta* was initially regarded as somewhat difficult to harvest for seed because the plant dries slowly after the seed matures (McKee, 1952). Modern, high-powered headers should handle it readily. McKee (1952) recommends mowing and drying in the windrow before threshing, although he believed that a header could be used on thin, short stands.

Seed yields. A good seed producer.

Cultivars. Seed was first offered for sale in Florida in 1945 from work begun at Gainesville in 1931. Two distinct types were initially used, one a large, late-maturing strain and the other a smaller type that matured a month earlier (McKee, 1952). No cultivar name has been noted for either line.

Pests and diseases. Resistant to root knot nematode (*Meloidogyne* spp.) and most insects and diseases.

Main attributes. Free-seeding and regenerating readily each year, hairy indigo's tolerance of low soil pH and fertility coupled with its resistance to root knot nematode make it an attractive legume for poor sandy soils in Florida.

Main deficiencies. A rather low-quality forage, the quality of which declines even further as growth advances.

Main references. McKee (1952); Kalmbacher, Hodges and Martin (1980).

Lablab spp.

Lablab purpureus (**L.**) **Sweet** (See Colour plate XXI)

Synonym. Dolichos lab-lab.

Common name. Rongai dolichos, lab-lab bean (Australia), poor man's bean, Tonga bean (England), lubia (the Sudan), batao (Philippines), hyacinth bean (Brazil), frijol jacinto (Colombia), quiquaqua, caroata chwata (Venezuela), poroto de Egipto (Argentina), dolique lab-lab, dolique d'Egypte (France), fiwi bean (Zambia), chicarros, frijol caballo (Puerto Rico), gallinita (Mexico), frijol de adorno (El Salvador), wal (India).

Description. Summer-growing, rampant and vigorously twining herbaceous annual or short-lived perennial. Stems robust, 3 to 6 m, leaves trifoliate;

311

leaflets broad ovate-rhomboid, 7.5 to 15 cm long, thin, acute at apex, almost smooth above and short-haired underneath. Petioles long and slender. Inflorescence lax, fascicled, of many-flowered racemes on elongated peduncles. Flowers white (in 'Rongai') or blue or purple, on short pedicels. Pod 4 to 5 cm long, broadly scimitar shaped, smooth and beaked by the persistent style, containing two to four seeds. Seeds in 'Rongai' buff or pale brown coloured, ovoid, laterally compressed, with a linear white conspicuous hilum, 1.0 cm long × 0.7 cm broad (Barnard, 1967).

Distribution. Widespread throughout the tropics, especially in Africa as a food crop.

Season of growth. A summer-growing annual, biennial, or short-term perennial.

Temperature for growth. Requires warm temperatures for good growth. Does not grow rapidly till December in southeast Queensland, Australia, when temperatures exceed 29°C. It shoots rapidly in the spring from old plants. Minimum temperature for growth is about 3°C (Murtagh and Dougherty, 1968). It is more tolerant of cold than velvet bean (*Mucuna pruriens*). Its frost tolerance is low. It usually seeds late and so early frosts affect it. In Georgia, United States, and Queensland, (Downes, 1966), a breeding programme for earliness of flowering and seed production is in progress.

Latitudinal limits. It extends south to beyond latitude 30°S. It is cultivated in Buenos Aires, Argentina (Burkart, 1952).

Altitude range. Sea level up to 2 000 m (Crowder, 1960), but it prefers the lower elevations.

Rainfall requirements. Used for a food crop in rainfall as low as 400 mm with summer incidence and where deep soils are available. Prefers a rainfall in excess of 750 mm but not above 2 500 mm.

Drought tolerance. It is quite drought-tolerant when established (Luck, 1965b).

Tolerance of flooding. Very poor; it will not grow in wet soils (Luck 1965b; Wilson and Murtagh, 1962).

Soil requirements. Extremely tolerant of soil texture, growing in deep sands to heavy clays, provided drainage is good. It will grow in a wide range of pH, from 5.0 to 7.5. Salinity reduces the plant population and produces chlorotic leaves (Wilson, personal communication).

Rhizobium *relationships*. Lablab does not easily nodulate with native strains of rhizobia, and although it is often not inoculated it is preferable to treat the seed with the cowpea strain CB 756 (Norris, 1967). Diatloff (1967) recorded poor growth on poor sandy soils in southeast coastal Queensland, where

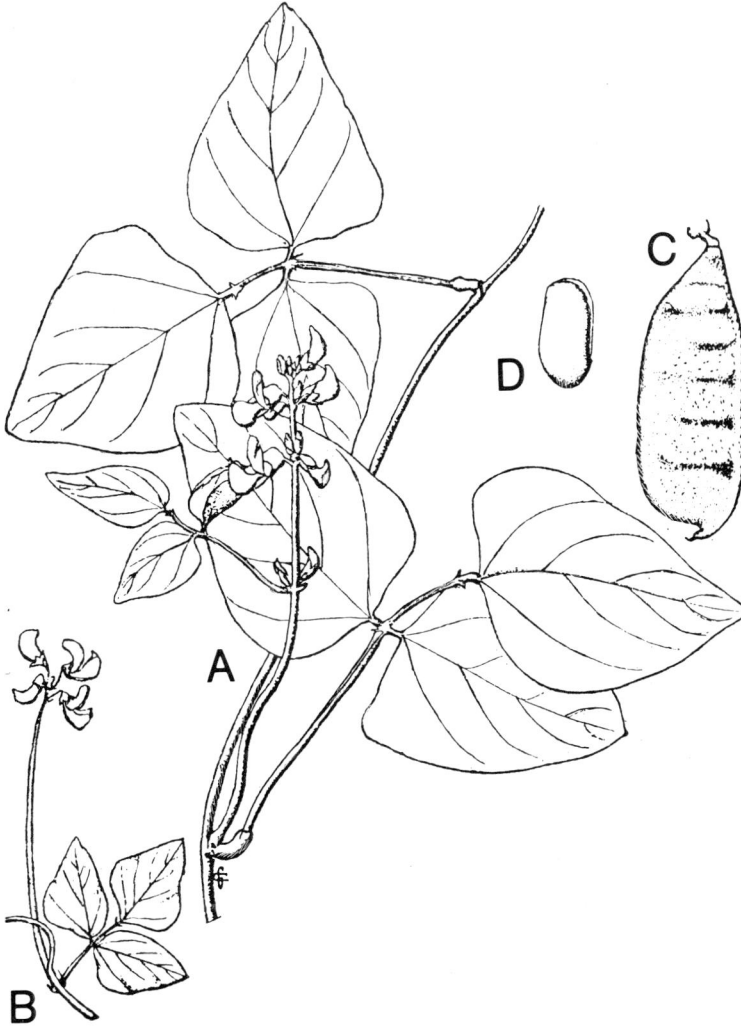

Figure 61. *Lablab purpureus.* **A**-Habit **B**-Flowering stem **C**-Pod **D**-Seed (**Source:** Burkart, 1952)

313

uninoculated plants yielded 203 kg/ha of dry matter compared with 1 611 kg/ha inoculated. Only three out of 25 virgin soils gave good growth without inoculation of seed. Cloonan (1963) found that crown nodules on lablab were pink at four weeks, dark pink at six weeks, and black at 12 weeks and still active. He suggested that this feature might be used as a diagnostic check on successful strain inoculation.

Ability to spread naturally. Will not spread naturally.

Land preparation for establishment. Lablab performs best when drilled into a well-prepared seed bed, but it can establish by broadcasting into roughly ploughed or cultivated land if the seed is covered to some extent.

Sowing methods. It is drilled in 1-m rows into a prepared seed bed, or broadcast onto rough seed beds. In Brazil, it is commonly drilled in with maize at planting or when the maize is 15 cm high, using 20 percent by weight of lablab seed and 80 percent maize, in alternate rows 80 cm apart (Schaaffhausen, 1966).

It does not establish well in natural pastures unless they are cultivated. It can be sod-seeded into pastures (McAdam and Swain, 1969) with adequate fertilizer and preferably inoculated. In Brazil, it is sometimes broadcast from horseback into *P. maximum* pastures where the deep red latosolic surface is loose (Horrell, personal communication). Drilled rows allow interrow cultivation for early weed control. It is often sown with maize and sorghum in alternate rows for silage.

Sowing depth and cover. The seed germinates from sowing as deep as 10 cm, but it is usually sown at 2.5 to 5 cm and harrowed. Hand planting by dibbling in the seed or using a one-row hand machine is also practised.

Sowing time and rate. Five to 7 kg/ha drilled, with a heavier rate of 8 to 10 kg/ha broadcast. Sown in early summer, it will yield up to three grazings; later summer planting yields only one grazing.

Number of seeds per kg. 3 300 to 4 290. Percentage of hard seed is very low.

Seed treatment before planting. Not necessary to break dormancy. Inoculation with a cowpea type is advisable. Pelleting is not necessary unless to protect rhizobia, in which case rock phosphate should be used. Seeding with a neutral fertilizer will also protect the rhizobia. For insect and disease control, treat seed with dieldrin or endrin prior to sowing (to protect from bean fly).

Nutrient requirements. Generally in fertile soils, no fertilizer is necessary. In poor sandy soils, use 250 to 500 kg/ha molybdenized superphosphate and some potash if needed.

Toxicity levels and symptoms. There is evidence of an adverse effect of salinity on lablab (Wilson, personal communication).

314

Response to photoperiod and light. It is a short-day plant. In southeast Queensland, flowering commences in May; but if unfrosted, flowering may continue through the winter into the spring. Cv. Rongai is later flowering than other types. It is sensitive to day length and flowers best with less than 11 hours of daylight, but it requires ample sunlight. In New South Wales, Australia, Murtagh and Dougherty (1968) got full light interception with a canopy of pure lablab. If grown with tall grasses or crops, it can climb to the light.

Compatibility with grasses and other legumes. It is usually sown alone or in widely spaced maize or sorghum rows because of its slow early growth and short life.

Ability to compete with weeds. Excellent when once established, but its early growth is slow and so it should not be subject to weed competition at this stage.

Tolerance of herbicides. No reference in the literature to this effect. Being a plant with broad-leaved, tender foliage, it is probably highly susceptible to herbicide damage.

Vigour of seedling, growth and growth rhythm. Its seedling is vigorous, a little slower growing than cowpeas early in the season. It is easy to establish because of the large seeds. Murtagh and Dougherty (1968) showed leaf yield increasing rapidly in midsummer and continuing up to the first frost (mid-June), whereas stem growth continued into the spring despite the frosted leaves. With a dense growth, the lower leaves are shed. These are lost to grazing, but form an excellent mulch and provide nitrogen on decomposition. In indicating growth rhythm of dolichos, cowpea and velvet bean in northern New South Wales, Wilson and Murtagh (1962) clearly show that the lablab retains its foliage, and therefore its feed value, much later into the winter than the other two (see Figure 62).

Nitrogen-fixing ability. In relation to yield. Besides nodulation from nitrogen fixation, it also supplies large amounts of nitrogen by leaf decay. At São Paulo, Brazil, it is estimated that it provides 220 kg/ha of nitrogen (Lambert, personal communication). Parbery (1967b) obtained dry matter yields up to 44 832 kg/ha in 287 days at the Kimberley Research Station, Australia, which contained 6 279 kg/ha of protein, unfertilized with nitrogen, indicating its extensive nitrogen-accumulating ability.

Response to defoliation. Will not stand heavy grazing of stems, but if only the leaf is taken it will provide two to three grazings in a season. Neme (1970, unpublished) advises that the plant should not be below 25 cm and recovery will take four to five months to give a second cut in Brazil.

315

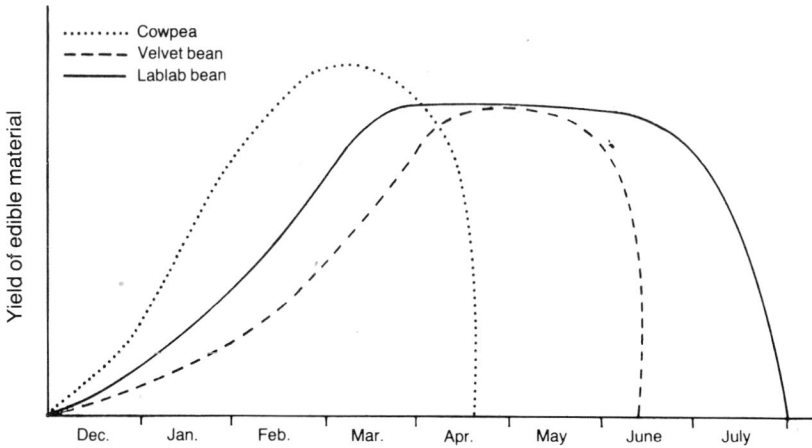

Figure 62. Growth rhythm of legumes in New South Wales, Australia

Grazing management. Graze first in about ten weeks from sowing to remove the leaf only and then remove the animals to allow further leaf to develop (Hamilton, 1969). Do not turn hungry animals onto a sole diet of lablab or bloat may occur, especially with young regrowth. Use a mixed grass/legume diet if possible or spray the material before grazing with an antibloat agent.
Response to fire. It will not tolerate fire.
Breeding system. The flowers are mainly cross-pollinated. Chromosome number $2n = 22$. Patil (1958) has dealt with anthesis and pollination in the field.
Dry- and green-matter yields. Crowder (1960) reported 25 tonnes/ha of green material after four to six months growth in Colombia. In Brazil, 40 tonnes/ha were obtained, and 35 tonnes of mixed maize and dolichos. Van Rensburg (1967) obtained 5 438 kg DM/ha at Mount Makulu, Zambia, of 23.38 and 11.5 percent crude protein for the first and second cut, giving 870 kg of crude protein per hectare. In another experiment, 35.64 percent of the dry matter of a crop yielding 3 374 kg DM/ha was consumed at grazing. The dry matter of the whole crop consisted of 69 percent stem of 8.25 percent protein, and the leaf DM per hectare was 1 031 kg/ha, containing 17.63 percent crude protein. Murtagh and Dougherty (1968) averaged a yield of 4 035 kg DM/ha from three sites on the north coast of New South Wales (lat. 28°50'S), leaf yield averaging 2 094 kg/ha. They felt that a ceiling leaf yield of 2 200 kg/ha

316

can be expected in that environment. Neme (personal communication) calculated that 1 ha of lablab could furnish 1 500 kg protein/ha.

Suitability for hay and silage. Lablab makes excellent hay if the leaf is preserved. The stem is difficult to dry and must be conditioned mechanically to hasten curing. Thurbon, Byford and Winks (1970) made hay of lablab in north Queensland. The material was mown, crushed (conditioned), windrowed and, when dry enough, baled with a pick-up baler.

Ryley (1966) recorded good silage made from lablab alone in trench silos in Queensland. Skerman (1958b) made excellent silage with a mixture of lablab and sorghum, lifting the protein of the sorghum from 4.5 percent alone to 8.1 percent with a 1:2 lablab/sorghum mixture and to 11 percent with a 2:1 mixture. It is often grown with maize for ensiling. Morris and Levitt (1968) recorded the intake and digestibility of lablab silage. The material was ensiled immediately after harvesting, after wilting for two days with and without 3 percent molasses, and after wilting for three days. All silages were satisfactory and readily eaten by sheep.

Value as standover or deferred feed. Excellent if there are no frosts. It flowers late and carries a large body of feed into the winter. Even if frosted, if it has set seed the pods do not dehisce and so there is good feeding value in them alone. In Brazil, the lablab crop planted with maize is fed off with the old maize residues after the maize harvest.

Feeding value. It is excellent for bridging the gap between summer and winter grazing crops and pastures (Luck, 1965b).

● Chemical analysis. French (1937) reported 11.74 percent crude protein, 37.67 percent crude fibre and 39.47 percent carbohydrates with 2 percent CaO, 0.42 percent P_2O_5, 0.36 percent Na_2O, 1.69 percent K_2O and 0.13 percent Cl in lablab hay at Mpwapwa, Tanzania. Some of the leaf had been lost in making the hay. Luck (1965b) reported 25 to 26 percent crude protein in the leaf of lablab cv. Rongai compared with 18 to 23 percent for velvet bean, and the stem crude protein was 9 to 11 percent. Neme (1970, unpublished) recorded 22.17 percent crude protein and 27.44 percent crude fibre on a dry-weight basis. Thurbon, Byford and Winks (1970) found lablab hay to have a protein content of 11 to 14 percent. For weaner calves, it can be cut within ten weeks to give a higher protein; for mature animals it can be cut later. Digestibility of the dry matter of the young plants (77 days) was 61.3 percent and for old plants (140 days) 48.6 percent, and for crude protein 66.4 and 61.7 percent respectively. Morris and Levitt (1968) reported 24.9 percent dry matter, 2.3 percent nitrogen, 30.1 percent crude fibre, 42.6 percent carbohydrates, 1.5 percent Ca, 0.3 percent P and 4.6 percent sugar in green lablab before

317

ensiling. In the resulting silage made from nonwilted and wilted material plus molasses, there was no significant difference in composition. The average dry matter was 38.2 percent, nitrogen 2.4 percent, crude fibre 30.9 percent, carbohydrates 37.9 prcent, Ca 1.8 percent, P 0.3 percent and sugar 1.0 percent with final pH ranging from 4.2 to 4.6.

● Digestibility. Of the organic matter in the silage, digestibility was 49.1 percent (nitrogen 58.4 percent, crude fibre 55.3 percent and nitrogen-free extract 9.2 percent) for cattle. With sheep, digestibility figures were higher. The voluntary intake of sorghum/lablab silage was directly related to the proportion of lablab in the silage.

● Palatability. Green lablab is not usually eaten for up to four days (Murtagh and Dougherty, 1968), after which the cattle become used to it and then eat it readily. The palatability of the hay (French, 1937) and silage (Ryley, 1966) for sheep has been recorded. Sheep ate the silage readily at approximately 1 kg/head/day.

Toxicity. A sole ration of lablab caused a "feedy" flavour in milk, similar to that from clovers and lucerne. Pasteurization and/or homogenization rendered milk acceptable (Hamilton, Fraser and Armitt, 1969). A case of bloat in cattle eating a sole diet of lablab was reported by Hamilton and Ruth (1968).

Seed harvesting methods. Often hand-picked in the tropics. It can be directly headed when the seed is ripe and standing or twining fine-stemmed crops, or it can be mown, cured in the field and subsequently threshed.

Seed yield. Up to 1 000 kg/ha in Brazil and Bolivia. Davies and Hutton (1970) give an average figure of 500 kg/ha. Its seed yield is best at elevations of 1 200 to 1 800 m in Colombia. It does not seed very well in Venezuela.

Minimum germination and quality for commercial sale. Minimum germination of 75 percent, with a maximum of 10 percent hard seed and purity of at least 97.5 percent in Queensland. The seed is germinated under cover at 25°C (Prodonoff, 1968).

Cultivars. Cultivar Rongai originally came from Kenya as CPI 16883. A much earlier-flowering cultivar, cv. Highworth, was introduced to Australia as CPI 20212 from southern India. It has high seed yield coupled with adequate foliage DM production. It has purple flowers and black seeds (those of 'Rongai' are white and light brown).

There are numerous cultivars in the tropics. Selection No. 697 performs well in Brazil.

Diseases. The plant is attacked by numerous diseases throughout the world. In Australia, cultivar Rongai is fairly disease-free. A stem rot caused by

318

Sclerotinia sclerotiorum may attack the plant under wet conditions (Wilson and Murtagh, 1962).

Pests. Colbran (1963) found that the roots of *Lablab purpureus* were attacked by the nematodes *Helicotylenchus dihystera*, *Meloidogyne hapla* and *M. incognita*. It is also attacked by leaf-eating insects.

Main attributes. Its late maturing habit allows it to grow well into the autumn to provide feed between the normal summer species and winter species (e.g. oats). Its large seed allows easy establishment. It provides a high yield of dry matter and is drought tolerant. It is a good pioneer crop to prepare land previously infested with *Axonopus* and *Cynodon* grasses for sowing grass/legume mixtures (Cassidy, 1968).

Main deficiencies. Its short life, low palatability of the stems and its susceptibility to frost.

Performance. At Fazendo el Prata, São Paulo, Brazil (Lambert, personal communication), lablab raised the milk yield by 1.5 kg/day after two days. French (1937) successfully fed lablab hay to sheep at Mpwapwa, Tanzania. In Brazil, on rotation pastures with lablab, pigeon pea and grasses, 47 bulls gained an average of 40 kilograms per head in 63 days (Schaaffhausen, 1966). Hamilton (1969) obtained 9 to 13 litres milk/head/day from cows grazing pure dolichos.

Main references. Luck (1965b); Morris and Levitt (1968).

Lotononis spp.

Lotononis bainesii **Baker** (see Colour plate XXII)

Common name. Lotononis or Miles lotononis.

Description. Herbaceous perennial with slender stoloniferous stems, irregularly branched, up to 1 to 1.5 m long. Plants form a dense pasture up to 60 cm high, or under heavy grazing a more prostrate herbage. Leaves are smooth, digitately trifoliate and borne in groups of three to five at the nodes; leaflets linear, sessile with narrow bases and slightly rounded mucronate points. One or more leaves at each node are large, the central leaflet being about 0.8 to 2.5 cm wide and 2.5 to 5 cm long; the other leaves are much smaller, having a central leaflet 3 mm broad and 1.25 cm long. Petioles 6 mm to 5 cm long. A taproot is developed as well as numerous secondary roots, and similar root systems develop from the nodes of the stolons. Inflorescences are racemes, which may be contracted to dense umbellate heads with 8 to 23 flowers. Flowers small, yellow; pod linear-oblong, 8 to 12 mm long, many-seeded, shattering at the base to release seed. Seeds cream-yellow to magenta-rose, obovoid and asymmetrically heart-shaped, laterally compressed (Barnard, 1967). It was introduced into Australia from the Worcester Veldt Reserve in South Africa by Miles (1952), as CPI 16833.

Distribution. In southern central Africa, in the interior near the Tropic of Capricorn (Oliver, 1868, quoted by Norris, 1958a).

Season of growth. A summer-growing perennial with considerable winter greenness.

Temperature for growth. About 13.5 to 21°C is optimum. Ludlow and Wilson (1967) found that lotononis had a similar dry matter yield, 20 percent lower growth rate and half the leaf area when grown at 20°C compared with 30°C. It probably has a lower growth limit at about 7.5°C. It is quite frost-tolerant (the best of the tropical legumes in this regard) and recovers well in the spring. Remained green in winter at Beerwah, southeastern Queensland, Australia, through minimum grass temperature of − 3.3°C (Bryan, 1961) and down to − 6.5°C at Armidale, New South Wales, Australia.

Latitudinal limits. Bryan (1961) lists its growth limits at 1°N to 33.39°S. It occurs naturally between 23 and 33.39°S. Growth is probably limited to above 600 m.

Rainfall requirements. It requires at least 875 mm annual rainfall and has been most successful between 1 125 and 1 625 mm (Bryan, 1961). It is fairly drought-tolerant — more so than the *Desmodium* spp. and almost as much as siratro. If often survives as seed. It is also tolerant of temporary waterlogging. It survived two floods, one of five days' duration in January and another of six days in March (Wright, 1964).

Soil requirements. It is best adapted to sandy soils such as lateritic podzolic, low-humic gley, red-yellow podzolic and meadow podzolic soils (Bryan, 1961). It will, however, grow on heavy clay (Wright, 1964). In sand culture, it has grown and nodulated quite successfully at pH 4.0.

Rhizobium *relationships.* Lotononis is unable to nodulate with local rhizobia;

it requires an extremely specific type which develops red colonies on culture medium (Norris, 1958a). The current Australian recommendation is CB 376. This culture is ineffective with *L. angolense*, which nodulates with another strain, CB 1323.

Ability to spread naturally. Where natural pasture is kept short, lotononis can spread naturally; Wright (1964) reported that it invaded the tall *Paspalum urvillei* pasture at Coolum in southeastern Queensland. Lotononis seed is also spread in the dung of cattle that graze the plant.

Land preparation for establishment. Will only establish satisfactorily if sown on a fine, well-prepared compact seed bed. Seed is so small that it needs a weed-free planting medium.

Sowing methods. Seed is usually broadcast on the surface. If a roller is available, the seed bed can be rolled both before and after planting; otherwise, a very light harrowing is necessary. The plant will also establish readily from cuttings and turves. A turf 15 cm square is a suitable planting material, each piece being placed about 2.5 m apart in rows 2 m apart, to allow interrow cultivation (Wright, 1964). It can be introduced by oversowing into an *Imperata* pasture after burning but generally it is not established easily in natural pasture. Bryan (1968) found that oversowing failed in *Paspalum commersonii* pasture. Seed must not be sown deeper than 5 to 6 mm. It is fairly cold-tolerant. It is best sown in the autumn when evaporation will not be so severe and soil moisture will be in better supply. The seeding rate is 0.5 to 1 kg/ha.

Number of seeds per kg. 3 300 000. (A seed weighs 0.26 mg, of which 55 percent is seed coat.) The percentage of hard seed is fairly high. Hand-harvested seed has a hard-seed content up to 86 percent. Purple seed has 58 percent hard seed, 14 percent green seed. Purple and brown seed keep their viability better than yellow and green seed (Poulsen, 1966).

Seed treatment before planting. To break dormancy: (*a*) treat with concentrated sulphuric acid (sp. gr. 1.8) for 20 minutes, wash and dry (Prodonoff, 1968); (*b*) hot-water treatment higher than 35°C lowers germination, temperatures between 20 and 30°C improve germination (Poulsen, 1966). Inoculation is absolutely necessary; pelleting is unnecessary; insect and disease control are usually not necessary.

Nutrient requirements. Lotononis responds to fertilizer, although it has moderate demands for phosphorus. On light sandy soil, it should be given a complete mixture of fertilizer. For establishment on poor soil, give it 750 kg/ha molybdenized superphosphate, 125 kg/ha muriate of potash with small amounts of copper and zinc. Annual dressings should thereafter be 250 kg/ha superphosphate plus 125 kg/ha muriate of potash.

321

● Calcium. Norris (1966) has shown that *Lotononis bainesii* nodulated successfully in sand culture at pH 4.0 and in the presence of only 5 ppm of calcium. Its calcium uptake is 1.1 to 1.6 percent of the dry matter (Andrew and Hegarty, 1969). However, in areas where it has been established with grasses and other legumes (such as with white clover in the *wallum* country in Queensland), 600 kg/ha of lime are applied in the establishment year.

● Phosphorus. Andrew and Robins (1969a) determined the critical percentage of phosphorus in the top growth at the immediately preflowering period as 0.17 percent, and lotononis yielded 55 percent of its maximum yield at 250 kg/ha superphosphate. Its maximum yield was attained above 500 kg/ha. It was shown to be almost as effective as *Stylosanthes humilis* in extracting its phosphorus needs from low-fertility soils. Andrew and Robins (1969b) further showed that *Lotononis bainesii* and *S. humilis* were the most effective in producing nitrogen per unit mass of soil, especially on low phosphorus soils. The percentage of nitrogen in the tops of *Lotononis bainesii* increased from 3.29 percent with 125 kg/ha superphosphate to 3.46 percent at 250 kg/ha and 3.91 percent at 1 000 kg/ha.

● Potash. Andrew and Robins (1969c) determined the critical percentage for potash in the plant tops as 0.90 percent. It yielded only 23 percent of its maximum yield without the addition of potash, and required more potash than siratro.

Lotononis and *Desmodium intortum (aparines)* are usually high in potassium (Andrew and Robins, 1969d). *Lotononis bainesii* growing in a soil with adequate potash had a potassium concentration of 1.08 g DM/kg, and a deficient plant recorded only 0.41 g DM/kg in the tops. The initial effect of potassium deficiency in *L. bainesii* is the appearance of a generalized interveinal chlorosis of the leaflets of the older leaves, commencing as a blotching of the leaflets, particularly toward the tips and margins. With increasing deficiency, the chlorosis intensifies, particularly toward the tips of the leaflets, which finally develop light brown, irregular shaped necrotic spots which enlarge and coalesce to form large brown areas with some distortion around the necrotic areas. Finally, necrosis encompasses the entire leaflet and death ensues. The petioles remain unaffected for some time but eventually absciss, and even the stipules of the affected petioles develop necrosis. In general, potassium deficiency symptoms are more severe on the central upright stem of the plant than on the lateral stems. Under conditions of extreme deficiency, even the freshly expanding leaves are chlorotic and malformed and have pinched tips (Andrew and Pieters, 1970a).

● Sodium. Andrew and Robins (1969b) found that the sodium content of *L.*

bainesii was high in comparison to other tropical legumes and increases with fertilization with sodium dihydrogen phosphate at the expense of potassium.

Toxicity levels and symptoms. Manganese — the initial effects of manganese toxicity are interveinal chlorosis in the young leaves and chlorotic shoots. The interveinal chlorosis is general. With increasing toxicity, irregularly shaped brown areas occur on both leaflet surfaces; these subsequently become necrotic. At the same time, younger expanding leaves are small, chlorotic and malformed by unusual leaflet shape, distortion of the edges, and intense puckering of the central portion of the leaflet. This is accompanied by outward curvature of the leaflet margins and a downward curving of the distal portion of the leaflets. Brown-coloured necrotic areas occur on the petiole and stems. Restricted growth gives rise to strong secondary growth which is chlorotic at emergence (Andrew and Pieters, 1970b). Lotononis is tolerant of excess manganese and the toxicity threshold value is 1 320 ppm (Andrew and Hegarty, 1969).

Vigour of seedling, growth and growth rhythm. Seedling vigour is not good and the plant is rather slow to start. Growth rhythm of *Lotononis bainesii* has been determined by Whiteman (1969) and is shown in Figure 63. (Lotononis reaches its peak growth in October in southeast Queensland, lat. 27°22'S.)

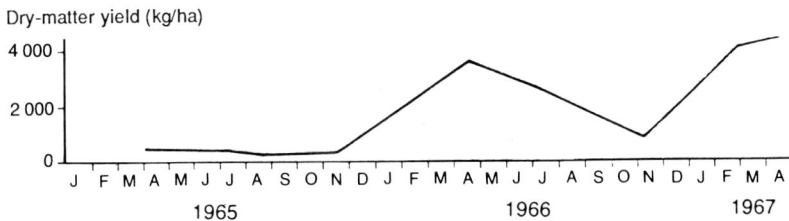

Figure 63. Growth rhythm of *Lotononis bainesii* in southeast Queensland, Australia

Response to photoperiod and light. Flowers mainly in the spring, with a secondary blooming in late summer (Bryan, 1961; Wright, 1964). It is very sensitive to light and will not tolerate excessive shading. Growth ceases when radiation is less than 11 000 langleys for 28 days. This explains its slow growth in winter and in cloudy periods in late summer.

Compatibility with grasses and other legumes. Combines well with *Paspalum*

323

commersonii, P. plicatulum and *P. dilatatum*, and especially well with *Digitaria decumbens* (pangola grass), as each species can withstand heavy grazing. Because of its unpredictable performance, it should be sown with another legume in mixtures, e.g. *Desmodium intortum* or *Macroptilium atropurpureum*. It will grow with tall grasses such as *Setaria anceps* if they are defoliated frequently. It is less compatible with thickly matted Kikuyu grass (*Pennisetum clandestinum*). It is compatible with *Macroptilium lathyroides, Trifolium repens* and *Centrosema pubescens* (Bryan, 1961).

Ability to compete with weeds. Low in the seedling stage and is susceptible to shading by tall weeds.

Tolerance of herbicides. No record in the literature.

Nitrogen-fixing ability. It averaged 200 kg N/ha with grass over 5 years at Samford, southeastern Queensland, and in the last year transferred 15.4 kg N/ha to the grass. Bryan (1961) showed that its addition to a pasture of *Paspalum commersonii* and *P. plicatulum* increased the nitrogen content of the grass by 14 percent.

Response to defoliation. New stolon development accompanies grazing and so it is able to withstand grazing pressure. Jones, Davies and Waite (1967) and Whiteman (1969) found that it did not persist with high yields for more than a few years, but Bryan (1968) found it more persistent at Beerwah in southeastern Queensland.

Grazing management. It should be kept fairly closely grazed to encourage its runners to root down and reduce damage by little-leaf and *Rhizoctonia*. Its combination with pangola grass where the two have similar grazing requirements is sound. Some shade from associated grasses in the heat of summer is beneficial. However, it should not be grazed so severely that the crowns are eaten. Under grazing conditions in mixed pastures in southeast Queensland, the grazing rotation is two weeks' grazing, four weeks' rest.

Response to fire. Recovers readily.

Breeding system. A tetraploid with 2n = 36 (Byth, 1964). Flowers are cleistogamous and self-pollinated.

Dry- and green-matter yields. At Rodd's Bay, Queensland, 1 850 kg DM/ha were harvested after four months' growth. At Beerwah, the yield was 5 050 kg DM/ha/year. In grazing trials at Beerwah with 6 weeks' rest between grazings, yields have ranged from 25 kg/ha in a dry early spring to 2 673 kg DM/ha in mid-summer. With pangola grass (*Digitaria decumbens*), with 8 weeks between grazings, it has yielded 250 kg/ha to 1 912 kg/ha, associated with corresponding grass yields of 1 564 kg/ha to 2 340 kg/ha at an annual grazing rate of 1 beast/0.80 ha (Bryan, 1961). Jones, Davies and Waite (1967) harvested at

Samford successive yearly dry matter yields from lotononis in association with grass of 3 850 kg/ha, 2 937 kg/ha, 3 751 kg/ha and 1 078 kg/ha, representing 33.5, 43.3, 27.9 and 12.9 percent of the total dry matter respectively. Under irrigation at Perth, Western Australia, in a Mediterranean climate (lat. 32°S), Roberts and Carbon (1969) obtained average yields at two sites of 8 268 kg/ha/annum.

Suitability for hay and silage. No hay appears to have been made with lotononis, but if care is taken to preserve the leaf the hay should be of high quality. Catchpoole (1970) made excellent silage from lotononis cut at 2.5 to 4 cm above ground level without the addition of molasses. The pH of the silage was below 4.2 with a low content of volatile constituents, and small losses of dry matter and nitrogen during storage. The material was chaffed for ensilage. The fermentation followed the usual lactic acid sequence, but development of lactic acid and attainment of the final pH were slow. The final silage had 25.6 percent dry matter with 15.37 percent protein on a dry-matter basis.

Value as standover or deferred feed. It is excellent feed in winter because of its greenness and tolerance of frost. However, its total yield is small in midwinter, usually less than the frosted species (Jones, 1967a). Green material at this time has a food value in excess of its summer contribution on a weight basis.

Feeding value. Cattle prefer lotononis to any other tropical legume because of its extreme palatability.

● Chemical analysis. The results of 14 analyses showed that the crude protein content of the dry matter reached 25.9 percent (Bryan, 1961), with a mean value of 18.3 percent. The full analyses obtained by Milford (unpublished) were: crude protein 19.3 percent, crude fibre 27 percent, ether extract 4 percent, nitrogen-free extract 41.6 percent, ash 8.1 percent and moisture 74 percent.

● Digestibility. Dry matter and organic matter, 60 percent; crude protein, 74 percent; and crude fibre, 47 percent. Details of other analyses are listed in Appendix 1.

Toxicity. No toxicity observed, although the lush growth will cause a temporary taint in milk from dairy cows grazing it.

Seed harvesting methods. Lotononis is harvested by hand-picking the ripe seed heads for small samples or by mowing the green material when about 80 percent of the crop has purple seed. This will usually contain 50 percent fully mature purple seed, 30 percent soft purple seed and 20 percent green, yellow and brown seed. The mowing can be done with an auto-scythe pushing the cut material onto a tray, or with a tray attachment to the power take-off of a

Figure 64. Tray attachment to a three-point linkage mower blade for harvesting *L. bainesii* for seed and for holding and spreading cuttings by hand (**Photo:** C. Wright)

tractor, so that the material can be collected in a heap. The material is left in the swath, windrow or heap for three to four days. Then thresh with a high drum speed without an air blast and with the sieves closed. Lotononis has a sticky sap and after threshing the seed will heat in heaps — up to 60°C in four hours — and therefore should be dried immediately.

Seed yield. Only about 33 to 55 kg/ha.

Minimum germination and quality required for sale. A minimum of 50 percent germination and 93 percent purity with a maximum of 45 percent hard seed is required in Queensland. The seed is germinated at 20 to 30°C.

Cultivars. There is only one cultivar, cv. Miles.

Diseases. It is especially susceptible to little-leaf, more so if it is grazed lightly in late summer (Bryan, 1961), and also to attacks by *Rhizoctonia solani* when in lush growth. Grazing helps reduce these hazards.

Pests. Colbran (1963) found that *Lotononis bainesii* roots were attacked by the nematodes *Helicotylenchus dihystera* and *Meloidogyne hapla*.

Main attributes. Ability to grow on poor soils; winter greenness; high palatability; ability to stand high grazing pressure; high drought tolerance; resistance to the *Amnemus* weevil.

Main deficiencies. Unpredictable growth cycle; lack of bulk; susceptibility to disease; low seedling vigour; low seed yields (consequently, seed price is high).

Performance. Bryan (1968b) tested several legumes in complex pasture mixtures under grazing at Beerwah, southeastern Queensland (lat. 27°S, rainfall 1 630 mm, altitude 15 m). Lotononis was a good pioneer species, but almost disappeared and then recovered. In a mixed pasture, it contributed 100 kg DM/ha and persisted. It is recommended as a constituent of *wallum* pastures because of its winter and spring contribution.

Main reference. Bryan (1961).

Macroptilium spp.

Macroptilium atropurpureum **(DC) Urb.** (see Colour plates XXIII, XXIV)

Synonym. *Phaseolus atropurpureus* DC.
Common names. Siratro. The unselected parents are commonly known as atro.
Description. Deep-rooting perennial with trailing pubescent stems which may root anywhere along their length, especially in moist clay soils but rarely in drier sandy soils. Leaves pinnately trifoliate, dark green and slightly hairy on the upper, silvery and very hairy on the lower, surfaces. Lateral leaflets ovate, obtuse, about 4 to 6 cm, often asymmetrically lobed. The inflorescence is a raceme; peduncle 10 to 30 cm long, with 6 to 12 flowers crowded at the apex, deep purple with a reddish tinge near the base of the petals. Pod straight, about 7.5 cm long and many-seeded. Pods dehisce violently when ripe. Seeds from light brown to black, flattened ovoid in shape, $4 \times 2.5 \times 2$ mm (Barnard, 1967).
Distribution. Siratro was bred by E.M. Hutton (Queensland, Australia) from two ecotypes of *M. atropurpureum* from Mexico — one from the Vera cruz waterfront and the other from near Matlopa in San Luis Potosi. *M. atropurpureum* occurs naturally in Central and South America.
Season of growth. Summer-growing perennial with greatest growth in mid-summer to autumn in southeast Queensland.
Temperature for growth. Optimum, about 26.5 to 30 °C with average daily minimum temperatures above 21°C. Growth was poor at a day/night temperature range of 21/16°C and 18/13°C and maximum dry matter was produced at 30/25°C and 27/22°C in a long day.

Ludlow and Wilson (1970) found that siratro gave only 24.5 percent of its yield, had only 43 percent of the growth rate and only 14.3 percent of the leaf area at 20°C as at 30°C. Jones (1967b) and Whiteman and Lulham (1970) found that growth of siratro ceased at 14°C.

328

Frost tolerance. Winter frosts cause severe defoliation, but survival is one of the highest among the tropical legumes. Jones (1969) found that it survived a cold winter at Samford in southeastern Queensland when the terrestrial minimum temperature reached –8.35°C.

Latitudinal limits. About 30°N and S. At latitude 28°S, growth is very slow at elevations higher than 610 m.

Altitude range. In Kenya, it grows at elevations up to 1 600 m, but the temperature must be more than 15.5°C (Jones, personal communication).

Rainfall requirements. It requires at least 615 mm and preferably more than 850 mm. It does not thrive in high rainfall regions above 1 800 mm.

Tolerance of flooding. Extremely drought-tolerant by reason of its deep-rooting habit. In summer droughts, large leaves are shed and small leathery leaves produced until conditions are more favourable (Davies and Hutton, 1970). It is not tolerant of flooding.

Soil requirements. Thrives on a wide range of soils, except poorly drained ones (Davies and Hutton, 1970). Will grow in soils ranging from deep sands and loams to light clays (e.g. podzolic, deep latosolic and alluvial soils). Grows over a range of pH from 4.5 to 8.0. Kretschmer (1966) got best growth in Florida from an application of 2.2 tonnes of lime, which lifted the pH from 4.5 to 6.1. It is one of the best of the tropical legumes under moderately saline conditions.

Rhizobium *relationships*. It nodulates freely with native rhizobia, but seed should be inoculated at sowing with inoculum of the cowpea type. The current Australian strain recommendation is CB 756 (Date, 1969). Van Rensburg (1967) found that siratro had frequent large nodules on the taproots and laterals which were very easily dislodged. Whiteman and Lulham (1970) found that under heavy grazing there is evidence of nodule loss and replacement, as the grazed plants produced larger numbers of smaller nodules. Nodule weight peaked at the end of March (lat. 27°22′S) and then there was a rapid shedding of nodules, with few present during the winter.

Ability to spread naturally. Under favourable conditions, siratro will spread naturally, but it spreads more readily with some preliminary cultivation.

Land preparation for establishment. Establishes best in a well-prepared seed bed, and has been successfully established, though not so rapidly, in roughly prepared ground, in the ashes of a burn from forest debris or from *Imperata* grassland (Douglas, 1965), and by sod-seeding. In roughly prepared seed beds, the seeding rate should be doubled.

Sowing methods. It is preferably drilled into a well-prepared seed bed, but can be broadcast from the ground or by aerial seeding. If conditions are favourable, it can be oversown into existing pastures. Downes (1966) established it by oversowing into natural grassland in north Queensland. It can be sod-seeded into natural pastures with superphosphate and also oversown into burnt *Imperata* grassland with superphosphate. The seed is viable after passage through an animal and germinates in dung pats.

Sow at 1.5 to 2.5 cm into prepared seed beds and cover with a harrow or with a roller. Sow from spring to late summer. Whiteman and Lulham (1970) found that the best time to sow for yield in southeastern Queensland was in December. If if is sown in early summer, it is more likely to escape bean-fly attack at the seedling stage. Van Rensburg (1967) found that sowing distances of 30, 60 and 90 cm made no significant difference to subsequent yield and regrowth in Zambia. Jones and Andrew (1967) found increasing yields by increasing sowing rates from 0.46, 1.84 and 3.38 kg/ha and increasing phosphorus rates from 250 kg/ha to 1 500 kg/ha. Middleton (1970) found seedling density proportional to the sowing rate. Seed is usually sown at 2 to 8 kg/ha.

Number of seeds per kg. 80 000. Percentage of hard seeds in commercial samples is from 40 to 70 percent.

Seed treatment before planting. To break down dormancy: (*a*) scarify mechanically; (*b*) treat with concentrated sulphuric acid (sp. gr. 1.8) for 25 minutes, wash and dry (Prodonoff, 1968). Inoculation is not necessary but preferable. Pelleting is not necessary unless to protect rhizobia; pellet with rock phosphate (Norris, 1967). For insect and disease control, treat the seed with 6 cc of dieldrin 15 percent emulsifiable concentrate per kg seed (Jones, 1965).

Nutrient requirements. Siratro responds in podzolic and solodic soils to molybdenized superphosphate up to 500 kg/ha for maximum yields (Truong, Andrew and Skerman, 1967) but gives a yield increase from dressings of 125 kg/ha. On fertile soils, no fertilizer will be necessary for a number of years. There is need for balanced fertilization; Jones (1967a) showed that high phosphorus plots had 2 percent siratro, while high phosphorus-high potassium plots had 20 percent siratro. An annual application of 125 to 200 kg superphosphate and 125 kg potassium chloride should maintain production in this species on sandy soils.

● Calcium. Siratro is not sensitive to calcium deficiency, but on acid soils molybdenum may be unavailable and addition of calcium to raise the pH released molybdenum; thus, it is really a molybdenum effect. Kretschmer (1966) got a direct response to lime up to 2.2 tonnes/ha in Florida, United States. Truong, Andrew and Skerman (1967) found that Ca gave no response, but that 125 and 250 kg/ha of superphosphate increased yield over nil treatment and maximum yield was at 500 kg/ha.

● Magnesium. Siratro is relatively high in magnesium (Andrew and Robins, 1969b), and addition of calcium phosphate to the soil increased magnesium at the expense of potassium.

● Molybdenum. It responded to molybdenum on a solodic soil (Truong, Andrew and Skerman, 1967).

● Nitrogen. At about 100 kg/ha, siratro practically disappeared from a pangola grass mixture grown on a sandy soil in Florida (Brohlman, personal communication). Parbery (1967a) got no response to 100 kg N/ha in the Kimberleys, northern Australia. Jones (1965) recorded 28 percent, 5.7 percent and a trace of siratro after four years' growth with Nandi setaria to which dressing of nil, 250 kg and 750 kg/ha of urea had been applied.

Henzell *et al.* (1968) found that when grown separately, siratro and Rhodes grass each took up an equal amount of nitrogen, but when grown together only one-third was taken up by the siratro and two-thirds by the Rhodes grass.

● Phosphorus. The critical percentage for phosphorus in the plant tops of siratro is 0.24 percent (Andrew and Robins, 1969a). At 250 kg/ha superphosphate, siratro yielded 54 percent of maximum yield at 1 320 kg/ha superphosphate when grown on a gley soil at Samford, Australia. No response to phosphorus was obtained on an alluvial soil derived from basalt.

● Potassium. The critical percentage of potassium in the plant tops was determined by Andrew and Robins (1969c) to be 0.75 percent. It yielded 38 percent, at the nil treatment, of the yield at its maximum response at 185 kg/ha. In the presence of high phosphorus, potassium may become deficient (Jones, 1966). Jones (1966) increased the percentage of siratro in a mixed pasture with Nandi setaria from 12 to 16 percent with adequate phosphorus to 42 percent with adequate P and 185 kg/ha muriate of potash. Andrew and Pieters (1970a) obtained healthy growth in siratro plants containing 1.30 g/kg of K in the tops on a dry-matter basis, and 0.42 percent in a plant showing potassium deficiency. Deficiency symptoms commenced as a necrotic spotting on the lower leaves of the plant. This was not preceded by rust-coloured spotting, as in *M. lathyroides*. The necrotic spots were of pinhead size, irregular in shape and placed interveinally toward the margins of the leaflets. They were mid-brown in colour, surrounded by a pale chlorotic region in an otherwise normal green leaf, visible on both surfaces of the leaflets but with a sunken appearance, particularly on the lower surfaces of the leaflets. With increasing severity, the edges and tips of the leaflets became chlorotic and some of the necrotic spots enlarged and coalesced to give a marginal necrotic effect, particularly near the tip of the leaflets.

331

In some cases, the interveinal necrotic spotting was not evident, but the symptom commenced as marginal necrosis and interveinal chlorosis of the lower leaves. In severe cases of deficiency, affected leaves abscissed, and this effect progressed toward the younger portion of the plant. In this species, there was a suggestion that affected leaves tended to remain in a "sleeping" position during daylight (Andrew and Pieters, 1970a).

Toxicity levels and symptoms. Manganese — the toxicity threshold value in the dry matter of the tops was 810 ppm (Andrew and Hegarty, 1969). Siratro only gave 9 percent of its maximum yield in the presence of high manganese concentrations. As manganese content in the solution increased, there was a large reduction in the total uptake of nitrogen by the plant. However, *M. atropurpureum* had a high concentration of manganese, reaching a maximum of 5 590 ppm.

In tests to ascertain the toxicity symptoms of excess manganese in siratro, Andrew and Pieters (1970b) found that there were two initial symptoms of manganese toxicity. Firstly, young growth was interveinally chlorotic and, secondly, older leaflets showed brown, rust-coloured spots. Leaves slightly affected at emergence retained the interveinal chlorosis effect to maturity and also exhibited increasing numbers of brown spots, which appeared at the extremities of the veinlets and adjacent to the secondary veins, particularly on the underside of the leaflets. In young plants, this was often accentuated on the primary leaves and their petioles. As toxicity increased, emerging shoots were severely chlorotic, the leaflets showing an interveinal effect, but this was not as definite as in *M. lathyroides*; the main and secondary veins were pale green in colour toward the base of the leaflets but the remainder of the leaflet was almost devoid of chlorophyll. In severely affected plants, the young emerging leaflets had very little chlorophyll, numerous brown areas occurred adjacent to and on the main and secondary veins, and in severe cases puckering of the leaf surface occurred, usually associated with necrosis of the previously mentioned brown areas. This resulted in epinastic curvature of the petiolules and subsequently of the leaf petiole. In the older leaves of the plant, there was an increase in leaflet thickness. Following cessation of growth at the primary shoot, secondary growth was initiated, but this in turn was similarly affected by the toxicity.

Vigour of seedling, growth and growth rhythm. Excellent seedling vigour and can establish readily from shattered seed in an existing pasture.

The plant grows vigorously in the hot weather and is most productive in midsummer. Growth rhythm from an irrigated and fertilized pasture mixture of Rhodes grass and siratro, uncut, is shown in Figure 65.

332

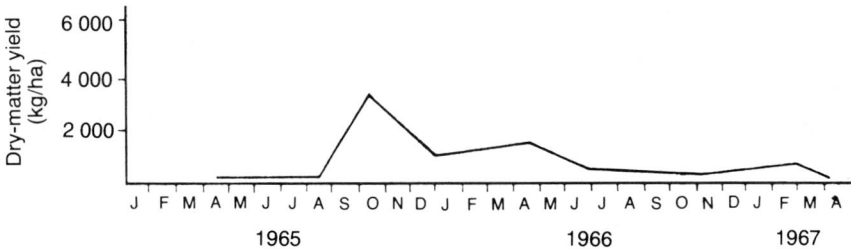

Figure 65. Growth rhythm of an uncut siratro/Rhodes grass pasture

Response to photoperiod and light. Flowering occurs in short and long days; best at 24/19°C, 27/22°C and 30/25°C. Flowers in 60 to 70 days in southeastern Queensland and in 57 days in the Kimberley district, Australia (Parbery, 1967a). It will grow reasonably well in the shade, but prefers ample sunlight and has the capacity to climb tall grasses in mixtures.

Compatibility with grasses and other legumes. It is compatible with Rhodes grass, buffel, green panic, guinea grasses and setarias. Middleton (1970) found siratro more competitive with *Setaria sphacelata* than it is with *Desmodium intortum*. In Florida, Kretschmer (1966) found that it grew better with pangola grass than with *Pensacola bahia* and *Setaria anceps*.

Ability to compete with weeds. Siratro competes well with weeds if allowed to develop to a stage where it can smother them. Douglas (1965) successfully established siratro on burnt blady grass (*Imperata cylindrica*) country by oversowing it with 370 kg/ha superphosphate containing 0.03 percent molybdenum. By the end of the fifth month, it was overtopping the blady grass.

Tolerance of herbicides. Bailey (personal communication) found that siratro was one of the most susceptible of the tropical legumes to 2,4-D. It should not be used on this species.

Nitrogen-fixing ability. Siratro fixes a good deal of nitrogen — about 100 to 175 kg/ha/year. Kretschmer (1966) found that introducing siratro at 1.1 kg/ha on a 25-cm grid raised the crude protein level of pangola grass from 4.7 to 7.1 percent in Florida, and at 11 kg/ha of seed the protein content was 11.2 percent for the mixture and 5.8 percent for pangola grass alone. He estimated that siratro contributed 55 to 138 kg N/ha/year. Jones, Davies and Waite (1967) found that the selections of *M. atropurpureum* produced as much nitrogen as grass fertilized with 187 kg/ha/year, but produced only as much dry matter as grass fertilized with 100 kg N/ha. Henzell *et al.* (1968) found that 43 to 50 percent of the nitrogen in siratro plants at three weeks of age was taken

333

Figure 66. Pastures based on *Macroptilium atropurpureum* cv. Siratro developing from minimum soil preparation in open forest country. Limited grazing has permitted the legume to climb the trees

up from the soil, the remainder coming from symbiotic nitrogen fixation, but at 15 weeks only 2 to 4 percent was coming from the soil.

Response to defoliation. Jones (1967a) found that siratro did not persist when cut to a height of 3.75 cm every four weeks. Under grazing by sheep and cutting treatments, Whiteman (1969) found that frequent defoliation by sheep or cutting at 5 cm steadily reduced the yield of siratro, grazing reducing survival more than cutting. Jones (1967a) further found that cutting to 15 cm maintained the vigour of the siratro in a setaria/siratro sward.

Grazing management. Siratro should be lightly grazed at all times. Livestock will eat the runners back toward the crown, which should be protected from overgrazing. The concept that "leaf begets leaf" is valid for siratro, and grazing to 15 cm maintains the stand. In thinning stands, siratro should be shut up to allow it to seed and shatter the seed, so that the stand can be improved by new seedlings. In this way, it will also climb over dominant grass and weeds and suppress them. Stobbs (1969j) found that a rotational grazing system of two weeks' grazing-four weeks' rest maintained the best botanical composition and equalled the weight gain obtained with continuous grazing.

Response to fire. Recovers well, new growth appears from the crown and new seedlings germinate.

Breeding system. Self-pollinated; chromosome number 2n = 22.

Dry- and green-matter yields. Roe and Jones (1966) obtained 3 394 kg/ha siratro in a mixture with Nandi setaria, the mixture yielding 12 200 kg/ha at Gympie, Queensland, in 1962/63, and 1 094 kg/ha out of a total of 4 873 kg/ha in 1963/64. Van Rensburg (1967) obtained an average dry-matter yield over the two years 1965-66 in Zambia of 7 960 kg/ha. In Florida, siratro yielded 11 610 kg DM/ha/year when grown with pangola grass (Kretschmer, 1966).

Suitability for hay and silage. Can be made into hay only with difficulty because of the heavy loss by leaf drop, leaving stemmy material and very young shoots. Catchpoole (1970) found that ensilage of siratro without added molasses was never successful, but that the second harvest material in a season was better than the first. Satisfactory silage was made by adding 8 percent molasses with the first cut and 4 percent with the second cut. Two percent molasses markedly improved an otherwise poor silage but was insufficient for a stable product.

Value as standover or deferred feed. Not very valuable because of leaf shedding. Jones (1967b) found that it lost over 75 percent of its dry matter and over 80 percent of its nitrogen on average over two winter periods (1962-63).

Feeding value. Quite palatable; it is readily eaten by livestock.

● Chemical analysis. Milford (1967) gave the analyses of mature siratro at the seed-shedding stage, leafy but 20 to 30 percent of the leaf dry. He recorded figures of 35 percent dry matter, 16.8 percent crude protein, 33.4 percent crude fibre, 1.2 percent ether extract, 38.8 percent nitrogen-free extract and 9.8 percent ash.

● Digestibility. The digestibility figures were 50.4 percent for dry matter, 53.4 percent for organic matter, 67.6 percent for protein, 50.9 percent for fibre and 50.6 percent for the nitrogen-free extract. The intakes of dry matter and digestible dry matter were 37.5 and 18.9 g/kg live weight/day respectively. Other analyses for siratro are given in Appendix 1. Minson and Milford (1966) showed that the mean energy value of the digestible dry matter for siratro was 4.2 times higher than for pangola grass, while that of the digestible organic matter was 8.2 percent higher.

● Palatability. Although siratro is palatable, Stobbs (1969k) found that cattle grazed *Panicum maximum* first and allowed the siratro in the mixture to become dominant.

335

Toxicity. None reported in livestock feeding.

Seed harvesting. Siratro does not seed prolifically in districts where it thrives as pasture. Therefore, although pastures may be harvested for seed, it is better in the long term to seek specific seed producing districts. These should have a very dry and frost-free winter. From one to four crops may be produced each dry season, depending on temperature and rainfall patterns and irrigation use.

Use of insecticide over the flowering period is necessary for heavy yields. Each crop may be harvested as it ripens (if hand harvesting or using a small header); or a single end-of-season header harvest may be taken, followed by suction harvesting. The latter system produces very high yields with minimum labour input, but requires sophisticated management and machinery.

Pods shatter on ripening, and harvest time should be chosen to forestall this. Hand picked pods will shatter during drying. Both hand- and header-harvested seed need drying. Sun drying on tarpaulins is adequate (Hopkinson, personal communication).

Seed yields. A single vigorous flower flush may produce over 1 000 kg/ha of seed, but seldom is more than 200 kg/ha harvested from a single crop. Direct heading of locked up pasture yields far less than this. Single header harvests of several irrigated flushes have yielded up to 700 kg/ha commercially, and suction harvesting generally recovers 100 to 400 kg/ha.

Maximum germination and quality required for sale. Seventy percent minimum germination with a maximum hard-seed content of 100 percent and a minimum purity of 97.5 percent in Queensland. The seed is tested for germination at 25°C (Prodonoff, 1968).

Cultivars. There is at present only one commercial line of siratro available, though an active breeding programme by the CSIRO Division of Tropical Agronomy (Australia) will soon lead to the release of cultivars for special purposes.

Diseases. Siratro is attacked by *Rhizoctonia solani* under very wet conditions (Kretschmer, 1966; Dunsmore and Ong, 1969), and a case of attack by a powdery mildew was recorded in a late-sown crop at Campinas, Brazil. It is relatively resistant to little-leaf, but is attacked occasionally. It is severely attacked by an orange-coloured rust (*Uromyces phaseoli*) in high rainfall areas of Guatemala (Rodriguez, personal communication).

Jones, Alcorn and Rees (1969) reported death of siratro plants from the attack of violet root caused by *Rhizoctonia crocorum*. A growth of reddish-brown to purple fungal mycelium occurred over the top 20 cm of taproot, accompanied by decay of internal tissues where the advancing internal margin of the rot sometimes showed a reddish band.

Pests. Colbran (1963) found that siratro was attacked by the root nematode *Helicotylenchus dihystera*, but was resistant to *Meloidogyne javanica javanica* and *Radopholus similis*. He therefore recommended it (Colbran, 1964) in conjunction with green panic as a suitable cover crop for control of nematodes in banana plantations. The bean fly (*Melanagromyza phaseoli*) will attack seedlings up to three to four weeks of age, but it can be prevented by seed treatment (Jones, 1965). Meloid beetles, which may prevent flowering in the tropics, can be controlled by a DDT spray. The plant is resistant to the *Amnemus* weevil. In Florida, the bean leaf roller (*Urbanus proteus*) attacks siratro in late summer and autumn (Kretschmer, 1966).

Main attributes. It is productive under a wide range of soils; easy to establish; drought-resistant; combines well with a wide suite of grasses. It is promiscuous in *Rhizobium* requirement; has high seedling vigour; is very palatable.

Main deficiencies. Low cold tolerance and comparatively low seed yields.

Performance. Siratro is performing well in the medium rainfall areas of the tropics such as at Campinas (Brazil), Serere (Uganda), Makulu (Zambia), Ukirigiru (Tanzania), Queensland and Northern Territory (Australia) and Panama. Stobbs (1969h) obtained a mean live-weight gain of 432 kg/ha/year from a *Panicum maximum/Macroptilium atropurpureum* pasture at Serere, Uganda. The highest percentage of legume was maintained in the sward under continuous grazing at 7 beasts/ha, but weed invasion was also highest. Stobbs concluded that over long periods rotational grazing is necessary to maintain a satisfactory sward. The inclusion of siratro in the mixture considerably extended the quality of the herbage during the dry season.

Main references. Hutton (1962); Jones (1966).

Macroptilium lathyroides **(L.) Urb.** (see Colour plate XXV)

Synonyms. *Phaseolus lathyroides* L. *Phaseolus semi-erectus* (L.) DC, *P. crotalaroides* Mort.

Common names. Phasey bean, Murray phasey bean (Australia), frijol de monte (Venezuela), frijol de los arrozales (Colombia), wild pea bean (Hawaii).

Description. Herbaceous annual or short-lived perennial, erectly branching, 0.5 to 1 m high under normal conditions, lower parts becoming somewhat woody; when grown under shade, in association with tall grasses or sown

337

early, assumes a twining habit and may attain 1.2 m. Branches terete, clothed with long deciduous reflexed hairs. Leaflets ovate or lanceolate, 3.5 to 7.5 cm long. Inflorescence, semierect racemes about 15 cm long borne on axillary peduncles up to 25 cm; pedicels short. Flowers red-purple. Pods subcylindrical, 7.5 to 10 cm long and 3 mm wide, slightly curved, approximately 20-seeded. Seeds oblong or rhomboidal, slightly compressed, about 3 mm long, mottled light and dark grey-brown (Barnard, 1967). It regenerates from seed under favourable conditions but Bryan (1968a) found that it was unreliable in this regard at Beerwah, southeast Queensland.

Distribution. Originated in tropical America and is now widespread and naturalized in the tropics, including Colombia, Venezuela, Paraguay, Belize, Panama, Brazil, the Caribbean (Jamaica, Antigua, St. Vincent, Cuba) and Australia.

Temperature for growth. Whiteman (1968) found that the optimum temperature for growth was 30/25°C. Minimum temperature was not recorded. It is susceptible to frost, but has usually seeded well ahead of them. Paltridge (1955) recorded occasions when it survived the winter and provided new growth in the following summer. Jones (1969) recorded 21 percent survival of plants at Samford after a winter in which the lowest temperature reached −8.3°C.

Latitudinal limits. About 23°N to 28°S.

Altitude range. It grows from sea level to 1 800 m in Colombia (Crowder, 1960).

Rainfall requirement. It is adapted to a wide range of climates with a rainfall from 475 to 3 000 mm, although insect damage may be severe in the higher rainfall regime and its growth will depend on summer storms and wetter depressions in the lower rainfall areas.

Tolerance of drought and flooding. It escapes drought by its free-seeding habit. Phasey bean is fairly tolerant of waterlogging and frequently grows in the drains (or water-tables) on the edge of highways in areas where it is adapted.

Soil requirements. Phasey bean is adapted to a wide range of soils, from deep sands to heavy clays, including lateritic types. It tolerates both acid and alkaline soils. It is fairly tolerant to saline conditions (Fretes, Samudio and Gay, 1970).

Rhizobium *relationships*. Nodulates freely with native rhizobia. If inoculation is desired, use the cowpea strain, the current Australian recommendation being CB 756 (Date, 1969) and, in Argentina, strains C5 (from the United States) and 95 (of Australian origin) (Batthyany, 1970).

Ability to spread naturally. Under moist conditions, it will spread well naturally in both sandy and clay soils if competition is not too severe.

Land preparation for establishment. Phasey bean grows best on well-prepared seed beds but will give less satisfactory establishment with less cultivation.

Sowing methods. Seed can be drilled, or broadcast from the ground or from the air. In the poorly drained flood-plain soils of the Burdekin River in north Queensland, Seton (1962) suggested that phasey bean seed should be drilled into the top of planting ridges (15 cm high and 1 m apart) and vegetative grass material (e.g. Para grass, planted two to three months later) between these ridges. Irrigation water is then applied between the ridges so that the legume seed is moistened only. Oversowing into natural pastures is usually not very successful. It can be sown at any time during spring and summer when moisture is adequate, using 1 to 3 kg seed/ha. Sow on the surface or no deeper than 1.25 cm and cover lightly with a harrow.

Number of seeds per kg. 119 000 (ranging from 88 000 to 154 000). Percentage of hard seed is fairly low.

Seed treatment before planting. To break dormancy, treat the seed with concentrated sulphuric acid (sg. gr. 1.8) for 20 minutes, wash and dry (Prodonoff, 1968). Inoculation is not necessary. Pelleting is not necessary unless to protect the rhizobia (use rock phosphate; Norris, 1967). For insect and disease control, treat the seed with 2.2 g acid equivalent (13.2 ml of 15 percent emulsifiable concentrate) of dieldrin per kg of seed before planting to control bean-fly damage to seedlings (Jones, 1965).

Nutrient requirements. On fertile alluvial soils or heavy clays, it need not be fertilized. On poor sandy soils, use 250 kg/ha molybdenized superphosphate. On lateritic gravels in Sri Lanka, Paltridge and Santhirasegaram (1957) found responses to Ca, N, P and K, and on cinnamon sand to P, Ca, Mg, K and B.

● Calcium. *Macroptilium lathyroides* gave 68 percent of its maximum yield in the absence of calcium, and its maximum yield at 995 kg/ha. In the absence of calcium, N, P and K had an adverse effect on its root growth. It responded to lime by increasing its nitrogen content from 1.7 percent with nil to 3.8 percent with 740 kg/ha calcium carbonate (Andrew and Norris, 1961). Paltridge and Santhirasegaram (1957) found small responses to lime on lateritic gravels and sand in Sri Lanka. The calcium uptake by phasey bean was found by Andrew and Hegarty (1969) to be 1.5 to 1.9 percent of the dry matter.

● Copper. Phasey bean gave 46 percent of its maximum yield in the absence of copper, and so is relatively insensitive to low copper status in the soil. Copper concentration in the plant was relatively high, being higher in the roots than in the tops (Andrew and Thorne, 1962). Santhirasegaram (1967) recorded 57 percent of the maximum yield in the absence of copper in a sandy soil in Sri Lanka. Copper deficiency in this species does not show specific symptoms. It causes reduced plant growth, with a slight change in colour of the younger portions of the plant from the natural mid-green or greyish-green, to pale green. Progressively, these portions become yellowish green and the younger, fully expanded leaves absciss. The leaf abscission is not preceded by interveinal chlorosis or any form of necrosis. Immediately prior to abscission, the leaves tended to fold down against the stem of the plant, the condition resembling partial wilting. In severe cases, after several leaves had dropped, the growing shoots became necrotic and the plant commenced axillary growth, which also became necrotic (Andrew, 1963).

● Molybdenum. 't Mannetje, Shaw and Elich (1963) found that phasey bean responded to molybdenum and that the molybdenum had a residual effect in the soil for three to four years. Santhirasegaram (1967) also obtained a response to molybdenum on sandy soils in Sri Lanka.

● Phosphorus. Phasey bean gave 55 percent of its maximum yield at 250 kg/ha superphosphate, with its maximum at 1 132 kg/ha. The critical percentage of phosphorus in the dry matter of the leaves was determined by Andrew and Robins (1969a) as 0.20 percent. Increasing phosphorus supply in the form of monocalcium phosphate increased the magnesium content but decreased the potassium content.

● Potassium. Santhirasegaram (1967) obtained 50 percent of the yield of phasey bean at first harvest and only 12 and 4 percent at the second and third harvest when potash was eliminated — compared with the yields from a complete fertilizer mixture. Andrew and Robins (1969c) found the critical percentage for potassium in the dry matter of the tops to be 0.75 percent. It gave only 24 percent of its maximum yield in the absence of potash. Added potassium had little effect on the concentration of other cations. Andrew and Pieters (1970a) showed that deficiency symptoms commenced on lower leaves as a rust-coloured spotting on the upper surfaces of the leaflets. Individual spots were asymmetrical in shape and mostly interveinally placed toward the margins of the leaflets. In this phase, there were no signs of blemish or discoloration on the lower leaf surfaces, nor was there any associated chlorosis or necrosis. With increased severity, the rust-coloured areas became necrotic with light brownish-grey centres and dark brown peripheries. Necrosis was evident on both surfaces of the leaflet, but, in addition, the lower surface of the leaflet was sunken at the necrotic areas. Following this phase, there was a general marginal interveinal chlorosis of the lower leaves, but a halo of green tissue remained around the necrotic spots. Chlorosis intensified, followed by necrosis of the leaflet margin, particularly toward the distal end of the leaflets. Finally, severely affected leaflets abscissed, the central leaflet of each leaf falling before the lateral ones, resulting in an upright, single-stemmed plant, abscissed leaves in the lower portion, necrotic interveinally chlorotic leaves in the central portion, and an otherwise normal top portion with reduced growth rate. The authors found the healthy plant to have 1 g/100 g potash in the dry matter of the tops. In the deficient plant, the concentration was 0.57 g/100 kg.

- Sulphur. Santhirasegaram (1967) recorded only 28 percent of the yield of phasey bean growing on sandy soils in Sri Lanka compared with the yields obtained from a complete fertilizer.
- Zinc. Russell (1966) found that the yields of phasey bean growing on a solodic soil in southeast Queensland were severely depressed by liming above pH 6.3 in the absence of zinc; the addition of zinc gave small yield increases.

Toxicity levels and symptoms. Manganese — phasey bean is subject to manganese toxicity in high manganese soils. Andrew and Pieters (1970b) showed that the first sign of toxicity in this species was a light interveinal chlorosis of the growing shoots. With increasing severity, these leaves upon expansion continued to show the interveinal chlorosis effect, but new shoots were more severely chlorotic. As toxicity progressed, leaflets moderately affected by chlorosis also showed brown necrotic areas at the extremities of, and along, the secondary veins; these brown areas appeared to be more developed on the underside of the leaflets than on the upper side. Associated with this effect, the younger shoots were severely chlorotic and when fully expanded showed puckering of the leaflet surface, usually raised above the upper surface. In addition, the necrotic spots at the vein tips coalesced, and those that affected leaflets showed edge and tip necrosis with outward and downward curling of the leaflet lamina, while the base of the leaflets remained interveinally chlorotic with some brown necrotic spots along the veins. In the severe form, new shoots were almost devoid of chlorophyll, were puckered and malformed, and abscissed prematurely, while brown necrotic patches occurred on the stems and petioles. Axillary growth followed this phase but it was also chlorotic upon emergence. Aluminium — Russell (1966) reported aluminium toxicity in *Macroptilium lathyroides* growing on a solodic soil of low calcium status in southeastern Queensland. By adding lime he was able to eliminate the toxicity.

Response to photoperiod and light. It is day-neutral. Capable of climbing tall grasses, it is not unduly suppressed by absence of light except in the seedling stage.

341

Compatibility with grasses and other legumes. Compatible with Para grass (*Brachiaria mutica*) under swampy conditions; on well-drained land it grows well with *Panicum maximum*, *Setaria anceps* and *Paspalum commersonii*.

Ability to compete with weeds. It competes favourably with weeds under fertile, moist conditions.

Tolerance of herbicides. No record.

Vigour of seedling, growth and growth rhythm. It has a vigorous seedling and establishes readily. Grows vigorously in the warm moist summer period.

Nitrogen-fixing ability. Paltridge (1955) analysed its effect on a companion grass, *Paspalum commersonii*, at Lawes, Queensland (lat. 27°30'S), growing in a heavy black clay soil. Grown as a sward, the legume increased the yield over grass plus 100 kg N/ha by 77 percent. In terms of the total yield of nitrogen, Paltridge estimated that the inclusion of phasey bean into *P. commersonii* would be the equivalent of adding 800 kg/ha sulphate of ammonia.

Sen and Paul (1959) found that the nitrogen content of the soil after three years' growth of phasey bean was the same in the first 30 cm of the profile and had increased 10 and 15 percent respectively in the 60- and 90-cm zone. No fertilizer data were given.

Response to defoliation. Paltridge (1955) found that *M. lathyroides* did not persist with continuous grazing. Heavy grazing including woody parts severely reduces its vitality.

Grazing management. It should only be lightly grazed to encourage persistence. Only leaves and fruiting branches should be removed by the grazing animal. Paltridge (1955) found that it should be rested for six to eight weeks during the summer growing season, as it is selectively grazed early in the growing period. A light cultivation of the pasture helps annual regeneration.

Response to fire. Does not survive fire, but the seed will germinate afterwards when conditions are suitable.

Breeding system. It is closely self-pollinated (Hutton and Beall, 1957). A relative humidity of 95 percent or more at 28°C was necessary for vigorous pollen germination and pollen tube growth. The flowers do not "trip". Temperatures between 23 and 30°C are satisfactory (Hutton, 1960).

Dry- and green-matter yields. Paltridge (1955) obtained a mean yield of 5 439 kg DM/ha from a mixed pasture of *M. lathyroides* with *Paspalum commersonii* and 4 115 kg/ha from the grass alone. Crowder (1960), in Colombia, recorded a production of 15 tonnes/ha of green material when cut at a height of 1.25 to 1.5 m.

Suitability for hay and silage. Makes quite good hay if an effort is made to conserve the leaf, handling the material when it is supple. Robertson (1971) made excellent silage with *Sorghum almum* mixed with phasey bean in Uruguay and fed it successfully to Brahman cows and calves.

Value as standover or deferred feed. It contributes very little to winter feed, as it is susceptible to frost and defoliates after frosting.

Feeding value. Has a high feeding value, which declines after frosting.

● Chemical analysis. Milford (1967) determined the analyses for phasey bean shown in Table 14.6.

TABLE 14.6 **Compositional analysis of *Macroptilium lathyroides***

Clearing cut	Harvest date		Dry matter	Crude protein	Crude fibre	Ether extract	N-free extract	Ash
					(%)			
January 1957	2-12 August 1957	(1)	48	7.6	32.7	3.1	53.0	3.5
December 1957	26 April-6 May 1958	(2)	24	17.1	27.0	3.1	48.2	4.7
	14-24 June 1958	(3)	30	12.2	37.0	1.9	45.0	4.1
	25 July-4 August 1958	(4)	82	12.8	43.4	2.6	38.1	3.1
December 1958	19-29 April 1959	(5)	21	19.2	32.3	3.5	37.3	7.6
	1-10 August 1959	(6)	25	16.5	32.4	3.5	40.7	6.8

● Digestibility. Figures were as shown in Table 14.7.

The mean intake of 19.6 g/kg $W^{0.75}$ was higher than the autumn means for subtropical grasses. The highest intake was before leaf shedding and intake fell by 37 percent after frosting.

● Palatability. Stock prefer to graze *M. lathyroides* after the seed has set; young plants appear to be less palatable (Young, Fox and Burns, 1959).

Toxicity. It had been reported as poisonous to horses, but Paltridge (1955) found no toxicity with horses at Lawes and chaffed green material had no effect on dairy cows when fed for ten successive days. It did not produce any milk taint.

Seed harvesting methods. Its shattering habit makes it difficult to obtain full seed yield. It can be direct-headed when a fair percentage of the pods are ripe to ripening and the material subsequently dried where seed from dehiscing pods will not be lost. Subsequent threshing in a stationary thresher and winnowing to clean the seed are required.

Seed yield. Two hundred to 250 kg/ha.

TABLE 14.7 Digestibility figures for *Macroptilium lathyroides*

	Dry matter	Organic matter	Crude protein	Crude fibre	Ether extract	Nitrogen-free extract	Starch equivalent	Total digestible nutrient	Intake of dry matter	Digestibility of dry matter	Nitrogen retention
	(%)	(%)	(%)	(%)	(%)	(%)	(%)	(%)	(g/kg $W^{0.75}$/day)	(g/kg $W^{0.75}$/day)	(g/day)
(1)	48.0±1.1	50.3	52.8	32.1	27.9	62.8	32.3	51.9	35.9±2.6	17.2±1.6	−0.58
(2)	62.0±0.2	64.7	78.0	41.9	48.8	74.8	47.8	65.0	43.7±1.9	27.1±1.5	+3.95
(3)	57.5±3.2	58.8	59.2	43.2	44.5	69.4	35.1	57.5	28.4±0.8	16.3±1.4	+0.26
(4)	41.7±3.6	43.4	57.5	35.0	37.5	49.0	17.5	43.4	24.5±0.8	10.2±0.7	+0.55
(5)	60.4±0.8	64.2	79.0	55.4	69.4	63.9	41.9	62.4	37.8±5.8	22.8±3.2	+3.41
(6)	59.9±1.3	61.2	75.1	44.2	61.1	69.4	39.5	58.9	40.2±0.7	24.1±0.6	+1.99

Key: (1) Most leaves shed, stems green, not frosted. (2) Fully mature, seed shed, twining very leafy. (3) More advanced stage of maturity than the previous cut (2). (4) Frosted, all leaves shed, mainly dried stem, some thick green stem. (5) Flowering, some immature seed, height 1.7 m, very leafy. (6) Few flowers and seed pods, height 1.7 m. twining, very leafy.

Source: Milford, 1967.

Minimum germination and quality for commercial sale. Seventy percent germination with a purity of 98 percent and a maximum hard-seed content of 10 percent in Queensland. Seed is germinated at 25°C.

Cultivars. There is only one cultivar, cv. Murray, which is more robust and vigorous than other ecotypes so far tested. Some more prostrate new ecotypes are under test.

Diseases. Susceptible to *Phaseolus* virus 2 (Wilson, 1958). Shows field resistance to little-leaf (Hutton and Grylls, 1956).

Pests. Seedlings are severely attacked by bean fly (*Melanagromyza phaseoli*). Nematodes attack the roots in sandy soils, the main species being *Meloidogyne incognita* (Koford and White) Chitwood.

Main attributes. An excellent pioneer species giving rapid growth, nitrogen fixation and early feed in pasture mixtures in the establishment year. Can be grazed without damage to the permanent legumes, which helps in early weed control (Bryan, 1968a). Is adaptable to wet conditions and to poor soils. Seeds heavily. Andrew (1964) has found it particularly suitable as an indicator plant in nutrition studies.

Main deficiencies. It is only an annual or biennial and does not persist in pastures at Beerwah, southeastern Queensland (Bryan, 1968a). Is susceptible to bean-fly and nematode attack. Because of seed shattering, it is difficult to obtain an effective seed harvest.

Performance. It has not been widely tested as a major constituent of tropical pastures. Bryan (1968a) and Ebersohn (1969) stressed its value as a pioneer species to give an early return on the capital invested in pasture improvement. At Rodd's Bay (central Queensland), Bryan and Shaw (1964) obtained a live-weight gain of 404 kg/head and 90 kg/ha on *Paspalum plicatulum*/ phasey bean pastures in $7^1/_2$ months of grazing in the first year, stocked at 1 beast/1.8 ha. The yield from native pasture alone was 185 kg/head and 19 kg/ ha stocked at 1 beast/4 ha.

Young, Fox and Burns (1959) found that it produced up to 495 kg DM/ha in the second autumn in a grass/legume mixture at Gayndah, southeast Queensland, and produced high-quality fodder in late autumn. However, it did not persist longer than two years and gave no winter grazing because of defoliation by frost.

Main reference. Paltridge (1955).

Macrotyloma spp.

Macrotyloma axillare (E. Mey.) Verdc. (see Colour plates XXVI, XXVII)

Synonyms. Dolichos axillaris E. Meyer.; *Clitoria vividiflora* Boulton.

Common name. Archer axillaris (Australia).

Description. A trailing and twisting legume with hairy stems and bright green ovate leaflets 3 to 5 cm long and 3 cm wide, slightly pubescent but glossy. Racemes usually of three flowers in the leaf axils on short stalks. Flowers greenish yellow, 1.2 to 1.5 cm long. Pods slightly curved, hairy, 3 to 5 cm long and 6 mm wide, seven- to eight-seeded. Seeds mottled light and dark (Davies and Hutton, 1970).

Distribution. Widespread in tropical Africa, reaching Natal in the south, also in Madagascar and Sri Lanka (Hutchinson and Dalziel, 1954).

Season of growth. Short-lived summer perennial which grows early in the spring before most other summer legumes.

Temperature for growth. It is very heat-tolerant and has a high degree of cold tolerance (Luck and Douglas, 1966). It is susceptible to frost but will recover quickly and well (Downes, 1966).

Latitudinal limits and altitude range. Probably about 25°N and S. Grows from sea level to 1 230 m in Panama.

Rainfall requirements. It requires more than 1 000 mm and is tolerant of high rainfall in Panama up to 3 000 mm.

Drought tolerance. Quite good; produced well on alluvial and basalt soils in the dry season in north Queensland, Australia (Downes, 1967). Has a very strong root system. Farinas (1966) reported it to be drought tolerant in the Philippines.

Tolerance of flooding. Has no tolerance to waterlogging (Farinas, 1966).

Soil requirements. Has a wide tolerance, from sands to clays — as long as they are well drained. It requires a pH in excess of 5.5. Is fairly tolerant of salinity.

Rhizobium *relationships.* A promiscuous species; nodulates with native

346

rhizobia. It is an advantage to inoculate with the cowpea strain, the Australian one being CB 756 (Norris, 1967).

Ability to spread naturally. Spreads quite well naturally (Downes, 1967). Some seed passes undigested through the animal and will germinate in dung pats.

Land preparation for establishment. Prefers a good seed bed, but will establish on roughly prepared areas.

Sowing methods. Drill or broadcast the seed. Farinas (1966) states that it will grow from cuttings. It has been established in pangola and *Brachiaria* grass swards, but is soon beaten by grazing and plant competition.

Sowing time, rate, depth.Sow in late spring to summer, at 1 to 2 kg/ha as a pure stand, 0.5 to 1 kg in mixtures, at a depth of 1 to 2.5 cm, lightly covered.

Number of seeds per kg. About 120 000. Percentage of hard seeds is low.

Seed treatment before planting. Not required to break dormancy. Inoculation is not required, but preferable. Pelleting necessary only to protect rhizobia; pellet with rock phosphate (Norris, 1967). Insect and disease control is not necessary.

Nutrient requirements. It responds to phosphorus. Tutt and Luck (1969) at Cooroy, in coastal Queensland, fertilized it on establishment with 370 kg/ha molybdenized superphosphate (0.06 percent Mo), with a similar dressing a little later to encourage it to smother *Eupatorium*. Normally the annual maintenance fertilizer would be 120 to 250 kg/ha superphosphate. Nitrogen at 100 kg/ha suppressed growth (Parbery, 1967a).

Response to light. Grows very well with tall-growing grasses.

Compatibility with grasses and other legumes. It grows well with *Setaria*, *Paspalum* and *Panicum* spp. and with *Chloris gayana*. Also combines well with *Desmodium* spp., *Glycine* and siratro (Luck and Douglas, 1966).

Ability to compete with weeds. Excellent. Tutt and Luck (1969) found that it successfully controlled the shrubby *Eupatorium* in southeast coastal Queensland if allowed to grow vigorously from June to September.

Tolerance of herbicides. Tolerates 2,4-DE as a post-emergent spray at rates of up to 2.5 kg/ha active ingredient.

Vigour of growth, growth rhythm. It has a fairly strong seedling. It is a vigorous, summer-growing plant.

Nitrogen-fixing ability. It appears to fix ample nitrogen.

Response to defoliation. Moderate for grazing and trampling (Farinas, 1966).

Grazing management. Tutt and Luck (1969), using it to smother *Eupatorium* in winter and early spring (June-September), grazed it from September to mid-February and conserved it for the former period.

347

Normally, the plant should be grazed lightly until it becomes established. If protected while seeding, seed will shatter and later germinate to thicken the sward.

Rotational grazing to a height of 15 cm will maintain its productivity.

Response to fire. Recovers well from the passage of fire (Downes, 1966).

Breeding system. Self-fertile. Chromosome number 2n = 20.

Dry- and green-matter yields. In the Kimberley area of northern Australia, Parbery (1967a) obtained 15 509 kg DM/ha without added nitrogen and 11 260 kg/ha with 100 kg/ha of added nitrogen on Cunnunurra clay under irrigation.

Suitability for hay and silage. No record of hay making has been found, but if care is taken to preserve the leaf, it should make good hay. No record of silage being made has been found in the literature.

Value as standover or deferred feed. Fairly good in frost-free areas, as leaf will remain on the plant (Downes, 1966); it is useful for winter feeding, as it is more palatable at maturity.

Feeding value. Chemical analyses and digestibility: it contains 11.9 percent crude protein in full growth, declining to 6.2 percent when seeding (Parbery, 1967a). Palatability: many reports claim that *M. axillare* is not initially palatable (Snook, personal communication), but that when cattle become used to it they find it very palatable (Luck and Douglas, 1966).

Toxicity. Nontoxic, but its bitter taste tends to deter stock. They become used to it as it matures.

Seed harvesting methods. Pods are set just above ground level, which makes seed harvesting difficult. The seed pods ripen on the stem. If cut and put on a tarpaulin or a concrete floor, the stems take up to a month to dry.

Seed yields. Seed production is poor because of sparse flowering. Davies and Hutton (1970) give seed yields of 200 to 560 kg/ha, but 100 to 150 kg/ha is more common.

Minimum germination and quality for commercial sale. Minimum of 60 percent germination, a maximum of 10 percent hard seed, and 97.5 percent purity in Queensland. It is germinated under cover at 25°C after cutting the seed coat (Prodonoff, 1968).

Cultivars. There is only one cultivar, cv. Archer, registered in Australia.

Diseases. It is attacked by little-leaf, but in the wet areas of Colombia, Costa Rica, Guatemala and Panama it is remarkably free from diseases and pests, much more so than desmodium and siratro.

Pests. It suffers little from pests, probably because of a little astringency. It has some tolerance to the *Amnemus* weevil.

Main attributes. Drought and disease resistance and its vigorous growth in summer. Valuable grazing in early winter and early spring.

Main deficiencies. It is not very resistant and there is difficulty in seed production.

Performance. It has not been tested in grazing trials as yet, but its drought resistance and ability to smother weedy species are well known. Chatterjee and Singh (1966) have found it promising at the Agricultural Research Institute at Ranchi, India, and it has also shown preliminary promise in Ghana, Guatemala and Panama.

Main reference. Luck (1965a).

Macrotyloma uniflorum (Lam.) Verdc.

Synonyms. Previously (erroneously) called *D. biflorus; D. uniflorus*.

Common name. Leichhardt biflorus.

Description. Climbing annual herb with slender, slightly hairy stems. Three oblong leaflets, blunt at the apex, the terminal one 1.8 to 2.5 cm long, the lateral ones very unequal sided. Flowers yellow or greenish yellow, one to three on very short pedicels in the axils of the leaves. Pods linear-oblong, slightly curved, sessile, 2.5 to 5.5 cm long, smooth or slightly hairy, six- to eight-seeded, tipped with a persistent style 0.6 cm long (Andrews, 1952).

Distribution. Widespread in eastern Africa, southern Africa, India, Malaysia, the West Indies and Australia.

Temperature for growth. A summer-growing annual, it requires hot moist weather for maximum growth. Ludlow and Wilson (1970) obtained only 8 percent of the dry matter, 25 percent of the growth rate and 4.5 percent of the leaf area at 20°C as that yielded at 30°C. It is completely intolerant of frost, but usually seeds before the frosts and regenerates in the summer from seed.

Latitudinal limits and altitude range. About 28°S latitude; about sea level to 1 000 m altitude. Grows well in the central Tanzanian plateau at about 1 000 m.

Rainfall requirements. Has a rainfall range of 500 to 2 500 mm but is more common in the drier areas.

Drought and flooding tolerance. Very good drought tolerance; seeds early and hays off. Will not tolerate flooding.

Soil requirements. Adapted to a wide range of soils from granitic sands to latosols and heavy clays. Its pH range is about 6.0 to 7.5. It is fairly tolerant of salinity.

Rhizobium *relationships*. A promiscuous legume, it usually nodulates with native rhizobia. It is preferably inoculated before sowing with the cowpea

349

Figure 67. *Macrotyloma uniflorum*

strain, the present Australian recommendation being CB 756 (Norris, 1967).
This strain produces black nodules on *Macrotyloma uniflorum* and *Lablab
purpureus* (Cloonan, 1963).

Ability to spread naturally. Good; it occurs spontaneously in old cultivations
and along roadsides in the drier areas of Tanzania.

Land preparation for establishment. Does best in a well-prepared seed bed
but will establish with little ground disturbance.

Sowing methods. Seed can be drilled or broadcast. Can be sown with the early
cultivation of maize so that it can mature with it and provide feed with the
maize residues after harvest. It can be easily established in natural pastures
(Downes, 1966). Staples (1966) showed that it could be established by drilling
in 15-cm rows on burned, disced land. Sow at 1 to 1.5 cm and lightly cover.
Sow in spring to early summer — late summer planting is unsatisfactory. Sow
1 to 3 kg/ha.

Number of seeds per kg. 75 000. Percentage of hard seed is not high.

Seed treatment before planting. Not necessary to break dormancy. Inoculation is not necessary, but desirable. Pelleting is not necessary.

Nutrient requirements. It will grow in many soils without fertilizer, especially
in old crop land. For best results, sow with 130 to 250 kg superphosphate per
hectare.

Photoperiod and light response. A short-day plant, it flowers in autumn. It
can climb tall grasses, crops and weeds.

Ability to compete with weeds. Good — its vigorous growth tends to smother
them.

Vigour of seedlings, growth and growth rhythm. Grows vigorously in summer,
seeds early and then drops its leaves in autumn to early winter.

Nitrogen-fixing ability. Not tested.

Grazing management. It is essentially a dry-season feed reserve in the form of
pods hanging from tall vegetation. The pods are indehiscent and retain the
seed (Downes, 1966). It does not appear to be very palatable when green. Do
not graze during the summer, but allow it to seed and then utilize it during
winter and spring.

Response to fire. Easily killed by fire.

Breeding system. Self-fertile; chromosome number 2n = 20.

Dry- and green-matter yields. Staples (1966) records 6 600 kg DM/ha/year, of
which the seed may contribute 2 200 kg.

Suitability for hay and silage. Collected in early autumn with care taken to
preserve the leaves, it would make good hay. Sown with sorghum or maize for
support, it should make useful silage.

Value as standover or deferred feed. Excellent as seed pods.

Feeding value of the plant. No analytical data are available.

Toxicity. None reported.

Seed harvesting methods. At present, by hand, but with a companion crop grown for seed purposes, it could be easily harvested mechanically.

Seed yields. Heavy yields recorded by Staples (1966) — about 2 200 kg/ha.

Cultivars. Only one — cv. Leichhardt, which was introduced from Kenya as CPI 26260. Commercial seed no longer available.

Diseases and pests. Mainly pod rot in wet situations. Pod borers can cause some damage, and rodents harvest the seed.

Main attributes. Rapid summer growth, heavy seeding habit, and drought tolerance.

Main deficiencies. Short life and autumn and winter defoliation.

Main reference. Staples (1966).

Mimosa spp.

Mimosa pudica L. (see Colour plate XXVIII)

Synonyms. *M. hirsuta* HBK; *M. pudibunda* Willd.

Common names. Sensitive plant (Australia), sien-sien (Suriname), hila hila (Hawaii).

Description. A prostrate to semierect herb, 0.3 to 0.8 m tall, often forming a small compact bush about 0.6 m tall; branches glabrous or pubescent, heavily armed with broad-based, black-tipped recurved thorns. Leaves small, sensitive, bipinnate; leaflets about 16 pairs; petioles up to 4 cm long. Inflorescence of axillary headlike spikes, flowers pinkish, sessile. Pods usually in a dense cluster, 10 to 15 mm long and 4 mm wide, flat, segmented, somewhat constricted between the articulations, one-seeded segments (Woodson and Schery, 1950).

Distribution. Native to tropical America from southern Mexico to Central America, West Indies, Hawaii, Fiji and Australia.

Characteristics. It is very common in the high rainfall areas of Fiji and Hawaii, where it is grazed by livestock in areas in which it is difficult to establish other legumes — e.g. in the humid and fog-prone windward slopes of high mountains (Ripperton and Hosaka, 1942). In Fiji, it is a useful forage for dairy cows in the wet zone in the winter, but in the dry zone near Sigatoka it is a troublesome weed, germinating prolifically wherever the ground is disturbed.

Mucuna spp.

Mucuna pruriens (L.) DC var. *utilis* (Wight) Burck.

Synonyms. *Stizolobium deeringianum* Bort.; *S. aterrimum* Piper and Tracey; *S. niveum* Kuntze.

Common names. Velvet bean (Australia, United States, southern Africa), pica-pica (Venezuela), frijol terciopelo (Latin America), Bengal bean (India).

Description. Vigorously growing, trailing vine (grown mostly for green manuring or temporary pasture), slender, slightly ridged, and extending over 6 m in length (grown on supports, they may grow to over 10 m). Leaves large and smooth, the terminal leaf being rhomboidal-ovate and the lateral ones oblique, 20 to 25 cm long and 7.5 to 12.5 cm wide. Flowers borne in long racemes, white with purple tinge. Pods, 10 to 14 in a cluster, are borne singly, 10 to 12.5 cm long, curved, with a greyish-white pubescence of short, silky hairs. Seeds are black (Mauritius bean) and glossy when mature, 1.2 to 1.5 cm long and 0.9 to 1.1 cm broad with raised white hilum half as long as the seed. Each pod contains three to five seeds (Paul, 1951). Velvet bean seeds are commonly mottled.

353

Distribution. Probably native to southern Asia and Malaysia, now widely distributed in the tropics.

Characteristics. Velvet beans require a hot moist climate for maximum growth, and are usually used as cover crops where there is a long frost-free growing season during the wet months. They grow from sea level to 2 100 m in Kenya — apparently the altitude limit. Rainfall range of the species is from 650 to 2 500 mm. Tolerates a wide range of soils, from sands to clays, and will grow on soils of appreciable acidity. Nonspecific in its *Rhizobium* requirements. Mes (1959) found that a low night temperature of 10°C was more harmful to the growth of the plants when they were dependent on *Rhizobium* than when nitrogen was applied. Nitrogen fixation, increased with increasing night temperatures, was poor below 18°C.

Velvet beans will grow in roughly prepared land, provided the seed is covered or rain falls soon after seeding. Better results are obtained by sowing into a prepared seed bed. For a green manure crop, the seed is usually broadcast onto ploughed land following the ploughing out of the previous crop. It can be sown in furrows or dibbled in at 1 × 1 m spacings with two seeds per hill. It can also be sod-seeded into existing pastures (Doherty, 1963b). Seed is sown in early summer at 22 kg seed/ha. About 1 000 seeds weigh 1 kg. The crop is usually grown alone for green manure and for silage, because sown with maize or sorghum it tends to pull the crop down, making harvesting difficult.

Velvet beans are slow to start and, if drill-sown, one or two interrow cultivations improve early development. When established, the crop smothers weeds effectively. King, Mungomery and Hughes (1965) stated that a crop of velvet beans yielded 17.4 tonnes of green material per hectare in north Queensland, and that its nitrogen content was 331 kg/ha, equivalent to 1 615 kg sulphate of ammonia per hectare. Doherty (1963b) obtained a yield of 11 176 kg/ha of green matter from velvet beans sod-sown into a Rhodes grass/green panic pasture at a seed rate of 22 kg/ha and fertilized with 264 kg/ha molybdenized superphosphate in coastal Queensland, Australia.

When grazing the crop, the animals are turned in after the pods are mature, some 170 to 220 days in Hawaii (Takahashi and Ripperton, 1949). These authors obtained 19 tonnes green forage and 3.85 tonnes seed per hectare in Hawaii with a variety named Lyon. Because of its dense matted growth, velvet bean is difficult to harvest and cure for hay. French (1935, 1937) made small quantities for experimental feeding of sheep at Mpwapwa, Tanzania. It is grown fairly extensively as a silage crop, but as it is hard to harvest, flail-type harvesters do a better job than mowing machines. It is better to grow velvet beans and the companion crop separately and mix them when ensiling.

Axtmayer, Hernandez and Cook (1938) had the plant analysed when in flower together with the pods. The plant contained 15.65 percent crude protein, 34.47 percent crude fibre, 1.21 percent calcium and 0.16 percent phosphorus. Pods contained 16.36 percent crude protein, 27.95 percent crude fibre, 1.02 percent calcium, and 0.18 percent phosphorus. French (1935, 1937) found that velvet-bean hays contained 17.87 and 12.95 percent crude protein and 10.88 and 7.12 percent digestible protein, respectively. Krause (1911) stated that velvet bean was the least palatable of the forage legumes he had tested. Crops are harvested by hand-picking the pods or mowing the crop, drying and subsequently threshing.

Numerous cultivars are in use. The Mauritius bean is a black-seeded type used in green manuring, while the Australian velvet bean has a large mottled seed. A variety J52 yielded 1 980 kg DM/ha in Tanzania, followed by 'Somerset' with a yield of 1 780 kg/ha. The cultivar Somerset has been found useful in South Africa, Tanzania, and Queensland, where it has superseded 'Mauritius'. Other cultivars include 'Stringless', 'Osceola', and 'Bunch' (Queensland), J54, J77, 'Local White' and 'Local Black' in Tanzania. The latter produces green foliage long after 'Somerset' and others have died.

In north Queensland, Sturgess and Egan (1960) isolated the fungus *Phytophthora dreschleri* from wilt-affected crops (it had come from southern Africa in seed contaminants). All velvet bean varieties are affected, but cowpeas are resistant.

The main attribute of velvet beans is their long growing season in frost-free environments, which enables them to protect the soil throughout the wet monsoonal season.

355

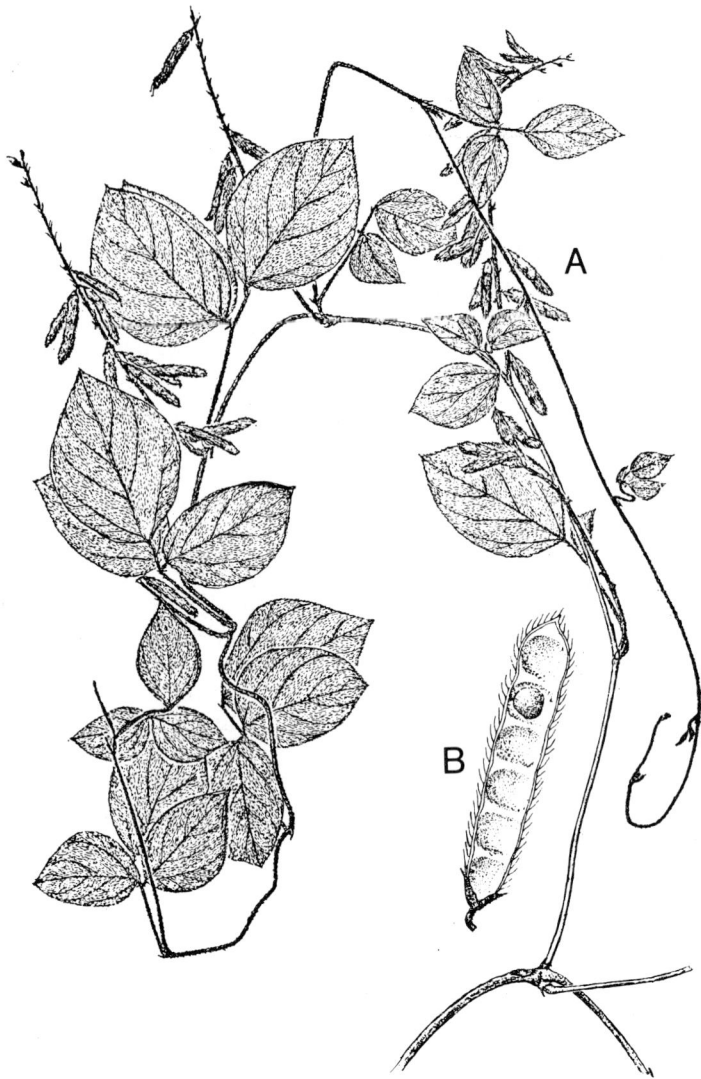

Figure 68. *Neonotonia wightii.* **A**-Fruiting branch **B**-Pod and seed

356

Neonotonia spp.

Neonotonia wightii (Arn.) Lackey (see Colour plates XXIX, XXX)

Synonyms. Notonia wightii Arn.; *Glycine javanica* L.; *G. wightii* (Arn.) Verdc.

Common names. Glycine (Australia, Kenya); soja perene or perennial soybean (Brazil, Colombia); fundo-fundo (Tanzania).

Description. Herbaceous perennial with strong taproot and trailing, climbing and twining stems. Stems slender and well branched, and under grazing may arise from a crown below the soil surface; runners frequently root at the nodes and are moderately hairy. Leaves, pinnately trifoliate with ovate leaflets 5 to 10 cm long and 3 to 6 cm wide, with short hairs on both surfaces, and small triangular stipules. Elongated racemes from 4 to 30 cm long bear clusters of white or violet flowers 5 to 8 mm long. In some varieties the flowers become yellow or orange-yellow. Pods are hairy, straight or slightly curved, 1 to 4 cm long, about 3 mm wide, with three to eight seeds. Seeds vary in size, shape and colour, depending on variety. The pods differ from those of *Teramnus* in that the tips are straight while *Teramnus* pods have a distinctly curved tip.

Distribution. Neonotonia wightii belongs to the subgenus Glycine, which is entirely African in origin. It is found in the East Indies, tropical Asia, Ethiopia, through east and central Africa and down to southern Africa, where it occurs in the warmer parts of the Transvaal, Natal and east Cape Province.

Season of growth. A summer-growing perennial, but the cultivar Tinaroo makes some useful winter growth under frost-free conditions at Kairi, north Queensland, Australia (lat. 17°17′S), until May.

Temperature for growth. Wutoh, Hutton and Pritchard (1968c) showed that the optimum temperature for growth and seed production was a day/night regime of 27/22-16°C. High root temperatures affect growth (Tow, 1967).

Growth slowed down at 16°C and ceased at 13°C. At Campinas, Brazil (lat. 23°S), there was little growth during June (12.8°C minimum), July (11.4°C) and August (13°C). Leaf shedding occurred in the Burdekin Delta, Queensland (lat. 19°S) at 1.5°C (Allen, 1961a, b).

Frost tolerance and regrowth after frosting. Frost causes leaf shedding, but the plant has some degree of frost tolerance. Tolerance to cold is greater in the Tinaroo cultivar than in the cultivars Cooper and Clarence in Australia. Glycine is more tolerant to frost than *Centrosema pubescens*, and leaf shedding is less than with *Pueraria phaseoloides*. Glycine grows vigorously from the stems and crowns in warm weather (when moisture is adequate).

Latitudinal limits. Extends in Australia to approximately 29°S latitude (Wollongbar Agricultural Research Station, lat. 28°50'S).

Altitude response. In Kenya it grows from the lowlands up to 2 450 m at the latitude of the equator. In Colombia, it occurs from sea level to 1 800 m.

Rainfall requirements. Best adapted to areas of summer rainfall of 750 to 1 500 mm, but newer varieties are expected to move into drier areas. Does not perform as well in areas of higher rainfall. Outstanding performances are recorded at Atherton, Queensland (1 420 mm); Campinas, Brazil (1 200 mm); Arusha, Tanzania (1 000 to 1 500 mm); and Palmira, Colombia (1 200 mm). The Tanzanian variety Moshi (T2063) grows naturally at Itigi (rainfall 600 mm).

Drought and flooding tolerance. Reasonably drought-tolerant (in Australia the cultivar 'Cooper' is more tolerant than 'Clarence' and 'Tinaroo'). It grows slowly during dry spells but recovers quickly when favourable conditions resume. A well-established plant has a deep persistent taproot (see Colour plate XXXI). Tolerance to flooding is low; it requires good drainage.

Soil requirement. Performs best in deep, freely drained latosolic soils derived from basic igneous rocks, on self-mulching black soils and fertile alluvial soils. It is not suited to acid podzolic soil or solodic soils. It prefers free-draining loams to clays of basaltic or alluvial origin. Prefers a pH above 6.5, but can grow at pH 6.0 if adequate lime is present. Souto and Döbereiner (1968) reported manganese toxicity at pH 5.1 to 5.7, which disappeared on liming to pH 6.5.

Glycine has been shown to be reasonably tolerant of salinity — more so than other tropical legumes. Although salinity severely inhibits growth, nodulation and nitrogen fixation, it does not affect nitrogen and phosphorus content. Nitrogen-fertilized plants are more tolerant than nodulated plants. Salinity tolerance varies with varieties and is inherent (Gates, Haydock and Little, 1966; Gates, Haydock and Claringbold, 1966).

Rhizobium *relationships*. Use the cowpea type — the current Australian culture is CB756 (1970). In Brazil (1970) cultures SFS288 and SFS404 are used. Kennedy (1962) found that glycine nodulated with native rhizobia from *N. wightii, G. maxima, G. tomentella, G. tabacina, Centrosema pubescens* and *Vigna unguiculata*.

Ability to spread naturally. Not good.

Land preparation for establishment. The seed bed should be well prepared with as complete destruction of the original vegetation as possible.

Sowing methods. Drill in the seed with standard seed drills, a "Planet Junior" seeder for small areas, or a maize planter. When using a maize planter, watch for blockage of the planting plates by the gummy inoculated seed (Murtagh and Wilson, 1962). Seed may also be broadcast. Roll after planting. In pure stands in Brazil, it is sown at 2.5 kg/ha in rows 0.5 m apart with a group of 20 seeds placed in holes 0.5 m apart in the row.

Generally not oversown into existing pastures without some cultivation. Murtagh (1963) successfully sod-seeded it into a *Paspalum/Axonopus* sward suppressed with herbicide containing 27.7 percent acid equivalent of the sodium salt of 2,2-DPA and 12.5 percent acid equivalent amitrole, seeded at 5 kg/ha with adequate neutralized fertilizer.

Sowing depth, time and rates. Sow at 1 to 2 cm and lightly cover with harrow or roll. Murtagh (1970) proved that glycine needs relatively cool temperatures for the first two days of germination. Laboratory germination is seriously impaired at temperatures above 37°C. Sow in early summer at 0.5 to 3 kg/ha depending on cost of seed and desired rate of establishment. Six kg/ha are used in Colombia, 2.5 kg/ha in Brazil.

Number of seeds per kg. 90 000 for flat-seeded Zimbabwean type to 330 000 for fine violet glycine ssp. *micrantha* (Bogdan, 1966). Percentage of hard seed is quite high.

Seed treatment before planting. To break dormancy: scarify with special machine (Neme, 1966b); scarify in cement mixer with stones; or treat with concentrated sulphuric acid (sp. gr. 1.8) for 25 minutes, drain seed for two minutes, wash thoroughly in water and dry (Prodonoff, 1969); or treat with $24N$ or $36N$ sulphuric acid for seven minutes, then wash (Black, 1968); or put seed in boiling water for one minute (Naveh, 1966) or let stand overnight (Lotero, personal communication). Inoculation: use *Rhizobium* strain CB756 (Australia), SFS288 or SFS404 (Brazil). Pelleting: lime pelleting gives a response in Brazil in high manganese soils (Döbereiner and Aronovich, 1966). Lime-pellet only with soils with pH below 4.5 or on high manganese soils (Norris, 1967). Insect control: to control ants, dust with lindane at a

359

strength of 5.3 g/kg dry seed. As lindane affects *Rhizobium*, the inoculated seed should be lime pelleted (Gartner and Fisher, 1966).

Nutrient requirements. Glycine requires sufficient calcium, phosphorus, sulphur and molybdenum. On fertile soils, no fertilizer may be required. On previously cropped land on the Atherton Tableland, Queensland, 450 kg/ha molybdenized superphosphate are used for establishment (Gartner and Fisher, 1966) on red latosolic soils. The amount of molybdenum required for glycine is about twice that for other tropical legumes (Mears and Barkus, 1970) and 312 g per ha sodium molybdate are desirable (Luck and Douglas, 1966). The annual fertilizer dressing for maintenance should be 250 kg single superphosphate per hectare.

● Calcium. A concentration of 1.6 to 2.8 percent of the dry matter of the tops was measured by Andrew and Hegarty (1969). Glycine responded to lime in the presence of molybdenum (203 g/ha) (Mears and Barkus 1970).

● Phosphorus. The critical level for P in the dry matter of the leaves at the immediately preflowering stage is 0.23 percent (Andrew and Robins, 1969a). Glycine yielded 42 percent of its maximum yield at the equivalent of 250 kg single superphosphate per hectare. Its maximum yield was at 1 350 kg/ha.

● Potash. Deficiency symptoms of potash show when the K percentage of the dry matter at the immediately preflowering stage is 0.80 percent (Andrew and Robins, 1969c). The first visual sign of potash deficiency in this species was peripheral necrotic spotting on the lower leaflets of the plants, irregular in shape, interveinal, and equally visible on both surfaces of the leaflet. Centres of the spots were light brown surrounded by darker brown and circled by an area of chlorotic tissue. Necrotic areas increased in number and size with increasing severity of deficiency and finally coalesced to give a necrotic leaflet margin at the tip and backward along the leaflet margins; there was a tendency for necrotic leaflet margins to curl upward and inward. Old leaves at the base of the plant were extremely necrotic but did not show curling effect. Fully necrotic leaves became papery white in colour but there was little leaf abscission even at that stage (Andrew and Pieters, 1970a).

Toxicity levels and symptoms. Manganese — the toxicity threshold value for Mn is 560 ppm (Andrew and Hegarty, 1969). Glycine is more sensitive to excess Mn than white clover. In seedling growth the first visual symptom of toxicity was slight interveinal chlorosis of the young fully expanded leaves; however, the chlorosis was initially restricted to the marginal areas of the leaflets, giving the appearance of blotches on the leaflet margin, particularly toward the tips. At the same time, small irregularly shaped brown spots occurred on the surface of the primary leaves. With intensification of toxicity, interveinal chlorosis extended to the midrib area and brown markings appeared. Young shoots were pale in colour and showed uniform interveinal chlorosis. In older plants, the brown markings were on or adjacent to the veins, with prominence on the underside of the leaflets, and became purplish in colour. In plants suffering from severe toxicity, leaf puckering on the newer fully expanded leaves occurred mainly toward the base of the midrib of the leaflets, the raised portions being above the upper leaf surface and usually associated with severe interveinal chlorosis. Leaflet margins were not usually puckered but showed an undulating margin. Newly emerging leaves were fully formed but were severely chlorotic, accompanied by a bronze colour over the full leaf. There were no visual effects on the petioles or stems (Andrew and Pieters, 1970b).

Aluminium — glycine is no more tolerant to excess than white clover (Andrew and Robins, unpublished).

Vigour of seedling, growth and growth rhythm. Glycine is slow to start because of nodulation difficulties. In Zambia, it does not gain prominence until the second year (van Rensburg, 1967). In Brazil, it is often hand weeded during early establishment. After a slow start, it becomes quite vigorous. In Colombia, the plants cover the soil in two months after seeding.

Response to photoperiod and light. Glycine is a short-day plant. Cultivars Clarence and Cooper are early maturing types; cv. Tinaroo is a late-flowering type.

Response to light is fair; it grows with medium-height grasses, though it does better in pure swards. Basic light and CO_2 responses are given by Ludlow and Wilson (1970).

Compatibility with grasses. Combines well with *Panicum maximum, P. maximum* var. *trichoglume, Setaria anceps, Chloris gayana, Melinis minutiflora* and *Pennisetum purpureum*. In *Digitaria decumbens* it lasts two years at Campinas, Brazil, and at least three years in the fertile Cauca Valley of Colombia. At Atherton, north Queensland, it is sown at 11 kg/ha with Rhodes grass at 3.3 kg/ha (Tow, 1967). In Brazil, it is sown at 0.5-m intervals at 3 kg/0.40 ha in three rows between colonial guinea rows sown 2 m apart; with elephant or napier grass, at 0.5-m intervals at 6.5 kg/ha in one row between two rows of grass spaced 0.5 m apart; with pangola grass in alternate rows 1 m apart, the glycine planted at 20 seeds 0.5 m apart; with molasses grass the glycine is planted on the square with 20 seeds at 0.5 m apart and the molasses grass subsequently broadcast over the glycine (Menegario, 1964).

Ability to compete with weeds. Early growth suffers from weed competition, but when established can suppress weeds, even *Imperata cylindrica*. Weeds can be suppressed with a slasher and, thereafter, glycine can assume control. Savage (1970) found that cutting at 15 cm every nine weeks gave weed-free swards at Redland Bay, Queensland, on red latosolic soils.

Tolerance of herbicides. Bailey (1970) found that tolerance of glycine to 2,4-D and 2,4-DB improved with age, but 2,4-D is too damaging to use for overall spraying before glycine is three to four months old, and 0.82 kg of acid equivalent per hectare should not be exceeded. A check to growth must be expected. 2,4-DB is somewhat more selective than 2,4-D, and 1.1 kg of acid equivalent per hectare can be used at five weeks of age if other means of weed control are unsatisfactory. At three to four months of age, 2.2 kg of 2,4-DB acid equivalent per hectare can be used with safety. Remember, however, that 2,4-DB is slow to kill many weeds.

Glycine shows useful tolerance to diquat and this may be used at 140 g diquat cation per hectare from five to eight weeks of age, and 280 g per hectare on established glycine. However, seedling glycine must be healthy at the time of spraying; otherwise severe damage may result.

Nitrogen-fixing ability. At Palmira, Colombia, glycine pastures fix 160 kg N/ha/year (Lotero, personal communication).

At Campinas, Brazil, it has been shown to fix 170 to 290 kg/*alqueire*/year or about 70 to 120 kg/ha (Menegario, 1964). In Kenya, Gethin-Jones (1942) found that if fixed 165 kg N/ha/year for four years, and on another East African site it fixed 175 kg N/ha (= 875 kg sulphate of ammonia) annually for the first five years and then 110 kg N/ha (= 550 kg sulphate of ammonia) per year for the following four years as the soil nitrogen percentage rose. Colman, Holder and Swain (1966) recorded that it fixed 156 to 203 kg N/ha over three seasons, but in a dry season at the same station, Mears (1967) recorded only 16.5 kg N/ha. Edye (1967) showed that in fertile black earths, nitrogen fixation by glycine is low.

Response to defoliation. Grazing or cutting to 5 cm over a two-year period reduced the stand to 12 to 15 percent (Whiteman, 1969) but it survived better than *Macroptilium atropurpureum*, *Desmodium uncinatum* and *Lotononis bainesii*. Grazing every four weeks at Wollongbar, New South Wales, Australia, reduced plant numbers compared with eight- and ten-week intervals. Savage (1970) recorded highest yields from cutting at 3.75 cm every nine weeks. More frequent cutting at this height reduced yields. Cutting at 5 cm caused a significant depression of yield at Lawes (lat. 28°S, rainfall 680 mm) on heavy black soil (Santhirasegaram, Coaldrake and Salih, 1966). Left ungrazed, the plant's leaves eventually drop and the production of new growth is considerably delayed (van Rensburg, 1967).

362

Grazing management. Glycine must be allowed to become established and cover the ground before animals are allowed to graze the pasture. Gartner and Fisher (1966) recommend the following schedule:

● In the first year, graze the pasture often enough but only long enough to remove the grass canopy and allow light to reach the legume. Cattle should not have time to graze seedlings. Weeds should be carefully controlled by slashing. Under normal conditions, guinea grass/glycine pastures should be ready to graze seven to eight weeks after sowing.

● In the second season, the legume is well established and it can be safely grazed back winter and spring. It is then shut up at the beginning of the wet season to build up leaf area.

● If pasture is unused in summer, grass dominates; if unused in winter, legume dominates. Graze rotationally in the warm wet months, when growth is fast, and graze continuously in winter. Stock at 60 to 120 adult cattle/ha when grazing.

● If pasture is to be saved for winter grazing, graze lightly in summer and spell during the autumn. If it is to be saved for dry spring or early summer grazing, it should be grazed heavily in the previous summer and allowed to build up green forage from autumn onwards. With deferred grazing, there is a danger of late winter frosts, so graze low-lying frost-prone areas early in the winter and use the slopes later. If the pasture becomes frosted, graze it immediately before palatability declines and leaves fall.

Menegario (1964) recommends the following grazing heights in Brazil:
    Pure glycine pastures — graze to ground level from 30 cm;
    Glycine + *Melinis minutiflora* — graze mixture from 30 to 10 cm;
    Glycine + *Digitaria decumbens* — graze mixture from 30 to 10 cm;
    Glycine + *Pennisetum purpureum* — graze mixture from 40-50 to 10 cm; and
    Glycine + *Panicum maximum* cv. Coloniao — graze mixture from 50-60 to 10 cm.

Response to fire. Moderate. Van Rensburg (1967) reports that, even at the height of the dry season, it produces green growth after burning. This would be governed by the store of subsoil moisture and how well the plants were established.

Breeding system. Self-pollinated and cleistogamous (Hutton, 1960) but some cross-pollination occurs (Bogdan, 1966; Hutton, 1970b). Chromosome number 2n = 22 (diploid) and 2n = 44 (tetraploid).

Dry- and green-matter yields. It is not very productive at low altitudes. Menegario (1964) recorded 120 tonnes of green material per *alqueire*, or

363

50 tonnes/ha/year, and Lovadini and Miyasaka (1968) 8 to 10 tonnes DM/ha at Campinas, Brazil.

At Wollongbar, New South Wales, Colman, Holder and Swain (1966) recorded up to 4 510 kg DM/ha, and over a three-year period an average of 3 850 kg DM/ha/year. The total yield of Kikuyu grass/glycine in this same pasture was 7 744 kg DM/ha.

Suitability for hay and silage. It is widely used for hay in Brazil. It is cut at the early flowering stage, dried and baled. The moisture content of the hay should be reduced to 12 percent (Lovadini and Miyasaka, 1968).

At Atherton, north Queensland, successful silage has been made with a mixture of green panic (*Panicum maximum* var. *trichoglume*) and glycine. The yield of green matter was 9.8 tonnes per hectare. Sixty kg of molasses per tonne of green material were added during the ensilage process. The final pH of the silage was 4.2 (Barker and Kyneur, 1962). Menegario (1964) records successful silage made from glycine combined with pangola grass, elephant grass and Coloniao guinea grass, and it is frequently ensiled in a mixture with 20 percent of fodder (sugar) cane.

Value as standover or deferred feed. Stands in the field fairly well, but drops its leaves when frost occurs. Provides good winter grazing in Brazil (Menegario, 1964).

Feeding value. Glycine is valuable pasture for cattle and pigs (Kyneur, 1960) and makes good hay and silage.

● Chemical analysis. Van Rensburg (1968) recorded 16.25 and 12.38 percent crude protein from young shoots and leaves respectively in Zambia, while Bogdan (1966) found 20.4 percent crude protein in the dry matter of the whole plant and 26.5 percent in the leaf in Kenya. An analysis of glycine hay and the digestibility of its components in Campinas, Brazil (Peixoto, de Moraes and Próspero, 1966), is given in Table 14.8.

● Digestibility. Compares favourably with best-quality lucerne and alfalfa hay, with equal digestibility of protein, superior use of fat and fibre and slightly lower digestibility of carbohydrate content. Higher digestibility than *Centrosema* and *Lablab purpureus*.

The digestibility of *Neonotonia wightii* at Wollongbar, New South Wales, according to Holder (1967) is given in Table 14.9.

Holder (1967) concluded that digestibility varied widely with the age of the pasture and the moisture regime under which the pastures were grown, and assessed that the digestibility of the organic matter would be about 10 units lower than a temperate legume at an equivalent stage of growth.

● Palatability. Good, improving as it becomes more vigorous in midseason;

364

TABLE 14.8 **Compositional analysis of *Neonotonia wightii* hay and digestibility of its components**

Component	Analysis	Digestibility
	(%)	
Dry matter	88.94	75.38
Protein	12.14	71.77
Fibre	30.39	54.55
Ash	2.25	44.43
Non-nitrogenous substance	35.87	68.91
Fat	7.59	
Ca	0.88	
P	0.28	

TABLE 14.9 **Digestibility of *Neonotonia wightii* at different stages of growth at Wollongbar, New South Wales**

Stage of growth	Crude protein	Digestibility	
		Dry matter	Crude protein
		(%)	
February – leafy	18.9	61.7	80.3
May – flowering	20.2	59.6	79.3
June – post-flower	14.9	55.7	72.2
July – seeded and frosted	12.9	56.2	66.6

maintains palatability into the autumn (Lychatchynsky and Steenmeyer, 1968 — unpublished).

Toxicity. Although oestrogenic substances are present in *Neonotonia wightii*, no breeding troubles have been reported (Colman, Holder and Swain, 1966).

Seed harvesting methods. In the Cauca Valley, Colombia, seed is hand harvested. In Brazil, the seed crop is mown, left in the field to dry for one to two days and then threshed through a stationary thresher. In Queensland, seed is harvested directly with an "All-crop" harvester when the tops of the plants are drying off. A two-year-old crop is used for seed, as the first year's crop is

365

generally weedy and has to be slashed with the slasher blade set high. Glycine is not easy to harvest due to the bulk of green material and uneven maturity.

Seed yields. For seed production, glycine can be sown in 1-m rows and cultivated interrow to control weeds (Cowdry, 1960). Wutoh, Hutton and Pritchard (1968a) found that a day/night temperature regime of 27/22 to 16°C was most suitable for seed production and growth.

At the IRI research station, Matão, Brazil, 1 000 kg seed/hectare have been harvested; at Fazenda da Prata, 750 kg/ha. In the Burdekin Delta, north Queensland, 500 to 750 kg/ha were harvested from an irrigated crop (Allen 1960), while at Kairi in north Queensland, 330 kg/ha can be obtained from an ungrazed pasture and 110 kg/ha from a crop which had previously been cut for hay earlier in the season (Kelly, 1963).

Minimum germination percentage required for sale. Germinable seeds, 60 percent, with a maximum of 10 percent hard seed, and 97.5 percent purity in Queensland, 1966; 60 percent germination (Colombia).

Cultivars

● <u>Australian</u>. 'Tinaroo', 'Cooper' and 'Clarence' (Barnard, 1967).

1. Cv. Tinaroo. Diploid 2n = 22; 132 000 to 143 000 seeds per kg. Stems develop brown coloration in the epiderms with age, are moderately hairy. Young shoots lack pigment. Stem hairs are semierect to appressed and point toward the base. Leaflets ovate-acute, almost glabrous at times; lower surface has a fine venation. Flowers are creamy white with obscure small violet streaks on the lower part of the standard. Seeds olive-green to light brown, occasionally mottled. Growth from seed is slower than 'Clarence' and 'Cooper' during the first year; it commences to flower in early to mid-June and seed matures in September (in frost-free areas). It forms a high percentage of hard seed. The original seed came from Kenya, was proved in Queensland and the cultivar released in 1962. It was proved and developed for commercial release at Atherton, north Queensland.

2. Cv. Cooper. Botanically, var. *moniliformis* (Hochst. ex A. Rich.) F.J. Herm. It is diploid 2n = 22; 150 000 to 160 000 seeds per kg. Stems are branched, slender and stoloniferous, pubescent. Stem hairs are white and reflective and produce an ash-silvery appearance. The inflorescence is strongly interrupted. Flowers white, with pink-violet streaks on the standard. Approximately five-seeded, hairy, constricted pods; seeds light brown. This

Figure 69. Harvesting *Neonotonia wightii* hay in Brazil (**Photo:** L.A.C. Lovadini)

cultivar can be distinguished from 'Clarence' and 'Tinaroo' by its ash-silvery vesture, its skew-shaped leaflet and the constrictions in its pods.

It flowers in Queensland in late April; matures four to six weeks ahead of 'Tinaroo', a little later than 'Clarence', and because of its late flowering it continues growth later into the autumn than 'Clarence'. It is more drought resistant and tolerates excess moisture better than either of the other two and is adapted to slightly less rainfall. It is more productive in its first year of growth.

The original Australian seed came from Kongwa in Tanzania in January 1959 as CPI 25702; it was proved and developed at Lawes, Queensland, and was released commercially by the Queensland Pasture Liaison Committee in August 1962.

3. Cv. Clarence. Botanically, var. *claessensii* (De Wild) Haumann. A tetraploid 2n = 44; 158 000 to 169 000 seeds per kg. Stems are coarser, less well branched than 'Cooper' or 'Tinaroo' and less stoloniferous: brown pigmentation of epidermal tissues and pronounced hairs over the whole plant, extending to young shoots. It is more pubescent over the whole plant. Both surfaces of the asymmetrical leaflets are hairy. Veins of the lower surface are prominent and rusty brown coloured. Interrupted inflorescence; flowers have obvious pink-violet marks on the standard. Pods are dark brown, hairy. Seeds olive-green to dark brown.

It is a little earlier flowering than 'Cooper', maturing seeds before frosts in northern New South Wales (lat. 29°S) and southern Queensland (lat. 27°S). It is considered more suitable than the other varieties for areas subject to early frosts. It is the first variety to start growth in the spring. The original seed came from southern Africa and, after testing at Grafton and Wollongbar in New South Wales, was released commercially in 1962. It was proved and developed on red latosolic soils developed on basalt at Grafton, 970-mm rainfall, latitude 29°42′S.

● Kenyan. 'Kenya white glycine' and 'Kenya violet glycine' (Bogdan, 1965).

1. 'Kenya white glycine' (K51394) - tetraploid. Leaflets of medium size to large, subacute, terminal leaflets mostly 4 to 6 cm long. Racemes medium to long, mostly over 15 cm. Flowers 7 mm long, white, often with a greyish-violet spot on the standard; occasionally the whole flower is slightly bluish in colour. After flowering, the corolla often turns yellow or orange-yellow in colour. Pods are straight, hairy, with the hairs directed toward the apex. Seeds 3 mm long, oblong, brown to black in colour. The weight of 1 000 seeds varies

from 5.5 to 7.0 g. Cross-pollination frequently occurs. A vigorous variety, with numerous leaves and satisfactory seedling qualities; is recommended over other types in Kenya. The original seed was collected by Strange from wild plants near Eldoret, Kenya.

2. 'Kenya violet glycine' (K51393) - diploid. Leaflets small, acute, the terminal ones mostly 3 to 4 cm long. Stems fine. Racemes 5 to 10 cm long. Flowers 5 mm long, violet in colour, not turning yellow after flowering. Pods slightly curved, hairy, hairs directed toward the base of the pod. Seeds 2 mm long, reddish-brown in colour. The weight of 1 000 seeds is about 2.5 g. It is a very uniform variety, apparently self-pollinating and belongs to the subspecies *micrantha* (Hochst ex A. Rich.) F.J. Herm. Although it is a small variety, with fine stems and small leaves, in mixed swards it may be only slightly less vigorous than K51394. Seeding qualities are good. The original seed was collected by Strange near Eldoret, Kenya.

● Brazilian. IRI No. 1 (SP1) is more tolerant of the high manganese content in some of the latosolic soils at Campinas, Brazil, and, under such conditions, outyields 'Tinaroo'. If manganese is not excessive, cultivars will outyield IRI No. 1 (Souto, 1969).

● Tanzanian. M218 goes into a drier climate of 700 to 800 mm at Kilosa (Rijkebusch, 1967), and cv. Moshi grows naturally at 600 mm at Itigi.

● Others. Cultivars under test worldwide include the following.

1. 'Choma' — at CSIRO, Australia; Turrialba, Costa Rica; and Marandellas, Zimbabwe.

2. 'Citrus' — at CSIRO, Australia; Beltsville, Maryland, United States; Maracaibo, Venezuela; and Fort Jameson, Zambia.

3. 'Kingsize' — at Beltsville, Maryland, United States; and Maracay, Venezuela.

4. 'Nelsfruit' — at CSIRO, Australia; Marandellas, Zimbabwe; and Beltsville, Maryland, United States.

5. 'Norval's Special' — at CSIRO, Australia; Marandellas, Zimbabwe; and Beltsville, Maryland, United States.

369

6. 'Nyasaland' — at Marandellas, Zimbabwe.

7. 'South African' — at CSIRO, Australia; Lilongo, Malawi; and Ilonga, Tanzania.

Diseases. In northern New South Wales, glycine is attacked by *Cercospora* leaf spot and *Sclerotinia sclerotiorum*. It is often attacked by *Xanthomonas* and *Pseudomonas* during wet weather in Zambia (van Rensburg, 1967). Bogdan (1966) reported small yellow rust spots on the leaves and stems caused by *Synchytrium dolici*; and *Rhizoctonia solani* can affect it in wet weather in Queensland.

Pests. The *Amnemus* weevil (*Amnemus quadrituberculatus*) attacks the roots of the plant in northern New South Wales (Braithwaite, Jane and Swain, 1958) and southeastern Queensland (Roe and Jones, 1966). A *Bruchus* weevil attacks seed at Kitale, Kenya (Bogdan, 1966). Seed dusting with Ceresan or other mercurial dust will protect seed supplies for planting.

Main attributes. An excellent legume for suitable soils, palatable and persistent, and combines well with grasses; will compete with *Imperata cylindrica* and *Pteridium* in north Queensland; produces a heavy crop of seed.

Main deficiencies. Slow establishment and nodulation; high percentage of hard seed; susceptibility to the *Amnemus* weevil and to frost.

Performance. 'Tinaroo' glycine greatly increased the productivity of dairying pastures at Kairi, north Queensland, on red loam (latosolic) soils derived from basalt. *Neonotonia/Panicum maximum* pastures yielded 2.09 tonnes/ha green weight, while pure *P. maximum* pastures yielded only 1 109 kg/ha green matter (Kyneur, 1960). At the same site, Edgley (1962) reported an increase in milk yield of 22 percent when cows were transferred to glycine/ green panic pastures from lucerne/Rhodes grass pastures.

At Campinas, Brazil, glycine significantly increased the protein percentage of *Pennisetum purpureum* from 10.69 to 14.56 percent, and *Melinis minutiflora* from 3.06 to 4.37 percent. Colman, Holder and Swain (1966) lifted dairy production from 2 700 kg butterfat per year with existing low-yielding matted *Pennisetum clandestinum, Paspalum dilatatum, Axonopus repens* pastures to 4 000 kg with an improved Kikuyu/*Neonotonia wightii* mixture, of which glycine contributed an average of 45 percent of the dry matter.

Main references. Allen *et al.* (1961); Gartner and Fisher (1966); Lovadini and Miyasaka (1968); Menegario (1964).

Psoralea spp.

Psoralea australasica Schltdl.

Synonyms. P. patens sensu auct., non Lindl.; *P. dietrichiae* Domin.

Common names. Native verbine; spreading scurf pea (Australia).

Description. An erect or semi-erect perennial, soft-wooded, shrubby legume, 0.5 to 2.5 m tall. Stems usually densely clothed with soft, spreading or downward-turned hairs. Leaves trifoliate. Leaflets ovate, the margins denticulate, glabrous to sparsely hairy above, usually pubescent below, at least on the veins; terminal leaflet 2.5 to 4 cm, sometimes 1.5 to 7.5 × 1.5 to 2 (0.5 to 4) cm. Inflorescence an axillary raceme of almost sessile flowers in groups of three, widely spaced along the 7- to 10- (5- to 23) cm-long rachis. Peduncle 7 to 16 (3 to 23) cm long. Flowers small, 6 to 7 mm long. Corolla pink, purplish or bluish, clearly longer than the calyx. Calyx densely pubescent, the hairs black, brown, grey, white or mixed, thinning with age, the lowest lobe only slightly longer than the others, not as long as the corolla. Pods silky-pilose, less enclosed in the calyx than those of *P. patens* (Lee, 1980).

Distribution. Native verbine is found in the central districts of Queensland, Australia, reaching the coast near Rockhampton, and in western districts; in western New South Wales, Australia, with many occurrences of semi-erect plants; and in South Australia from the coast near Adelaide to central Australia, on more alkaline soils or in depressions, rarely on sandy soils and then probably only when these overlie silts. Its main habitat is heavy clay soils.

General features. Lee (1980) pointed out that there had been considerable confusion within the *P. australasica-patens-pallida* complex in the past. *P. pallida* N.T. Burbidge was not recognized, being considered part of *P. eriantha* (syn *P. patens*). *P. australasica* was referred to as *P. patens* and *P. patens* referred to as *P. eriantha*. References to *P. patens* prior to Lee (1980) refer essentially to *P. australasica*.

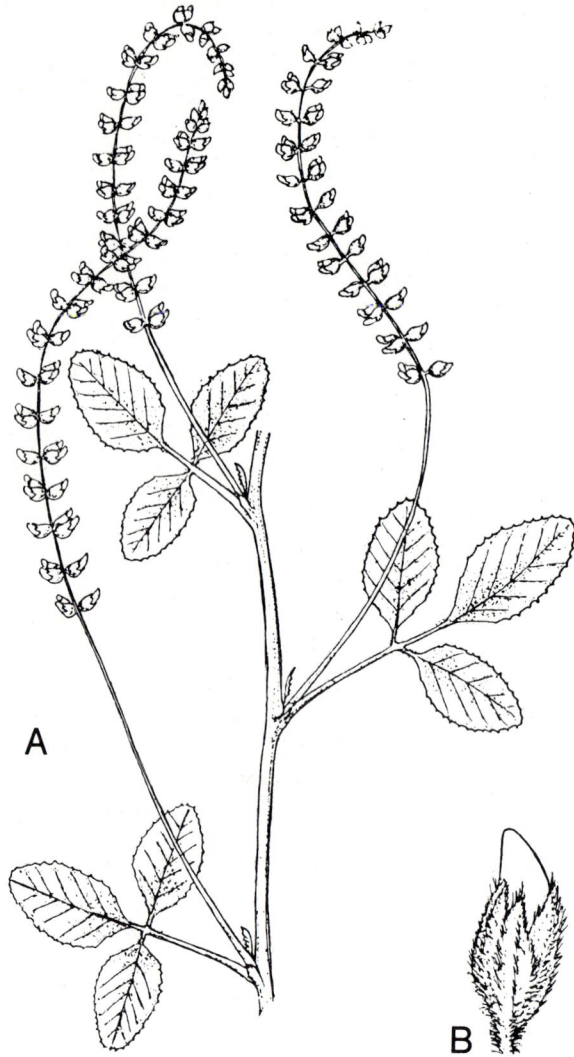

Figure 70. *Psoralea australasica.* **A**-Habit **B**-Bud

372

Native verbine has been found in areas subject to flooding, on loam to clay loam soils in western New South Wales. It has been found to be very palatable and a good fodder plant, probably greatly reduced by heavy stocking. Beeston (1978), on the other hand, lists it as of moderately low palatability on the alluvial plains and eucalypt woodland areas around Blackall, central western Queensland. In those areas, it vegetates the overgrazed southeast corner of paddocks on the Mitchell grass (*Astrebla* spp.) downs where stock graze into the southeast winds. Native verbine is not known to be toxic. It carries nodules (Bowen, 1956).
Main reference. Lee (1980).

Psoralea patens Lindl.

Synonyms. *P. eriantha* sensu auct., non Benth.; *P. discolor* Domin.
Common names. Woolly-flowered scurf pea, bullamon lucerne (Australia).
Description. A prostrate, rarely erect, soft-wooded, perennial, shrubby legume. Stems several to many, radiating from a central taproot, usually densely clothed with softly spreading hairs. Leaves trifoliate. Leaflets basically ovate, the margins denticulate, glabrous or hoary above, usually more pubescent below. Terminal leaflet 2 to 3 (1 to 4) cm long × 1 to 1.5 (0.7 to 2) cm wide.
Inflorescence an axillary raceme of almost sessile flowers, in groups (usually of three), continuously and usually compactly arranged along the 2- to 3- (1- to 9-) cm-long rachis. Peduncle 2 to 7 (sometimes 1 to 14) cm long. Flowers small, 6 to 7 cm, sometimes to 8 mm long. Corolla pink, purplish or bluish, slightly longer than the calyx. Calyx densely pubescent, hairs white or pale, always dense, the lowest lobe longer than the others and as long as the corolla. Pods more or less enclosed in the calyx, silky-pilose (Lee, 1980).
Distribution. The typical form occurs in all Australian states except Tasmania. An atypical form, with more or less densely tomentose leaflets, occurs in a broad band from the central Northern Territory southeast across the Simpson Desert to western New South Wales. This species is usually found only on sandy soils, but often on sites with favourable moisture relations (Lee, 1980).
General features. This is the plant usually referred to as *P. eriantha* Benth. prior to the recent clarification of the taxonomy of the *P. patens* complex by Lee (1980). It is the plant generally referred to by Skerman (1957), de Lacy and Britten (1970), Kerridge and Skerman (1968) and other University of

373

Queensland Department of Agriculture workers. The chromosome number is 2n = 20.

Characteristics. A summer-growing perennial; also gives some winter green growth (even though the tips of the leaves may be damaged by frost). Growth rhythm appears to be more closely related to temperature than rainfall. It is found mainly at altitudes of 300 to 600 m in rainfall regimes of 450 to 750 mm. It is very drought-tolerant and has a deep root system (Figure 3, Chapter 2). Skerman (1957) found that the taproot was still 1.25 cm thick at a depth of 1.5 m and that the roots reach a depth of up to 5 m in deep sandy soils. It occurs most abundantly on sandy levee soils of medium phosphate level (90 ppm available P_2O_5) and also on heavy clays. De Lacy and Britten (1970) found that it required a soil pH of a least 6.5. The plant is highly specific in its *Rhizobium* requirements, and two strains, CB762 and CB362, have been isolated from plants in Queensland, Australia, the former being more effective. There is up to 84 percent of hard seed. Dormancy can be broken by scarification or by treatment with sulphuric acid (Kerridge and Skerman, 1968).

P. patens grows quite well in association with native *Aristida* and *Heteropogon* species of grasses. Slashing of the taller grasses improves its growth and weakens that of the grasses. When established, it successfully competes with local weeds and can smother them with its foliage. The plant is particularly tolerant of defoliation and will regenerate even if the taproot of an established plant is cut 15 cm below ground level on sandy levee soil. Kerridge and Skerman (1968) found that plants survived and grew well even if cut to ground level every six weeks, but performance was better if the plants were cut at the second node at the longer cutting interval of 12 weeks. The additional foliage also improved frost tolerance in winter. It is tolerant of fire, with new shoots quickly arising from the crown.

The plant is very palatable to both cattle and sheep (and also to rabbits), and seed-eating birds harvest a lot of seed. Analyses of the air-dried whole plant revealed 9.6 percent moisture, 15.4 percent crude protein, 3.8 percent fat, 27.6 percent crude fibre, 35.6 percent nitrogen-free extract, 8.0 percent ash, 1.3 percent CaO and 0.41 percent P_2O_5 (Skerman, 1957). In pot tests, Kerridge and Skerman (1968) found that plants cut at the second node at 12-week intervals yielded as much dry matter as lucerne (alfalfa) cut every six weeks at a height of 2.5 cm.

Flowering occurs in western Queensland (lat. 26°25'S) over an extended period from late September to May. Ripe seed is shed under the plant and is difficult to retrieve except by suction. Its deep-rooting habit, ability to regenerate from a severed taproot, persistence under severe defoliation and

374

Figure 71. *Psoralea patens,* showing vigorous prostrate growth produced by a single plant

maintenance of green leaf after frosting make this plant valuable for the semiarid regions of the Australian tropics. De Lacy and Britten (1970) are endeavouring to breed superior types with a more erect growth habit to facilitate seed collection.

Subsequent work at the University of Queensland has suggested highest annual yields are obtained at longest intervals between harvests, especially from the more erect forms (Gutteridge and Whiteman, 1975; Britten and de Lacy, 1979). Frequent defoliation also markedly reduced the number of primary and secondary shoots, suggesting that survival under grazing may be a more important selection criteria than high yielding ability (Gutteridge and Whiteman, 1975). Yields in the field over the first six months after planting were equal to those of lucerne and higher than those of siratro (Britten and de Lacy, 1979).

Main references. Lee (1980); Skerman (1957).

Pueraria spp.

Pueraria phaseoloides (Roxb.) Benth. (see Colour plates XXXII, XXXIII)

Synonyms. *P. javanica* Benth.; *P. phaseoloides* var. *javanica* (Benth.) Hook.
Common names. Puero (Australia), tropical kudzu (most of the tropics).
Description. Vigorous twining and climbing, slightly woody, hairy perennial legume, deep-rooting and rather slender. Its main stems are about 0.6 cm in diameter and may extend for 5 to 6 metres. They may root at the nodes and from the nodes a number of lateral or secondary branches are formed. These intertwine and may result in a tangled mass of vegetation 60 to 75 cm deep within eight to nine months of sowing. The young shoots are densely covered with brown hairs. The leaves are large and trifoliate, borne on petioles 5 to 10 cm long covered with ascending hairs. Leaflets are thin, triangular-ovate and

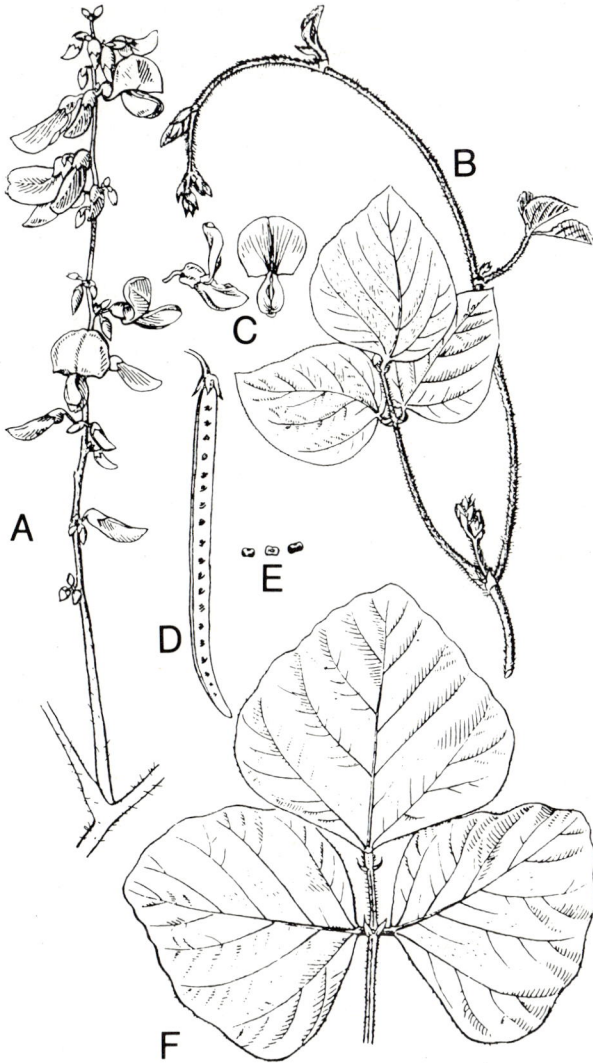

Figure 72. *Pueraria phaseoloides.* **A**-Inflorescence **B**-Branch **C**-Flowers **D**-Pod **E**-Seeds **F**-Leaf (**Source:** Malaya Planters Bulletin)

377

very shallowly lobed. Small mauve to deep purple flowers are borne in scattered pairs in axillary racemes about 15 to 30 cm long on peduncles about 12.5 cm long. The pod is straight, or slightly curved, linear, cylindrical, 7.5 to 8.5 cm long, thinly clothed with stiff adpressed hairs, black when mature and containing 10 to 20 (usually about 16) seeds, oblong to squarish with rounded corners, brown to brownish black, about 3 mm (Barnard, 1969).

Distribution. It is native to southeast Asia, Malaysia and Indonesia and is now widespread throughout the wet tropics.

Temperature for growth. Optimum about 15°C. Minimum about 12.5°C. Ludlow and Wilson (1970) obtained only 8.3 percent of the dry matter, 24 percent of the relative growth rate and 4 percent of the leaf area at 20°C as was produced at 30°C, indicating that puero is essentially a species for the humid tropics. It developed chlorotic leaves at the lower temperature. It is easily killed by frost.

Latitudinal limits. It extends to about 23°S, but for best performance should be grown in equatorial regions down to 17.5°S.

Altitude range. Essentially a plant of low altitudes, it generally grows below 600 m, but in Tanzania it reaches 1 000 m and in Colombia 2 000 m (Crowder, 1960). Altitudes above 1 200 m are too high in Kenya.

Rainfall requirement. Grows best in a rainfall exceeding 2 500 mm or in swampy land in areas of lower rainfall. In Tanzania, it grows in a minimum rainfall of 850 mm as a cover crop in sisal but grows better at 1 160 mm at Mlingano, Tanzania (Hopkinson, 1969). At 2 000 mm it is difficult to control in cocoa.

Tolerance of drought and flooding. Not drought tolerant. In prolonged dry periods it sheds its leaves but survives at Mlingano, Tanzania (Hopkinson, 1969). Kannegieter (1966) found it to be drought resistant in the forest zone of Ghana. It is one of the best tropical legumes for tolerance to waterlogging and nodulates freely in very wet soils; it can stand short periods of flooding.

Soil requirements. Has a wide range in soil adaptability, from sands to clays, although it does not grow well in tight heavy clays. Does well on sands and clays in Suriname, on latosols in Tanzania and north Queensland, Australia. Loustalot and Telford (1948) found a pH of 4 to 5 to be best. Fe deficiency showed at a pH of 6 to 8. N production was greatest at a pH of 4. Landrau *et al.* (1953) quote good growth at pH 4.5 on a lateritic soil, and at pH 4.6 to 5.1 in a clay. Best growth at pH 5.5 was recorded by Smith and Chandler (1951) but other co-workers increased growth by liming from pH 5.3 to 6.5 and, in the greenhouse, from pH 5.2 to 7.5. Molybdenum release may have been responsible. It is not tolerant of salinity.

Rhizobium *relationships*. A promiscuous species; nodulates with the cowpea type of *Rhizobium*, strain CB756 in Australia. However, Bowen (personal communication) obtained nearly double the yield from inoculated plants (as compared with the uninoculated control).

Ability to spread naturally. Spreads mainly by runners and in this way colonizes widely on suitable soils with adequate rainfall.

Land preparation for establishment. As early growth is slow, seed should be sown into a weed-free seed bed. It responds to good seed-bed preparation, which controls weeds by cultivation after the initial ploughing. Can also be established readily in the ashes of a forest burn.

Sowing methods. Seed is usually broadcast or drilled in. In Sri Lanka it is hand planted (15 to 20 seeds every 3 m mixed with the top 2.5 cm of soil). Drill in rows 1 m apart. Can also be propagated by cuttings 0.7 to 1 m long planted at two per point on a 1- to 2-m grid (Schofield, 1941). Establishment can be achieved by oversowing into existing pasture if the pasture is disced or burnt beforehand. Usually, however, it will not establish in grass, but grass will establish in the legume (Santhirasegaram, personal communication). It is best sown in midsummer to coincide with the wet season. Sow at 1 to 2 kg/ha in mixture. Rijkebusch (1967) recommends 3 to 6 kg/ha scarified or 8 to 10 kg/ha unscarified seed for sowing down the centre of sisal rows which are 3.5 m wide. Sow at 1.5 cm and roll or harrow.

Number of seeds per kg. 81 400 to 88 000.

Percentage of hard seeds. Eighty percent (Colombia) to 95 percent (Venezuela).

Seed treatment before planting. To break dormancy: (*a*) treat with concentrated sulphuric acid (sp. gr. 1.8) for 20 minutes, wash and dry (Rijkebusch, 1967; Prodonoff, 1968); (*b*) put in hot water at 50 to 70°C for several hours and allow to cool (Wycherley, 1960); (*c*) immerse in glycerine at 50°C for one hour — this increased germination from 10 to 50 percent (Wycherley, 1960); or (*d*) use infra-red lamp irradiation — Philips Infraphil Type 13373F/479 (150 watts) for one hour or Osram I.R.R. 4892 (250 watts) for two hours (Wycherley, 1960). Inoculation is advisable but not necessary. Pelleting is usually not necessary. Insect and disease control are usually not required.

Nutrient requirements. Dirven and Ehrencron (1969) have intensively tested the nutrient requirements of *Pueraria phaseoloides* and have published coloured photophaphs of the deficiency symptoms. They found that the lowest yields were with rain-water alone and minus phosphorus. Minus Ca and Mg reduced yields by 72 and 84 percent respectively and the plants did not recover when fertilized with a complete mixture four months later. Minus K,

Na or N reduced yields by 50 percent. Nodulation was lacking without calcium, fair with complete fertilizer and minus K, P, Mg or Na; good with rainwater, and very good with minus N. After complete fertilization, nodulation was very good in the minus Na, good in the minus N, Mg and in the rain-water treatments, and fair in all the others. In the absence of Ca, P or Mg, root systems developed slowly.

● Boron. Landrau *et al.* (1953) found that boron applied as 33 kg/ha of borax had no effect on yields or nodulation in an acid lateritic clay of pH 4.4.

● Calcium. Watson (1960) found that the Ca content of the leaf, stem and nodules was increased by liming to pH 6.0, with a smaller increase with liming to pH 7.0. The side-effect of liming, however, was to release molybdenum, especially at pH 7.0, the Ca and Mo contents rising appreciably. Liming also decreased the Mn content markedly. *Pueraria phaseoloides* is the most successful leguminous cover crop in sisal (*Agave sisalana*), which is a luxury consumer of calcium and is usually heavily limed. Loustalot and Telford (1948) found that lack of Ca caused rotting of the root system. The chlorophyll of the leaves faded along the margins and between the main veins, and the green was replaced by a buff pigment around the midrib. The area immediately adjacent to the main vein was unusually dark green. The leaves dropped before they became necrotic. The plants gave 57 percent of their maximum yield in the absence of Ca. Calcium deficiency appeared early and very few nodules were produced, some of which were decayed. Dirven and Ehrencron (1969) found that the omission of calcium resulted in less well-developed plants. Only a few new leaves and vines were formed. The root system was found to be small and there were no nodules. About three months after germination, dark yellow patches slowly spread to the leaf margin and base. The veins and the leaf top remained green for a considerable time, but eventually the entire leaf yellowed. Necrosis developed in the interveinal spaces, even after the first yellowing. As soon as the leaves had entirely yellowed, the tissue of the leaf tip or leaf margin began to die off. A brown coloration of the veins was observed on the top and underside of some of the leaves. Young leaves also exhibited distinct deficiency symptoms.

● Magnesium. Dirven and Ehrencron (1969) found that magnesium deficiency produced very weak plants with fairly small leaves and a poorly developed root system.

Necrosis occurred dispersedly over laminae and also at the tips and margins of the leaves. When the plants were supplied with a complete nutrient solution there was a slow and incomplete recovery.

● Nitrogen. Loustalot and Telford (1968) found that absence of N produced no symptoms but yields were only 55 percent of their maximum. Nodules were very large and moderately numerous. Landrau *et al.* (1953) found that puero responded to nitrogen at 275 kg/ha for the first cutting at four months from planting when nodule numbers were small; thereafter, no response was obtained. A similar response to early and also late nitrogen applications was recorded by Hopkinson (1969). Parbery (1967a) obtained a response to 100 kg N/ha in the Kimberley area of northern Australia. Evidently the puero nodules do not become active fixers of nitrogen until about four months after germination.

● Phosphorus. Loustalot and Telford (1948) obtained increased yields with puero from applications of phosphorus. In the absence of P, no symptoms were visible for two months after planting and then the basal leaves turned yellow and abscissed. The root system was abnormally large, with a low top-to-root ratio, and there were very few nodules, which were also small. In the absence of P, puero plants produced 87 percent of the yield of dry matter with complete fertilizer in pot tests. P had a highly beneficial effect (at 100 kg/ha P_2O_5) on poorly drained soils and there was a positive P × K interaction. The plants growing without P on these soils were chlorotic and small. Grof (1966) obtained linear responses to phosphorus with puero up to at least 110 kg/ha of P_2O_5, the response by puero in dry-matter yields being significantly higher than by centro and stylo. Dirven and Ehrencron (1969) found that P deficiency resulted in stunted growth. Only a few new leaves and vines were formed and the leaves were small and stiff, olive-green in colour, and in some cases the leaf margin was wavy in the middle. After application of complete nutrient solution, puero made a rapid recovery.

● Potassium. Loustalot and Telford (1948) found that K deficiency produced partial chlorosis and/or necrosis between the veins. The basal leaves were affected first, later the apex. Yellow areas at the margins of the leaves and between the veins extended irregularly inward. Entire leaf margins became chlorotic and then necrotic. In the absence of K, the yields were 87 percent of the maximum yield from complete fertilizer, and nodules were abnormally large and moderately numerous. Landrau *et al.* (1953) found that puero competed unsuccessfully with Merker grass (*Pennisetum purpureum*) for small K supplies in the soil. Chlorotic symptoms appeared on the margins of the legume leaves. Normal leaves were found to contain 2.20 percent of K in the dry matter, while chlorotic leaves contained 1.28 percent, and the N and P percentages were also low. Other workers have found that potash is deficient when the dry matter of the tops is less than 1.60 percent.

● Sodium. Dirven and Ehrencron (1969) found no sodium deficiency symptoms in puero supplied with a minus Na nutrient solution, but the top leaves were noticeably large. They stated that the significance of sodium deficiency was obscure, since few puero leaves contain less than 0.2 percent Na in the dry matter.

Toxicity levels and symptoms. Manganese — none described, though Watson (1960) found that the addition of lime markedly reduced the manganese content of the plant.

Vigour of seedling, growth and growth rhythm. In the seedling stage, has only moderate vigour and should have little competition. Grows slowly for three to four months in Colombia and then grows well, probably coinciding with active nodulation. Once established it is very vigorous, quickly smothers weeds and will climb trees and fences.

Response to photoperiod and light. A short-day plant; it flowers about 180 days from seeding in north Queensland, and in 126 days in the Kimberley district of north Australia (Parbery, 1967a). It tolerates partial shading, such as the edges of rain forests and in plantation crops such as coconuts. In Sri Lanka, it is the best legume under coconuts, growing with *Brachiaria miliiformis* (cori grass), but the trees should be at least 25 years old to provide enough light at ground level, with 150 to 180 trees/ha.

Compatibility with grasses and other legumes. Grows well with cori grass, molasses grass, guinea and elephant (napier) grass, but cannot persist with *Brachiaria decumbens* or pangola grass. As a cover crop it is often sown with centro and calopo. Calopo dominates the cover in the first year, then puero becomes dominant and finally *Centrosema* persists (Wilson and Lansbury, 1958).

Ability to compete with weeds. It is one of the best tropical legumes for smothering weeds; hence, its wide use as a cover crop in sisal (Hopkinson, 1969) in Tanzania, and in smothering nut grass (*Cyperus rotundus*) in Venezuela.

Tolerance of herbicides. From greenhouse work, 2,4-DB, ametryne, linuron and 2,2-DPA were promising preplant herbicides that could be safely used before sowing *Pueraria phaseoloides* (University of the West Indies, 1963). Riepma (1965) found that puero was adversely affected by both pre- and post-emergent application of neburon. The pre-emergence treatment was 2.2 to 4.5 kg active ingredient per hectare applied one day after sowing; the post-emergence application was made six to eight weeks after sowing. The growth of puero was impaired on sandy soil but not on clay soil at the 4.5 kg/ha rate. Increasing the seeding rate helped overcome the adverse effect. Hopkinson and Breitenstein (1969) found that 2 kg acid equivalent/ha of MCPA killed the runners of puero.

Nitrogen-fixing ability. Schofield (1945) found puero to be a better contributor of nitrogen to the soil than calopo, centro and stylo. After 18 months' growth, it was ploughed into the soil. The nitrogen content of a similar soil under bare fallow was 34.4 ppm, and in the soil into which puero was ploughed it was 171.8 ppm compared with 71.7 for centro, 66.7 for calopo and 54.5 for stylo. Hopkinson (1969) found that puero used as a green cover in sisal increased the fibre yield by 26 percent and equalled the yield with puero plus nitrogen. Rijkebusch (1967) found that puero had the same effect with sisal as the application of 635 kg N/ha in Tanzania. Oke (1967b) showed that puero fixed 9.3 mg N/plant/day, compared with 3.8 mg for *Calopogonium*, and transferred 92 percent of the fixed nitrogen to the plant tops, compared with 87 percent for calopo. Bruce (1967) found that an elephant (napier) grass/puero pasture added 143 kg N/ha/year to the top 15 cm of soil and raised the protein content of the grass by 7.1 percent. In puero stands, puero is self-mulching and adds considerable nitrogen by mineralization of leaf fall (Horrell, 1958).

Response to defoliation. It is moderately tolerant of defoliation, and recovers well after lenient grazing. Vicente-Chandler, Caro-Costas and Figarella (1953) found that cutting at 25 cm instead of 10 cm favoured puero in a molasses grass/puero mixture and gave better rooting and drought resistance.

Grazing management. Should be leniently grazed at all times to maintain the botanical composition of the pasture, as it is very palatable when selectively grazed. If it dominates the pasture mixture, grazing pressure can be increased.

Response to fire. Little tolerance.

Breeding system. Self-fertile; chromosome number 2n = 22.

Dry- and green-matter yields. Grof (personal communication) obtained 9 607 kg DM/ha in north Queensland from three cuttings made up of 3 684 kg/ha at the end of the wet season in June, 2 483 kg/ha at the second cut in the cool dry season in September, and 3 440 kg/ha from a third cutting in the wet season in January. Payne *et al.* (1955) obtained an average of 4 180 kg DM/ha/year over three years at Sigatoka, Fiji, 62 percent being obtained in the wet season, 38 percent in the dry. In Suriname (lat. 4 to 6°N), 30 to 35 tonnes of green fodder per hectare per year have been harvested. Vicente-Chandler, Caro-Costas and Figarella (1953) obtained dry-matter yields up to 22 896 kg/ha with a molasses grass/puero pasture with a protein content of 10.39 percent.

The contribution of puero was 9 141 kg DM/ha, with a protein content of 16.35 percent.

Suitability for hay and silage. It has been made into hay successfully in Colombia. It can stand two to four cuttings per year to give a hay yield of 4 tonnes/ha/year (Crowder, 1960). Cabrera and Rivera-Brenes (1953) prepared silage from a mixture of tropical kudzu and *Pennisetum purpurascens* (Merker grass) in Puerto Rico, and fed it to dairy cows with no beneficial effect compared with feeding green Merker grass, but the protein contents of two grass/legume silage samples were only 6.35 and 4.71 percent, while the green Merker grass had a protein content of 6.08 percent. No proportions of legume to grass were cited. It also makes good silage mixed with sorghum (one-third legume, two-thirds sorghum) and with elephant grass.

Value as standover or deferred feed. In frost- and fire-free areas, it is excellent as standover feed for the dry season.

Feeding value. It is a valuable fodder plant and has given excellent results in the wet tropics.

● Chemical analysis and digestibility. Reyes (1955) found that puero contained 8.4 percent fibre and 3.65 percent protein in green material with 22.59 percent dry matter. In the Kimberley district, Parbery (1967a) reported 11.6 percent protein, and Bermudez *et al.* (1968) reported 19.9 percent protein in Colombia.

● Palatability. Puero is very palatable. At South Johnstone, north Queensland, the palatability rating is puero > *Vigna hosei* > stylo > centro > *Desmodium heterophyllum* > calopo (Barrau, 1953).

Toxicity. None reported.

Seed harvesting methods. Seed is often harvested by hand in the tropics. It can be machine-harvested directly in the field, but low yields result because of the uneven maturity of the pods.

Seed yields. Yields are often affected by the legume pod borer. With full insect control, 330 kg/ha can be obtained by hand harvesting, 55 kg/ha by machine.

Minimum germination percentage and quality for commercial sale. Fifty percent germination, with a maximum of 10 percent hard seed and a purity of 93.5 percent in Queensland. Seed is germinated at 25°C under cover (Prodonoff, 1968).

Cultivars. At present there is only the one commercial line available in Australia. Cultivar IAC in Brazil has been reported by Souto (1969) to be more heat-tolerant than other types.

Diseases and pests. It is remarkably free from disease. Leaf-eating caterpillars cause damage in ungrazed plots; pod borers interfere with seed production.

Main attributes. Its compatibility with guinea and elephant (napier) grasses and its longer season of growth and higher yield than centro; high palatability; one of the best of the tropical legumes in nitrogen production; smothers weeds effectively.

Main deficiencies. Slow establishment; sensitivity to overgrazing; fire and drought intolerance; difficulties in seed production.

Performance. Puero has been an excellent cover crop to prevent soil erosion and to contribute nitrogen in plantation crops in the tropics for many years. Pereira *et al.* (1954) found that it gave very good protection of the soil for three years, but that its effect on structure was only transient. Hopkinson (1969) and Rijkebusch (1967) found its nitrogen-fixing ability of great value in sisal. Teitzel (1969b) reported that puero-based mixed pastures were some of the most productive under grazing in the wet tropics of north Queensland.

Vicente-Chandler, Caro-Costas and Figarella (1953) obtained 550 kg/ha live-weight gain with a molasses grass/puero mixture.

Main references. Dirven and Ehrencron (1969); Loustalot and Telford (1948); Schofield (1944).

Pueraria thunbergiana (Sieb. and Zucc.) Benth.

Common name. Kudzu.

Description. Spreading perennial vine; woody, strong, hairy, with many shoots developing from the crown. Leaves trifoliate, leaflets ovate and angularly lobed, 4 to 8 cm long, leathery, smooth or nearly so on upper surface, densely covered with soft greyish hairs on the lower; the leaf stalk hairy. Flower dull purple-red with yellow spot near the base, about 0.6 cm long, in a pendant raceme. Seed pod thin, 5 to 7.5 cm long, about 0.6 cm wide, covered with fairly stiff spreading brown hairs. Seed oval or oblong, about 0.3 cm long, dark, shiny (Hosaka and Ripperton, 1944).

Distribution. Native to East Asia, it is now widespread in the tropics.

Characteristics. Best adapted to a warm, moist climate but, because of the stored plant food in the roots, can stand protracted drought when once established. It grows from sea level to 900 m in Hawaii. Kudzu is a fast-growing plant and produces long, prostrate branches which root at many of the joints if the soil is moist and contact good. New plants are established in this manner. In the southern United States it is grown for grazing and cut feed, especially on poor eroded soils. It takes one year to establish. Quite a useful forage, retaining its palatability throughout the growing season, but new pasture legumes are likely to replace it for grazing purposes (Hosaka and Ripperton, 1944).

Rhynchosia spp.

Rhynchosia minima (L.) DC

Synonyms. *Dolichos minimus* L.; *Glycine rhombea* Schum. et Thonn.
Common names. Rhynchosia (Australia); least rhynchosia (United States); burn mouth vine (Barbados).
Description. A glabrous to somewhat pubescent, perennial, twining or sub-erect legume. Stems slender, numerous, 80 to 120 cm long. Stipules 2 to 3 mm long, lanceolate. Leaves trifoliate. Leaflets rhomboid, ovate or suborbicular to broad-acute, apex acute or rounded, 0.5 to 3 cm long × 0.5 to 3 cm wide, glabrescent to velvety. Inflorescence a lax 6- to 12-flowered raceme, 5 to 10 cm long. Calyx 3 to 4 mm long; five acuminate lobes. Corolla yellow, 1 cm long. Standard 5 to 7 mm long. Keel as long as the standard, wings shorter. Pods 1 to 1.5 cm long × 0.4 to 0.6 cm wide, black when ripe, two seeded, oblong, finely pubescent, slightly impressed between seeds, shortly beaked. Seeds monocoloured, black or brown, *c*. 3 mm long, with a short hilus.

There are numerous varieties. Those in southeastern Queensland are var. *minima* pods with short fine hairs only, and var. *australis* (Benth.) C. Moore (syn. *R. australis* Benth.) with short fine hairs and long tubercular-based hairs on the pods (Stanley and Ross, 1983; Andrews, 1952; Pulle, 1976; Gooding, Loveless and Proctor, 1965).
Distribution. Rhynchosia is an almost cosmopolitan plant on the heavier tex-tured soils of the tropics and subtropics. It is widespread in the Sudan and East Africa, and is also prominent from the Syrian Arab Republic to India (on black cotton soils in Mysore State) and Nepal. It also occurs from the

Figure 73. Natural plants of *Rhynchosia minima* following cultivation

southern United States (South Carolina to Florida and Texas) to Mexico. Rhynchosia is common in Panama and Venezuela and extends to Argentina. It is also native to all mainland sates of Australia, on heavy clay soils, particularly those high in lime and phosphorus; it also grows in association with *Astrebla* spp. (Mitchell grasses) and *Dichanthium sericium* grasslands in the 450- to 750-mm rainfall belt.

General features. Palatability of *R. minima* appears to vary widely from place to place. It is probable that it varies with the wide range of different ecotypes that exist. Hassell (1945) states it is eaten readily in Queensland when young but becomes rather fibrous and coarse when mature, while Beeston (1978) lists it as highly palatable in the Blackall District of central west Queensland and Boyland (1974) as moderately palatable in the far southwest of Queensland. Bogdan (1949) says its slight scent prevents it from being completely eaten in the flowering stage in Kenya. In Kenya, *R. minima* is more readily eaten by cattle than sheep (Edwards and Bogdan, 1951). However, Shukla, Ranjhan and Katiyar (1970) showed a daily consumption of 3.5 kg per 100 kg

387

bodyweight by rams in India, and considered it palatable to sheep as well. Crude protein was only 15.1 percent and crude fibre 45.9 percent, with 60 percent digestibility of dry matter, 54 percent of crude fibre and 64 percent of crude protein. The animals showed positive balance for N, Ca and P, and gained 0.32 kg per day. Studies with fistulated sheep at Toorak Research Station, on the northern end of the Queensland Mitchell grasslands, showed that *R. minima* was an important component of the "other species" in the pasture. These, while rarely exceeding 10 percent of the forage on offer, at times accounted for 60 percent of the diet consumed (Lorimer, 1978).

There are no known toxicities.

In Barbados, *R. minima* was among the most frequent plants on soils with a salt concentration of 1.6 meq/100 g, a concentration that prevented establishment of 30 other species (Eavis, Cumberbatch and Medford, 1974).

R. minima is a very hardy and widely adapted plant, to date much neglected. Many lines are lowly productive but others grow much more vigorously. The species warrants a thorough examination for pasture cultivars, especially for harsher, heavy-textured soils.

Rhynchosia sennaarensis Hochst. ex Schweinf.

Synonym. *R. flavissima* (Hochst.) Hochst. ex Bak.
Characteristics. An almost glabrous twiner. Central leaflet broad-ovate, 2.5 to 4 cm long; lateral leaflets very unequal sided, both surfaces glabrous, or the lower slightly grey-silky when young and dotted with golden glands. Yellow, 1.5-cm flowers in lax, 6- to 12-flowered racemes, 5 to 10 cm long on spreading peduncles. Pod gradually narrowed to the base, 2 cm long, glabrous when mature. It is indigenous to the Sudan (Andrews, 1952) and to Masailand, Tanzania. In nursery plots at Kongwa, Tanzania, it grew well to a height of 0.6 m and persisted under the low rainfall conditions (575 mm). Anderson and Naveh (1968) listed it as a promising legume for the semi-arid livestock areas. In cutting trials at three sites in northern Tanzania, type No. T1963 averaged 12 078 kg DM/ha/year from an average of four cuts per year. It is not as palatable as glycine, but is eaten and is more persistent.

Rhynchosia sublobata (Schumach.) Meikle

Synonyms. *Glycine sublobata* Schumach.; *R. melanosperma* Klotzsch.; *R. caribaea* sensu Baker in part, non (Jacq.) DC.
Common name. Munkolo (Zambia).

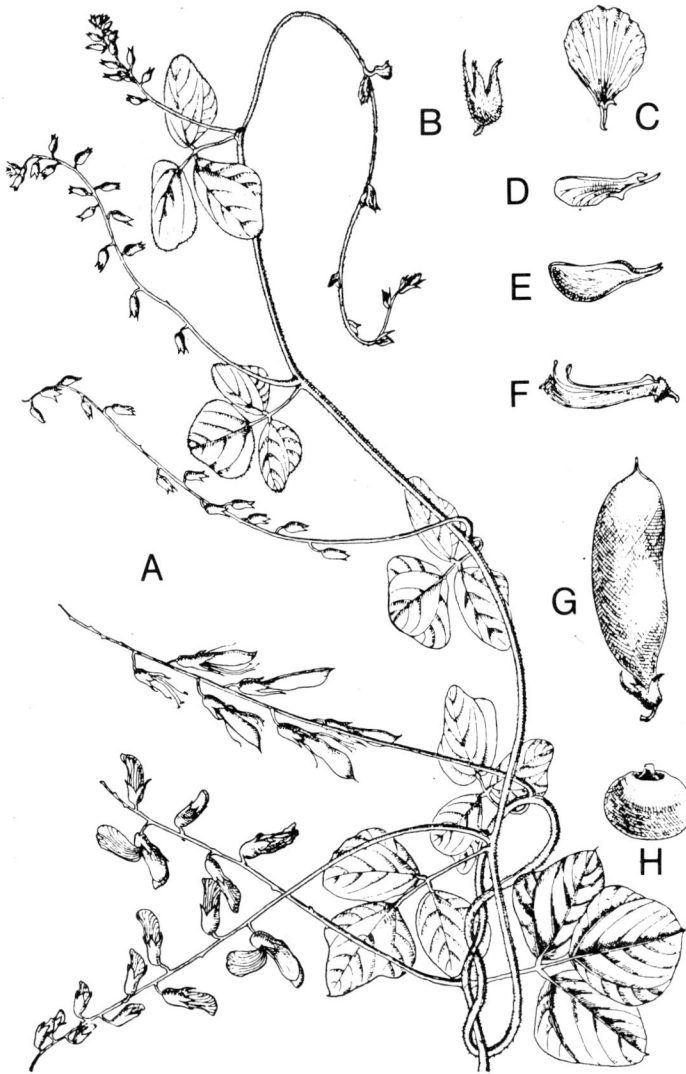

Figure 74. *Rhynchosia sublobata.* **A**-Habit **B**-Calyx **C**-Standard **D**-Wing **E**-Keel **F**-Stamens and gynoecium **G**-Pod **H**-Seed

Description. A perennial, prostrate or climbing legume with stems up to or more than 2 m long, growing from a long, stout, woody root-stock. Stems glabrescent to densely pubescent. Three leaflets, rhomboid, elliptic or ovate, 1.8 to 8 cm long, 1.8 to 8.5 cm wide, rounded at both ends, often slightly lobed at the base, the laterals on the outer side glabrescent to pubescent, rarely velvety. Petiole 1.2 to 6 cm long. Petiolules 1 to 2 mm. Stipules lanceolate, 5 to 6 mm long × 2 mm wide, hairy. Inflorescence lax. Rachis 3.5 to 20 cm long. Peduncles 2 to 14 cm long. Pedicels 1.5 to 4 mm long. Flowers 14 to 16, opening in the early evening. Calyx pubescent and glandular, tube 3 mm long, lobes linear-lanceolate to oblong-triangular, 2.5 to 6 mm long. Standard deep yellow or cream with a reddish-brown or purplish veining outside, 0.8 to 2.2 cm long. Wings yellow. Keel greenish yellow, usually purplish at the apex. Pods oblong-falcate, inflated very characteristically when ripe, 2.1 to 3 cm long, 0.8 to 1.3 to 1.3 cm wide, narrowed to the base, shortly pubescent or puberulent and glandular but without long hairs. Seeds dark brown, mottled black or entirely black, rounded-reniform, 4.3 to 5 × 3.5 to 3.8 mm, 2.5 to 3.2 mm thick, microscopically roughened (Gillett, Polhill and Verdcourt, 1971).

Distribution. R. sublobata is native from West Africa (Senegal) to Namibia; from central Africa to Zaire; and from the Sudan and Somalia to the Transvaal and Swaziland. It probably also occurs on Great Comoro Island and Madagascar (Gillett, Polhill and Verdcourt, 1971).

General features. The main habitats of munkolo are grasslands with scattered trees and woodlands, or abandoned cultivations from sea level to 1800 m in elevation (Gillett, Polhill and Verdcourt, 1971). It is widespread in the natural grasslands over large areas of southern and central Zambia, where it is regarded as a valuable native legume (Craufurd and Prins, 1979). It is frost and fire resistant, regrowing from the larger stems and from below ground during September/October (the end of the dry season), during which time it is freely grazed.

Parts of the areas in Zambia in which it grows suit also glycine and siratro. In spaced plants and swards, growth habit and protein levels were similar to those of glycine and siratro (Craufurd and Prins, 1979). The weight of 1 000 seeds varied from 45.0 to 71.0 g (mean = 59.6 g), compared with 14.0 g for siratro and 7.8 g for glycine. Seed production appears to require bee pollination (Craufurd and Prins, 1979).

Main reference. Craufurd and Prins (1979).

Stylosanthes spp.

Stylosanthes capitata Vog. (see Colour plates XXXIV, XXXV)

Description. An erect, ascendent to decumbent, much-branched perennial. Stems to 1 m high, woody toward the base, with short, dense, white hairs and some scattered bristles, the pubescence generally more dense below each node. Leaflets oblong to elliptic, to 30 mm long and 15 mm wide but usually less, densely villous on both surfaces, seven to nine pairs of conspicuous veins. Petioles 3 to 6 mm long, densely villous. Rachis 1.0 to 3.5 mm long. Sheath of stipules 7 to 9 mm long, densely villous, longer than the subulate teeth, several nerved. Inflorescence a thick spike, capituliform, about two-thirds as wide as high, to 35 mm long, many flowered, on peduncles usually 5 to 7 cm long. Bracts with a single, very reduced leaflet. Sheath often purplish, 8 to 12 mm wide, conspicuously 11 to 17 nerved, copiously softly pubescent. Axis rudiment 5 to 7 mm long in fruit, very long ciliate. Outer bracteoles 1 to 3 mm long, ciliate. Inner bracteoles two, 2.0 to 2.5 mm long, much narrower than the outer, lobes mostly acute and sparsely ciliate, about 2.5 mm long. Standard obovate, 5 to 7 mm long. Wings obovate, 4 to 5 mm long, auriculate at the base. Keel petals 3 to 4 mm long, falcate and auriculate. Loment to 2.5 mm broad, reticulate nerved. Both articulations usually fertile but either can sometimes be abortive, upper about 3.5 mm long and glabrous, lower some-what shorter and glabrous or sparsely pubescent. Beak uncinate, about 1 mm long, glabrous or with a few short stiff hairs on the inner face (Mohlenbrock, 1957).

Distribution. Native in a discontinuous pattern from southeastern Brazil to Venezuela. Restricted in distribution compared with *S. humilis* and *S. guianensis*. CIAT collections range from 22°S in Brazil to 10°N in Venezuela (Grof, Schultze-Kraft and Muller, 1979), and from lowlands to 1 000 m in elevation.

Soil requirements. Often found in treeless and open woodland savanna habitats. It thrives in the extremely acid (pH 4.5), infertile, sandy Cerrado soils, often in association with *S. scabra*, but is absent from the savannas on the western side of South America. On soils of medium acidity (pH 5.5 to 6.5), it fails to nodulate and dies.

Rainfall requirements. Grows from semi-arid tropical areas in Bahia, Brazil, to the subtropical conditions of Mato Grosso and Minas Gerais, Brazil, to the humid tropical coast of Pernambuco, Brazil. Rainfall range is from 500 to more than 1 500 mm annually.

Breeding systems. Self-pollinates readily in isolation, but outcrossing of up to 20 percent has been recorded in mixed populations (Miles, 1983).

Seed yields. Some accessions are extremely heavy seeders, producing yields of up to 1 000 kg/ha.

Rhizobium *requirements*. Highly specific.

Cultivars. There is wide variability and many natural ecotypes. One cultivar, Capica, a mixture of five accessions, CIAT 1315, 1342, 1693, 1728 and 1943, has recently been released by CIAT in Colombia. It is hoped that its wide base will give it wide adaptation and some anthracnose control (Edye, Grof and Walker, 1984).

Pests and diseases. Variation in reaction to anthracnose is present, with some lines resistant to a wide range of races, and to stem borers and budworms (Lenne and Calderon, 1984).

General features. A very variable species in many morphological characters, including size and shape of the inflorescence, habit of growth, hairiness of the stems and leaves, and colour of the stems.

Its adaptation to the low phosphate, acid, sandy soils of the Cerrado give it considerable promise in South America. Because of its many similarities to *S. scabra*, it could also have considerable value in Australia, especially in view of its high levels of anthracnose resistance. It appears well suited to low-input, extensive grazing systems. To date, it has not received the attention it deserves in Australia because of *Rhizobium* difficulties.

A prolific seed producer, with a high level of hard seed in inflorescences that are readily grazed over the dry season, *S. capitata* can thus be spread widely by livestock. It also regenerates readily from seed.

Main references. Grof, Schultze-Kraft and Muller (1979); Edye, Grof and Walker (1984).

Stylosanthes fruticosa (**Retz**) **Alston** (see Colour plate XXXVI)

Synonyms. *S. mucronata* Willd.; *S. flavicans* Bak.

392

Description. Copiously branching ascending shrub or undershrub; branches densely clothed with short yellowish pubescence. Leaflets oblanceolate, narrowed to both ends, long mucronate at the apex, 9 to 18 mm long, prominently nerved, both surfaces nearly glabrous. Flowers in dense oblong terminal heads. Pod with two articulations, about 6 mm long, both faces and remains of style densely silky (Andrews, 1952). Beaks 1.5 to 3 mm long, and the plant has evenly pubescent stems. (*S. fruticosa* has the beak of the loment 3.5 to 4 mm long and unilaterally pubescent stems; Mohlenbrock, 1963).

Distribution. Africa: the Sudan, Nigeria, Kenya, Uganda, Tanzania, Zambia, Mozambique, Zimbabwe and South Africa.

Characteristics. It is a perennial which may behave as an annual in the subtropics; occurs from sea level to 1 800 m. Commonly grows on sandy soils derived from granite, sandstone or coral, of low phosphorus status. Adapted to rainfall from as low as 284 mm (occurring over a four-month rainy season, from June to September, with the following eight months completely dry, in the Kordofan Province of the Sudan) to 1 000 mm on the coast at Dar-es-Salaam, Tanzania.

'T Mannetje (1965) found the plant to be day-neutral in its flowering response, though there was a tendency to flower more rapidly in the shorter day. Greater dry-matter production occurred under a 14-hour day length. It flowered in 52 days at a temperaure range of 30/25°C for a ten-hour photoperiod, and in 68 days under a 14-hour photoperiod. Temperatures of 35/30°C prohibited flowering. It appears to be promiscuous in its *Rhizobium* requirement but Norris (personal communication) has isolated a strain CB2618 from nodules brought from the Sudan. In Kordofan Province it combines well with the tall-growing *Andropogon gayanus* (Skerman, 1966), and in Tanzania with *Hyparrhenia* spp. and *Heteropogon contortus*. In areas protected from grazing and intermittenly mown, it forms a dense sward (e.g. Dodoma airstrip). In southern Queensland, Australia (lat. 27°30'S), it has seeded well ahead of frosts, and although some plants have been killed, others survive frosts of 3°C. Seed falls on ripening and germinates immediately under the mother plant if moisture and temperature are favourable. Seed kept in storage for 18 months had 90 percent hard seed in the pod. When dehulled and scarified, 60 percent of the seed germinated within ten days at 26.5°C.

The plant is eaten readily and grazed heavily by stock in the Sudan and Tanzania (Skerman, 1970). Preliminary tests show it to be promising on sandy soils in northern Queensland and on red latosols in south Queensland, and to respond to phosphorus. Dry-matter production from small plots was

6 000 kg/ha from rows sown 1.5 m apart. The dry material consisted of 38 percent stem, 19 percent leaf, and 42 percent inflorescence, of which one-third was seed (i.e. 14 percent of the plant weight was seed). There were up to 28 seeds per inflorescence.

All accessions in Queensland have proved extremely susceptible to anthracnose (*Colletotrichum gloeosporioides*).

Stylosanthes guianensis (Aubl.) Sw. (see Colour plates XXXVII, XXXVIII)

1. *Var.* guianensis

Synonyms. *Trifolium guianense* Aubl.; *Stylosanthes hispida* Rich.; *S. guyanensis* var. *subviscosa* Benth.; *S. gracilis* H.B.K. var. *subviscosa* (Benth.) Burkart; *S. pohliana* Taub.

Common names. Common stylo, stylo (Australia, Malaysia); alfalfa do nordeste, trifolio, mangericão do compo, saca-estrepe (Brazil); alfalfa del Brazil (Colombia); Brazilian lucerne, tarbardillo (Venezuela); tropical lucerne (Malaysia).

Description. Erect summer-growing herbaceous perennial with branching upright stems up to 1 m tall, which may become more prostrate under grazing. Stems hairy, becoming woody at the base with age; leaves pinnately trifoliate with elliptic leaflets 15 to 55 mm long and 7 to 13 mm wide; sticky in some ecotypes; petiole 6 to 15 mm long. Inflorescence of several spikes of a few flowers crowded into terminal heads; spikes sessile in unifoliate bracts and hairy; no axis rudiment; flowers yellow; pod hairy with one fertile joint and a very small beak. Seeds yellowish brown, averaging 1.75 mm long (Barnard, 1967), flat sided, tightly enclosed in a brown hull which can be removed by light threshing. Main taproot extends to 1 m. Runners root downwards but are ineffective (Gilchrist, 1967). At three months, 83.7 percent of the roots were in the top 20 cm of profile, 11 percent from 20 to 40 cm, 3.4 percent from 40 to 60 cm, 1.3 percent from 60 to 80 cm, and 0.4 percent had reached from 80 to 100 cm (Blouard and Thuriaux, 1962).

Distribution. Native to Latin America (Mohlenbrock, 1963), occurring mainly in the northern states of Brazil, from São Paulo and Rio de Janeiro northwards. Now widespread in the tropics and naturalized in many countries.

Temperature for growth. Prefers high summer temperatures. Adapted to frost-free conditions; continues active growth to 15°C (Allen and Cowdry,

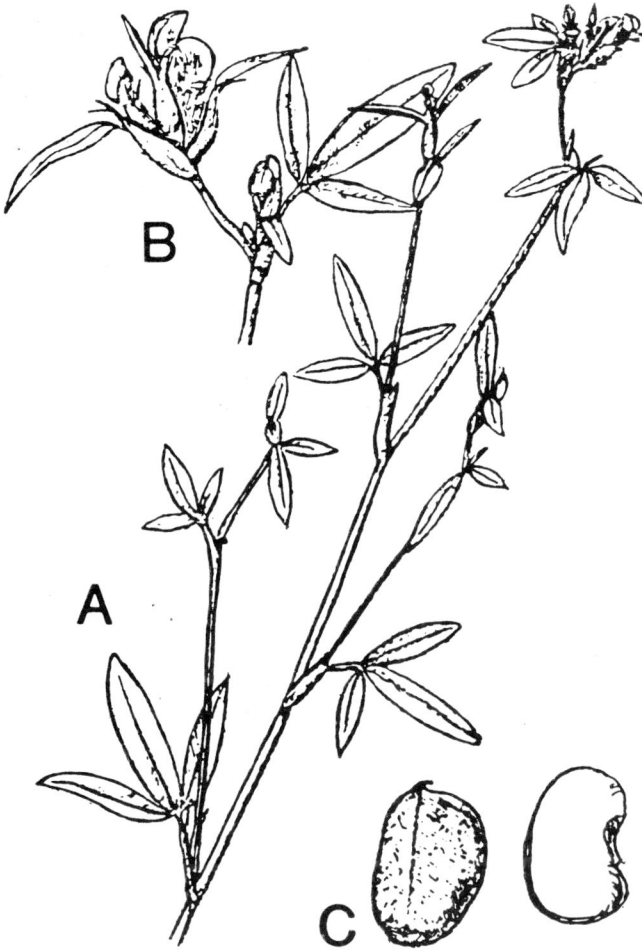

Figure 75. *Stylosanthes guianensis*. **A**-Branch **B**-Flower **C**-Seed

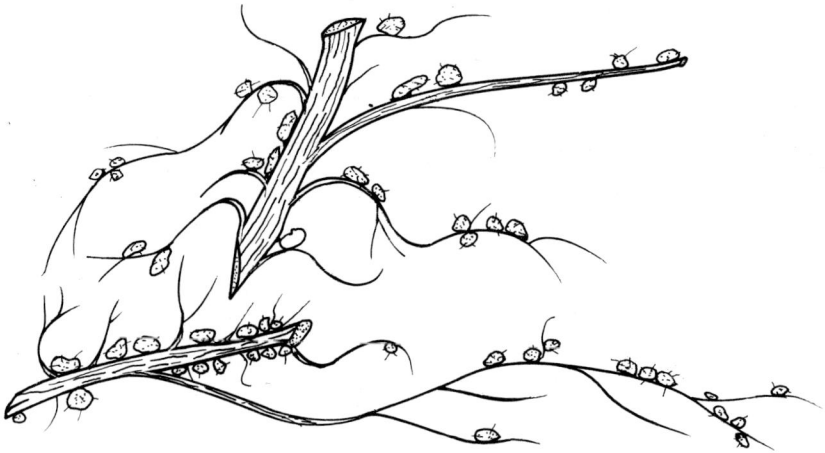

Figure 76. *S. guianensis,* root nodulation

1961b); defoliates at 0°C and plants are killed at −2.5°C (Boelcke, 1964). Tops are cut by heavy frosts. In the subtropics, it will grow on elevated slopes above the frost line (Gilchrist, 1967).

Latitudinal limits. About 23°N and S. Grows well at Campinas, Brazil (lat. 22°45′S) and in Laos (lat. 19°N).

Altitude range. In Brazil it occurs from 200 to 1 000 m; in Colombia from sea level to 2 000 m (Crowder, 1960). At Mareeba in north Queensland it grows at 450 m. It is not found below 650 m in Costa Rica, but occurs from 650 to 1 000 m (Kretschmer, unpublished).

Rainfall requirements. In Queensland, it is adapted to a rainfall range from 900 to 4 000 mm (Davies and Hutton, 1970); in Uganda above 500 mm (Stobbs, personal communication); in Brazil in the 1 000 to 1 700 mm range; in Chad at 1 100 mm.

Tolerance of drought and flooding. Has good drought tolerance, even in areas subject to frost. It will tolerate temporary waterlogging (Rijkebusch, 1967), but will not grow in swamps (Gilchrist, 1967).

Soil requirements. Does well on the coarser textured soils, but not so well on heavy clays. It grows on tropical latosols, gleys, loams and sandy podzolic soils. Does not do well on fine-textured montmorillonitic clays; prefers well-drained open-textured soils. Can tolerate highly acid soils (Davies and Hutton, 1970), and nodulates at pH 4.0. It is not very tolerant of salinity.

Rhizobium *relationships*. Inoculates freely with native rhizobia in the soil, although it nodulates better in the second year. On new land it is advisable to inoculate the seed. At Pitanguerias, Brazil, plants inoculated with the Australian strain of cowpea type, CB756, grew better than those inoculated with a local strain. Nodules are abundant, of small to medium size on the tap-roots and laterals, 30 per 5 cm on the taproot, and 15 per 5 cm on the laterals (van Rensburg, 1967).

Land preparation for establishment. For drill-sowing on a prepared seed bed, the soil should be ploughed in early spring and worked down to a fine tilth with disc harrows, finishing with a peg-tooth harrow. Allow weeds to germinate and then apply a preplant 2,4-D amine spray at 0.55 kg acid equivalent per hectare (280 ml of 50 percent 2,4-D/ha) on old cultivation and sow one week later with a minimum of soil disturbance (Gilchrist, 1967). For cheaper introduction, it can be sown without land preparation or with only one cultivation.

Sowing methods. Where necessary (e.g. on erosion terraces or ridges on slopes), it can be planted by stem cuttings (Schofield, 1941; Vivian, 1959; Nwosu, 1960).

Stylo has been proved to be the best legume to establish in *Imperata cylindrica* (blady, lalang or sword grass) grassland. Vivian (1959) stated that stylo cuttings can be established in this grass by digging individual holes 1 to 2 m apart, adding 50 g rock phosphate per hole and planting three to five stylo cuttings per hole with at least three nodes buried under the damp soil. Risopoulos (1966) established it in Zaire in *Imperata* grassland by broadcasting at the rate of 3 kg seed/ha after a single passage of a "Rome plough", after a light scarification of the soil with a disc harrow or after passage of a brush cutter. His findings are shown in Table 14.10.

TABLE 14.10 **Number of seedlings/1 000 m² required to establish *Stylosanthes guianensis* in *Imperata* grassland using different land preparation methods**

Land preparation method	Broadcast	After light cultivation with harrow	After passage of brush cutter
	Seedlings/1 000 m²		
Without burning; immediate grazing	1 565	1 744	2 678
After burning; immediate grazing	1 606	2 538	1 477
After burning; ungrazed for the rest of the year	1 237	2 505	3 474

He found that it was essential that the newly sown area be fenced off from cattle for eight to nine months and that fire be excluded, to allow for good establishment. Other successful methods are to burn the *Imperata* and scatter seed and superphosphate over the area (Blouard and Thuriaux, 1962), or to feed the seed to cattle and give the cattle access to young *Imperata* regrowth when it is palatable after a burn. The grazing of the *Imperata* reduces the competition for the stylo. Stobbs (1969k) found that seed collected from faeces and fistula samples were viable and in fact had an even higher germination rate. Much seed is spread in this way by the grazing animal (Stobbs, 1969k; Foster, 1961).

For seed production and for silage in combination with sorghum or other forage, it can be drill-sown in alternate rows spaced at 0.6 m (Risopoulos, 1966). For general pasture purposes, seed is sown by aerial or ground broadcasting or drilling into seed beds prepared by a single preliminary cultivation, or into ashes or natural pasture.

In Laos, Shelton and Humphreys (personal communication) have successfully established stylo in the rice crop up to day 35 from rice seeding. If left later than this, the rice shades out the developing stylo. The stylo may need hand weeding. When the rice crop is harvested, the stylo becomes a useful nitrogen-fixing pasture ley in a shifting system of rice cultivation.

Oversowing into natural pastures is quite successful, particularly on sandy soils in areas of adequate rainfall, as it the case of *Imperata* grassland. Van Rensburg (1967) succeeded in oversowing it into *Hyparrhenia*, Rhodes and star grass pastures in Zambia; Wendt *et al.* (1970) into *Hyparrhenia rufa* pastures in Uganda by broadcasting seed and superphosphate just before the rains. It can also be sod-seeded into pasture (Miller and Rains, 1963).

Sowing depth, cover, time, rate. Sow into a cultivated seed bed at a depth no greater than 1.5 cm and lightly cover with a "Cambridge"-type roller, harrow, or a bush dragged over the area. Sow at the start of the rains at 0.5 to 2.0 kg/ha.

Number of seeds per kg. 264 000 to 352 000. Percentage of hard seeds, 30 to 75 (Risopoulos, 1966).

Seed treatment before planting. Mechanically harvested seed does not require scarification to break dormancy (Rijkebusch, 1967; Gilchrist, 1967). Otherwise: (*a*) soak for 25 min. in water at 55°C (Risopoulos, 1966) or 85°C for 2 min. (Gilchrist, 1967); (*b*) scarify mechanically with scarifier or rice polisher (Blouard and Thuriaux, 1962); or (*c*) treat with concentrated sulphuric acid for 10 min. (Gilchrist, 1967). Pelleting is not necessary unless to protect rhizobia, when it should be pelleted with rock phosphate (Norris,

398

1967). For insect and disease control, dust with Fernasan D (Wendt *et al.*, 1970).

Nutrient requirements. Stylo is efficient in extracting phosphorus from the soil and is often not fertilized, but it responds to dressings of 125 to 250 kg/ha of superphosphate. At high levels of P, stylo responds to additions of copper (Grof, 1966), and heavy dressings of muriate of potash can cause chlorine toxicity. Horrell and Newhouse (1966) at Serere, Uganda, on a low fertility soil improved the yield of a stylo/*Hyparrhenia* pasture by 153 percent with added P, 197 percent with added S, and 243 percent with added P and S. Risopoulos (1966) in Zaire established it with 200 kg/ha dicalcic phosphate, 100 kg/ha ammonium nitrate and 50 kg/ha potassium sulphate, with good results for at least two years.

● Boron. Deficiency shows up as yellowing of the leaf tip, little-leaf and unthrifty growth (Stobbs, personal communication).

● Calcium. Stylo gave 64 percent of its maximum yield in the absence of added calcium, the maximum yield being at 740 kg/ha. It was depressed at higher levels. The percentage of calcium in stylo tops is higher than in temperate legumes because of its ability to extract Ca from the soil. In the nil treatments, the percentage of Ca was almost double that of other leguminous species (Andrew and Norris, 1961), indicating its high efficiency in extracting its necessary calcium from low calcium soils.

● Copper. Stylo gave only 7 percent of its maximum yield in the absence of copper and is sensitive to low copper supply. However, it has a high copper uptake per unit weight of root tissue and the Cu content of seed is 20.1 ppm (Andrew and Thorne, 1962). Eleven kg/ha of copper sulphate should be applied to deficient soils (Gilchrist, 1967). With copper deficiency in stylo, the first symptom is interveinal chlorosis of the younger fully expanded leaves, and a change in habit of growth from semiprostrate to upright. Tip necrosis of the leaflets follows, with subsequent twisting and then abscission of those immediately below the growing shoot. The growing shoot is retarded and shows some chlorosis. The final appearance is a stem devoid of leaves, except for a few old leaves at the base of the plant and one or two young chlorotic leaves at the top. Axillary growth takes place from the base of the plant (Andrew, 1963).

● Nitrogen. Parbery (1967a) obtained a positive response by stylo to 100 kg/ha N on Cunnunurra clay in northern Australia, but a depression of growth with a similar application to Cockatoo sand. Risopoulos (1966) used 20 kg N/ha as ammonium nitrate in Zaire with success.

● Phosphorus. Lack of phosphorus causes yellowing of the leaf tips in seed-

399

lings, small leaves, generally unthrifty growth and eventual death of young plants (Gilchrist, 1967).

● Potash. Andrew and Pieters (1970a) obtained healthy growth in stylo when the potassium concentration was 1.53 g/100 g dry weight, but deficiency symptoms at 0.55 percent. Symptoms commence as dark brown spots on the leaflets of the mid- to lower-positioned leaves; spots are irregular in shape and the pattern of placement and leaflets are not uniformly affected. Initially they are restricted to the upper leaflet surface and are in greatest concentration interveinally near the leaflet tips. This condition is followed by a general chlorosis of the leaflet, particularly the distal half and, subsequently, the brown spots become necrotic, resulting in leaflets having varying degrees of intensity of deficiency. Thus, a leaflet may have a brown necrotic tip, the necrosis merging into severely chlorotic tissue grading to light chlorosis with normal green colouring at the base of the leaflet. As necrosis proceeds further, particularly when half or more of the leaflet is affected, there is a tendency for the distal half to roll upward. With increased severity, all three leaflets become necrotic and may absciss from the petiole, which itself may develop brown necrotic spots and eventually absciss. The stipules of affected petioles may also show necrosis. The end result is a plant with a few fully expanded leaves at the growing tip in varying stages of chlorosis and necrosis, and a bare stem devoid of leaves and petioles. The primary stem of the plant is more affected than the secondary stems (Andrew and Pieters, 1970a). Stylo will respond to potash only in the presence of adequate copper (Teitzel, 1969a).

● Sulphur. Horrell and Court (1965) found that S gave the greatest response on the poor sandy soils of Serere, Uganda. Sulphur was used at 33 kg/ha, along with P at 104 kg/ha.

Vigour of seedling, growth and growth rhythm. Stylo is slow to establish (Risopoulos, 1966). Oke (1967a) explains this in terms of nodulation: he found it slow-growing up to six weeks and then rapid, the increased growth coinciding with the increase in nodule size. The older nodules of stylo were more effective than the young ones. Its growth during the wet season is good. It grows vigorously to a height of 1 to 1.5 m. If uncut, it collapses and new shoots arise from the pendant stems. In this way it has a valuable suppressive effect on weeds.

Response to photoperiod and light. A short-day plant, it requires a day length of less than 12 hours for flowering, with best results at ten hours ('t Mannetje, 1965). Does not flower well in the subtropics. It flowers in June-August (north Queensland), March-May (Kelantan, Malaysia), June-July (Colombia) and January-February (northern Nigeria). Common stylo is only moderately tolerant of shading. It grows under coconuts in Tanzania, though with a more erect habit. Its yield is reduced more than 75 percent when grown under closely planted oil-palms (Fodder crops and fallows in Dahomey, 1965b). It will grow with the tall *Panicum maximum* if well managed. Among tall tussock-forming grasses, it often fails to persist because cattle avoid the grass tussock and trample the legume — thus, the trampling factor may be more important than light competition in its disappearance from a mixture (Schofield, 1945).

Compatibility with grasses and other legumes. Stylo combines well with colonial guinea grass (*P. maximum*) in north Queensland (Davies and Hutton, 1970) and in Brazil; molasses grass and *Brachiaria ruziziensis* in Madagascar (Granier, 1966) and Zaire (Risopoulos, 1966); and *Hyparrhenia rufa*, Rhodes grass, *Panicum maximum* and *Setaria* at Serere, Uganda (Stobbs, personal communication). It grows with pangola grass in Brazil and north Queensland. Schofield (1945) found that stylo was shaded out by *Brachiaria decumbens*, *B. brizantha*, *P. maximum* and *P. coloratum*, lasted two years with *Brachiaria mutica* and persisted with Kikuyu (*Pennisetum clandestinum*) and *Paspalum dilatatum*. Ordinary guinea grass (*P. maximum*) offers it intense competition in summer (Gilchrist, 1967). If the pasture is short, it is compatible with puero, centro and siratro.

Ability to compete with weeds. When it is established, it competes very successfully with weeds and can invade natural grassland. It is aggressive because of its low early palatability and heavy seeding habit (Horrell, 1963).

Tolerance of herbicides. Bailey (1964, 1965a) has shown that stylo has good tolerance to 2,4-D; from about six weeks of age, 1.65 kg of acid equivalent per hectare can be used. Because stylo is well adapted to low fertility soils where bush regrowth is often troublesome, it is well to know the effect of 2,4,5-T, used in timber control, on this legume. Between 50 and 75 percent reduction in stylo can be expected 12 to 18 months after spraying with 2,4,5-T at 1.1 kg/ha acid equivalent, mixed with water. Provided the associated grass is not too dense and stylo has set seed, regeneration occurs from both seed and surviving root stocks. Obviously, 2,4,5-T should be used in stylo pastures only in exceptional circumstances.

Nitrogen-fixing ability. Unfertilized stylo/*Hyparrhenia* pastures at Serere, Uganda, yielded equivalent to grass receiving 165 kg N/ha (Horrell and Newhouse, 1966). In north Queensland, soil nitrogen under a bare fallow was 34.4 ppm, and in soil which had stylo ploughed in after 18 months of growth the nitrogen content was 54.5 ppm (Schofield, 1945). Grass/legume swards containing stylo and centro fertilized with phosphorus and sulphur gave live-weight gains equivalent to animal production from grass swards receiving 157 kg N/ha/year (Stobbs, 1969). In Nigeria, stylo fixed 4.6 mg N/day, compared with 14.5 mg for *Cajanus cajan* and 10.3 mg for *Centrosema pubescens*, and 98 percent of the fixed N was transferred to the plant (Oke, 1967a).

Response to defoliation. Heavy grazing is detrimental. Grof and Harding (1968) found that harvesting at 18-week frequency caused the lower stems and crown to become woody, with an almost complete loss of stand. There are very few growing points on these plants as they mature. Stylo persisted under grazing at eight-week intervals at Sigatoka, Fiji (Payne *et al.*, 1955). Root-stocks die if cut after two years' ungrazed growth, as the base of the plant is very woody then (Risopoulos, 1966). Cutting lower than 20 cm also affects it (Vivian, 1959). Grazing sheep do less damage to stylo as they tend to pluck the leaves (Tuley, 1968).

Grazing management. Stylo should be lightly grazed in the first year after six to eight weeks, to promote tillering and prevent it from becoming woody (Risopoulos, 1966). Long grass should be prevented from shading the stylo. Rotational grazing, one week on and four to eight weeks off, is best. Cattle graze the leafy material first and then move lower down with each successive grazing until the woody parts are consumed and damage results.

Response to fire. It is advisable to exclude fire from common stylo, as dense stands may be weakened. Light fires or fires passing when the ground is wet are less severe and may even stimulate it. Hot fires when the soil is dry will kill it, and two successive fires will virtually eliminate it. Seeds survive, however, and fire stimulates new seedling growth by breaking seed dormancy.

Breeding system. Self-fertilized; chromosome number 2n = 20.

Dry- and green-matter yields. Otero (1952) recorded yields of 15 to 20 t/ha of green material, in Brazil; Risopoulos (1966), 35 t/ha/year in Zaire; and Granier (1966) 43 t/ha on high ground and 70 t/ha after one year on low ground in Madagascar. At Sigatoka, Fiji, Payne *et al.* (1955) recorded 4 180 kg/ha/year of dry matter averaged over three years; and van Rensburg (1967) reported 4 600 kg/ha/year in Zambia. Gilchrist (1967) gives yields of 11 000 kg DM/ha in north Queensland. Blouard and Thuriaux (1962) recorded no significant differences in yield from four cutting regimes over 24 months in Zaire. The average annual yield from four cuts over 24 months was: cut every three months at 15 cm, 7 281 kg/ha/year; at 25 cm, 6 785 kg. Cut every $4^1/_2$ months at 15 cm, 6 845 kg; at 25 cm, 6 529 kg.

Suitability for hay and silage. Makes good hay, containing 14 to 16 percent crude protein. Only one cut should be taken annually in north Queensland, in late summer to autumn at a height of at least 20 cm. The stubble after seed threshing has 5 percent crude protein and can be hammer-milled for roughage feeding — the seed left in it is then distributed by the livestock (Gilchrist, 1967). Nwosu (1960) cut it three times a year in Nigeria, taking only the top 45 cm of crop (cut at 45-cm height), dried it in the field for seven days and hammer-milled it into feeding meal containing 17.17 percent crude protein — at a cost of about 19.8 U.S. cents/kg. In well-established stands in Malaysia, cutting has little effect for up to four years, but thereafter yield declines and at six years its economic production is finished (Vivian, 1959). The hay should be handled as little as possible to preserve the more nutritious leafy portion.

Risopoulos (1966) sowed sorghum and stylo in alternate rows 0.6 m apart and with this mixture made excellent silage of 6 to 7 percent crude protein. The first year the proportion of stylo to sorghum is 1:3, the second year the mixture is balanced and from the third year there is a pure stylo crop. Cattle ate stylo silage whether ensiled with 1 percent salt, 1 percent molasses, 1.5 percent molasses, 2 percent molasses or with no additive. One percent molasses gave a pleasant odour to the silage.

Value as standover or deferred feed. It is excellent as standover feed as its palatability is high at this stage. Sillar (1969b) showed that cattle could be fattened during the three normal dry months in north Queensland on a pure diet of standing stylo.

Feeding value.

● Chemical analysis and digestibility. Chemical analysis figures for stylo are listed in Appendix 1. Crude protein figures range from 12.1 to 18.1 percent for the whole plant. Van Rensburg (1967) recorded 7.9 percent for the stem and 13.6 percent for the leaf. Crude fibre figures range from 21.7 to 37.7 percent. Milford (1967) recorded 11.8 percent crude protein with a digestibility of 52.6 percent, and 37.7 percent crude fibre with a digestibility of 42.2 percent. The digestibility of the dry matter was 48.4 percent and voluntary intake of dry matter 33.9 ± 4.3 g/kg live weight/day, the lowest of the tropical legumes tested. The plant material tested was green, fairly stemmy, unfrosted and seeding freely, growing at Lawes, Queensland. Risopoulos (1966) in Zaire recorded a range of 15 to 17 percent crude protein, 33 to 40 percent crude fibre, 0.1 to 0.2 percent P, 0.8 to 1 percent Ca, 1.2 to 1.8 percent K, 0.3 to 0.8 percent Mg, a trace 0.02 percent Na and 0.1 to 0.8 percent Cl in the dry matter. Nwosu (1960) in Nigeria prepared a meal with 17.17 percent crude protein and 28.9 percent crude fibre.

● Palatability. It is relatively unpalatable in the early stages of growth. Schofield (1945) found that it was not grazed by cattle which had no experience with the legume but, after confining the stock for two days to pure legume alone, they ate it readily. Nwosu (1960) also observed this lack of early palatability and suggested that it was due to the harsh hairs on the plant. In stall-feeding green stylo, he found that the material had to be wilted before feeding (to remove the turgidity of the hair) to make it palatable to the cattle. In Brazil (Lychatchynsky and Steenmeyer, unpublished), palatability tests showed stylo to be the least palatable of eight tropical legumes early in the summer, but among the most palatable later on. Stobbs (1969k) observed intake of stylo over a period. Over six weeks, 27.8 percent of stylo of a pasture containing 29.1 percent of this legume was consumed. Animals preferred to eat *Hyparrhenia rufa* when introduced to the mixed pasture, and intake of stylo was significantly lower on the first day of grazing. When herbage was plentiful, mainly the leafy parts of stylo were selected, but as the amount of herbage declined, complete florets, which contained viable seed, were eaten. On the last few days of grazing, a higher proportion of the stemmy parts of the legume was consumed.

Toxicity. None has been reported, but it has a toxic effect on succeeding cotton crops at Serere, Uganda (Horrell and Newhouse, 1965). Leaf and stem exudate depressed seed germination of *Pennisetum typhoides* in Petri dishes. At Ngeta, Zambia, van Rensburg (1968) found that it reduced subsequent

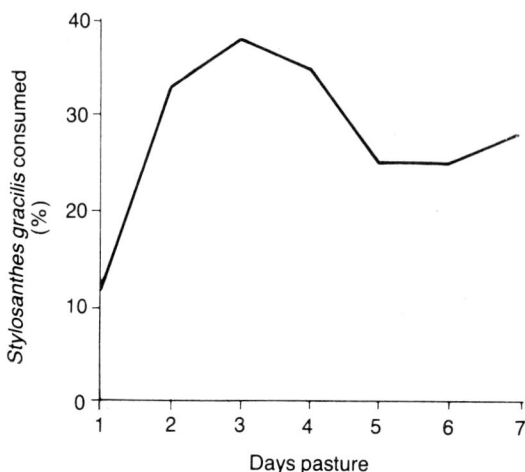

Figure 77. Mean percentage of stylo intake during the grazing period (**Source:** Stobbs, 1969k)

cotton yields by 30 percent and also cottonseed germination. The toxic effect was confirmed by pot tests in which stylo leaves were incorporated in the soil. Rijkebush (1967) also observed an adverse effect on subsequent crops of sisal in Tanzania.

Seed harvesting methods. As it matures over an extended period, some loss of seed is unavoidable owing to immaturity or shedding when seed is mechanically harvested. The crop is harvested with an all-crop harvester with a peg drum, a high cylinder speed being employed in combination with a full set of concaves. The comb is set 20 to 25 cm below the top of the sward. A second harvest a few days later, using a lower comb height, recovers more seed. The seed is cured by drying, during which some after-ripening of seed takes place. The seed sample has testa colouring varying from black to yellow; the paler seeds germinate better (Cowdry and Verhoeven, 1961). It is not advisable to let it grow for two years for a seed crop and then cut it to the ground in harvesting, as the woody root-stocks will die (Risopoulos, 1966).

Seed yields. Seed production is heavy, but the small pods shed on ripening so that only a fraction of the seed crop can be mechanically harvested. Risopoulos (1966) recorded 100 to 200 kg/ha in Zaire, and at the IRI Research Institute at Campinas, Brazil, the yield is about 100 kg/ha. In north Queensland, yields average 90 to 100 kg/ha, reaching up to 330 kg/ha (Gilchrist, 1967).

Minimum germination percentage and quality required for sale. Minimum germination 40 percent, maximum hard-seed content 10 percent, and purity of at least 96.5 percent in Queensland. Seed stored in sacks deteriorates after one year, but in sealed containers retains its viability longer (Risopoulos, 1966). Blouard and Thuriaux (1962) obtained 26 percent germination in three-month-old seed, 73 percent after eight months, 52 percent after 10 months, 12 percent after 36 months and only 7 percent after 41 months. In the field, stylo seeds kept germinating continuously over three years of recording (Tuley, 1968).

Cultivars. Cv. Schofield was introduced into Australia from Brazil in 1933 and was the major cultivar until it collapsed with anthracnose (*Colletotrichum gloeosporioides*) in 1978. Cv. Cook, from Colombia in 1965, flowers earlier and gives better cool-season yields; it also currently maintains some fields tolerance of anthracnose. 'Endeavour', from Guatemala in 1965, collapsed to anthracnose just as it was beginning to be widely used in 1974. Cv. Graham was released in 1979 as an earlier flowering, more persistent type. It still retains some field tolerance to anthracnose.

In Brazil, three cultivars — Deodora I, Deodora II and FAO 13821 — and in Senegal, one cultivar — N-6399 — are available. In addition, in Brazil, EMBRAPA is in the process of releasing a *tardio* line called cv. Bandeirante.

Diseases. Anthracnose has been reported in Brazil under wet conditions (Otero, 1952). *Corticium* and *Rhizoctonia solani* attack it under wet conditions in Zaire (Blouard and Thuriaux, 1962), and *Diplodia* in Malaysia (Vivian, 1959). Hutton and Grylls (1956) listed it as susceptible to little-leaf. Generally, however, stylo is one of the most disease-free of the tropical legumes. Grazing management can limit damage by pathogens.

Pests. A variety of insects attack the plant and reduce seed yields under wet conditions in Panama and Bolivia. The caterpillar of the moth *Lamprosema diemenalis* Gueen. did some slight damage. Colbran (1963) recorded the nematode *Meloidogyne hapla* on the roots of stylo. In cast Nigeria it is recommended in an arable crop rotation because it is not susceptible to the particular nematode affecting these crops (Tuley, 1968).

Main attributes. Where fertilizer costs are high it offers one of the best opportunities to raise productivity of natural grassland because of its low phosphorus requirement. An adaptable nonclimbing legume, it grows in poor soils, is easily established by oversowing, continues to increase in palatability and persists into the dry season, when it is most needed. Its phosphorus requirement is only 0.17 percent, compared with 0.24 percent for siratro (Andrew and Robins, 1969a).

Main deficiencies. It is frost susceptible; will not stand heavy grazing; can reduce the yield of subsequent crops (seed tends to shatter on ripening, thus reducing yields); has a relatively low protein content and tends to become woody.

Performance. The inclusion of the legumes *Stylosanthes guianensis* and *Centrosema pubescens* into a *Hyparrhenia* grassland at Serere, Uganda, increased animal production by 11 to 49 percent (Stobbs, 1969). Over a three-year period, rotationally grazed oversown *Stylosanthes guianensis* pasture at Serere produced a mean of 254 kg/ha/year live-weight gain without fertilizer and a mean of 450 kg/ha/year where single superphosphate was applied (Stobbs, 1969). A *Hyparrhenia*/stylo pasture under low-frequency rotational grazing (i.e. seven days grazing and 28 days resting) and high-frequency rotational grazing (i.e. $3^{1}/_{2}$ days grazing and 14 days resting) gave equally high live-weight gains per hectare — 740 kg/ha/year at Serere (Stobbs, 1969i). This pasture showed better animal returns in the dry season than a *Hyparrhenia*/centro pasture which gave the same yearly live-weight gain. Both swards appeared tolerant of both grazing regimes. As a cover crop, stylo competes with rubber for water in Sri Lanka, but in Malaysia competition is reduced by cutting the stylo periodically for feeding (Vivian, 1959). It grows well under coconuts in Tanzania, though more erectly.

Agronomically quite a different plant to the subtropical fine stem stylo.

Stylosanthes guianensis (Aubl.) Sw.

2. *Var.* **intermedia** *(Vog.) Hassler.*

Synonyms. S. montevidensis Vog. var. *intermedia* Vog. and others according to 't Mannetje (1977) and (1984), but Williams *et al.* (1984) regard it as *S. hippocampoides* Moh., and could well be right. Agronomically fine stem stylo is quite a different plant to the var. *guianensis* stylos.

Common name. Fine stem stylo.

Description. A semiprostrate perennial with a strong taproot. Differs from common stylo in having a well-developed crown with buds both below and above ground level. It also has finer stems and smaller leaflets. The narrow deep green leaflets are 1.5 to 3.5 cm long, 3 to 5 mm wide and have few hairs. It has 4 to 20 small yellow flowers subtended by leaflike bracts in small compact spikes. Flowers are mainly terminal but sometimes occur in the leaf axils. The light brown pods are flattened, single-seeded, 3 mm long and 2 mm wide with a minute coiled beak; conspicuously fine-veined without hairs. Pods fall as they ripen. Seeds are yellowish brown (Davies and Hutton, 1970). Very little root development occurs at the nodes of prostrate stems except under high moisture conditions, and even then rather poorly (Stonard, 1968). It was introduced to Australia by W. Hartley, as CPI 11493, from sandy loam at Asunción, Paraguay.

Distribution. It is found in Paraguay, Argentina, northern Uruguay and southern Brazil, extending to latitude 33°S.

Season of growth. Comes away earlier in the spring than other legumes, grows through the summer with peak growth in November and in the absence of frosts grows into the winter. It is perennial in habit and has a longer growing season that *S. humilis.*

Temperature for growth. Optimum approximately 27 to 29°C. It withstands temperatures up to 43°C. Although frosts affect tops, crowns of established plants have survived temperatures of −10°C. In the absence of frost it grows through the winter.

Latitudinal limits. It is used in latitudes 24 to 30°S; its performance elsewhere is not known.

407

Figure 78. *Stylosanthes guianensis* spreading through *Imperata cylindrica* near Cooktown, north Queensland, Australia (lat. 15°S)

Altitude range. In Queensland, it is grown at 100 to 130 m, in Paraguay it occurs at about 200 m (Fretes, Samudio and Gay, 1970).

Rainfall requirement. Medium — from 625 to 875 mm in Queensland. In Paraguay it occurs in a rainfall zone of 1 300 to 1 500 mm.

Tolerance of drought and flooding. Has excellent drought-resisting qualities (deep taproot, small leaves); it is unable to withstand flooding.

Soil requirements. Prefers sands and sandy loams, but will establish on basalt loams. Does not establish on heavy self-mulching clays. Has a wider range of adaptability than common and Townsville stylos. It grows best in soils of pH 6.5 to 6.7. Its tolerance of salinity is unrecorded.

Rhizobium *relationships*. Although it may nodulate weakly with some local rhizobia and those of the cowpea type, it requires a specialized *Rhizobium* for best growth. The current (1970) Australian recommendation is CB1552. The nodules are small and irregular in shape and are found mainly on the taproot — seldom on the laterals (Stonard, 1968).

Ability to spread naturally. Quite good in suitable soils. At Brian Pastures, Queensland, it has spread on granitic sands with the aid of grazing by cattle, hares and wallabies (Stonard, 1968).

Land preparation for establishment. Does not need elaborate land preparation; burn the pasture in the spring, then cultivate lightly.

Sowing methods. Seed is broadcast after light cultivation. Oversowing into natural pastures is successful; the area should be continuously stocked after seeding to reduce competition from the grass. Do not sow deeper than 1 to 1.5 cm and lightly cover. Sow in midsummer (during the rainy season) at 2 to 5 kg/ha.

Number of seeds per kg. 770 000 (Davies and Hutton, 1970). There is a high percentage of hard seed in the pod (Stonard, 1968), which is an advantage in perpetuating the sward under unreliable rainfall conditions.

Seed treatment before planting. To break dormancy: (*a*) scarify mechanically; or (*b*) immerse in hot water at 80°C and allow to cool for 40 min.; or (*c*) treat with concentrated sulphuric acid for 10 min., wash and dry. Inoculation is necessary. Pelleting is not needed unless to protect the rhizobia, when rock phosphate should be used (Norris, 1967). If seed-harvesting ants are troublesome, dust with dieldrin, 13.2 cc of dieldrin 15 percent emulsifiable concentrate (or 2.2 g acid equivalent per kg seed).

Nutrient requirements. Grows well even on soils of low fertility, but responds to phosphorus and molybdenum.

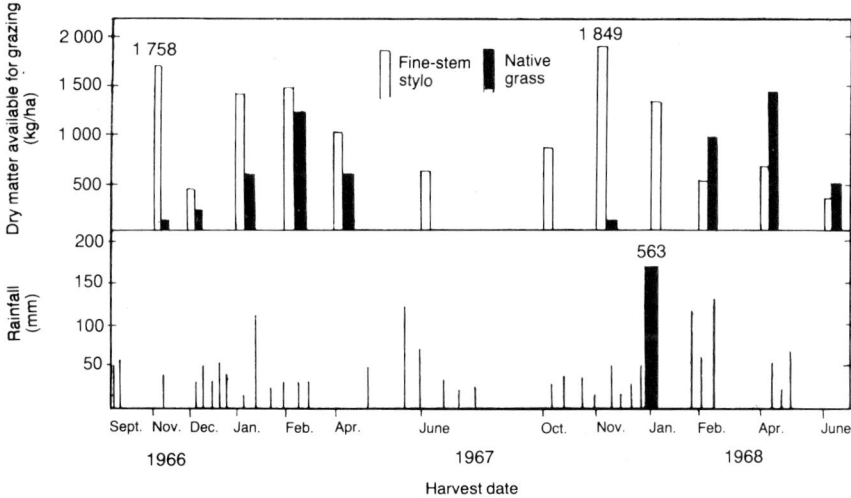

Figure 79. Yield of *Stylosanthes guianensis* over a three-year period in southeastern Queensland, Australia

409

● <u>Phosphorus</u>. Fine stem stylo, like *S. humilis*, is very efficient in extracting phosphorus from the soil and grows quite well on granitic sands at Brian Pastures, Queensland, with an available P content of 35 ppm. It responded to 250 kg of molybdenized superphosphate per hectare, which increased total native grass/fine stem stylo yield by 23 percent and the protein content of the grass by 0.47 percent, but not that of the legume.

Vigour of seedling, growth and growth rhythm. Seedlings quite vigorous, but light should be available to them for quick establishment. It has a long growing season (September-April in southeastern Queensland). Its yield over nearly three years is given by Stonard and Bisset (1970) in Figure 79.

Response to photoperiod and light. Fine stem stylo flowers over a long period; grows best in pasture mixtures where there is ample light provided by burning or heavy grazing.

Compatibility with grasses. Combines well with *Heteropogon contortus* and with buffel (*Cenchrus ciliaris*) and Rhodes (*Chloris gayana*) grasses.

Ability to compete with weeds. In its adapted low rainfall environment, it competes successfully with weeds.

Tolerance of herbicides. Herbicide responses of *Stylosanthes* spp. reported by Stonard (1968) are given in Table 14.11.

Nitrogen-fixing ability. Evidence of effective nodulation is provided by Stonard (1968): yield from unfertilized native pasture was raised by 17 percent and from similar pasture fertilized with 250 kg/ha molybenized superphosphate by 23 percent when fine stem stylo was grown with it.

Response to defoliation. Can stand close grazing; a profusion of short leafy shoots develops from the crown, giving the plant a higher leaf to stem ratio.

Grazing management. Can be grazed throughout the year. Stock it heavily

TABLE 14.11 **Response of *Stylosanthes guianensis* to various herbicides**

Herbicide	Application rate	Effect on fine stem stylo
2,4-D	0.55 kg AE/ha	Nil
2,4-DB	2.20 kg AE/ha	Nil
2,4,5-T	2.20 kg AE/ha	Leaf scorch, plants recovered
Diquat	1.37 kg ion/ha	Leaf drop, plants recovered
Paraquat	0.44 kg ion/ha	Leaf drop, plants recovered
Dalapon	11.00 kg/ha	Prolonged yellowing of leaves, plants recovered

Figure 80. A single plant of fine stem stylo (*Stylosanthes guianensis* var. *intermedia* cv. Oxley)

after burning native pasture and introducing stylo seed, and again in the autumn to control excessive grass growth.

Response to fire. Survives fire much better than common stylo because of its buried crown, and its seed-shattering habit and hard-seededness allow it to regenerate from seed.

Breeding system. Self-fertilized; chromosome number 2n = 20.

Dry- and green-matter yields. In unfertilized natural pasture, Stonard (1968) obtained 637 kg/ha of fine stem stylo in a total yield of 1 735 kg of mixed natural pasture/fine stem stylo pasture, and 950 kg/ha when fertilized with 250 kg/ha molybdenized superphosphate in a total yield of 2 065 kg/ha. The highest dry-matter yield of legume in a grass/legume mixture was 5 044 kg/ha/year over two harvests when fertilized at 23 kg/ha of P; unfertilized legume yielded a maximum of 2 027 kg/ha (Stonard and Bisset, 1970).

Suitability for hay and silage. No records of conservation are documented. The plant is quite prostrate and would be difficult to cut.

Value as standover or deferred feed. It is valuable in the absence of frosts.

411

Figure 81. A fine stem stylo sward in native pasture in southern Queensland, Australia

412

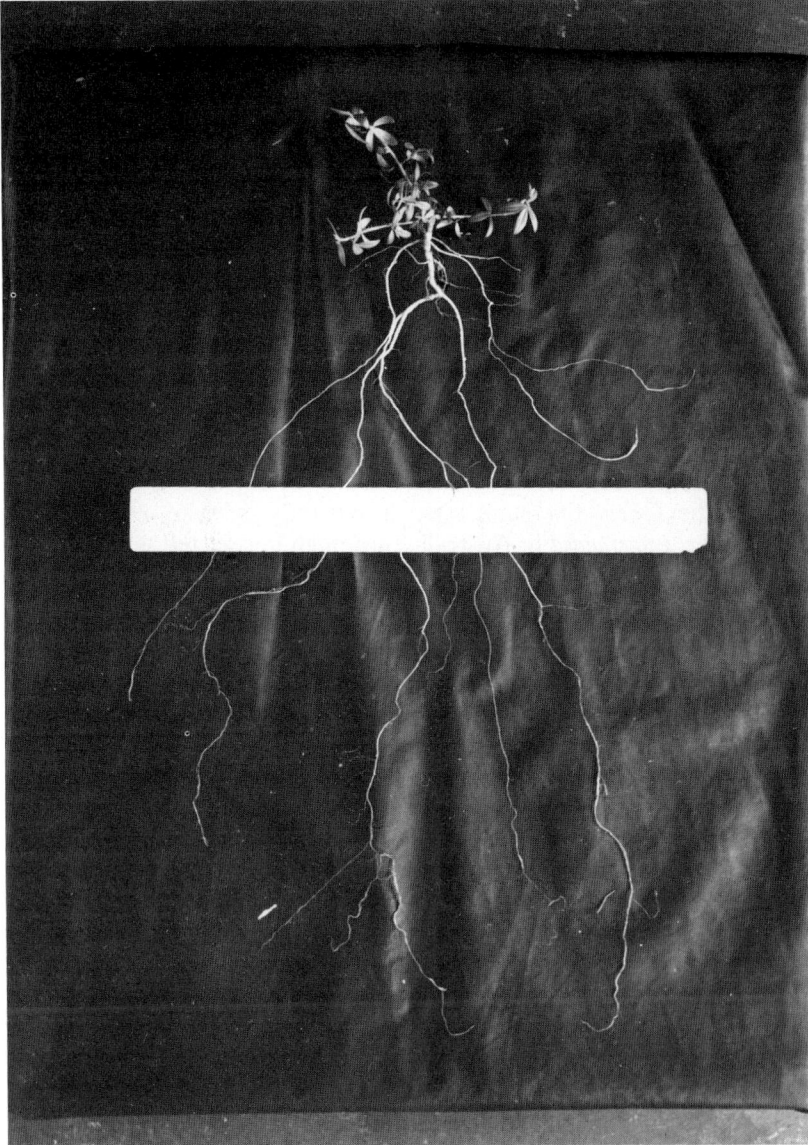

Figure 82. The well-developed root system on a fine stem stylo plant, regenerating early in the season

413

Feeding value. It is well grazed but no data on live-weight gains are available. Chemical analyses and digestibility are not available. It is a little unpalatable when young, but improves as it matures.

Toxicity. None recorded.

Seed harvesting methods. Ripe seed is exserted from the flower head, making harvesting difficult. The whole plant can be harvested with a rotary lawn-mower fitted with a grass catcher when 30 percent of the seed is ripe. The material is then dried and threshed (Stonard, 1968). Otherwise, the seed can be allowed to fall and be sucked from the ground with a vacuum harvester (Bisset, 1968). If the crop is grown for seed production, it can be grown in rows, and, at flowering, plastic strips are laid under the plants to catch seed when it falls on ripening. Seed-harvesting ants may take some seed.

Seed yields. The average yields of seed is 300 to 500 kg/ha, but up to 1 000 kg/ha have been obtained.

Minimum germination and quality required for commercial sale. Not available.

Cultivars. The Oxley fine-stem stylo is the only cultivar of *Stylosanthes guianensis* var. *intermedia* in Australia, although unselected material is also available. Several more ecotypes are under test but have not been released for commercial planting.

Diseases and pests. Only seed-harvesting ants have been recorded.

Main attributes. Extreme tolerance to grazing and fire; drought and some frost resistance; efficiency in extracting calcium and phosphorus from the soil; compatibility with grasses. Strong ability to naturalize on suitable country.

Main deficiencies. Difficulty of seed harvest; specificity of *Rhizobium*; lack of bulk.

Performance. Dry-matter yields have been given.

Main reference. Stonard (1968).

Stylosanthes hamata (L.) Taub. (see Colour plates XXXIX, XL)

Synonyms. *Hedysarum hamatum* L.; *Stylosanthes procumbens* Sw.; *S. eriocarpa* Blake.

Common names. Caribbean stylo (Australia); pencil flower, mother segal (West Indies); tebeneque (Venezuela).

Description. A herbaceous annual to short-lived perennial with a non-determinate, much-branched, semi-erect growth habit. It can reach 0.75 m.

Stems spreading to ascending, with a dichotomously branching habit, smooth, except for a very fine line of white hairs down one side, but no bristles. Leaves trifoliate with long, narrow, shiny leaflets, the latter 4.5 to 5.3 times as long as they are wide. Stipules 4 to 6 mm long, bidentate, adnate to the base of the petiole, hairs on the sheath and teeth. Inflorescence an oblong spike, greater than 20 mm long, with 8 to 14 small yellow flowers on a long stem. Outer bracts have hairs on the edges, one outer and two inner bracts and an axis rudiment subtending each flower. Loments have two articulations, both usually fertile, the lower pilose, the upper glabrous. Beak slightly curled, equal to or shorter than the upper articulation, total length of upper pod including beak 6 to 7 mm, a few hairs on the underside of beak. Seeds medium to dark brown, 2 to 2.5 mm long, unsymmetrically reniform, radicle ends fairly prominent (McKay, personal communication).

Distribution. S. hamata occurs mainly in the West Indies, in the Caribbean area generally, and in coastal areas of Venezuela and Colombia bordering the Caribbean. It also occurs in coastal areas of southern Florida, United States. It is occasionally reported from Honduras and various parts of Brazil but these may not be native populations, especially those growing in isolated areas in southern Brazil.

Season of growth. A summer-growing plant. Cv. Verano flowers in 65 to 75 days after planting, thus approximating the early- to mid-season-flowering types of Townsville stylo. However, unlike Townsville stylo, vegetative growth and flowering can continue throughout the rest of the season.

Temperature for growth. 'Verano', at least, is a heat-loving plant, and in Queensland, Australia, does not perform well south of the Tropic of Capricorn.

Frost tolerance and regrowth after frosting. Aerial growth is cut even by light frosts but the crowns will survive moderate frosts.

Latitudinal limits. S. hamata grows between 12 and 28° latitude. It is doubtful that the plant is native to areas outside these limits (Williams *et al.*, 1984).

Altitude range. S. hamata is essentially a low-altitude, maritime plant, lacking the cold tolerance necessary for growth at higher elevations.

Rainfall requirements. 600 to 1700 mm per year, with 700 to 900 mm ideal. *S. hamata* requires a pronounced dry season for optimum growth.

Drought tolerance. 'Verano' is certainly more drought tolerant than Townsville stylo and is sufficiently drought hardy to survive the long dry season in the dry tropics of northern Australia. To ensure persistence of the crown, however, considerable leaf fall may occur.

415

Tolerance of flooding. Certain forms have quite good flood tolerance (Brolmann, 1978).

Soil requirements. Soil requirements are fairly specific for each ecotype. 'Verano' is suited to the infertile, acid, sandy-surfaced solodics that suit Townsville stylo but with some tolerance of slightly more alkaline soils, such as the duplex texture-contrast brigalow (*Acacia harpophylla*)/dawson gum (*Eucalyptus cambageana*) soils of central Queensland. Other forms, in Florida and Antigua, have thrived on alkaline soils varying from coarse coral beach sands to relatively heavy clays (Burt *et al.*, 1983).

Rhizobium *relationships.* 'Verano' is promiscuous, nodulating freely with a wide spectrum of native cowpea rhizobia in Australian soils. Other ecotypes are much more specific and lack of suitable *Rhizobium* strains has held back the testing of some lines.

Ability to spread naturally. The rhizobially promiscuous 'Verano' spreads readily and widely, the hooked upper seed pod being spread readily by stock. The more specific types may be slower in spread, depending on how quickly appropriate rhizobia can travel.

Land preparation for establishment. Requirements are similar in most respects to those of Townsville stylo. The better the seed bed and the more complete the removal of competing plants, the more successful will be the establishment. However, both strip planting and oversowing into burnt native pasture have been quite successful in northern Australia.

Sowing methods. Because only half the seed pods carry a beak and the hook is not as long and complete as that of Townsville stylo, *S. hamata* seed tends to flow more freely than that of Townsville stylo. As a result, an even wider range of sowing methods can be used. Most seed is distributed either from the air or by ground spreading from fertilizer spinners, in the latter case mixed with a small quantity of fertilizer as a carrier.

Oversowing into natural pastures. Provided the native grass growth is not strong and the impeding cover is removed by burning, quite successful establishment is possible. Where stronger competition exists, light cultivation will be desirable.

Sowing depth and cover. Surface sowing is possible, but shallow coverage of the seed enhances establishment success, especially on soils with a tight-setting surface.

Sowing time and rate. In the seasonally dry tropics, where break of the seasonal drought is reliable, the ideal sowing time is just before the beginning of the next wet season. In northern Queensland, this is from December to early January. Sowing rates of 1 to 4 kg of seed/ha are used.

416

Number of seeds per kg. Cv. Verano has about 270 000 seeds in the pod and 450 000 without pods.

Percentage of hard seed. For 'Verano', hard-seed levels are high immediately after seed set. Seed softens more slowly than that of Townsville stylo, though high surface soil temperatures hasten softening (McKeon and Mott, 1984).

When planted on bare burnt surfaces, commercial seed will soften quickly if planted some weeks before the break-of-season rains. However, if planted into a moist conventional seed bed, such as is used for seed production, seed should be scarified before sowing. In some circumstances, half of the seed is treated and half left hard. The hard seed softens naturally and gives an insurance against a false start to the first growing season. Scarification can be by dry heat (Mott, 1979), hot water or dehulling.

Nutrient requirements. Like all *Stylosanthes* spp., Caribbean stylo has the ability to forage strongly for phosphorus in the soil. As a result it can grow on quite low phosphate soils. It will nevertheless respond strongly to applied phosphate, establishment in particular being aided by a small quantity of superphosphate. Cv. Verano responded to applied P in the same degree as Townsville, Cook and endeavour stylo (Jones, 1974).

Responses have also been widespread to molybdenum and sulphur applications. Plants in calcareous soils apparently have much higher phosphorus requirements for adequate growth.

Compatability with grasses and other legumes. Cv. Verano is no more tolerant of shading by strong grass growth than is Townsville stylo, but, because of its rather more vigorous growth, it can keep up above the grass for longer. As a result, it combines well with *Urochloa mosambicensis* and *Bothriochloa pertusa*, two grasses planted or often naturalized in areas where 'Verano' is planted in north Queensland.

Ability to compete with weeds. Caribbean stylo competes better with weeds than does Townsville stylo, because of its more vigorous growth and weak perenniality.

Tolerance of herbicides. Like all the *Stylosanthes* spp., 'Verano' appears remarkably tolerant of a wide range of herbicides. For seed production, Trifluralin (40 percent w/v) at 1.4 to 2.0 litres/ha is recommended preplant, followed by 2,4-D amine (50 percent w/v) at 1 litre/ha of the product after the plants are 5 cm high and 2 litres/ha of product on established crops. Alachlor, dinoseb, paraquat, dimethaline and aciflourfen all seem to be effective herbicides for use in seed crops, but actual recommendations are not yet available (English and Hopkinson, 1984).

Grazing management. As with Townsville stylo, management, through

417

adjustment of stocking rate, should aim at suppression of the associated grass should this show signs of dominating the stand, especially early in the growing season. Heavy grazing is readily tolerated and reduces the competition from taller-growing grasses while the stylos form a dense, ground-hugging mat of foliage. Perennation of individual plants could also be favoured by close grazing, inasmuch as plants left ungrazed throughout their growing season tend to behave as annuals.

Response to fire. 'Verano' has no tolerance of dry-season fire, perenniating plants being readily killed. However, hard seed is softened by the fire and the stand is generally replaced at the beginning of the next growing season.

Breeding systems. 'Verano' is essentially a self-pollinating, cleistogamous plant, pollination occurring within the flower before it opens. There is, however, a low level of cross-pollination that can occur even with related species such as *S. scabra*. Unlike most *S. hamata* ecotypes, which are only diploids with 2n = 20 chromosomes, 'Verano' is a tetraploid line with 2n = 40.

Suitability for hay and silage. Like Townsville stylo, *S. hamata* makes excellent hay so long as it is cut before leaf fall commences. If left too late, it can give a rather stemmy hay.

Value as standover or deferred feed. Unlike Townsville stylo, 'Verano' continues vegetative growth after flowering begins, and will continue to grow while seasonal conditions remain suitable. This gives a longer season of nutritious feed in some years, but once the soil dries out, leaf fall can be rapid and standover feed of little value, consisting mainly of bare stems.

Feeding value. Protein levels are reasonable and *in vitro* digestibilities are in the order of 60 to 65 percent. The phosphate levels of the forage however depend on the status of the soil in which the stylo is growing. On low phosphate soils, low levels of forage phosphorus can be expected, especially toward maturity.

Toxicity. None recorded.

Seed harvesting methods. Direct heading is the usual method with 'Verano' in north Queensland, but suction harvesting can sometimes be used under very intensive management conditions. For direct heading, correct timing is essential. Flowering should have ceased or nearly so, and some seed should already have been shed. Most seeds should be no longer green but brown (English and Hopkinson, 1984).

Seed yield. Yields of up to 700 kg/ha have been obtained for seed in the pod in north Queensland for 'Verano'. Normal commercial yields are in the range of 300 to 600 kg/ha.

Minimum germination and quality of seed for commercial sale. The Queensland standard is 40 percent minimum germination plus a maximum of 20 percent hard seed deemed to be germinable. Minimum pure seed content is 90 percent.

Cultivars. 'Verano' is currently the only commercial cultivar. Released in Australia in 1973, it was collected in 1965 at Maracaibo Airport, Venezuela (10°N lat., elevation 10 m, MAR 563 mm), but it is believed that other cultivars of diploid types are under development in Florida and the West Indies. 'Verano' could in fact be a separate species (Stace and Cameron, 1984).

Diseases. To date, 'Verano', while showing some susceptibility to anthracnose (*Colletotrichum gloeosporioides*) in the greenhouse, has remained, for all practical purposes, resistant or at least tolerant to the ravage of anthracnose in the field. It is also susceptible to botrytis head blight (*Botrytis cinerea*) in high-rainfall areas and to rhizoctonia web blight (*Rhizoctonia* sp.), but these cause little damage in the main 'Verano' areas, which are located in drier districts.

Main attributes. Essentially a bigger, better, short-lived perennial form of Townsville stylo with, to date, anthracnose tolerance and a better competitive ability with native grasses in northern Australia.

Main deficiencies. 'Verano' is very much a tropical plant, performing poorly at higher latitudes and especially in the subtropics.

Performance. 'Verano' has replaced Townsville stylo with the latter's collapse to anthracnose. Its greater competitive ability, drought resistance and wide soil adaptation were already giving promise of its being a complementary species to Townsville stylo before the appearance of anthracnose.

Main references. McKeague, Miller and Anning (1978); Burt *et al.* (1983); English and Hopkinson (1984).

Stylosanthes humilis **H.B.K.** (see Colour plates XLI, XLII, XLIII)

Synonyms. *S. sundaica* Taub.; *Astyposanthes humilis* H.B.K.

Common names. Townsville stylo, Townsville lucerne (Australia); Magsaysay lucerne (Philippines); wild lucerne, alfalfa selvagem, alfalfa de Townsville (Brazil).

Description. A self-regenerating, self-fertile summer annual or short-lived perennial legume with trifoliate leaves and branched stems, ascending or prostrate; can reach a height of up to 0.7 m. Leaflets lanceolate, narrow and pointed. Fruit, a biarticulate pod terminated by a persistent style that gives a beaked appearance. Each loment contains one seed. Six or more pods are

419

produced in each seed head. Under heavy grazing or when plant stems rest on moist soil, adventitious roots are formed on the stems several inches away from the taproot.

Distribution. Native to northeast Brazil, Venezuela, the drier Caribbean littoral, Costa Rica and the Chitre region of Panama. It is now widespread in the tropics.

Temperature for growth. Over a range of 15/10°C to 27/22°C an increase of 2 to 7°C doubled the growth rate. Whiteman (1968) considers optimum growth temperatures to range between a 27°C and 33°C day temperature. Cameron (1967a, b) suggested an optimum temperature for growth as ± 30°C for day temperature and ± 25°C for night temperature. He also found that night temperatures below 25°C and day temperatures below 30°C inhibit dry-matter production. It is affected by frost, but usually sheds ripe seed before it occurs. Frosts prevent seed production in the subtropics.

Latitudinal limits. In Australia it grows between 11 and 28°S, in the Americas from 23°N in Mexico to 14°S in Brazil. Cameron (1965) found that flowering was negatively correlated with latitude and positively with rainfall.

Altitude range. Occurs from sea level (510 m at Rodd's Bay, Queensland) to up to 1 500 m in Burma (Snook, personal communication).

Rainfall requirement. In Queensland, it requires more than 550 mm. Davies (1966) suggests a range of 635 to 1 778 mm. Late maturing types can grow from 550 to 1 020 mm. Snook (personal communication) found that it did well in Burma under an annual rainfall of 2 500 mm. Winter rain is detrimental to its performance as it tends to cause secondary fungal infection on overwintering standing hay. Fisher (1970) found it relatively insensitive to within-season droughts.

Tolerance of drought and flooding. It survives drought because of its annual habit and heavy seeding capacity. Outside of its rainfall range (above), it should be replaced by newer low-rainfall ecotypes of *Stylosanthes*. Miller (1970) suggests that in areas of uncertain rainfall *S. humilis* may survive in-season drought better under a low phosphorus status than if ample fertilizer is applied. Humphreys (1967) says that it does not tolerate waterlogging, but Tiver (personal communication) found that it tolerated waterlogged conditions, but not swamps in the Northern Territory, Australia; Farinas (1966) in the Philippines, Snook (personal communication) in Burma, and Kretschmer (1968) found that *S. humilis* would survive intermittent flooding but not waterlogging, and did well where there was a high water table in Florida, United States.

Soil requirements. Prefers sands and sandy loams, but will grow on heavier

420

types such as the rice lands in Burma (Snook, personal communication), where rainfall is 2 500 mm, falling during seven months, and on heavy rice soils on the Adelaide River, Northern Territory (Northern Territory, 1961). It prefers the soil to be slightly acid. It will nodulate at pH 4.5, or pH 4.0 if the calcium supply is adequate. It has fair tolerance of salinity.

Rhizobium *relationships*. Does not need inoculation, as it nodulates freely with native rhizobia of the cowpea type in the soil. If it is desired to inoculate the seed, the current Australian inoculant recommended is CB756.

Ability to spread naturally. Excellent; cattle and other grazing animals quickly spread the seed and it will gradually cover the ground. This factor enables seed to be planted in strips to economize on labour and seed (Graham, 1963b).

Land preparation for establishment. Competing tall plants must be removed by heavy grazing, burning or discing. A well-prepared seed bed gives quicker and better establishment but is more costly. Cultivation of strips in natural pasture for seeding is cheaper, and aerial seeding into natural pasture is quite effective but gives slower establishment. Norman (1961) found ploughing to 15 cm, then discing level and broadcasting the seed to be the most effective at Katherine, Northern Territory. For aerial sowing into natural pasture, burning or heavy stocking of the natural grassland immediately before the rains is all the land preparation required. Establishment in contour strips by ground cultivation and seeding about one-quarter of the area is another common practice which succeeds best if the timber has been previously ringbarked.

Sowing methods. Townsville stylo is either drilled into a well-prepared or roughly prepared seed bed (Strachan, Lambert and Finlay, 1967) or broadcast from ground machines or from the air. Establishment in cultivated strips has been described and illustrated (see Figure 10). With aerial seeding, it is best to divide the ground area into blocks by cutting boundaries with a bulldozer and have flag men to guide the aeroplane. Sowing height is 120 m and the seed used either hulled or dehooked for freer flowing. Seed gives three times the establishment that is obtained from pods. Land should be burnt or stocked heavily before seeding and stocked heavily after seeding to reduce competition from tall grasses. Stocker and Sturtz (1966) found that burning the native *Sorghum intrans* pasture in the middle of the wet season, during the period of reliable rainfall, and aerial seeding *S. humilis* at the rate of 2.3 kg/ha pods with 250 kg/ha superphosphate gave good establishment on lateritic red earths in the Northern Territory. For good establishment, the soil should be moist after seeding for two to four days (Sillar, 1969); Norman

421

(1961) suggests that moisture content in the top 15 cm of soil should be above wilting point for five to six days.

Oversowing into natural pastures is also successfully accomplished. Establishment is slower and it may take up to three years to cover the ground adequately, but it is very convenient. Miller (1967) states that a period of four consecutive wet days with at least 25 mm of rain in the first day and 75 mm over the whole period would be favourable for establishment — such conditions occurred 15 times in nine years at Katherine, Northern Territory.

Sowing depth, cover, time and rate. Broadcast on the surface. In drill-sowing into prepared seed beds, a light harrowing can be given. Sow just before the rainy season, 4 to 6 kg/ha; the plant stand is linearly related to seeding rates (Miller, 1967).

Number of seeds per kg. 396 000 to 484 000. Percentage of hard seed, 17 to 99 percent at harvest. Hard-seededness of 74 to 99 percent in July had declined steadily to 31 to 49 percent by November (Cameron, 1965, 1967). About 10 percent germination a year is satisfactory in the field. Graham (1963b) recorded germination of fresh seed as 2 to 35 percent.

Seed treatment before planting. To improve handling: seed is preferably dehulled or at least debeaked or dehooked so that it will flow more freely through seeding machinery. The hooked pods can be diluted with sand for sowing. To break dormancy: treatment is not usual, as hard seed is an advantage under the irregular climatic conditions in which it is normally sown. Scarification by light abrading in a hammer-mill improves germination. Pelleting is not usual. If the seed is to be pelleted, use rock phosphate as a base. Harty (1967) found that superphosphate at the rate of 308 kg/ha inhibited germination of abraded seed, and so a neutral fertilizer or pelleting should be used with heavy superphosphate dressings. Insect and disease treatment is not usual.

Nutrient requirements. Townsville stylo is one of the most efficient of the tropical legumes in extracting its calcium and phosphorus requirements from the soil. It will grow in soil with as little as 3 to 10 ppm available phosphorus, and so will establish in most soils without P, although it performs better with additional P. It is also tolerant of high manganese and aluminium. It is usual to apply 130 kg/ha molybdenized superphosphate at establishment and about 65 kg/ha straight superphosphate annually thereafter.

● Calcium. Townsville stylo gives 64 percent of its maximum yield in the absence of calcium (Andrew and Robins, 1969). The calcium uptake is 1.8 to 2.5 percent of the dry matter (Andrew and Hegarty, 1969). Shaw, Gates and Wilson (1966), analysing the response by applying superphosphate, con-

422

cluded that the calcium of superphosphate had little effect. However, the calcium concentration of *S. humilis* is the highest of all the tropical legumes analysed by Andrew and Robins (1969), being 67.5 percent of the total calcium compared with the lowest, 48.4 percent, for *Desmodium intortum*.

● Nitrogen. Kretschmer (1968) found that as little as 27.5 kg/ha N applied after each clipping of a pangola grass/*S. humilis* pasture eliminated *S. humilis* after three clippings. Andrew and Robins (1969) found the N content of *S. humilis* low at 3.28 percent compared with *Desmodium intortum*; the highest, at 4.23 percent. In association with Rhodes grass, Vallis *et al.* (1967) found that, with added N, Rhodes grass took up 20 times as much N in the first nine weeks and eight times as much as the *S. humilis* between nine and 13 weeks. At five weeks, they estimated that 47 percent of the N of the legume growing with the grass had come from the soil. At nine weeks, the cumulative uptake of N by the legume was only 6 percent and at 13 weeks only 3 percent of its total N yield.

● Phosphorus. The critical percentage of phosphorus in the plant tops was determined in the field by Jones (1968) as 0.16 to 0.17 percent and the latter figure was confirmed by Andrew and Robins (1969). The plant produced its maximum yield when 500 kg/ha superphosphate were added to the soil. The ability of *S. humilis* to absorb greater quantities of phosphorus from a soil with low available P has been shown by Andrew (1966), who also found that the uptake of P by excised roots of *S. humilis* from dilute solutions per unit time was greater than that of *Macroptilium lathyroides*, *Desmodium uncinatum* and *Medicago sativa*. Shaw, Gates and Wilson (1966) showed a fivefold increase in yield from the application of PK and Mo. Phosphorus in the presence of Mo also increased N content. In the field, the yield of dry matter was increased by a factor of 2.4 by superphosphate in the presence of Mo, and the yield of nitrogen grew by a factor of 3.3. They found that the P in superphosphate increased dry matter and that S increased the N content. Heavy dressings of P in the absence of S led to a leaf tip necrosis. Heavy dressings of P also reduced the K content below the critical level and increased the Na and Mg levels. They concluded that *S. humilis* had a high degree of efficiency in profiting from limited reserves of an essential element by substitution of another. Jones (1968) found that for dry-matter production from a solodic soil in north Queensland, an initial dressing of 370 kg/ha superphosphate gave higher dry-matter yields over three years than yearly dressings of 133 kg/ha, but annual dressings gave high N and P levels. Norman (1959) increased nodulation by 293 percent with a dressing of 370 kg/ha superphosphate over the nil treatment. Fisher (1970) achieved maximum yields at 625 kg/ha superphosphate but 90 percent of the maximum was obtained with 375 kg/ha and 75 percent with 250 kg/ha.

Although adequate dressings of phosphorus are required to give maximum production of Townsville stylo, too much phosphorus will encourage the legume to such an extent that it will eliminate the associated grass and expose the area to a shortage of dry matter (Woods and Dance, 1970). Fertilizer application should be adjusted to keep about 50 percent legume and 50 percent grass. At sowing, 250 kg/ha superphosphate and 62 kg/ha yearly thereafter gave 90 percent of what was considered to be the maximum yield at Katherine, Northern Territory (Norman, 1965b).

● Potash. The critical percentage of potassium in the plant tops was determined by Andrew and Robins (1969) as 0.60; and *S. humilis* gave 47 percent of its maximum yield at low levels of potassium. Gates, Wilson and Shaw (1966) quoted K percentages of 0.35 to 0.55 percent in the tops of plants showing potash deficiency symptoms. Andrew and Pieters (1970a) showed a normal plant to have 1.22 g of K per 100 g of dry matter and a deficient one to have 0.30 g K/100 g dry matter. In plant analyses under differing degrees of K fertilization, *S. humilis* was found to have low concentrations of K (11 percent compared with *D. intortum*, which was the highest of eight tropical legumes, with 29.1 percent). It had high concentrations of Ca, Mg and Na and low concentrations of N and P.

Deficiency symptoms for potash are first evident on the leaves of the middle portions of the plant. They commence as a series of rust-coloured spots of pinhead size, irregular in shape, largely interveinal, but in some cases also on the veins, spread at random over the leaflets. They may or may not be associated with chlorotic leaflet tips, which may be preceded by a bronze coloration on the upper leaf surface. With increasing severity, some of the larger of the rust-coloured areas become necrotic and are discernible on both surfaces of the leaflet. This condition intensifies and is followed by severe leaflet tip and marginal chlorosis, culminating in severe necrosis of the affected areas. The necrosis encroaches toward the base of each leaflet, and in some cases ends up as fully necrotic leaves; each leaflet curls inward to give a ball effect. On complete necrosis and death, the leaflets and petioles absciss although sometimes the petioles remain green in colour and intact. Axillary growth is similarly affected. In very severe forms, the unfolding leaflets at the growing tip become slightly chlorotic and are puckered and pinched at their tips (Andrew and Pieters, 1970a).

Response to photoperiod and light. *S. humilis* is a short-day plant. With a day-length of under ten hours, its growth habit is prostrate, and from 12 to 14 hours it is erect. It needs less than 14 hours' daylight to flower; 't Mannetje (1965) found eight to ten hours best. Dry-matter yields, however, were greater with long days (Downes *et al.*, 1967). Townsville stylo requires adequate light for full development and close grazing provides this. Sillar (1967) found that reduction to 0.74 daylight reduced yield by 47 percent, and a reduction to 0.38 daylight caused 33 percent mortality. Tall grass depressed its growth.

Compatibility with grasses and other legumes. Does not do well with tall grasses. It associates with *Heteropogon contortus* naturally in Queensland. Selection and breeding for plants more tolerant to tall grasses are in progress. In the Northern Territory, Australia, it associated well with *Cenchrus ciliaris, C. biflorus, Urochloa bulbodes* and *U. mosambicensis* (Norman, 1967a). If P is applied to mixtures of grass/legume, the response in the first year is mainly by the grass. In Florida, Kretschmer (1968) has had outstanding success with pangola grass/*S. humilis* mixtures. It is compatible with angleton grass (*Dichanthium aristatum*) in north Queensland (Onley and Sillar, 1965). In the first year, *S. humilis* is compatible with one other legume, but not two; in the second year, it is strongly competitive with a second legume.

Tolerance of herbicides. Generally, there is little need to use herbicides. Grazing management is much more important. For the selective control of seedling *Hyptus suaveolens* in Townsville stylo pastures in the Northern Territory, 2,4-D has been used successfully, the rate of application being between 125 and 250 g/ha acid equivalent (Bailey, personal communication).

Seedling vigour. Seedlings are only reasonably vigorous, but rapid root development (see Figure 83), which Torssell *et al.* (1968) measured as 50 to 60 cm after the opening rain of only 38 mm, helps it to compete. Similar seedlings survived for another seven weeks without further rain.

Vigour of growth and growth rhythm. At Katherine, Northern Territory (lat. 14°3'S, rainfall 950 mm), seed germinated in the first week in October and there was slow early growth to the end of January and then rapid development until early April. Most of the dry matter was located in the stems, particularly in the lower layers. Stem development increased to a maximum at active flowering and seed setting in late March to early April. By contrast, there was a decline in leaf tissue mass in the lower layers as upper leaf mass and stand height increased. It was estimated that 35 percent of the total above-ground production was shed as leaf. Torssell *et al.* (1968) summarized growth rhythm as shown in Table 14.12.

Ability to compete with weeds. Under heavy grazing, Townsville stylo generally competes very successfully with weeds. If growth of associated weeds is rank, it competes less successfully.

425

TABLE 14.12 **Growth rhythm of *Stylosanthes humilis* in Northern Australia**

Season	Stage of growth	Growth features
Transition period between dry and wet season	Germination and establishment	Rapid root penetration after germination primarily with a vertical taproot. Shoot development slow. Possibility of seedling death as a result of water stress
Early wet season	Early growth	Additional germination and establishment. Continued root penetration, accompanied by increase in total root length, due to lateral root development in the topsoil. Increasing shoot development, the end of this period being marked by full light interception by the stand (if no water stress)
Late wet season and transition period between wet and dry season	Grand growth period including flowering and early seed setting	Growth rate high. If no water stress, full light interception at 15 to 20 cm below crop surface, leaf shedding from lower parts of the stem. Root length may decrease at end of period, which terminates by net assimilation decreasing to zero
Early dry season	Maturation and senescence	No net growth; decrease in above-ground material due to respiration or translocation to roots. The period terminated by the death of the plant
Late dry season	Drought survival (as seed)	Few physiological or chemical changes in standing dry material if no rainfall. May persist if soil water available to the plant

Nitrogen-fixing ability. S. humilis is able to fix moderate quantities of nitrogen. Andrew and Robins (1969) found that it was not very efficient in increasing its nitrogen content in relation to phosphorus concentration but, per unit of soil mass and within the time limits, it was one of the most productive of the tropical species, particularly at low phosphorus levels. In the Northern Territory, after four years Townsville stylo yielded 2.72 tonnes/ha; without legume but with 250 kg/ha sulphate of ammonia, it yielded 2.31 tonnes/ha (Northern Territory, 1964). Wetselaar (1967) found that the total amount of nitrogen added to the soil/plant system over three seasons at Katherine, Northern Territory, was 216 kg/ha, the figure for each of the second and third years being 91.3 kg/ha. Gates (1970) showed active nodulation and continuing fixation even in the presence of increasing nitrogen if the proper mineral

426

balance, especially between phosphorus and nitrogen, was maintained. Norman and Stewart (1964) found that a two- to three-year ley of Townsville stylo will provide enough nitrogen for the successive *Pennisetum typhoides* crop in the Northern Territory.

Response to defoliation. Will stand heavy grazing; the only adverse effect is in seed production near the end of the growing season. As the plant is grazed, new axillary buds develop.

Grazing management. The aim in effectively introducing and using *S. humilis* is to establish a light cover of the legume in the first year cheaply, and to encourage its subsequent development by grazing management. Heavy grazing in the early wet season, by reducing the vigour of the perennial grasses, allows the legume to develop. Heavy grazing in March-April reduces seed production (Norman, 1965a).

Response to fire. *S. humilis* can survive fires as seed, and the passage of fire stimulates germination. Burning of pastures should take place before the first spring or summer storms, as fire-susceptible seedlings will be present after the early rains (Sillar, 1969a).

Breeding system. Self-fertile; chromosome number 2n = 20.

Dry- and green-matter yields. The average yield of cultivated Townsville stylo

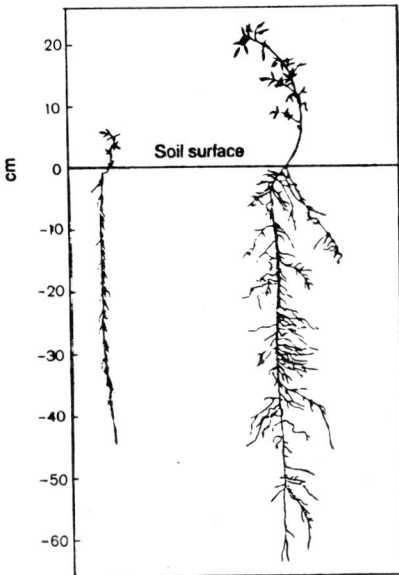

Figure 83. Root development of *Stylosanthes humilis* plants, fertilized with 50 mg/cm^2 of superphosphate mixed in the top 5 cm of soil. Grown in growth chamber for 33 days (left) and 43 days (right)

427

at Katherine, Northern Territory, has been 6 635 kg/ha over several years (Miller, 1967), at lat. 14°3'S, altitude 110 m and with an annual rainfall of 950 mm falling mainly from December to March. Norman (1959) obtained 5 610 kg/ha at 16 weeks from *S. humilis* fertilized with 132 kg/ha superphosphate with a crude protein yield at 12 weeks of 712 kg/ha. At Berrimah Experiment Farm in the Northern Territory, a *Melinis minutiflora/S. humilis* pasture yielded 3.82 tonnes/ha when fertilized with 1 232 kg/ha superphosphate. In Florida, Kretschmer (1968) obtained 9 295 kg/ha/year from pangola grass/*S. humilis* pasture, compared with 5 159 kg/ha from pangola grass alone.

Suitability for hay and silage. *S. humilis* makes quite good hay, the feeding value of which depends on fertilizer history. Shaw, Gates and Wilson (1966) showed that superphosphate fertilizer increased the N content in the field from 2.4 to 3.3 percent and proved that high-quality fodder can be obtained from *S. humilis* with adequate fertilizer. It yielded 2.4 tonnes of hay/ha/year over five years at Katherine, Northern Territory (Woods, 1969). Fully grown material is mown into windrows to dry within 24 hours and, depending on the weather, can be baled in three days or more. Too frequent cutting encourages weeds (Sillar, 1969). Little silage has been made from this plant. Where it grows erect among grasses, it is easier to handle and Kretschmer (1968) suggests that pangola grass/*S. humilis* would make good silage.

Value as standover or deferred feed. This is one of the main attributes of *S. humilis*. It is not very palatable in the young stage and the associated grass is usually eaten first. Its acceptability improves with age and the standing dry matter is sought after during winter and spring. The seed content also improves the food on offer.

Feeding value. Its chief advantage is its nutritive value in the winter and dry season, when grazing cattle gain weight. They lose weight during the "critical period" immediately after the onset of the rainy season, probably owing to a reduction in quality of the standing forage (Norman, 1967b), especially if this is the case with early maturing varieties that have reached maturity.

● Chemical analysis and digestibility. Few complete analyses of *S. humilis* have been recorded. Otero (1952) listed the average crude protein content of 14.2 percent with a range of 12.3 to 17.7 percent, crude fibre at a mean of 30 percent with a range of 25.5 to 36.2 percent, fat at 1.4 percent (range 0.7 to 2.5 percent), nitrogen-free extract 46 percent (range 41 to 49.7 percent), ash 8.3 percent (range 5.2 to 13.1 percent), calcium 1.7 percent and phosphoric acid 0.35 percent. Several other analyses are listed in Appendix 1. Crude protein figures range from a very low 5.6 percent to 21.4 percent — a reflection on soils, fertilizers and stage of growth.

428

● Palatability. *S. humilis* is not readily eaten in the young stages but increases in palatability as it matures. This protects the young growth and enables immediate and heavy stocking after seeding.

Toxicity. None observed.

Seed harvesting methods. For good seed production, grazing should cease toward the end of the season to allow ample seed to develop. Several methods of harvesting are available, the choice depending on financial outlay. Graham (1963a) outlined methods in use in Queensland.

● Sweeping, winnowing and cleaning.

Sweeping. The plants are allowed to mature fully with harvesting delayed until the seed has fallen and the plants themselves have begun to break up and wither away. This takes place about September-October, when conditions in the tropics are likely to be dry. The area is then swept with a rotor-broom such as that used for road sweeping. The broom sweeps sand, seed and crop residue into windrows for convenience in picking up. The material is then dumped at a convenient spot for winnowing.

Winnowing. The first step in cleaning the seed is to remove the maximum of extraneous matter from the bulk in one operation. Standard winnowers have been used for removing small stones, dirt and large trash from the collected sample, but with most of them the sieves are too small and the cleaning process is slowed down. The winnowing section of a harvester fitted with appropriate trays will do a similar cleaning job at a much faster rate. Where winnowing machinery is not available, a simple sieve can quite easily be constructed on the property. This consists of a wooden rectangular frame 1.85×0.77 m built of 10×2.5 cm wood and the bottom fitted with perforated zinc. The lower end is partly opened to allow the coarse material to be shaken off the sieve. This tray is suspended by hoop-iron from a special frame erected to support the shaker, or it can be suspended in a similar manner from the rafters of a shed. The movement of the sieve must be primarily in the horizontal plane; otherwise pieces of straw will tend to stand on end and fall through. The frame is tapped against a solid upright to facilitate screening. The desired rate is approximately 150 taps a minute. This can quite easily be converted to power operation by fitting a simple cam coupled to a small motor by a V pulley. A fan adjusted in a suitable position by trial and error should remove most of the straw that falls through the tray. Wind can sometimes be used to advantage by placing the tray in a position to take full advantage of it.

Cleaning. Townsville stylo seed will pass through perforated zinc but not through screen cloth of the same hole diameter. It would appear that where

Figure 84. Home-made cleaner and sieves for *S. humilis*. Some of these cylinders are hand-operated; the more modern ones use a small 1.5 horsepower engine. As the cage, which is mounted on rollers, slowly revolves, the material containing seed is fed into the revolving cylinder (**Photo:** T.G. Graham)

the diameter of the hole is less than the width of the solid area between the holes, provided the holes are large enough to take the seed, Townsville stylo seed will pass through without any trouble; but where the space between the holes is less than the diameter of the hole itself, the seed can attach itself to the sieve by means of its curious hooklike formation at the end of the seed. An ingenious device is used for the final cleaning of Townsville stylo seed (Figure 84). A cylindrical wire gauze cage from 0.6 to 1.0 m in diameter and about 1.8 m long is mounted on a frame which can be tilted at one end to gravitate the residue out at the other. The Townsville stylo seeds cling to the inside circumference of the cylinder, where they are removed by a scraper, either a spring-loaded rubber scraper or a revolving finned scraper. This scraper operates as the seed reaches the top of the cylinder. As the seed is scraped off, it falls into a chute which directs it to the far end of the cylinder. It is given considerable fall to facilitate the movement of the seed toward the outer end. However, Townsville stylo seed does not flow readily and consequently has to be scraped out at the lower end. This can be done while the

430

Figure 85. Horwood Bagshaw clover seed harvester for harvesting subterranean clover seed, barrel medick and Townsville stylo (**Photo:** Horwood Bagshaw, Adelaide)

machine is in motion. Extraneous matter remains in the cylinder until it reaches the far end, where it drops on the ground away from the clean seed chute. This is a positive method of cleaning Townsville stylo seed, and no other seed even of the same weight and size can become mixed with the sample. For rapid commercial handling of large quantities of seed, cylindrical rotary-screen cleaners have superseded other types.

● <u>Vacuum pick-up machine.</u> A home-made machine can be built which consists of a large fan mounted on three-point linkage and driven by the power take-off of the tractor. On the intake side of the fan a 1 m mouthpiece is fitted for collecting the seed. On the delivery end, the harvested material passes through a cyclone device which blows the dust through the top and delivers the seed and the remainder of the extraneous matter into a bagger or trailer. When the machine is in motion it rides on rubber wheels, which can be adjusted to maintain any desirable distance from the ground. The back of the intake chute is fitted with flexible sheet rubber which, fitting snugly on the surface, helps to make the vacuum more effective. Beneath the tractor and just in front of the vacuum intake a netting drag is fitted which teases the mat of vegetation lying on the surface and renders it easier to lift. Any form of roll-

431

ing or disturbing with chain harrows would serve a similar purpose (Graham, 1963a).

A more costly machine is manufactured commercially. It was designed to harvest subterranean clover seed but will deal effectively with Townsville stylo, and will clean the pods and straighten the hooks. Preliminary preparation of the land involves mowing and windrowing the overburden, and scarifying the surface between the windrows where the suction pick-up machine will operate (see Figure 85).

● Heading. Where the crop has been encouraged to grow upright by the application of superphosphate or where a dense crop is growing in association with grass, it is possible to head it with a header or all-crop harvester. This method may be rather wasteful of seed, since, as Townsville stylo does not seed uniformly, the optimum stage for harvesting has to be selected. This is usually at a point just before the crop lodges. It entails considerable care in the handling of green material prior to cleaning. This method has yielded up to 226 kg/ha in the Bundaberg district, Queensland. However, as seed collected from the ground presents far greater cleaning problems than does seed harvested from the standing crop, direct heading has largely replaced suction harvesting.

Seed yields. Average yield, about 330 kg/ha but yields up to 1 100 kg/ha have been obtained under good conditions.

Minimum germination and quality for commercial sale. Forty percent minimum germination and 96.5 percent purity is required in Queensland (Prodonoff, 1968).

Cultivars. In addition to commercial seed, there are three cultivars released for commercial use in Queensland: 'Paterson' (early), 'Lawson' (midseason) and 'Gordon' (late flowering). Several other ecotypes are being developed and tested. It is hoped that several cultivars will soon be available to fill specific ecological niches in the tropics.

Diseases. S. humilis is subject to few diseases. Van Rensburg (1967) reported a severe attack by *Corticium solani* in local patches at Mt. Makulu, Zambia. Badly damaged by anthracnose when this appeared in Australia in 1974.

Pests. Colbran (1963) reported that the nematodes *Meloidogyne javanica* and *Radopholus similis* attack *S. humilis*, but losses from this source are rare.

Main attributes. Adaptability to poor soils of low fertility due to its efficiency in extracting calcium and phosphorus and its tolerance of manganese and aluminium. Free-seeding, self-regenerating habit, and palatability increasing with age.

432

Main deficiencies. Lack of bulk; low nitrogen-fixing ability; tolerance of shade in the presence of tall grasses; its hooked seed.

Performance. Townsville stylo introduced into native pastures in central and northern Queensland and the Northern Territory in Australia, in Burma and in Tanzania has remarkably increased beef production per unit area. Sillar (1969) reports 182 kg/head/year live-weight gain in beef cattle and 91 kg/head in a drought year when grazing *S. humilis* pastures in north Queensland.

Shaw and 't Mannetje (1970) reported results with beef cattle grazing *S. humilis* for seven years at Rodd's Bay near Gladstone, Queensland (lat. 24°S), and Shaw (1969) has summarized the performance as presented in Table 14.13.

Cattle grazing natural pastures were usually not in a marketable condition at the end of the season, whereas those on the pastures containing Townsville stylo were marketable.

At Katherine, Northern Territory, Dr M.J.T. Norman has carried out numerous investigations into the use of *S. humilis*, the results of which were published during 1959-68. He found *S. humilis* to be particularly valuable as a dry-season feed and showed that the greater the time spent by cattle on Townsville stylo, the shorter the time from weaning to slaughter. Performances obtained by Norman (1968) from cattle grazing Townsville stylo and native pasture are given in Table 14.14.

Cattle wholly grazing sown pasture slaughtered at 438 kg/head at 30 months and averaged a live-weight gain of 0.46 kg/day. Cattle grazing half time on sown pasture and half on native pasture slaughtered at 300 to 384 kg/head and averaged a live-weight gain of 0.28 kg/day. Cattle grazing one-quarter of the time on sown pasture slaughtered at 247 kg/head and averaged a live-weight gain of 0.15 kg/day.

TABLE 14.13 **Weight gains and stocking rate over seven years of cattle grazing *Stylosanthes humilis* and native pasture**

Pasture type	Beasts/ ha	Live-weight gain/ ha/year (kg)	Live-weight gain/ animal/year (kg)
Native pasture	0.30	24.2	83.2
Native pasture + *S. humilis*	0.77	91.3	121.4
Native pasture + *S. humilis* 132 kg/ha Mo superphosphate	1.01	145.2	149.1

TABLE 14.14 **Weight gains of cattle grazing _Stylosanthes humilis_ and native pastures in various seasons**

Season	Sown pasture (_S. humilis_ mainly)	Native pasture
	(kg/head/day)	
First dry season	+ 0.077	− 0.087
First wet season	+ 0.70	+ 0.28
Second dry season	+ 0.10	− 0.30
Second wet season	+ 0.87	+ 0.33
Subsequently	+ 0.30	—

Norman and Stewart (1964) showed that _S. humilis_ could support 2.5 beasts per hectare during both the wet and dry season at Katherine. Average intake per day was 11.9 kg. The use of native pasture in conjunction with Townsville stylo provided an efficient and flexible year-round grazing system. _Main references_. Cameron (1967); Humphreys (1967); Kretschmer (1968); Norman (1968); Shaw and 't Mannetje (1970); Sillar (1969); Torssell, Begg, Rose and Byrne (1968).

Stylosanthes montevidensis Vog.

Synonym. _S. gracilis_ var. _rostrata_.
Description. Upright plant, 0.6 to 1.2 m high with yellow to creamy flowers; pod pubescent with beak shorter than the upper articulation; stem without viscid hairs, leaflets not punctate, and articulation is fertile (Mohlenbrock, 1963).
Distribution. Occurs in Brazil, Paraguay, Uruguay, Bolivia, Argentina and Colombia. The CSIRO Australian introduction is CPI 11492, from Uruguay.
Characteristics. A sparse perennial fodder plant. 'T Mannetje (1965) showed it to be a long-day plant which can be found dormant under photoperiods of eight and ten hours. It flowered in 68 days under a 12-hour regime and in 64 days with 14 hours. Dry-matter production in a 12-hour photoperiod was nearly seven times that in the 10-hour treatment.

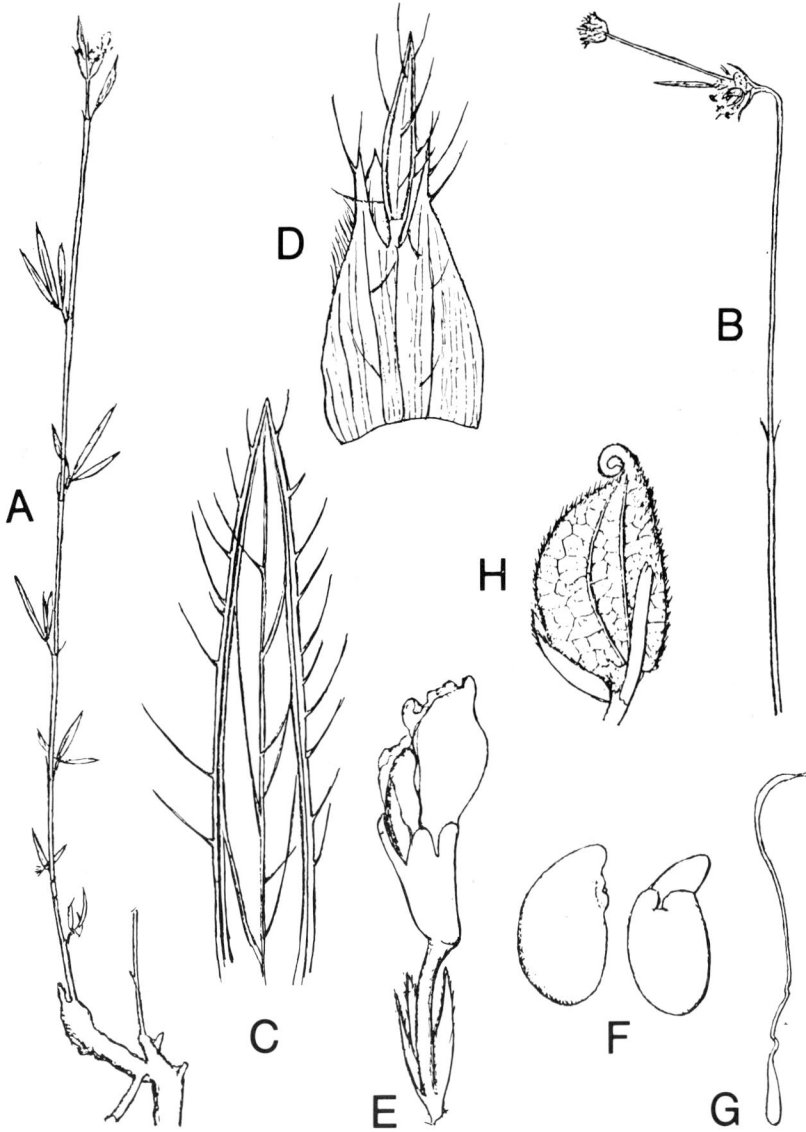

Figure 86. *Stylosanthes montevidensis.* **A**-Plant in flower **B**-Fruiting branch **C**-Portion of leaf **D**-Bract **E**-Flower **F**-Seed **G**-Carpel **H**-Fruit

***Stylosanthes scabra* Vog.** (see Colour plates XLIV, XLV, XLVI)

<u>Synonyms</u>. *S. diarthra* Blake; *S. gloiodes* Blake; *S. plicata* Blake.
<u>Common names</u>. Shrubby stylo (Australia), Capitan Juan and pata de terecay (Venezuela).
<u>Description</u>. A tall (up to 2 m), shrubby, perennial legume, erect to suberect, with strong, woody stems, usually densely hairy and viscid, making them sticky under hot, dry conditions. Stems vary in colour from pale green to dark blue-green, brown or reddish depending on the strain. Leaves trifoliate. Leaflets hairy on both surfaces, varying in shape from elliptic to oblong-lanceolate and in colour from pale green to dark green and dark blue-green. Stipules obovate, bristly hairy, 15 to 25 mm long including teeth. Inflorescence a broad, loosely formed and shortly elongate spike, occurring singly in the leaf axils. Fruit a biarticulating pod, the upper articulation carrying a short, slightly curved beak (residual stigma), and short viscid hairs. Seeds small (up to 2 mm long), pale to light brown, unsymmetrically reniform, the radical ends prominent (Mohlenbrock, 1957).
<u>Distribution</u>. *S. scabra* is a South American species, widely distributed across Brazil, especially in Minas Gerais, Pernambuco, Ceara, Piaui and Goias, to eastern Bolivia, and in a band across Venezuela and Colombia to Ecuador (Williams *et al.*, 1984).
'T Mannetje (1984) regards *S. scabra* as a problematic species that possibly should also include the presently separate *S. tuberculata* Blake and *S. nervosa* Macbr., and perhaps several others.
<u>Season of growth</u>. *S. scabra* is a summer-growing perennial with some strains possessing slight frost tolerance, giving them a long growing season. The tops are tolerant of light frosts and the crowns of quite heavy frosts.
<u>Latitudinal limits</u>. *S. scabra* occurs naturally at least as far south as the Tropic of Capricorn, while cv. Fitzroy has performed well as far south as 24°S, near Bundaberg in Queensland, Australia. Earlier-flowering lines should extend the limit further south if released there. At Lansdown Research Station, north Queensland, flowering dates varied from mid-January to mid-May (Edye, Grof and Walker, 1984).
<u>Altitude range</u>. Grows from sea level to *c.* 600 m.
<u>Rainfall requirements</u>. Requirements have yet to be fully determined. Upper limits will be determined by disease spectrum and resistances; the lower limit could be as low as 500 mm. *S. scabra* is currently planted to the 600-mm isohyet in Queensland.
<u>Drought tolerance</u>. *S. scabra* is very drought tolerant; plants that have

received no rain since the previous April will shoot with the rising tempera-tures in September in northern Australia's dry tropics. The very deeply penetrating taproot stands it in good stead in this regard. Leaf tissue is also able to survive much lower water potentials than other species before it dies (Williams and Gardener, 1984).

Tolerance of flooding. Cv. Seca, at least, is not very tolerant; its dry-matter production was only 38 percent of the control when subjected to artificial waterlogging (Williams and Gardener, 1984). Some variation between acces-sions exists (Edye, Grof and Walker, 1984).

Soil requirements. While best adapted to infertile, acid, sandy-surfaced soil, especially solodics, shrubby stylo is more versatile in its soil requirements than most stylos and at times will grow in heavier textured, less acid soils. Such results have been variable and planting on such soils is currently not recommended.

Rhizobium *requirements*. *S. scabra* nodulates readily with a wide spectrum of *rhizobium* strains. Specific inoculation is not required in northern Australia.

Ability to spread naturally. Spreads readily and widely on some soil types, especially sandy surfaced solodics, but spread on soils with hard-setting sur-faces has, at times, been disappointing.

Land preparation for establishment. In the seasonally dry tropics of northern Australia, shrubby stylos can be established satisfactorily by surface broad-casting into the burnt native grass swards. On these predominantly infertile soils, the competition to establishing seedlings is weak. On better soils, limited seed-bed preparation is necessary to reduce grass competition. Shrubby stylo seedlings tend to be weak and slow growing during the first sea-son.

Sowing methods. Suited to aerial seeding or ground spreading, depending on the sophistication of the farming system.

Sowing time and rate. In the dry tropics, *S. scabra* is sown just prior to the break of season at 1 to 2 kg/ha. Further south, in Queensland, it is generally sown somewhat later (into January), nearer the high-rainfall periods, and at 2 kg/ha. In central Queensland, Wildin and Middleton (1984) recommend burning in November-December and oversowing after a fall of at least 30 mm rain.

Number of seeds per kg. 400 000 to 500 000 in the pod and 600 000 to 800 000 clean seed.

Percentage of hard seed. Hard-seed levels are invariably high in freshly ripened seed. Little germination can be expected before these seeds either soften naturally or are treated.

Figure 87. An ungrazed plant of *Stylosanthes scabra* cv. Seca in mid-season

438

Figure 88. Branches and foliage of *Stylosanthes scabra* cv. Seca

439

Figure 89. A small plot of *Stylosanthes scabra* in northwest Queensland, Australia, after grazing late in the dry season

Seed treatment before planting. In the seasonally dry tropics of northern Australia, seed planted on the surface of a recently burnt area will soften rapidly under the influence of a hot sun. It is generally ready to germinate after the first rain. Further south, the rate of softening is slower, and it is normal to scarify half the sample by depodding prior to planting. This gives some seed ready for immediate germination while preserving some hard seed as an insurance against false germinations.

Nutrient requirements. While able to establish and persist on quite low phosphate soils, the shrubby stylos respond readily to applied phosphorus. Not only is dry-matter yield increased but phosphorus content of the forage available is also increased substantially. Sulphur and potassium responses are also regularly obtained.

Response to photoperiod. The flowering response of *S. scabra* to photoperiod is apparently unclear. 'Seca' was classified as a short-day plant in controlled environments (Cameron and 't Mannetje, 1977) but other factors also may operate in the field (Ison and Humphreys, 1984).

440

Figure 90. Stems of *Stylosanthes scabra*, heavily grazed only during the dry season

441

Compatibility with grasses and other legumes. Once established, *S. scabra* is a very strong competitor, able to persist with most companion species, including buffel grass (*Cenchrus ciliaris*), on the texture-contrast brigalow soils.

Ability to compete with weeds. Once established, *S. scabra* is well able to hold its own against most weed species, especially on the less fertile soils where shrubby stylos are most suited for use.

Tolerance of herbicides. Herbicides are not used in extensive field stands, but for seed production, trifluralin and 2,4-D are safe and widely effective in establishment. The shrubby stylos have the wide tolerance of herbicides common to most *Stylosanthes* spp.

Seedling vigour. Seedlings are very slow growing for the first season, but are particularly hardy and able to establish unless grass competition is strong.

Response to defoliation. Regrowth following grazing is normally from buds along the aerial stems, which, being hard and woody, are rarely removed. When mown, strong and rapid regrowth occurs from crown buds at or even slightly below ground level.

Grazing management. Normally, *S. scabra* is continuously but lightly grazed. Under extensive conditions, animals tend to ignore the shrubby stylos until late in the growing season.

Response to fire. Fire kills most aerial stems, but regrowth rapidly occurs from the low-set crown or from root tissue down to 38 mm below the soil surface at the beginning of the next wet season (Gardener, 1980). Mott (1982) found that three-year-old plants in northern Australia failed to survive at all.

Breeding systems. According to Stace (1982), shrubby stylo is normally a true breeding fixed heterozygote, an allotetraploid with 2n = 40 chromosomes, 98.2 percent self-pollinating and 1.8 percent outbreeding.

Dry- and green-matter yields. Dry-matter yields in cutting trials have varied from less than 1 tonne/ha on a solodic soil to more than 9 tonnes/ha on deep red earths and yellow earths in high-rainfall areas (Edye, Grof and Walker, 1984).

Suitability for hay and silage. *S. scabra* is not currently made into silage or hay, but regrowth on regularly harvested experimental plots suggests that, if cut early, before the stems begin to harden, useful hay could be made.

Feeding value. Few analyses are available, but Little, McIvor and McLean (1984) recorded a dry-matter digestibility of 70 percent for a 59 g/day intake of *S. scabra* grown with irrigation and 250 kg/ha superphosphate. They also record N content of 1.5 to 2.0 percent and P of 0.08 to 0.11 percent, both of which were little affected by superphosphate fertilization in mature herbage.

Toxicity. There is no record of toxicity to livestock, but acaricidal properties have been recorded for a number of accessions, including 'Seca' and 'Fitzroy' (Sutherst, Jones and Schnitzerling, 1982).

Seed harvesting methods. A perennial, the seed stands are retained for as long as possible, but usually require renovation after about four years as old plants tend to become moribund. Harvesting is by direct heading after the ripening crop has begun to shed seed. At this stage, the inflorescences are dry and beginning to disintegrate and the crop is less sticky than it would have been earlier. Suction harvesting is possible, but rarely used (J.M. Hopkinson, personal communication).

Seed yields. Mean seed yields for cv. Seca from nine crops at Walkamin, north Queensland, between 1973 and 1978 were 180 kg/ha off the header (yields ranged from 550 kg/ha to virtual failure) (Thompson and de Medeiros, 1981).

Cultivars. The collection available in Australia in the early 1970s contained four morphological/agronomic groups (Edye, Grof and Walker, 1984). Representatives of two of these, cvs Seca and Fitzroy, have come into use. 'Seca' is a late flowering, strongly erect, rather open bush with complete initial resistance to anthracnose (*Colletotrichum gloeosporioides*).

'Fitzroy', a denser, bushier and agronomically more attractive mid-season-flowering type, proved highly susceptible to anthracnose when it appeared and was restricted in use to lower-rainfall (less than 750 mm) habitats and higher latitudes (greater than 22°S). However, this is now breaking down. Earlier-flowering types are also available if anthracnose-resistant types cannot be located.

Diseases. Anthracnose is endemic in the natural habitats of the shrubby stylos in South America, but was first identified in Australia only in 1974, as cvs Seca and Fitzroy were about to be released. The present situation is extremely dynamic. In Australia, the only other serious disease is Botrytis head blight (*Botrytis cinerea*), which can be a problem in seed crops in years with overcast weather during flowering.

Pests. The larvae of a native *Cerambycidae, (Platyomopsis pedicornis)* (F), tunnelled into the lower main stems and killed all plants, usually after they had been weakened by fire, in several plantings in northwestern Queensland (Hall, 1980). In Colombia, stem borers of *Caloptilia* spp. attack the species (CIAT, 1978).

Main attributes. S. scabra is widely adapted and hardy, well suited to low-fertility soils and extensive stock-raising conditions, and possesses good drought, adequate fire and some frost resistance. The shrubby stylos were showing considerable promise before anthracnose arrived on the scene. There were still a number of unanswered questions, including their effect on

443

livestock performance on low-fertility soils in the absence of applied phosphate fertilizer.

Main deficiencies. The most serious defect of *S. scabra* is proving to be its susceptibility to anthracnose. It has also proved, on low fertility soils at least, to provide a very poor quality forage that is especially low in phosphorus. Palatability has also been low. The value of *S. scabra* is restricted to low-fertility soils and extensive land-use situations.

Main references. Burt *et al.* (1983); Edye, Grof and Walker (1984).

Stylosanthes viscosa Sw.

Synonym. *S. prostrata* M.E. Jones.

Common names. Sticky stylo (Australia); poor man's friend (Jamaica).

Description. An ascending and spreading or prostrate and matted, much-branched, perennial legume. Stems to 1 m long, densely pubescent with short viscid hairs. Leaflets to 25 mm long, 5 mm wide but usually much smaller, acute or obtuse, punctate beneath, shortly hairy or hispidulous, with two to four pairs of conspicuous veins. Petioles 2.5 to 5.0 mm long, hispidulous, viscid. Rachis 1 to 2 mm long. Sheath of the stipules 3.5 to 5.5 mm long, 1 to 2 mm longer than the teeth, hispidulous and viscid on the back, three- to five-nerved. Inflorescence a small, crowded ovoid spike, two- to five-flowered. Outer bracts usually trifoliate. Inner bract unifoliate, the sheath equalling or slightly exceeding the teeth, hispidulous and viscid on the back, five- to seven-nerved. Calyx tube 3 to 7 mm long, glabrous to pubescent. Standard suborbiculate, 4 to 7 mm long. Wings 4 to 5 mm long, auriculate, spurred within at the base. Keel petals 3 to 4 mm long, falcate. Loment to 2.5 mm long, shortly hairy, reticulately nerved. Beak short, less than half as long as the upper articulation, usually from one-third to one-quarter as long, shortly hairy, strongly uncinate, often coiled (Mohlenbrock, 1957).

Distribution. *S. prostrata* occurs from southern coastal Texas, United States, to Baja California and Sinaloa in Mexico, through Central America and some Caribbean islands into tropical South America and as far south as Paraguay, often on sandy soils.

General features. *S. prostrata* has proved to be one of the better adapted of the introduced *Stylosanthes* spp. in northern Australia. A frequent but erratically expressed lack of palatability has to date prevented the release of a commercial cultivar. In fact, it could confer weed potential on some accessions.

444

Its natural latitudinal range is 29°S to 29°N. It occurs on both acid and neutral soils, mostly in open scrub or woodland, or in fully exposed positions. The soils are generally well drained, sandy or loamy, composed mainly of sedimentary or granitic materials (Williams *et al.*, 1984).

Agronomically, *S. prostrata* is very similar to *S. scabra*, although taxonomically it is clearly a different species. Only a small fraction of the variation available has been tested to date and, with its wide range of adaptation, it should be much more closely examined, especially if the reasons for erratic palatability can be determined (Burt *et al.*, 1983).

Main references. Burt *et al.* (1983); Williams *et al.* (1984).

Teramnus spp.

Teramnus labialis (L.f.) Spreng

Synonyms. *Glycine labialis* L.f.; *Glycine abyssinica* Hochst. ex A. Rich.; *G. anonychis* Walp.; according to Hermann (1962), more than 12 other species are named within *Glycine*.

Common names. Rabbit vine, horse vine (Barbados).

Description. An extremely variable perennial, climbing or trailing to prostrate legume, sometimes with a woody rootstock. Stems 0.3 to 3 m long, slender, covered with adpressed to spreading white to ferruginous hairs, or glabrescent, occasionally rooting at the nodes. Leaflets rounded, elliptic, ovate, obovate or even narrowly oblong or lanceolate, 1 to 8 cm long, 0.5 to 4 cm wide, emarginate to acuminate at the apex, mostly rounded at the base, glabrous to densely covered with white or ferruginous hairs beneath. Petioles 0.9 to 4 cm long. Rachis 1 to 7 mm long. Petiolules 2 mm long. Stipules narrowly lanceolate, 2 to 3 mm long. Inflorescence slender and usually few flow-

445

Figure 91. *Teramnus labialis* (**Source:** Pittier, 1944)

446

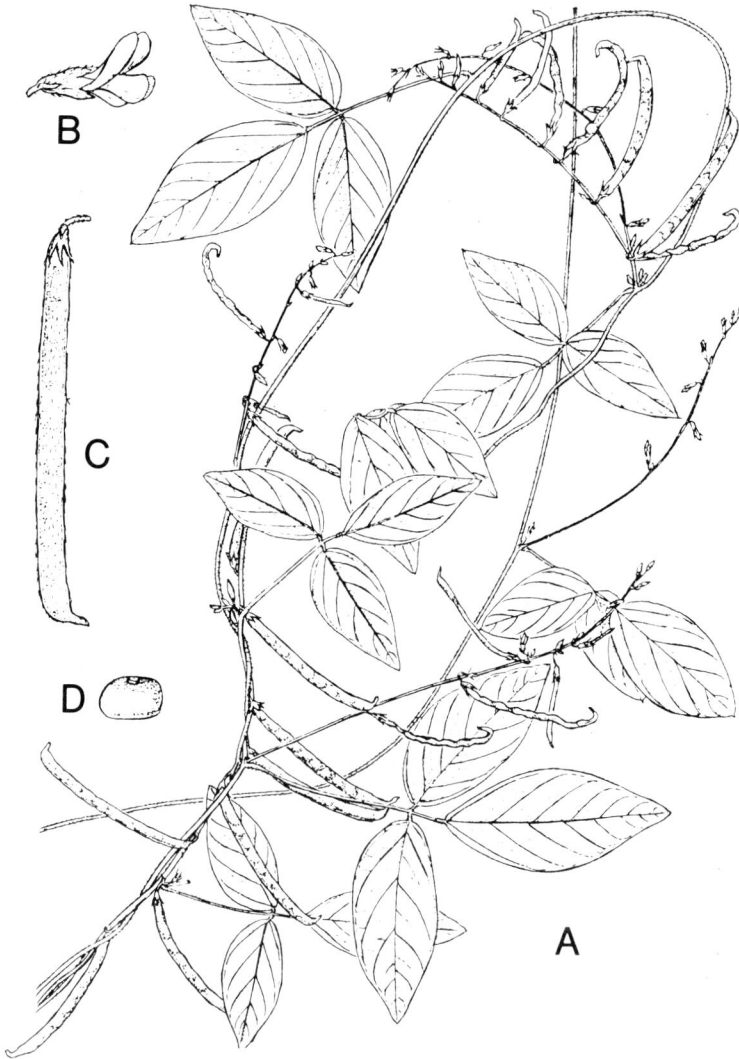

Figure 92. *Teramnus labialis* ssp. *arabicus.* **A**-Habit **B**-Flower **C**-Pod **D**-Seed

447

ered. Rachis 0.3 to 10 cm long. Peduncles 0.8 to 3 cm long. Pedicels mostly rather slender, 1.3 to 4 mm long. Bracteoles linear, about 1.5 mm long. Calyx tube glabrescent or hairy, ribbed, 1 to 3 mm long, lobes lanceolate, 0.8 to 3 mm long, acute, usually densely hairy. Standard white, pink or purplish, also described as yellowish or orange, obovate, 5 mm long, 3.5 mm wide. Wings pale mauve. Keel white. Pods linear, 2.5 to 6 cm long, 2 to 4 mm wide, glabrescent to densely covered with adpressed or spreading hairs, mostly rather sparse and not covering the surface, but rarely dense. Beak 2 to 3 mm long. Seeds yellow-brown to dark purplish-brown, oblong or almost cylindrical, smooth or covered with a granular encrustation, 2 to 3 × 1.2 to 2 mm. Hilum minute, aril slightly developed, white with a small scale-like extension (Gillett, Polhill and Verdcourt, 1971).

Distribution. At least two subspecies, sspp. *labialis* and *arabicus* Verdc., and several varieties exist. The species as a whole is distributed in tropical and southern Africa, India, Malaysia, Indonesia, the Philippines, New Guinea, the eastern Indian Ocean, the West Indies and Guyana.

In ssp. *labialis*, var. *abyssinicus* (A. Rich.) Verdc. occurs in three forms from 1 000 to 2 000 m elevation in Uganda, Tanzania, the Congo and Ethiopia, in grasslands, bushlands, cultivations and along the edges of roads, while var. *acutus* Verdc. occurs from the Congo to West Africa in evergreen forests and fringing forests from 1 100 to 1 200 m elevation.

Ssp. *aribicus* Verdc. occurs on the Arabian Peninsula and in the Sudan, Mali, Zambia, Zimbabwe, Principe, Mozambique, Mauritius, Reunion, Seychelles, the Comoros, the West Indies and Guyana, in grasslands with scattered trees, forests, clearings and bushlands from sea level to 1 000 m (Gillett, Polhill and Verdcourt, 1971).

The species occurs in the eastern highlands and prairie zone of Cuba (Menendez, 1982), Jamaica (Adams, 1972) and Barbados (Gooding, Loveless and Proctor, 1965).

General features. T. *labialis* is regarded as the most promising pasture plant in Cuba, where cultivar Semilla Clara is identified. There it maintains a good leaf to stem ratio and a high crude protein content (up to 20 percent) to maturity. It is well accepted by animals and persists well under grazing in either pure swards or mixtures. It can also be cut for green chop for up to two years without damaging the sward. Seed production is also good, up to 0.5 tonnes/ha, and seed germinates readily. Seed does not require scarification. Stands take six to eight months to become established.

No specific *Rhizobium* inoculum is required. It has been sown in Cuba at 6 kg/ha in rows 50 cm apart and no more than 3 to 5 cm deep. Productivity for beef and milk production has yet to be determined, as have fertilizer requirements (Menendez, 1982).

In the British Virgin Islands, *T. labialis* grows profusely in many pastures, where it persists well and produces considerable bulk without smothering the grasses (British Virgin Islands, 1963). In Barbados, it is regarded as a common weed(?) in sour-grass areas (Gooding, Loveless and Proctor, 1965), while it is common in thickets and on stony ground in Jamaica (Adams, 1972).

At Serere, Uganda, African ecotypes shed their leaves during dry periods and were killed off by the prolonged annual drought. They produced seed but did not reseed readily. They were not regarded as showing promise for pasture (Horrell, 1958).

Main references. Gillett, Polhill and Verdcourt, (1971); Menendez (1982).

Teramnus micans (Welv. ex Bak.) Bak.

Description. Fast, strong climber with white or pink flowers.
Distribution. West and East Africa (Sierra Leone, Cameroon, Zaire, Uganda, Angola).
Characteristics. This legume was observed growing strongly with *Cynodon* sp. on basaltic soils near Owen Falls, Uganda. It was well grazed by cattle and seeded well.

Teramnus uncinatus Sw. (see Colour plate XLVII)

Common names. Amendoim de veado, faveirà, and alfalfa Paulista (Brazil).
Description. A twining herb with angular, hairy stems and oblong-lanceolate leaflets slightly pubescent above. It resembles glycine, but the pod, which is 4 to 5 cm long and rusty-hairy, has a recurved beak (the beakage of the glycine pod is shorter and straight). Seeds of *T. uncinatus* are yellowish and oblong.
Distribution. Occurs from the West Indies, through Mexico, Central and South America, to the states of Minas Gerais, Bahia and São Paulo in Brazil.
Characteristics. A summer-growing annual or perennial preferring the hotter areas of the tropics, is sensitive to cold and becomes diseased under low temperatures. It flowers in 126 days in the Kimberley area of northern Australia (Parbery, 1967a).

It prefers a rainfall in excess of 750 mm; grows vigorously in the wet season (van Rensburg, 1967) but performs poorly in the dry season; Evans (1968), however, says that it has some drought and cold tolerance (CPI 25937). It has

449

a hard seed content of 68 percent (Otero, 1952). It has a low fertility requirement. Buller *et al.* (1970) obtained 2.45 tonnes DM/ha under low fertility conditions and 3.61 tonnes in a high fertility regime at Matão, Brazil, where it
was the poorest of eight tropical legumes, though better than *Desmodium
intortum* at the low fertility level.

Andrew and Pieters (1970a) found that *T. uncinatus* was severely affected
by potassium deficiency. The symptoms were first evident as very small necrotic spots occurring interveinally near the margins of the leaflets. In the early
stage, they were only evident on the upper surface of the leaflet but subsequently both surfaces were affected. With increasing severity, the necrotic
areas enlarged and the margins of the leaflets became chlorotic. This condition progressed toward marginal necrosis of the leaflet with full interveinal
chlorosis and prominent venation on the undersides of the leaflets.
Associated with this marginal necrosis the lateral margins of the leaflets
curled inward and in the very severe form a rolled tube effect was shown.
Severely affected leaves at the base of the plant abscissed and the partially
affected leaves attained a "sleeping position", with lateral leaflets folded
against each other. Parbery (1967a) found that it responded to 100 kg/ha nitrogen in northern Australia.

T. uncinatus establishes quickly and gives a good cover of palatable leafy
growth. It is resistant to trampling and nematodes, and Evans (1968) believes
that it may be a useful pioneer legume in pasture mixtures. Its recovery from
cutting is poor (van Rensburg, 1967; Parbery, 1967a); van Rensburg harvested 1 294 kg/ha dry matter in one cut at Mt. Makulu, Zambia, but it did
not give a second cutting. Parbery (1967a) harvested 7 076 kg/ha without
added nitrogen, and 12 146 kg/ha with 100 kg/ha applied nitrogen. The protein content was 17.6 percent. Otero (1952) gave the chemical analysis of the
air-dried material as 9.88 percent moisture, 13.17 percent crude protein, 2.18
percent fat, 35.68 percent crude fibre, 31.54 percent nitrogen-free extract and
7.55 percent ash.

Teramnus volubilis Sw.

Synonym. *Glycine oblonga* Benth.
Common name. Frijolito (Brazil).
Description. Closely resembles some forms of *T. labialis*, but the four-fid
calyx is distinctive. Leaflets elliptic or oblong, mostly glabrous above; pods
sparsely appressed, pubescent; seeds smooth.

450

Distribution. Central America, Colombia, Venezuela, Jamaica, Paraguay, Brazil, Ecuador.

Characteristics. It is vigorous, with a wide soil range and combines well with *Axonopus* and *Paspalum* grasses. Resists drought and trampling. A good forage for horses and cattle and contains 16.83 percent crude protein (Bermudez *et al.*, 1968).

Trifolium spp.

Trifolium africanum Ser.

Synonyms. *Lupinaster africanus* (Ser.) Ech. and Zey.; *Trifolium burchellianum* Ser. var. *africanum* (Ser.) Kuntze.

Common name. Erasmus clover.

Description. A perennial legume with stems mainly prostrate, rooting at the nodes of the lateral runners, pilose. Var. *africanum* has petioles usually free for more than 1 cm, although sometimes less than 1 cm in the upper leaves. Leaflets obovate to elliptic, ranging from less than twice as long as wide to between twice and four times as long as wide, pilose on both surfaces, apex acute, rounded emarginate. Inflorescences with many purple flowers, globose, about 2 cm wide. Peduncle up to 12 cm long. Pedicels pilose, erect after flowering. Calyx pilose with less than 15 nerves. Var. *glabellum* (E. Mey.) Harv. has glabrous leaflets and glabrous stems, while var. *lydenbergense* (Gillett) has glabrous stems and linear-lanceolate leaflets that are glabrous above and sometimes sparsely pilose below. The apex is acute, rounded or slightly emarginate, and is more than five times longer than wide ('t Mannetje, 1966).

451

Distribution. *T. africanum* and *T. burchellianum* are the only two *Trifolium* species indigenous to southern Africa. *T. africanum* var. *africanum* is mainly confined to the Witwatersrand and the eastern Transvaal highveld, that part of Natal above 500 m elevation and 750 mm rainfall, Lesotho, and adjacent areas of the Orange Free State and Cape Province. Var. *glabrellum* has a similar distribution but is more frequent in Natal and less frequent in the Transvaal, while var. *lydenbergense* is confined to the eastern Transvaal highveld, where the rainfall is about 750 mm (Jones, Strijdom and Theron, 1974).

General features. All areas of occurrence have summer-dominant rainfall and most frost in winter, some receiving isolated snowfalls. *T. africanum* occurs mostly in lightly grazed, cool, moist, high-altitude sites. At lower altitudes or lower rainfalls (to 650 mm), it occurs only in wet sites or on river banks (Jones, Strijdom and Theron, 1974). It appears that *T. africanum* occurs more widely on roadsides than in grazed veld and that it survives regular close mowing. *T. africanum* is adapted to acidic, highveld soils and is frost hardy, though not as hardy as *T. repens* (Jones, Strijdom and Theron, 1974).

As in the case of *T. semipilosum*, the growing points of the stolons are not confined to the soil surface, so that they can climb up to combat overshading to a greater extent than those of white clover. Swards develop numerous, small taproots connected by both surface and subsurface stolons (Jones, Strijdom and Theron, 1974).

Rhizobium *relationships*. *T. africanum* nodulates effectively with rhizobia not effective on European clovers. In addition, it nodulates at lower pH levels, although usually above pH 5.5, and nodulates and fixes nitrogen at higher temperatures than *T. pratense* and *T. repens*. It tolerates root temperatures of up to 40°C compared with 26 to 33°C for *T. pratense* and *T. repens* (Jones, Strijdom and Theron, 1974).

Nutrient requirements. *T. africanum* has been observed growing well on low-phosphate soils (Jones, Strijdom and Theron, 1974).

On acidic, P-fixing soils, *T. africanum* outyielded *T. repens*. Elimination of exchangeable aluminium from the soil by the use of lime significantly increased the yield of both species. *T. africanum* had twice the Ca content of *T. repens* on the limed soil. Lime also improved the ability of *T. africanum* to extract P from the soil (Hackland, Booysen and Somner, 1976).

Seed characteristics. *T. africanum* flowers from mid-spring to mid-summer. More than 90 percent of seed set is hard. There are about 700 000 seed/kg (range 450 000 to 1 100 000). It is cross-pollinating. The chromosome number is $2n = 32$ for var. *africanum* (Jones, Strijdom and Theron, 1974).

Pests and diseases. Red spider (*Tetranychus* spp.) and root knot nematode

452

(*Meloidogne* spp.) have been the main pests in cultivated stands but there are variations in reaction to them (Jones, Strijdom and Theron, 1974).
Toxicities. *T. africanum* contains no prussic acid.
Main references. Jones, Strijdom and Theron, (1974); Gillett (1952).

Trifolium burchellianum Ser.

Synonyms. *T. johnstonii* Oliv.; *T. brasileianum* Choiv.
Common names. Burchell's clover, Cape clover.
Description. A perennial clover with a variable habit and a strong taproot. Stems glabrous or nearly so, creeping and rooting at the nodes, or, less often, ascending. Leaves glabrous or nearly so, with a well-developed, free petiole. Free parts of the stipule triangular, tapering gradually to the tip, much shorter than the base. Leaflets cuneate-oblong, cuneate-obovate or cuneate-elliptic, less often oblong, emarginate or less often, truncate or rounded at the tips, up to 25 (50) × 19 mm, teeth rather small, main nerves 10 to 15 on each side at about 50° to the midrib. Inflorescence many-flowered, more or less globular, up to 3 cm across. Peduncle longer than the subtending leaf, pilose towards the top. Bracts 0.5 to 2 mm long. Pedicels stout, 2 to 2.5 mm long, more or less erect in fruit, glabrous. Calyx glabrous except for a few hairs at the margin. Tube 11-nerved, 2 to 3 mm long, teeth triangular for about 0.5 mm at the base only, then subulate, in all 4 to 7 mm long. Corolla purple. Standard 8 to 13 mm long. Blade of wings straight, more or less oblong, its auricle less than 0.5 mm long. Ovary glabrous or minutely scabid-punctate, usually two, occasionally three to four ovules. Pods about 5 by 3 mm. One to two seeds, dark brown, irregularly oval, about 2.1 by 1.8 mm (Gillett *et al.*, 1971).
Distribution. There are three subspecies. Ssp. *burchellianum*, one of only two *Trifolium* spp. indigenous to Africa, occurs in southern Africa, Lesotho and Angola and has $2n = 48$ chromosomes. It is found in lightly grazed, cool, high-altitude sites, in Lesotho up to 3 000 m elevation and in the fringing areas of the Orange Free State, Natal and Cape Province. It also occurs at lower elevations in southern Natal and over a wide range of Cape Province to the fringe of the Karroo, and occasionally into western Cape Province (Jones, Strijdom and Theron, 1974).
Ssp. *johnstonii* (Oliv.) Gillett has larger flowers and leaflets, and $2n = 96$ chromosomes. It occurs in Uganda, Kenya, Tanzania and southern Ethiopia, especially in moist places from 1 800 to 3 700 m elevation with annual rainfall

453

of 1 000 mm and over: upland grasslands, openings in evergreen forests, bamboo forests and moorlands. It is also found occasionally in the alpine zone. Ssp. *johnstonii* is the most common subspecies in East Africa.

Ssp. *oblongum* Gillett occurs in Kenya and southern Ethiopia in the undergrowth of forest margins (Gillett, Polhill and Verdcourt, 1971). Less commonly, it occurs in tall grass, where it is more upright.

General features. *T. burchellianum* is one of the more promising African species for possible domestication. It is a productive and palatable plant in suitable habitats, with a wide range of ecotypes owing to the wide distribution of the three subspecies. In the future, it will probably yield commercial cultivars. "Notes on Kenya agriculture" (1959) suggests that at least some ecotypes can be shy seeders and slow to establish. A minimum seeding rate of 2 kg/ha was recommended. As the growing points are not restricted to the soil surface, it can tolerate shade better than *T. repens*, being able to get up to the light to some extent (Jones, Strijdom and Theron, 1974).

The southern African forms appear to have greater ecological tolerance and genetic diversity than *T. africanum*. They are not as frost tolerant as *T. repens* and appear more frequently along roadsides than in grazed fields (Jones, Strijdom and Theron, 1974).

Rhizobium relationships. Effective nodulation is achieved by *Rhizobium* strains not suited to European clovers that nodulate at a lower pH, usually above pH 5.5 (Jones, Strijdom and Theron, 1974).

Nutrient requirements. Grows well in low phosphorus soils (Jones, Strijdom and Theron, 1974).

Seed characteristics. Flowering in the field occurs from mid-spring through mid-summer. More than 90 percent of seed set is hard and remains so. There are *c*. 700 000 seeds/kg. It is cross-pollinating (Jones, Strijdom and Theron, 1974).

Pests and diseases. Red spider (*Tetranychus* spp.) and root knot nematodes (*Meloidogyne* spp.) have been the two main pests in artificial cultivations in southern Africa. Rugose leaf curl has also been noted in Australia (Jones, Strijdom and Theron, 1974).

Feeding value. *In vitro* digestibility can be lower than for *T. repens*, but is still satisfactory. N content is similar to that of *T. repens*. In addition, *T. burchellianum* contains no prussic acid and has a low level of "fraction S" protein, so it should not have a high bloat potential (Jones, Strijdom and Theron, 1974).

Main references. Jones, Strijdom and Theron (1974); Gillett, Polhill and Verdcourt (1971).

454

Trifolium polymorphum **Poir.**

Common name. Polymorphous clover.

Description. A creeping, stoloniferous, perennial clover. It also produces rhizomes. Plants are pubescent, fine and low-growing. Inflorescences of two types, aerial and hypogeal, are produced. The aerial flower heads are small, with pink or rose coloured flowers that are readily eaten by livestock. Most of the seed is set by the small white, subterranean flowers, which occur in axillary, ovoid heads at many nodes. There is a main taproot and many adventitious roots from the nodes of the prostrate stems.

Distribution. Native to Buenos Aires and the northeastern provinces of Argentina, Uruguay and Rio Grande do Sul Province, Brazil. It is common on the sandy strip fringing the River Parana in Santa Fe Province, Argentina.

General features. A rather small, inconspicuous clover that is favoured by heavy grazing, especially by sheep, but disappears readily when shaded by taller-growing plants. The subterranean seed set particularly enhances its persistence under heavy grazing. While normally inconspicuous, it responds strongly to applied phosphate (Burkart, 1954), at times doubling its carrying capacity (Henry, 1954). It nodulates well with *Rhizobium* from *T. repens* and *T. alexandrinum* plants. It differs from most plants in its native habitat in being winter-spring growing, maturing in early summer (Hartley, 1949).

Main references. Burkart (1954); Henry (1954).

Trifolium rueppellianum **Fres.**

Synonyms. *T. subrotundum* Hochst. and Steud. in part; *T. goetzenii* auct., non Taub. ex Engl.; *T. preussii* Taub.

Common name. Rueppell's clover.

Description. An erect or occasionally prostrate annual. Stems glabrous, not rooting at the nodes. Leaves glabrous or nearly so. Stipules up to 15 mm long, with three to four nerves, united with the petiole for two-thirds of their length, tips entire, acuminate. Petiole only a few millimetres long and sometimes wholly adnate to the stipules in the upper leaves, up to 5 cm long in the lower leaves. Leaflets oval, oblong or obovate, rarely broadly-lanceolate, less than three times as long as wide, rounded truncate or, less often, emarginate at the tips, rarely acute, up to 24 mm long and 14 mm wide. Teeth strongly developed, up to 0.5 mm long, often with subulate tips, 10 to 15 main nerves on each side, at an angle of 40 to 50° to the midrib. Inflorescence more

455

or less globose, typically 15- to 30-flowered and *c*. 17 mm across, sometimes few flowered in stunted plants, especially at high altitudes. Peduncles pilose near the top, many times longer than the subtending petiole, bracts minute or linear-lanceolate, up to 3 mm long. Pedicels glabrous, pilose, up to 2 mm long, shorter than the calyx tube, suberect in fruit. Calyx glabrous, except occasionally for a few hairs at the margins; tube pale, rounded at the base, strongly 11 nerved, the commissural nerves almost always undivided to the top. Teeth abruptly narrowed near the base, subulate for most of their length, 2 to 5 mm long, sometimes exceeding the corolla. Corolla purple, rarely white, wings with a well-developed auricle. Pods 3 to 4 mm long by 2 mm wide. Two to three seeds, occasionally four to five, oval, brown, *c*. 1.5 × 1 mm (Gillett, Polhill and Verdcourt, 1971).

Distribution. There are two varieties.

• Var. *rueppellianum* occurs from Uganda, Kenya and Tanzania through eastern Zaire to Fernando Póo (Equatorial Guinea) and Cameroon, as well as in the Sudan and Ethiopia, mainly in upland grasslands, moorlands, tracks through upland forests and as a weed in cultivated areas. It usually occurs in rather wet places, from 1 600 to 3 600 m in elevation, with an annual rainfall of from 650 to 1 200 mm. It is the most usual form, common in seasonal swamps.

• Var. *minimiflorum* Gillett is known only from the Moshi District of Tanzania, where it occurs in grasslands from 1 000 to 1 800 m in elevation, with up to 2 000 mm annual rainfall (Gillett, Polhill and Verdcourt, 1971).

General features. Next to *T. semipilosum*, *T. rueppellianum* is the East African clover showing the most potential for domestication. Like *T. semipilosum*, it has a wide range of ecotypes, is easily grown from seed and reseeds itself readily, persisting well in association with perennial grasses in appropriate habitats. Seed yields of over 224 kg/ha have been harvested by direct heading, even though the pods shatter readily (Bogdan, 1956).

Palatability can be rather low when the plants are young but improves as they reach maturity (Strange, 1958). One form is cultivated in Ethiopia as forage for horses (Gillett, 1953).

Trifolium semipilosum **Fres.** (see Colour plate XLVIII)

Synonym. *T. johnstonii* auct., non Oliver.
Common name. Kenya white clover.
Description. A perennial legume with a strong taproot and prostrate, pilose

stems, often rooting at the nodes. Stipules whitish, the falcate-triangular tips usually longer than the base. Petiole pilose, much longer than a leaflet, except sometimes in stunted plants. Leaflets glabrous above, pilose at the margins and on the midrib beneath and also (except sometimes in var. *glabrescens*) on the undersides of the lower (outer) half of the two lateral leaflets. Inflorescence more or less globose, 10- to 25-flowered, about 2 cm across. Peduncle longer than the leaves, pilose. Pedicels pilose, reflexed in fruit, the outer up to 2 mm long, the inner up to 5 mm. Corolla white or pale pink. Standard 8 to 9 mm long. Pods 5 to 6 × 2.6 mm. Two to six seeds, dull yellow or light brown, sometimes mottled, irregularly discoid, about 1.5 mm across.

There are two varieties. Var. *semipilosum* Gillett, which now includes var. *microphyllum* Choiv. and var. *kilimanjaricum* Baker; and var. *glabrescens* Gillett (syn. *T. johnstonii* sensu Edwards, non Oliver and *T. repens* sensu Baker, non L.) (Gillett, Polhill and Verdcourt, 1971).

Distribution. Var. *semipilosum* occurs in Yemen, Ethiopia, Kenya and Tanzania, in upland grasslands, especially near the drier types of upland evergreen forests at 1 400 to 3 000 m elevation, usually in places where there is much mist and annual rainfall in the range of 550 to 1 400 mm.

Var. *glabrescens* occurs in southern Ethiopia, Uganda, Kenya, Tanzania and in Malawi, where it is perhaps not native. It occurs in rather moister habitats than var. *semipilosum* and is more variable. Sites are upland grasslands, often abundantly associated with *Pennisetum clandestinum* at 1 500 to 2 250 m elevation, occasionally 1 200 to 2 700 m, with an annual rainfall of 900 to 1 200 mm (Gillett, Polhill and Verdcourt, 1971).

Season of growth. *T. semipilosum* has a long growing season. It is somewhat less tolerant of frost than *T. repens* cultivars but shows more vigorous summer growth. The main growing season is spring and autumn, with better summer than winter growth.

Frost tolerance and regrowth after frosting. It is cut by heavy frosts, but resumes slow growth as soon as frosts cease.

Altitude range. Cv. Safari has grown from sea level to 700 m elevation in Australia.

Rainfall requirements. In Australia, cv. Safari has been useful in areas with between 800 and 1 600 mm mean annual rainfall.

Drought tolerance. Lines of *T. semipilosum* grown in various areas of Australia have all shown greater drought tolerance than control lines of *T. repens*.

Tolerance of flooding. Cv. Safari is regarded as less tolerant of waterlogging than *T. repens*, but in laboratory flooding studies by Whiteman *et al.* (1984)

457

it was placed in the second of four groups together with *Lotononis bainesii* and *Desmodium heterophyllum*, all of which developed medium to large adventitious roots under simulated flooding.

Soil requirements. *T. semipilosum* grows in soils of pH 5.0 to 7.7. It is also more tolerant of the aluminium ion than *T. repens* or *Medicago* spp.

Rhizobium *requirements*. Requires specific *Rhizobium;* CB 782 is used on cv. Safari in Australia.

Ability to spread naturally. One of the disappointments with cv. Safari to date in Australia has been its slow to nonexistent rate of spread from planted areas. This is possibly due to the lack of spread of the specific *Rhizobium*.

Land preparation for establishment. Seed is small and expensive and seedlings often grow slowly owing to rugose leaf curl. As a result, a well prepared, fine, firm, weed-free seed bed is necessary.

Sowing methods. Can be sown with conventional seed drills and combines, especially if fitted with a small seed box. Rolling after sowing is desirable.

Oversowing into natural pastures. Not particularly suitable for this method of establishment at this stage.

Sowing depth and cover. Sown on or near the surface and covered with trailing harrows, then rolled.

Sowing time and rate. In southern Queensland, Australia, sowing is generally in autumn, as rainfall then is more reliable. Further north, sowing immediately after the start of the wet season is recommended. Recommended rates are 2 to 3 kg/ha, but because of the high cost of seed, lesser rates are sometimes used in the hope the stand will thicken with age.

Number of seeds per kg. Cv. Safari ranges from 700 000 to 1 000 000.

Percentage of hard seed. Higher than for *T. repens*.

Seed treatment before planting. Normal mechanical harvesting and processing techniques appear to soften enough seed so that no special treatment is required. Hand-harvested samples may require scarification.

Nutrient requirements. *T. semipilosum* has a high nutrient requirement; in southeastern Queensland, up to 500 kg/ha Mo superphosphate is recommended at planting. For the next two years, 250 kg/ha of superphosphate a year is recommended, and subsequently this amount every second year.

Compatability with grasses and other legumes. As the stolons can scramble up through tall-growing companions and, unlike *T. repens*, are not confined to the soil surface, *T. semipilosum* combines better with the taller tropical pasture plants, and is more tolerant of lax pasture management for its maintenance in mixtures.

Figure 93. A sward of *Trifolium semipilosum* cv. Safari and *Pennisetum clandestinum*

Ability to compete with weeds. *T. semipilosum* is not as readily overtopped by taller weeds as *T. repens* but still needs careful handling.

Tolerance of herbicides. Appears tolerant of 2,4-DB at normal rates and of dalapon at 5 kg/ha.

Seedling vigour. Healthy seedlings are quite vigorous, but once infected with rugose leaf curl, vigour falls drastically. Survival depends on freedom from weeds and favourable growing conditions.

Vigour of growth and growth rhythm. Maximum growth rates are achieved in late spring and from late summer through autumn.

Response to defoliation. *T. semipilosum* withstands heavy grazing well, forming a dense, prostrate, short sward.

Grazing management. While able to climb to some extent, it still performs better if growth of the taller species is controlled. Heavy summer grazing for this purpose is therefore recommended. This ensures that the stolons stay close to the ground and gives rise to more adventitious roots from the nodes. These replace old crowns, which die after about one year.

459

Response to fire. Burning in spring appears to do little or no harm and can in fact soften hard seed and improve seedling regeneration.

Breeding system. Cv. Safari is cross-pollinating and self-sterile but does not cross with *T. repens*. Chromosome number 2n = 16.

Dry- and green-matter yields. Third-year dry-matter yields of 8 500 kg/ha have been recorded in the Bundaberg-Maryborough District of southeastern Queensland in wallum soils (Evans, 1967).

Suitability for hay and silage. There is no reason to expect that *T. semipilosum* will not make hay or silage as readily as *T. repens*.

Value as standover or deferred feed. Because of its long growing season and greater drought tolerance, *T. semipilosum* provides feed longer into difficult periods than *T. repens*. It does not, however, stand into frosty periods as well.

Feeding value. *In vitro* organic matter digestibility is in the order of 71 to 75 percent and crude protein around 25 percent, levels similar to those of *T. repens*. Sodium content can be as low as 0.05 percent.

Toxicity. There is no evidence of oestrogenic activity with *T. semipilosum*, but bloat can occur in cattle. However, the risks may be less than with *T. repens*.

Seed harvesting methods. Harvesting methods vary from collection of material with a forage harvester for thrashing through a stationary header, to prior windrowing and pick-up thrashing with the header, to direct heading of the standing crop. The seed heads of the main crop are carried well above the ground but those of succeeding crops are much closer to the ground. The machinery available often determines the harvesting technique. Harvested seed needs drying before final cleaning.

Seed yields. Yields of up to 400 kg/ha have been recorded from small areas, but commercial yields rarely exceed 300 kg/ha and are usually much lower than this. Two harvests per year are possible under some circumstances.

Minimum germination and quality of seed for commercial sale. The most recent standards in Queensland are minimum germination 60 percent, which can include the hard seed component; minimum pure seed 96.5 percent; and maximum other seeds, 2.5 percent.

Cultivars. Cv. Safari was released in 1973 by the Queensland Herbage Plant Liaison Committee. This derives from CPI 27218. received by CSIRO in 1960 from the Kitale Seed Co., Kenya. It is a representative of var. *glabrescens* Gillett.

At one stage, cv. Kabete 4, a mixture of four clones selected for vigour and persistence (Bogdan, 1965), was available in Kenya, but it is no longer believed to be available.

Figure 94. The foliage of *Trifolium semipilosum* cv. Safari

Diseases. The most serious problem is rugose leaf curl, which delays establishment of new stands. Plants that survive this phase show subsequent immunity to further infection. At times, slugs can be a problem, especially after prolonged wet weather, while occasionally *Amnemus* sp. weevils and root knot nematodes (*Meloidogyne* sp.) attack the roots. Unlike *T. repens, T. semipilosum*, is not attacked by clover rust or pepper spot.

Main attributes. Essentially a tropical white clover, *T. semipilosum* exhibits the same main attributes as *T. repens* and in addition can thrive in hotter and drier districts. It better tolerates tall growth of associated species, while retaining *T. repens'* tolerance of heavy grazing.

Main deficiences. The main deficiency of *T. semipilosum* is the slow and unreliable early growth of new stands due to rugose leaf curl syndrome. It is still only essentially a subtropical and high altitude, mild-climate tropical species, not adapted to hot, humid or lowland tropics.

Main references. Gillett, Polhill and Verdcourt (1971); Shaw and Quinlan (1976); Jones and Cook (1981).

461

Trifolium tembense Fres.

Synonyms. *T. umbellatum* A. Rich.; *T. goetzenii* Taub. ex Engl.; *T. calocephalum* var. *parviflorum* Choiv.; *T. subrotundum* Harms in part, non Hochst. and Steud. ex A. Rich.; *T. rueppellianum* Robyns in part, non Fres.

Common name. Temben's clover.

Description. An annual, glabrous or subglabrous, erect or prostrate legume, sometimes rooting at the nodes, especially when growing in water. Stipules united with the petiole for most of its length, the free part abruptly contracted to a fine point. Petioles as short as 5 mm in the upper leaves, where they may be wholly fused with the stipule; up to 5 cm long in the lower leaves. Leaflet elliptic or obovate, cuneate at the base, acute rounded, truncate or slightly emarginate at the apex, up to 17 mm long by 10 mm wide, teeth strongly developed with subulate tips, eight to ten main nerves on each side at an angle of 40 to 60° to the midrib. Inflorescence hemispherical, 3- to 16-flowered, up to 23 mm across. Peduncles up to 7 cm long. Pedicels up to 3 mm long, more or less erect in fruit. Calyx sparingly pilose at the margins, tube about 2 mm long, broadly campanulate, strongly 11-nerved, teeth about 5 mm long, gradually narrowing from the base. Corolla purple. Standard about 9 mm long, abruptly narrowed above the middle into an oblong, truncate, downward-curved tip. Pods glabrous, 5 to 7 mm long by about 3 mm wide. Seeds brown, oval, flattened, 2 × 1.3 mm (Gillett, Polhill and Verdcourt, 1971).

Distribution. *T. tembense* is found naturally from Zaire and Rwanda to Ethiopia, in wet places, sometimes in shallow water, and in upland grasslands, forests, moorlands and alpine areas, from 2 000 to 3 800 m in elevation. It occasionally occurs by bodies of water as low as 1 400 m in elevation. Mean annual rainfall in these areas is at least 1 000 mm (Gillett, Polhill and Verdcourt, 1971).

General features. *T. tembense* often grows with other *Trifolium* spp., which it resembles. It can easily be mistaken for *T. rueppellianum*, especially when the flowers are young, and may also interbreed with it (Gillett, Polhill and Verdcourt, 1971).

  T. tembense grows very fast from seed but is shorter lived (earlier maturing) than *T. rueppellianum* and was inferior at Kitale, Kenya (Strange, 1958). It is rather unpalatable until very mature. A seeding rate of 5 to 6 kg/ha is recommended ("Notes on Kenya agriculture", 1959).

  T. tembense is one of the numerous African *Trifolium* species that has shown some promise for domestication and continues to be tested from time to time.

Trigonella spp.

Trigonella suavissima Lindl.

Common names. Sweet fenugreek, Cooper clover, channel clover, Darling clover, Menindee clover, Australian shamrock (Australia).

Description. An annual with prostrate or ascending, usually hairless stems and leaflets. Leaves trifoliate. Leaflets broadly ovate or obcordate and more or less denticulate, about 12 mm long, the middle one on a longer petiolule. Petioles long and slender. Stipules semi-sagitate, deeply toothed. Four to eight flowers, small, yellow, in sessile axillary clusters or short, peduncled, umbellate heads. Calyx 5 mm long, the lobes lanceolate-subulate, rather rigid, fully as long as the tube. Standard longer than the calyx; wings and keel slightly shorter. Pods linear, curved, almost obtuse, 12 to 20 mm long, about 2 mm wide, opening in two thin reticulate valves, either flat or undulate.

Distribution. A native Australian species, *T. suavissima* is found along the flood plains of the interior rivers, growing prolifically on the rich alluvial soils after prolonged periods of inundation. It is rarely found on sands. *T. suavissima* is particularly prominent in the Channel Country of southwest Queensland.

General features. A strongly sweet scented herb, *T. suavissima* is quite palatable but somewhat ephemeral. It requires general flooding to induce massive germinations; local flooding will cause little or no germination. It grows particularly well after late summer-autumn or winter flooding. In the Channel Country, *T. suavissima* is considered the fattening feed, and a good clover year means a heavy supply of fat cattle (Skerman, 1947). It carries nodules (Bowen, 1956) but is fairly strain specific (Brockwell and Hely, 1966).

T. suavissima has long been recognized as a potentially useful plant for domestication but to date little effort has been directed toward this end.

Vigna spp.

Vigna hosei (Craib) Back.

Synonym. *Dolichos hosei* Craib.
Common name. Vigna (Far East).
Description. A twining or creeping legume, often forming a thick ground cover. Leaflets ovate-elliptic, on both surfaces thin, long hairs; terminal leaflet 3 to 7.5 cm × 2 to 5 cm; pod 1 to 2 cm long, black, one- to three-seeded. Besides producing normal flowers and pods, it also produces some which remain concealed under a dense carpet of half-decayed leaves, originating from the plant itself and which are borne on a 2.5- to 6-cm pale stalk; in a ripe state they are very yellow, finely and rather densely pubescent, one- to two-seeded, 1 to 1.5 cm long.
Distribution. Widespread in the humid tropics, native of north Borneo. Occurs from sea level to 1 100 m.
Characteristics. Perennial summer-growing legume, but has some active growth in the cool season. Requires 2 500 mm or more annual rainfall and has low tolerance of dry weather, but will stand flooding.

Bunting (1930) found that it gave good growth under coconuts as a green manure. It is adapted to a wide range of soil textures and prefers an acid soil of pH 4.9 or less (Grof, personal communication). It is nonspecific in its *Rhizobium* requirement, but preferably should be inoculated with a selected cowpea strain such as CB 756 (Norris, 1967).

Spreads very rapidly under natural conditions, but for establishment it is better to prepare a good seed bed and drill in the seed at 1 kg/ha at a depth of 1.25 cm and harrow or roll after seeding. Midsummer sowing is recommended. Seed has a hard coat and should be treated with concentrated sulphuric acid for ten minutes to break dormancy. It has fair seedling vigour and thereafter grows vigorously and suppresses weed growth. It fixes significant amounts of nitrogen and combines well with guinea, pangola and *Brachiaria* grasses.

464

Early grazing should be light, but once established it can tolerate heavy stocking. Fire and drought have serious effects on the swards. Flowers in about 180 days from seeding and is self-fertilized. Seed production is poor and only about 50 kg/ha can be harvested; a suction harvester is required. (Plants can be chemically defoliated before using the suction harvester). Leaf-eating caterpillars may attack the crop during its vigorous growth in midsummer.

Main attribute is its persistence in sward-forming grasses, having a stoloniferous habit of growth. Main deficiencies are poor seed-setting and difficulty of harvesting seed on a field scale.

Vigna lanceolata **Benth.**

Synonym. *V. suberecta* Benth.

Common name. Maloga bean (Australia).

Description. A twining or occasionally erect perennial legume. Taproot conspicuous, up to 40 cm deep. Stems glabrous or with few hairs. Stipules narrow, 2 to 3 mm long. Leaves trifoliate. Leaflets lanceolate, narrowly ovate, apex acute or somewhat obtuse, base entire or lobed, lateral leaflets usually lobed on one side only, 3 to 7 × 0.4 to 1.5 cm. Inflorescence up to 13 cm long; flowers greenish yellow in few clusters at the ends of the peduncle. Calyx short, about 2 to 2.5 mm long, the two upper lobes united into one very short and broad one, the lowest not as long as the tube. Standard 7 to 8 mm long. Keel broad, curved inwards, rather acute, not beaked. Pod 2 to 4 × 0.5 to 0.8 cm, glabrous or a few appressed hairs, nearly terete (Bailey, 1900; Stanley and Ross, 1984).

Distribution. *V. lanceolata* is native to tropical Australia and widespread south to western New South Wales, northern South Australia and Western Australia. It is particularly found on the fertile clay soils of such areas as the Darling Downs of southern Queensland.

General features. Together with several of the *Glycine* spp. and *Rhynchosia minima*, *V. lanceolata* is a legume of the fertile black soil downs, on which it can be quite prominent in favourable seasons. It is highly palatable and valued as forage in native pastures but can be a nuisance in cultivations on black soils. It nodulates under good conditions but the nodules are absent in exceptionally wet or dry conditions and after frosting in winter (Diatloff, 1967b).

465

Vigna luteola (Jacq.) Benth.

Synonyms. Incorrectly introduced as *Vigna marina* Merr. from Costa Rica to Australia; *V. nilotica* (Del.) Hook; *V. bukombensis* Harms.

Common names. Dalrymple vigna (Australia), gilibande, goko, masheke, mugulula, indolo, a kwakwa, kavuhivahi, toshimbo shimbo (Zaire).

Description. Sparsely hairy, short-lived perennial with a climbing and twining habit, rooting readily from the stems in contact with moist soil. Leaves trifoliate, leaflets ovate-lanceolate, 2.5 to 10 cm long and 1.5 to 5.0 cm broad, sparsely pubescent on both surfaces. Racemes axillary, many-flowered. Flowers yellow, 1.8 to 2.2 cm long. Pods pendant, thin shelled, with waxy margins when dry, up to 5 cm long and 0.5 to 0.6 cm wide. Zero to six, from brown to black in colour, 3 to 4 mm long and 2 to 3 mm wide, (Barnard, 1969). CPL 21347 was introduced into Australia from Costa Rica by W.W. Bryan and came to be called Dalrymple vigna.

Distribution. Occurs in wet tropical coastal areas of Africa, Central America, West Indies, southern United States and northern Australia (Grisebach, 1963).

Season of growth. Summer-growing. It disappeared after two years' grazing at Samford, southeast Queensland (Jones, Davies and Waite, 1967).

Temperature for growth. Optimum, about 20 to 30°C. Ludlow and Wilson (1967) obtained only 12.5 percent of the yield, 33 percent of the relative growth rate and 11 percent of the leaf area at 20°C compared with plants grown at 30°C. It gives quite good late autumn and early summer growth. It is very frost susceptible but regrowth will take place by established plants when warm weather ensues.

Latitudinal limits. Approximately 30°N and S.

Altitude range. It is essentially a plant of wet coastal areas.

Rainfall requirements. Prefers a rainfall of about 1 250 mm and above, but will give some useful growth down to 900 mm in wet situations.

Tolerance of drought and flooding. It has little drought tolerance and does not perform well under dry conditions. It is, however, one of the best legumes for wet conditions and one of the best pioneer plants in such situations.

Soil requirements. It does not appear to be very specific as to soil requirements, but friable soils allow it to root down readily. It is one of the few legumes which can tolerate saline conditions (Hutton, 1968).

Ability to spread naturally. It can be spread by seed in dung under favourable conditions of warmth and moisture, but it does not spread readily.

Land preparation for establishment. Requires a fully cultivated seed bed.

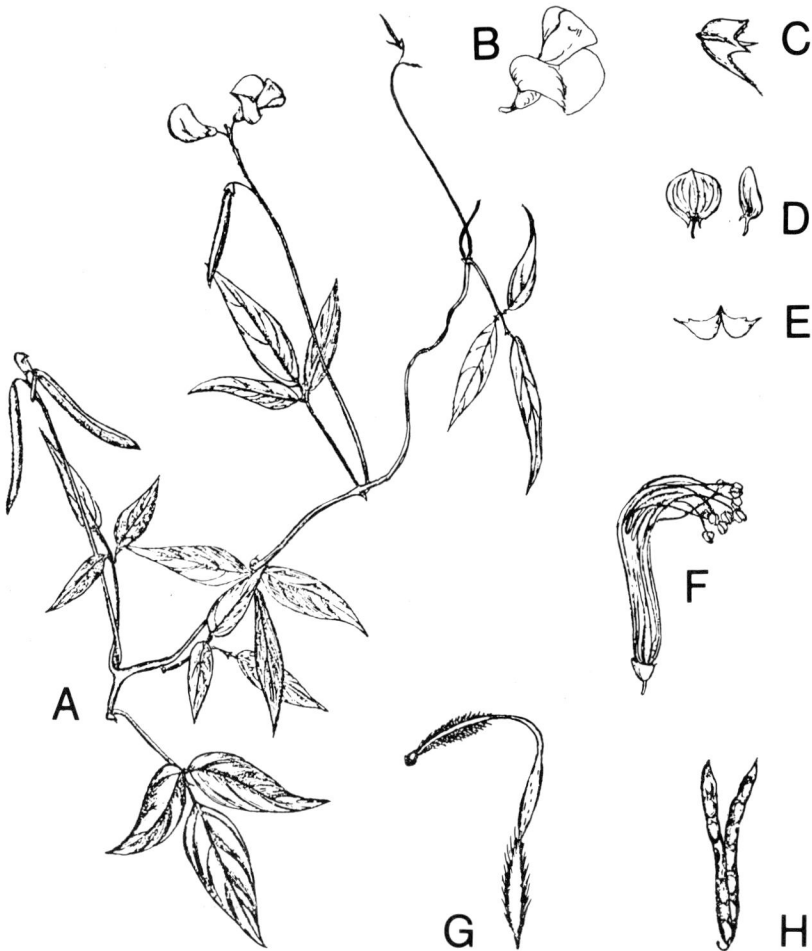

Figure 95. *Vigna luteola.* **A**-Flowering and fruiting branch **B**-Flower **C**-Calyx **D**-Standard and wing **E**-Keel **F**-Stamens **G**-Carpel **H**-Dehiscing pod

Sowing methods. Seed is preferably drilled into a well-prepared seed bed. Oversowing into existing pastures is not successful. Seed should be sown at a depth of 1 to 2.5 cm in spring to early summer, at 1 to 2 kg/ha in mixtures and at 3 to 5 kg/ha alone.

Seed treatment before planting. To break dormancy: a light mechanical scarification will assist germination, but normally no treatment is required. This legume is nonspecific in its *Rhizobium* requirement and need not be inoculated. However, inoculation with the cowpea type is advisable (the current Australian inoculum is CB756). If the seed requires pelleting to protect the *Rhizobium* from the treatment for insect control, rock phosphate should be used (Norris, 1967). To deter attacks by bean fly, treat the seed with 13.2 cc of 15 percent dieldrin emulsifiable concentrate per kg before planting (Jones, 1965).

Nutrient requirements. *V. luteola* requires adequate phosphorus and molybdenum. Use 250 kg/ha molybdenized superphosphate for establishment. Andrew and Robins (1969a) determined the critical percentage of phosphorus in the dry matter of the tops at the immediate preflowering period as 0.25 percent. *V. luteola* yielded 44 percent of the dry matter at soil dressing equivalent to 250 kg/ha superphosphate which it produced at its maximum production at 1 350 kg/ha.

When sodium dihydrogen phosphate was added to the soil in pots containing *V. luteola* and other legumes, *V. luteola* accumulated sodium more than any of the others, but the increased sodium uptake did not depress the potassium uptake as it did with other legumes (Andrew and Robins, 1969b). This would indicate that *V. luteola* has some tolerance to salinity.

Parbery (1967b) found that the addition of 100 kg N/ha depressed the growth of *V. luteola* on Cockatoo sand.

Vigour of seedling, growth and growth rhythm. It has very vigorous seedlings. Its rapid early growth is one of its strongest features (Davidson, 1966). It is a very vigorous grower, especially in its first year; thereafter, its performance is variable.

Response to light. It is day-neutral in its flowering habits and flowers throughout the year without interruption of its vegetative growth in frost-free situations.

Compatibility with grasses and other legumes. Grows quite well with *Paspalum* and *Setaria* species in moist situations.

Ability to compete with weeds. Has poor competitive ability in south Queensland. In frost-free, high rainfall areas in north Queensland, it competes well (Davidson, 1966).

468

Tolerance of herbicides. Bailey (1967b) found that 2,4-D amine caused severe leaf drop; 2,4-D ethyl ester, moderate leaf drop; and 2,4-DB and MCPB, slight leaf drop when sprayed on to 10- to 15-leaf stage *V. luteola*. MCPB spray caused 8.64 percent of leaf reduction; 2,4-DB, 18.4 percent; 2,4-D ethyl ester, 26.5 percent; and 2,4-D amine, 68.7 percent.

Nitrogen-fixing ability. It nodulates freely and effectively; Jones, Davies and Waite (1967) found that the nitrogen released when it died markedly improved growth of the associated grasses.

Response to defoliation. Did not survive close cutting in northern Australia (Parbery, 1967a) or heavy grazing (Davidson, 1966).

Grazing management. Do not graze the young swards heavily. Established plants will stand heavy grazing. Intermittent grazing is preferable to conserve this legume.

Response to fire. No record, but in areas to which it is especially adapted accidental fires would be infrequent.

Breeding system. Self-fertile; chromosome number 2n = 22. It flowered in 91 days at Kimberley Research Station, Northern Territory, Australia (lat. 15°17'S).

Dry-matter yields. At Eumundi, in southeastern Queensland, Australia (lat. 26°S), rainfall about 1 500 mm, it yielded 1 900 kg DM/ha in its first year — the highest of all the legumes tested — but only 375 kg/ha in its second year. At Kimberley Research Station, the best stand yielded only 550 kg/ha of dry matter, but on the upper Adelaide river it yielded 2 000 kg/ha (Parbery, 1967b).

Suitability for hay or silage. No record.

Value as standover feed. Poor. At Kimberley Research Station, it died out in 140 days on Cockatoo sand (Parbery, 1967b).

Feeding value. It is highly regarded as a pasture plant.

● Chemical analysis and digestibility. On Cunnunurra clay, Parbery (1967a) recorded 17.4 percent crude protein in the dry matter at flowering.

● Palatability. It is an extremely palatable plant; for this reason it must never be heavily grazed.

Toxicity. No record.

Seed harvesting methods. Owing to its indeterminate flowering habit, seed harvest is difficult. Pods are hidden by new growth before they can be picked, and so hand harvesting has been the only method employed to date. The pods shatter, but not as badly as those of siratro.

Seed yields. It is usually hand harvested, and no yields appear to have been recorded.

469

Minimum germination percentage and quality required for sale. In Queensland, all *Vigna* spp. are required to give a minimum of 70 percent germination with a maximum of 10 percent hard seed and 98.8 percent purity. They are germinated at 32°C after chipping the seed coat (Prodonoff, 1968).

Cultivars. There was only one commercial cultivar, 'Dalrymple', which is no longer readily available, but there are other local ecotypes.

Pests. It is attacked readily by leaf-eating caterpillars (e.g. *Prodenia* spp.). Bean fly (*Melanagromyza phaseoli*) will attack young seedlings (Jones, 1965).

Main attributes. High palatability; its ability to grow and nodulate in wet soils.

Main deficiencies. Rather short lived; susceptible to frosts and insect attack.

Performance. At Utchee Creek in north Queensland, it has persisted under grazing for five years and is still spreading and competing with weeds in this high rainfall area (about 3 750 mm). Jones, Davies and Waite (1967) obtained yields of 5 698 kg/ha in the first year and 10 934 kg/ha in the second year with a mixed *Paspalum plicatulum/V. luteola* pasture at Samford in southeastern Queensland. The legume contributed 43.4 percent of the dry matter and 65.1 percent of the nitrogen in the first year, and 31.1 and 58.7 percent in the second year. It did not persist into the third year.

In Zambia it behaved as an annual, reseeding itself during the second rainy season. It produced 4 660 kg DM/ha from two cuts per season, with an average protein content of 18.35 percent. The leaf contributed 781 kg DM/ha containing 13.13 percent crude protein, almost the same as the stems, which contributed 78 percent of the dry matter with a protein content of 13.25 percent (van Rensburg, 1967).

Main references. Barnard (1969); Davidson (1966).

Vigna oblongifolia A. Rich var. *parviflora* (Bak.) Verdc.

Synonym. *V. parviflora* Bak.

Description. A trailing, quick growing annual legume, 0.7 m high with pink or bluish flowers turning yellow; very variable in foliage.

Characteristics. It produces excellent leafy growth and cover, and yielded 5 507 kg dry matter per ha in Zambia. Produced no regrowth after cutting. It is capable of growing in Rhodes grass pastures, contributing a considerable amount of the herbage (van Rensburg, 1967).

Vigna parkeri **Baker** (see Colour plate XLIX)

Synonyms. V. gracilis auct., non (Guill. and Perr.) Hook.; *Dolichos maranguënsis* Taub.; *V. maranguënsis* (Taub.) Harms.

Common name. Creeping vigna (Australia).

Description. A perennial climbing or prostrate legume, sometimes forming mats 0.1 to 2 m long. Main root-stock rather slender but tough. Stems slender, often rooting at the nodes, sparsely to rather densely covered with mostly spreading hairs. Three leaflets, round, ovate or ovate-lanceolate, 1 to 8.8 × 0.8 to 5.4 cm, rounded to acuminate at the apex, rounded to subacute at the base, pubescent on both faces, margins densely ciliate; petiole 1 to 8.5 cm long. Two petiolules, about 3.5 mm long. Stipules persistent, lanceolate, 2.5 to 8 mm long, prolonged below the point of insertion and bilobed at the base. Two to five inflorescences, axillary, sometimes 10-flowered. Rachis 0.5 to 2.5 cm long, glandular-nodular. Peduncle 2 to 13 cm long. Pedicels 1 to 3 mm long. Calyx sparsely pubescent, tube 1.5 to 2 mm long, lobes deltoid, ovate or lanceolate, 1 to 1.5 mm long, the upper pair joined to form a more-or-less rounded lobe. Standard blue, turning purple-blue, yellow or white, usually paler at the base, oblate, 5 to 8.5 (12) × 5 to 8 (10) mm, glabrous. Wings blue, purple at the tips or yellow. Keel greenish with purple tip or yellowish, the beak not curved inwards. Pods linear-oblong, compressed 1.3 to 2 (0.9 to 3) cm long, 4.5 to 5.5 cm wide, pubescent or glabrescent, two- to five-seeded. Seeds grey to brown with black mottling, oblong-ovoid, 3 to 4 × 2 to 3 mm, 2 mm thick (Gillett, Polhill and Verdcourt, 1971).

In addition to ssp. *parkeri*, which is rare, there are two other subspecies in East Africa: *maranguënsis* (Taub.) Verdc. (syn. *V. maranguënsis* [Taub.] Harms.), and ssp. *acutifolia* Verdc. (Gillett, Polhill and Verdcourt, 1971).

Distribution. Ssp. *maranguënsis* occurs in Zaire, Rwanda and Angola to Ethiopia (?), generally from 1 000 to 2 460 m in elevation and sometimes as high as 2 700 m. It is extremely common in East Africa. Ssp. *acutifolia* occurs in Kenya, Tanzania, Zanzibar and Mozambique from sea level to 1 000 m in elevation (Gillett, Polhill and Verdcourt, 1971).

Habitat. V. parkeri occurs in grasslands, grasslands with scattered trees, thickets, forests and sometimes as a weed of cultivation.

Season of growth. V. parkeri is typical warm-season species, its growth pattern more closely resembling that of greenleaf desmodium than siratro in southeast Queensland, Australia.

Optimum temperatures. Optimum temperature for germination is 25°C, with germination delayed above 27°C and markedly reduced at 35°C. However,

471

there is little reduction down to 12.5° (Pentney, Whiteman and Sivasu-piramiam, 1984).

Frost tolerance and regrowth after frosting. Top growth is killed by even light frosts but stands regenerate readily from the root-stock. Early season growth is often rather slow, possibly owing to lack of moisture during this rather dry time of year in southeast Queensland (B.G. Cook, personal communication).

Rainfall requirements. Initially considered essentially a plant for wet areas, *V. parkeri* is now believed to be adapted to most areas that will carry greenleaf desmodium in southeast Queensland, that is areas with annual rainfall exceeding 1 200 mm and with good moisture-holding soils.

Drought tolerance. Established plants have little drought tolerance and are readily killed by dry conditions. Where these conditions occur only infrequently, regeneration from seed in the soil rapidly occurs. Soil seed reserves ranging from 50 to more than 1 000 seeds/m^2 (mean 450) have been measured (B.G. Cook, personal communication).

Tolerance of flooding. While *V. parkeri* is regarded as a useful plant for sea-sonally wet areas in southeastern Queensland, and has persisted for over 20 years in one area subjected to intermittent flooding, CQ 1374 was in the fourth and least tolerant group of 17 legumes tested for flooding tolerance in pots at the University of Queensland. In this group, no plants survived and no adventitious roots were formed when flooding was imposed at first flowering for 10 or 21 days. Others in this group were glycine, axillaris, silverleaf desmodium, centro and lablab (Whiteman *et al.*, 1984). The true picture is therefore unclear.

Soil requirements. The main soil requirement for *V. parkeri* appears to be a good moisture-holding ability, so adequate depth of the useful profile is important. *V. parkeri* persisted and spread from old trial plots on soils with pH as low as 4.6, acid and bicarbonate P levels as low as 7 and 13 ppm respec-tively, and exchangeable and replaceable K levels as low as 0.09 and 0.13 per-cent respectively in southeast Queensland (B.G. Cook, personal communi-cation).

Rhizobium *relationships.* *V. parkeri* nodulates with a wide range of *Rhizobium* strains. Eleven of 19 strains tested, including CB 756, gave yields equal to or in excess of the nitrogen-fertilized controls (Pentney, Whiteman and Sivasupiramiam, 1984).

Ability to spread naturally. Material planted during the 1960s has survived and spread away from the trial sites in a number of areas in southeast Queens-land. It has spread into areas carrying, among other species, *Imperata cylin-*

Figure 96. The foliage of *Vigna parkeri* cv. Shaw

drica and *Pennisetum clandestinum* (B.G. Cook, personal communication).
Land preparation for establishment. Naturally, a well-prepared seed bed will give the best establishment from sparse seed, but the plant has the ability to establish into lesser quality seed beds as well.

Oversowing into natural pasture. Experience over the last 20 years in south-east Queensland suggests that strip planting into either native or sown pastures on suitable country will eventually lead to excellent stands.

Sowing depth and cover. Shallow planting is indicated.

Sowing time and rate. 2 kg seed/ha. Plant at the break of the season.

Number of seeds per kg. 75 000 for CQ 1374 (B.G. Cook, personal communication).

Seed treatment before planting. Up to 96 percent hard seeds have been recorded in freshly harvested samples, so that seed scarification is essential (B.G. Cook, personal communication).

Nutrient requirements. Nutrient requirements have yet to be investigated, though the plant will persist and spread on fairly infertile soil. It will almost certainly respond strongly to fertilizer application.

Compatibility with grasses and other legumes. While the main stolons stay close to the ground surface, secondary branches from the leaf axils can climb up to the light through taller-growing species. *V. parkeri* is particularly compatible with the sward-forming species pangola and Kikuyu, but as long as the stand is not dense and undergrazed it is also able to persist in setaria stands. It is more akin to *Lotononis bainesii* cv. Miles and *Trifolium semipilosum* cv. Safari in its growth habits and species compatibility than to the more robust twining legumes (B.G. Cook, personal communication).

Seedling vigour. Stands are inclined to be slow to start, although in pots the plants established rapidly, forming 10 or more stolons per plant at 15 weeks from sowing. These stolons creep readily and root rapidly from the nodes (Pentney, Whiteman and Sivasupiramiam, 1984).

Nitrogren-fixing ability. Associated grasses generally show signs of nitrogen input, and nitrogen levels in the legume dry matter in excess of 3 percent have been recorded for CPI 25378 (B.G. Cook, personal communication).

Grazing management. Because of its growth habit, *V. parkeri* can tolerate somewhat heavier grazing pressures than the robust twining species such as siratro. It does not, however, tolerate as severe use as *Trifolium semipilosum*.

Dry- and green-matter yields. Under infrequent harvest regimes *V. parkeri* is not a high-yielding, bulky species, which is why it was initially passed over during selection of such species as greenleaf desmodium and siratro, but it is nevertheless able to contribute quite a useful yield.

Toxicity. None known.

Seed harvesting methods. Harvesting methods have yet to be developed. As *V. parkeri* is a rather low-growing plant and the pods shatter readily, it is likely that conventional header harvesting will still be possible from well-grown stands. As blue-flowered material does not flower until May/June, it will be necessary to plant in areas that remain frost free in early winter in order to achieve mature seed crops. The yellow-flowered lines have a much longer flowering period, beginning as early as October.

Pests and diseases. To date, insect damage has been minor in southeast Queensland, a leaf miner (*Acrocercops* sp.) sometimes being noted in autumn. *Anthracnose* devasted a number of accessions, but not CPI 25378, at Gympie in 1983. *Cercospora* leaf spot, *Sclerotium rolfsii* and the root knot nematode (*Meloidogyne javanica*) have been identified on CQ1374 but have been of no agronomic consequence (B.G. Cook, personal communication).

General features. Ssp. *maranguënsis* was long known to agronomists as

474

V. gracilis. It is not *V. gracilis* (Guill. et Pers.) Hook. (syn. *Dolichos gracilis* Guill. et Pers.), which comes from the region of the Senegal and Gambia rivers in eastern Senegal and western Mali. Ssp. *maranguënsis* is a fine legume, often forming natural mixtures with grasses on moist ground in East Africa and producing excellent grazing (Bogdan, 1977).

This plant was introduced into Australia in the 1950s and early 1960s as *V. gracilis* and widely tested before it was passed over for higher yielding plants such as desmodiums and siratro. However, these early plantings of a range of accessions have survived and spread in the Beerwah, Cooroy and Gympie districts of coastal southeast Queensland. As a result, since 1975 it has been under consideration as a plant for use in low lying moist areas in these above 1 200 mm rainfall districts. Its possible suitability to a wider range of habitats is also being appreciated.

Cultivar 'Shaw' was released by the Queensland Herbage Plant Liaison Committee in August, 1984. Previously it was identified as CQ1374, but is believed to be derived from CPI 25378, from Entebbe, Uganda, received in 1958.

Main references. Gillett *et al.* (1971); Pentney *et al.* (1984).

Vigna radiata (L.) Wilczek

Synonym. Phaseolus aureus Roxb.
Common names. Mung bean (Australia); green gram (India).
Description. An upright annual legume ranging in height from 15 cm to 1 m; average height of mature plant, 0.9 m. Branches freely, but not heavily foliaged. Leaves, stems and pods are slightly hairy. Junctions of branches and stems are stipuled. The first flowers appear seven to eight weeks after planting and the crop reaches maturity in 12 to 14 weeks. Pods borne at top of plant. Seeds, green and almost globular (Doherty, 1963a). Pods clothed in long, spreading, deciduous silky hairs.
Distribution. Native to India; is now widespread throughout the tropics, ascending from sea level to 1 850 m in the northwest Himalayan regions (Hooker, 1979).
Characteristics. Mung bean or green gram has long been a food crop in Asia. It is less known as a useful green manure crop. Recently it has become of interest in Queensland as a fodder crop. In its short growing season, *Vigna radiata* will outyield cowpea and velvet bean of the same age, although

475

maximum yields of the other two are greater. It is, therefore, a useful legume for early forage. It is adapted to a wide range of well drained soils, but is best on fertile sandy loams. On sandy soils of low fertility, 185 to 250 kg/ha molybdenized superphosphate will usually give adequate growth.

A good seed bed (as for maize or sorghum) should be prepared. The seed is broadcast or drilled in rows 16 to 35 cm apart, the usual seeding rate being 6 kg/ha drilled and up to 10 kg/ha broadcast. It can also be sod-seeded into existing pastures. Seed is preferably inoculated with the cowpea strain of *Rhizobium* before sowing. The first grazing can be given about six weeks after planting, before the flowers appear; two grazings are usually obtained. Green manure should be ploughed in when the plant is in full flower. Mung bean should be cut for hay as it begins to flower. The cut material should be conditioned to hasten drying. Doherty (1963a) obtained a yield of 1 872 kg/ha of green matter from mung bean sod-seeded into a Rhodes grass/green panic pasture at the rate of 11 kg/ha in 53-cm rows, fertilized with 264 kg/ha molybdenized superphosphate. Unfertilized pasture yielded only 623 kg/ha of green matter.

Vigna trilobata **Walp.**

Synonyms. *Dolichos trilobatus* L.; *Phaseolus trilobus* (L.) Schreb.; *P. trilobus* auct., non (L.) Aiton.

Common names. Phillipesara (India); mukni, jongli-math (Pakistan).

Description. An annual or perennial legume. Branches prostrate, diffuse, trailing, glabrous or pubescent. Stipules peltate, sometimes spurred, ovate, 4 to 15 mm long. Leaves trifoliate. Petiole 3.7 to 7.5 cm long. Leaflets 1.2 to 2.5 cm long, usually as wide, generally three lobed, lobes simple, various, oblong obtuse or subacute, glabrous to subglabrous. Inflorescence a few-flowered raceme. Peduncle 8 to 22.5 cm long; bracts deciduous. Bracteoles *c*. 3 mm long, ovate, below the calyx. Pedicels 2.5 mm long. Calyx 2.5 mm long, glabrous, teeth minute. Corolla yellow, 5 to 6.5 mm long. Pods cylindrical, 2.5 to 5.0 × *c*. 3 mm, glabrous to sparingly pubescent, 6- to 12-seeded (Ali, 1977; Saldanha and Nicolson, 1976).

Distribution. Native to India, Pakistan, Sri Lanka, Burma, Afghanistan and the Malay archipelago. It is probably fairly closely related to the moth or mat bean, *V. aconitifolia* (Jacq.) Marechal, with which it has at times been confused, (e.g. Whyte, Nilsson-Leissner and Trumble, 1953). It has also been suggested as the wild form of *V. aconitifolia* (Sampson, 1936).

General features. One of the outstanding features of this plant is its drought tolerance. One accession (CPI 28928) proved particularly hardy at Biloela, central Queensland, Australia, but the seed shattered readily and was cut by frost (Cameron and Mullaly, 1969).

V. trilobata is cultivated in India, Pakistan and the Sudan for short-term pasturage.

Vigna unguiculata (L.) Walp.

Synonyms. *Vigna sinensis* (L.) Savi; *Vigna sinensis* Endl.; *Vigna catjang* (Burm.) Walp. There is a difference of opinion regarding the valid name for this plant. Verdcourt of the Royal Botanical Gardens, Kew (personal communication), prefers *V. unguiculata*, with the several cowpeas throughout the world as cultivars.

Barnard (1969) has divided this variable species into three main groups:
● Var. *sinensis* — the common cultivated cowpea, with medium length, pendant pods, and medium-sized, kidney-shaped or roundish seeds;
● Var. *sesquipedalis* — the yard long or asparagus bean, which has long pendant pods which are inflated when green and shrivel when ripe, with elongate, kidney-shaped seeds; and
● Var. *cylindrica* or *catjang* (*Vigna catjang* [Burm.] Walp.) — which has short, erect pods and small oblong or cylindrical seeds.

Common name. Cowpea, occasionally southern pea (United States).

Description. Herbaceous annual with twining stems varying in erectness and bushiness. Leaves trifoliate, petioles 2.5 to 12.5 cm long. Central leaflet hastate, 2.5 to 12 cm long, smooth, lateral leaflets irregular. Flowers in axillary racemes on stalks 15 to 30 cm long. Pod pendulous, smooth, 10 to 23 cm long with a thick decurved beak and 10- to 15-seeded. Seeds 4 to 8 mm long, 3 to 4 mm broad, variable in size and colour (Barnard, 1969).

Distribution. The country of origin is uncertain. Vavilov (1951) thought it might be India, with secondary centres in China and Ethiopia. Recent workers believe it to be of central African origin. It is widespread throughout the tropics and most subtropical areas.

Season of growth. Summer-growing.

Temperature for growth. It prefers warm moist conditions, with a hotter climate than for maize or soybeans (Klages, 1942). Dart and Mercer (1965) found that a day temperature of 27°C gave optimum growth. It is sensitive to cold conditions (Johnson, 1970). Milford and Minson (1968) found that it

477

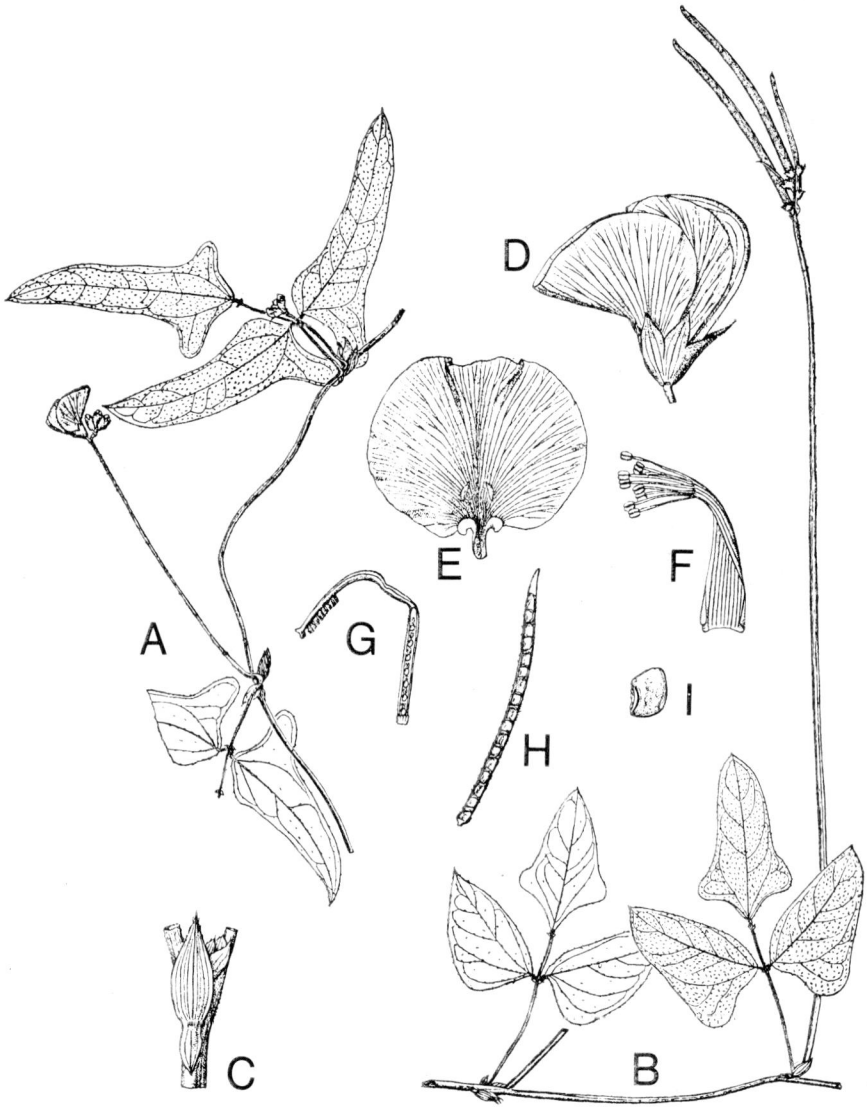

Figure 97. *Vigna unguiculata.* **A**-Flowering stem **B**-Fruiting stem **C**-Branch with stipules
D-Flower **E**-Standard **F**-Stamens **G**-Carpel **H**-Pod **I**-Seed

dropped from 33 to 14 percent of its leaves in winter. The crop is very susceptible to frost, and in frost-susceptible subtropical areas seed harvesting is usually deferred until frosts have killed and dried the top growth.

Latitudinal limits. Between latitudes 30°N and S.

Altitude range. Usually a low-altitude plant, but will grow quite well up to
1 500 m elevation.

Rainfall requirements. For forage purposes, a rainfall of 750 to 1 100 mm is
preferable. It will tolerate lower rainfall, but in high rainfall areas disease and
insect attacks increase.

Tolerance of drought and flooding. This species is one of the crop legumes
most tolerant to drought conditions and is used as a grazing crop and often as
a food crop down to 400 mm of annual rainfall in the Sudan. Flooding is usually fatal to cowpeas. They require well-drained soils.

Soil requirements. It is tolerant of a wide range of soil textures from sands to
heavy, well-drained clays. Heavy clays tend to encourage vegetative growth
at the expense of seed production. It adapts to a wide range of pH, but prefers
slightly acid to slightly alkaline soils. It has little tolerance of salinity
(Johnson, 1970).

Rhizobium *relationships*. The cowpea is nonspecific, but inoculation with a
selected cowpea strain of inoculum is an advantage. It nodulates freely with
native rhizobia. Johnson (1970) found that 45 kg/ha nitrogen depressed nodulation. Nodulation is also affected by high soil temperatures but deep planting
helps overcome this defect (Philpotts, 1967).

Ability to spread naturally. Does not spread in unprepared land.

Land preparation for establishment. Performs best if treated as a crop sown
on a well-prepared seed bed; will, however, establish quite well on a roughly
prepared seed bed from an initial ploughing or disc harrowing. The large seed
helps in establishment.

Sowing methods. Cowpeas may be sown broadcast on rough seed beds or be
drilled into well-prepared ground in rows 50 to 75 cm apart. Maize planters
with cowpea plates are usually used. Can be sod-seeded into pastures, provided soil disturbance is adequate and the crop is fertilized. It is often sown
in maize crops at the time of the last interrow cultivation. Seed is sown at from
2.5 to 7.5 cm, the latter being preferable, from spring to mid-summer where
frosts are likely; it is sown later in frost-free areas. Early sowings give higher
yields (Hendricksen, 1969). Sow 17 to 39 kg/ha drilled or 45 to 95 kg/ha
broadcast.

Number of seeds per kg. 4 180 to 8 800. Ezedinma (1965) found that seed size
did not affect performance. The percentage of hard seeds is low. Seed

will remain viable for up to three years if stored in a cool dry place, but normally seed from the last harvest should be used for the next planting.

Seed treatment before planting. Usually no preparation needed to break dormancy. Inoculation is not necessary as the cowpea is nonspecific. Pelleting is not needed. For insect and disease control, treat with thiram (1:200 by weight) and with aldrin to control bean fly before planting (Johnson, 1970).

Nutrient requirements. Cowpea grows well without fertilizer in the better soils. In soils of low fertility, it responds to phosphorus and potash and often some nitrogen. Up to 10 kg/ha of nitrogen and 40 to 70 kg/ha P_2O_5 and K_2O may be needed in low fertility soils (Johnson, 1970). There is a response to calcium where the pH is low but this may be a response to the released molybdenum.

Toxicity levels and symptoms. No specific toxicity effects recorded, though it does not tolerate salinity.

Response to photoperiod and light. The cowpea is usually a short-day plant or indeterminate in its flowering response. Time of sowing has little effect on plant growth, but later plantings flower earlier. Njoku (1958) stated that Nigerian types required a day length of less than $12\frac{1}{2}$ hours for flowering. It is tolerant of moderate shade and so grows with tall crops such as maize and sorghum.

Compatibility with grasses and other legumes. Usually grown as pure legume sward for forage or mixed with maize, sorghum or bulrush millet for green chop or silage. Does not compete with perennial grasses because of its annual habit. Pioneer crops of mixtures of cowpea and Japanese millet (*Echinochloa* sp.) have been used in coastal Queensland, ahead of permanent pastures.

Ability to compete with weeds. Can compete fairly well with low-growing weeds, but not with tall ones such as *Tagetes minuta*. It is preferably given one or two interrow cultivations if seed has been row-sown.

Tolerance of herbicides. Susceptible to MCPB (Gentner and Danielson, 1965). In the United States, trifluralin (Treflan) is used as either a preplant incorporated, preemergence or postemergence herbicide (Johnson, 1970).

Vigour of seedling, growth and growth rhythm. Quite a vigorous seedling; the young crop can be lightly harrowed during the heat of the day for seedling weed control. Under favourable conditions, cowpeas grow vigorously.

Nitrogen-fixing ability. Johnson (1970) estimates fixation at 84 kg/ha if rhizobia are effective. Denarie, Andreamanantena and Ramonjy (1968) found that cowpeas fixed 60 kg N/acre.

Response to defoliation. Should only be grazed to the stage of leaf removal.

Grazing management. Heavy grazing should be avoided. At no time should it be grazed or cut before flowering. With the 'Havana' type cowpea, Hendricksen achieved only 25 percent recovery when the plant was defoliated before flowering. For quick regeneration, leave four to six buds per plant which can commence regrowth immediately (Hendricksen, 1965).

Response to fire. Will not tolerate fire, but usually is too green to burn.

Breeding systems. Normally self-pollinated (Cobley, 1956), in dry areas, but cross-pollination occurs proportional to atmospheric humidity (Purseglove, 1968).

Dry- and green-matter yields. Milford and Minson (1968) found that cowpea yielded 2 310 kg DM/ha in summer and nil in winter, whereas *Lablab purpureus* yielded 4 950 kg DM/ha in summer and 4 070 kg DM/ha in autumn and early winter. Smith (1961) reported that a Poona cowpea/white *Panicum* pasture at Cooroy, Queensland, which had been irrigated and fertilized, yielded 27.15 tonnes of green material per hectare at the first grazing. Doherty (1963b) sod-seeded a mixture of mung bean, cowpea and velvet bean into a Rhodes grass/green panic pasture. Cowpeas fertilized with 265 kg molybdenized superphosphate per hectare planted in 53-cm rows at 9 kg/ha of seed yielded 29 075 kg green material per hectare.

Suitability for hay and silage. Cowpea makes quite good hay but care must be taken to preserve the leaf. The stem takes some time to dry and should be "conditioned" to hasten drying. It makes good silage if molasses is added to pure cowpea material at 30 to 40 kg per tonne. It is better mixed with sorghum or maize to provide sugar for fermentation, to add bulk and to prevent protein losses.

Value as standover or deferred feed. Can be utilized up to the first frost, after which the leaves drop. Not as tolerant of cool temperatures as *Lablab purpureus*.

Feeding value.

● Chemical analysis and digestibility. French (1935, 1937) fed cowpea hay to cattle at Mpwapwa, Tanzania. The crude protein contents were 13.01 and 12.8 percent, and the digestible crude protein 7.92 and 8.70 percent respectively. The seed contains 24 percent crude protein, 53 percent carbohydrates and 2 percent fat.

● Palatability. For the first day or two, cattle may shun the crop, but when they become accustomed to it they will eat it readily.

Seed harvesting methods. Usually by direct heading after frosts have killed the tops of the plants, especially in erect growing crops. For prostrate or trailing varieties, the crop can be mown when two-thirds of the pods are dry and rattle when shaken (in some varieties). The vines should be thoroughly dried before threshing by stationary or pick-up harvesters. Hand picking of pods a number of times during the season gives the highest yields; they can be threshed by flailing in bags or by machine.

Seed yield. Average yield is about 750 kg/ha but reaches as high as 2 800 kg/ha. Gill and Batra (1968) found that cowpeas gave the best seed yield with less than 200 to 250 mm rainfall during growth.

Minimum germination and purity required for commercial sale. A minimum germination of 70 and 98 percent purity, and a maximum of 10 percent hard seed are required in Queensland.

Cultivars. There are numerous local cultivars throughout the world to fill ecological niches. Many varieties are bred for resistance to prevailing diseases. Queensland varieties include 'Santiago', 'Havana', 'Malabar', 'Reeves', 'Cristaudo', 'Poona', 'Black' and 'Blackeye No. 5'.

Diseases. Many recorded — stem and root rots and some pod infections. *Phytophthora vignae* is most important in Australia (Purss, 1957), and breeding programmes aim at resistance to this disease. Fusarium wilt, septoria leaf spot and mildew are common.

Pests. Aphis, leaf-hopper and pod borer attacks in Hawaii (Takahashi and Ripperton, 1949); to nematodes and the cowpea witchweed (*Alectra vogeli*) in Zimbabwe (Johnson, 1970).

Main attributes. A quick-growing bulky leguminous crop of high protein content, it is a valuable catch crop for building soil fertility. It is drought tolerant and adapted to a wide range of soils.

Main deficiencies. It is an annual susceptible to frost. Thick stems make haymaking difficult.

Performance. Cowpea provides excellent grazing of high food value for dairy cattle and is also suitable for other livestock. Few records of actual grazing performance are available.

Main reference. Johnson (1970).

Vigna vexillata (L.) A. Rich.

Synonyms. V. *reticulata* auct. non Hook.; *V. capensis* Walp.; *V. hirta* Hook.

Description. Fairly strong twiner, usually with fusiform tuberous root; stems usually clothed with spreading silky hairs. Leaflets three, acute at the apex, obliquely and broadly cuneate to almost rounded at the base, the terminal one 7.5 to 15 cm long, all dark green and with appressed strong silky hairs on both surfaces. Flowers pink or purplish, turning yellow, 2.5 cm long, on two-

482

to four-flowered peduncles 7.5 to 30 cm long, keel prolonged into an uncurved beak. Pod recurved, linear, 7.5 to 9 cm long, silky (Andrews, 1952).

Distribution. It is common in the central Sudan, Zaire, Senegal, Sierra Leone, Liberia, Côte d'Ivoire, Togo, Nigeria, Republic of Cameroon, Angola, Ethiopia, Kenya, Tanzania, Malawi, Zambia, Colombia and Venezuela.

Characteristics. V. vexillata is an annual with good seedling vigour and good wet-season growth, and is effective in weed suppression. The seed shatters and it is not a good standover feed for the dry season. It is frost- and fire-susceptible (Downes, 1966). Although hairy, it is quite palatable (Bermudez, Ceballos and Chaverra, 1968). It spreads fairly well naturally and will appear seasonally in crop stubble. It did not perform as well as cowpea (*Vigna unguiculata*) at the Kimberley Research Station, in northern Australia (Parbery, 1967b). Its best yield of dry matter was only 309 kg/ha unfertilized and 525 kg/ha with 100 kg N/ha on the heavy Cunnunurra clay and 1 122 kg/ha on unfertilized Cockatoo sand. Added nitrogen at 100 kg/ha depressed the yield on Cockatoo sand. In Zambia, van Rensburg (1967) obtained 2 780 kg DM/ha.

The plant seeded in 91 days on Cockatoo sand and in 126 days on Cunnunurra clay. Milford (1967) found that the dry-matter intake of the actively growing flowering plant was one of the highest recorded among tropical legumes, but intake fell after the plants were severely frosted. The crude protein content of the dry matter at flowering was 20.3 percent, of which 80.9 percent was digestible.

It produced excellent leafy growth in Zambia and covered the ground quickly. It yielded 1 442 kg DM/ha but made no regrowth after cutting (van Rensburg, 1967).

Zornia spp.

Zornia diphylla Pers.

Synonyms. Z. *reticulata* Sm.; Z. *gracilis* DC; Z. *havensis* A. Rich.
Common names. Trencilla, zornia, barba de burro (El Salvador).
Description. A diffuse annual with two ovate-lanceolate pointed leaflets, flowers distant in peduncled spikes, pod joints two to seven, roundish, convex or reticulated on the sides; flowers yellow, concealed in the stipular bracts (Grisebach, 1963). The chromosome number is 2n = 20.
Distribution. All tropical countries from New Mexico (United States) to Brazil, the West Indies, Australia.
Characteristics. It is a native legume in the brigalow areas of Queensland, Australia, but is not very productive. Downes (1966) tested it at Parada, north Queensland (lat. 17°66′S, 480 m elevation, rainfall 900 mm) and found it very poor in establishment vigour and vigour of growth in the wet season. It did not recover from frost or fire.

Zornia glochidiata Reichb. ex DC

Synonym. Z. *diphylla* (non Pers.) Broun and Massey.
Characteristics. This legume is plentiful in the frost-free sandy areas south of the Sahara. In the Sudan, it occurs in a rainfall regime of 300 to 450 mm falling over four summer months with eight months' dry season. It provides some scant forage during the wet season and early in the dry season, but does not survive long into the dry season (Skerman, 1966). Valenza (1966) recorded that natural pastures in Senegal consisting of *Andropogon amplectens* and *Zornia glochidiata* were able to support one 200- to 300-kg beast per 2.5 hectares at the end of the rains, and one beast per 7.5 hectares in the dry season. The rainfall was under 520 mm per year.
Distribution. Occurs throughout tropical Africa.

484

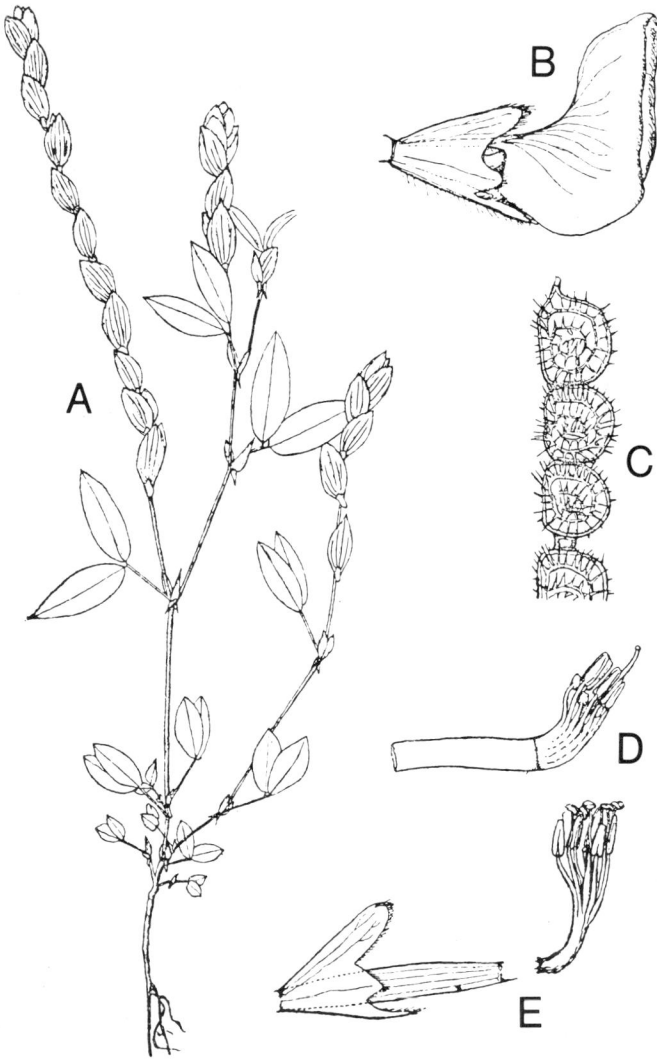

Figure 98. *Zornia latifolia.* **A**-Plant; *Zornia diphylla.* **B**-Flower **C**-Pod **D**-Pistils **E**-Stamens

Zornia latifolia Sm.

Synonyms. Z. *diphylla* auct., non (L.) Pers.; Z. *surinamensis* Miq.; Z. *gracilis* DC.; Z. *diphylla* var. *gracilis* (DC.) Benth.; Z. *pubescens* H.B.K.

Common names. Koemataballi (Suriname), tencilla, zornia, barba de burro (El Salvador).

Description. A perennial herb. Stems 20 to 50 cm long, glabrous or pubescent, with a prostrate growth habit and intense branching. Stipules lanceolate, striate, to 1 cm long. Leaves bifoliate. Leaflets lanceolate-oblong, acute at the apex, glabrous or pubescent, 1 to 4 cm long. Inflorescence a terminal peduncled spike, flowers alternate, 1 to 35 per inflorescence, bractlets stipuliform, to 1.5 cm long, nearly enclosing the flower. Calyx hyaline, 4 mm long, ciliate. Petals yellow, approximately 1 cm long. Pods two- to-eight jointed, shortly beaked, more or less spiny, pubescent, the inferior margin deeply crenate, the superior margin nearly straight, joints rounded, 2 to 3 mm long and wide (Pulle, 1976; Jutzi and Nosberger, 1984).

Distribution. Native to and widely distributed through tropical South America and north into the West Indies; possibly through Central America to southern Mexico and the United States. It is found as far south as Rio Grande do Sul in Brazil, and possibly northeastern Argentina. Its chief habitats are open fields and grassy areas. Z. *latifolia* has been naturalized as a weed in West Africa and the Congo, where it occurs in gardens and lawns, on roadsides, airfields and in savannas.

Mohlenbrock (1961) lists two varieties, var. *latifolia*, which is the widespread material, and var. *bernardinensis* (Chod & Hassl) Mohlenbrock which is confined to the savanna regions of Paraguay.

General features. Long confused with *Zornia diphylla*, an Old World plant based on *Hedysarum diphyllum* L., the true nature and extent of Z. *latifolia* still appear unclear. It has attracted attention in South America as a plant well adapted to the high-aluminium soils of the Brazilian Cerrado and Colombian Llanos. It has been under evaluation by CIAT since 1975 and has given consistently high forage yields and high forage quality on these toxic aluminium soils. One accession, CIAT 728, a Colombian ecotype, has been selected for possible development to cultivar status (Jutzi and Nosberger, 1984).

Active growth during the dry season is one of the important characteristics. In the Colombian Llanos Orientales, its dry-matter yield ranges from 617 to 4 917 kg/ha and protein yields from 60 to 801 kg/ha when grown in association with *Brachiaria decumbens* and *Andropogon gayanus* (CIAT, 1978). Z. *latifolia* has also formed a complete ground cover within 12 months when

486

planted in a low-density system (1 000 hills/ha) with initial fertilizer applied only to the hills (CIAT, 1978).

Breeding system. Flowering occurs sequentially from the basal to the terminal flowers of the inflorescence; flowers open for five to ten hours at anthesis. No insect tripping is needed for selfing but cross-breeding is possible. Fertilization is mainly autogamous, with a very low proportion of insect-dependent crossing (Jutzi and Nosberger, 1984).

Seeding features. The fruit is a one- to seven-articulated lomentum but most articulations are abortive. Disintegration occurs readily and pod shattering can begin as early as the third week after anthesis, before the seeds have matured.

CIAT 728 has a short-day flowering response, with a critical photoperiod of about 12.5 hours and optimum temperature for seed set ranging from 20 to 27°C. Higher temperatures are necessary for vigorous growth. Relatively high plant densities improved the seeding performance (Jutzi and Nosberger, 1984).

At the Cerrados Agricultural Research Centre, northeast of Brasilia, CIAT 728 yielded from 175 to 692 kg per ha per year over three years and was one of few, of the nine legumes tested, to yield seed each year. It had the highest mean annual yield. Seed-line maintenance was complicated by the presence of native lines of *Z. latifolia* (de Andrade, Thomas and Ferguson, 1983). In Jutzi and Nosberger's (1984) studies, seed yields of 400 to 700 kg/ha were recorded, with 90 percent hard seed at harvest and 770 000 seeds/kg.

Diseases and pests. Currently, the main limitation to the use of CIAT 728 in pastures is its susceptibility to disease. It is attacked by scab (*Sphaceloma zorniae*) and a virus-blackmould (*Meliola* sp.) complex, which is common in the native *Zornia* spp. in Brazil and causes leafrolling distortion and stunted growth. It was also attacked in seed stands by the bud worm (*Stegasta bosqueella*), which, however, is easily controlled by insecticides (de Andrade, Thomas and Ferguson, 1983).

Main reference. Jutzi and Nosberger (1984).

15. Leguminous browse

Dayton (1931) defined "browse" as the "shoots or sprouts, especially of tender twigs and stems of woody plants with their leaves, which are cropped (browsed) to a varying extent by domestic and wild animals". The term should be extended to include the fruit or pods, which are often more valuable than the foliage, especially if the browse shrub or tree is deciduous.

Browse in its natural habitat

In the process of free-range grazing, herbivorous animals frequently lightly nibble or browse the young foliage of palatable woody plants in addition to their intake of grass and herbs. This adds variety to their diet and helps maintain balance in the vegetation.

Herbivorous animals fall loosely into two classes: those that depend for their nutrition mainly on grass and herbs, and those that mainly browse woody plants and eat pods, with supplementation from grasses and herbage. This diversity in grazing habits helps to balance the plant-animal ecosystem. A third major factor, the soil, interacts with both plants and animals to form a soil-plant-animal ecosystem. In mixed vegetation, it is preferable that the balance between browsers and grazers be maintained, including game animals in the total biomass.

The advent of farmers with their commercial interests has led to specialization and segregation within the herbivores so that vegetation has been subjected to unnatural grazing pressures according to the choice of livestock and the aims of the production enterprise. A farmer is both a part and a manipulator of the ecosystem, but in reaching production goals often upsets the ecological balance, sometimes with disastrous consequences.

Animal feeding habits

Among game animals, the elephant eats both grass and woody plants. Lamprey (personal communication) has found in the Serengeti Game Reserve in northern Tanzania, where the grass cover is often insufficient for the animals' daily needs, that elephants are converting savanna woodland to grassland at the rate of 6 percent per year, as they seek food by pushing over the acacia and other trees.

The giraffe and the camel are well-known browsers and the eland and Grant's gazelle feed mainly on browse. Talbot and Talbot (1962) found that Grant's gazelle consumed in its diet 11 percent *Indigofera* browse and 4 percent leguminous pods and seeds. The impala feeds largely on the pods of *Acacia tortilis* (Lamprey, 1967); Talbot and Talbot (1962) found that 15 percent of its stomach content was made up of legume seeds and pods. Thomson's gazelle is mainly a grazer, but it also was found to have 2 percent of leguminous pods and seeds in its stomach contents.

The zebra, wildebeest, buffalo and the smaller antelopes feed mainly on grass.

Among domestic livestock, the horse and the sheep mainly utilize grass, other herbaceous plants, and low browse; the cow supplements its grass and herbage diet with some browse and pods; and the goat supplements its predominantly browse and pod diet with grass and herbage. French (1949) reporting trials at Mpwapwa, Tanzania, showed that plots grazed for six years only by cattle became open thickets liable to tsetse fly invasion, with poor grazing and many patches inviting erosion. Plots grazed exclusively by goats remained grassland with insufficient bush cover for tsetse fly development. However, there were insufficient goats to maintain a sufficiently heavy stocking during the flush season of growth. Goats did, however, assist in keeping browse under reasonable control and reduced labour needed for full control. Van Rensburg and Hornby (1948) summarized several years' work in these words: "On pastures obtained by clearing bushland, goats can be kept to advantage without fear of causing either pasture degeneration or soil erosion. They should *not* be regarded as bush-clearing agents. When present in reasonable numbers they will slow down the process of reversion to bushland, but the concentration necessary to completely check the processes is impractically high". Such heavy concentrations are usual around villages, and thus goats usually denude the browse or keep it under continuous suppression, with damage to other vegetation and to the soil. A compromise in browse use is often made by keeping a mixture of cattle and goats, usually at about a ratio

489

of four goats to one cattle beast, the proportion being varied according to the trend in botanical composition.

The delicate ecological balance

The vegetation in any natural semi-arid region is not only in a state of delicate equilibrium with climatic and edaphic factors, but also with its population of native fauna. Any increase in pressure from fauna numbers can lead to drastic changes for the worse in ecological balance.

There is also a delicate balance in land use between the demands of cultivation for food crops and the natural grazing land. The financial return per unit of cultivated land under crop is usually higher than that from grazed pasture, so whenever cultivation can be extended, however hazardously, natural pastures make way for the plough. There is also pressure, apart from money, for more food to relieve a usually hungry population, and so cropping moves into marginal lands, initially often as shifting cultivation (see Chapter 1), but tends to become permanent when the pressure of population exceeds the amount of available land. Erosion problems soon occur. It was estimated by Negi (1956) that about 100 000 hectares of land previously under cultivation had then been destroyed and turned into ravines in Uttar Pradesh, India, and an equal area under cultivation stood in danger of destruction.

The utilization of browse in semi-arid regions for the establishment of a commercial livestock industry often conflicts with the concept of conservation of natural resources. Many governments withhold land from grazing for the creation of national forests, parks, botanical reserves or recreation areas, and in enlightened lands laws are enacted to protect the lands from exploitation beyond safe limits.

The variable nature of the climatic factor in a semi-arid environment makes the definition of "safe limits" a rather arbitrary one, and few governments have at their disposal the necessary information from long-term research to accurately assess these limits. A good deal of intelligent guessing from the knowledge accumulated by experienced landholders often determines present policy.

Such experience in the developing countries in the tropics is severely limited, and so the problem of maintaining the ecological balance becomes more critical and urgent. In many semi-arid tropical countries, large areas of grazing lands have been so seriously damaged by mismanagement of an already fragile ecosystem that extensive rehabilitation is required.

The importance of browse

"It is a humbling fact for grass pasture experts to realize that probably more animals feed on shrubs and trees, or on associations in which shrubs and trees play an important part, than on true grass-legume pastures". This statement in the Commonwealth Agricultural Bureau Publication No. 10 (1947) sums up the importance of browse, and is particularly relevant in the drier tropical areas of the developing countries which lie beyond the existing limits of agriculture, or in communal grazing land within these boundaries.

French (1949) stated that in semi-arid scrublands, where grass growth is scanty, the nomad economy is based on the browsing of trees and shrubs by camels, goats and other animals.

The use of so much browse is a reflection of the relative aridity of many parts of the earth.

The browse plants of semi-arid regions, because of their adaptation to harsh environments by deep rooting habits, their leathery low-transpiration, xerophytic foliage and, often, their water-harvesting architecture (Slayter, 1965) are usually the only perennials to withstand the climatic conditions. Leguminous browse, because of its nitrogen-accumulating economy, is even more valuable in animal nutrition than other types. Chemical analyses reveal that browse has high nutrient potential, except for high fibre, and, by selective grazing, animals will choose a diet of higher nutritive value than normal sampling and chemical analyses reveal (Weir and Torrell, 1959; Hardison *et al.*, 1954). Consequently, browse is a valuable component of any environment. The absence of "top-feed" in grassland areas such as the *Astrebla* (Mitchell grass) grasslands of central western Queensland, Australia, (Skerman, 1958) can lead to disastrous livestock losses when the grass fails during drought.

Problems of the dry season, drought and insufficient water

The month-to-month, season-to-season, and year-to-year availability of food for the grazing animal is a major problem in all but the wet tropics, and even there the quality of the forage is affected by leaching of nutrients.

A large proportion of the tropical grazing lands have dry seasons lasting from two to eight months each year, ranging from a gradual change from wet to dry with intermittent showers to a sudden cessation of rain followed by rainless months. In many cases, the latter is preferable because the standing dry grass and herbage carrying over into rainless months are safe from deterioration by fungal attack and leaching when the above-ground portions of the plants are dead.

With the advancing dry season, grazing animals turn more and more to browse to satisfy their daily needs of food. The advantage of browse is its ability to maintain its feeding value into the dry season (Everist, 1969) if it is not deciduous in habit. Even in deciduous species, leaf fall is valuable feed for the smaller herbivores.

If the water supplies are adequate and well located, serious damage to browse need not occur. In semi-arid areas, however, sites suitable for water development are often limited, and so increasing numbers of animals converge on fewer and fewer watering points.

In a well-disciplined and prudent rural society, the approach of the normal dry season is the signal for disposal of surplus stock finished animals ready for slaughter, surplus males which will command a satisfactory price for finishing elsewhere, or females culled for poor quality or declining breeding performance. In this way, numbers can be reduced to more nearly match the available feed, and the surplus animals will be sold at their true value if disposed of before the dry season worsens. If the animals are held longer and dry conditions persist, both fleshing of the stock and enthusiasm of the buyers will decline.

It often happens, however, that the traditional social prestige attached to the ownership of large herds of livestock or the undue optimism that the drought will shortly break causes the retaining of the animals. Then, the concentration of stock around the diminishing water supplies is too great for the natural vegetation to support, and rapid deterioration of the grass, and later the browse, sets in near these waters. Also, as animals weaken in a worsening drought they become reluctant or unable to move far from water and the grazing pressure is further increased, with consequent further deterioration in grazing resources.

The size of the denuded grazing area will depend on the class of animals involved (Skerman, 1966). Camels and desert sheep graze farther from water and drink less frequently than cattle and goats. Within each species, the very young and the old, the pregnant and nursing females remain relatively close to water, while the stronger animals venture farther afield. In undeveloped countries the peasant village creates a grazing situation similar to that of the water supply, especially where animals return to bomas or kraals each night or are brought home for milking.

In both cases, grass and browse suffer severe grazing pressure. Grass usually disappears and is replaced by unpalatable and aggressive weeds, and browse is kept constantly defoliated — tree species are often kept at shrub height by such grazing.

Yield of browse

Few quantitative measurements of browse yield have been undertaken and accurate figures for production are difficult to find. Without such data the proper utilization of browse is, at best, arbitrary; considerably more attention should be paid to browse by pasture ecologists and animal nutritionists.

One of the difficulties is the lack of adequate techniques for browse measurement. This problem is being investigated widely in the United States, mainly in nonleguminous species.

Schuster (1965) found that total twig length was closely correlated with yield of deer browse plants, but the parameter for predicting yields was a combination of twig numbers and length.

Burrows and Beale (1970) have endeavoured to ascertain the yield of browse from mulga (*Acacia aneura*) in Queensland. Trees normally used for lopping for drought feeding have stem diameters in excess of 10 cm when measured 30 cm above ground level. A regression equation relating leaf yield to stem circumference has been developed which gives a satisfactory coefficient of correlation of $r = 0.985$. Testing of this equation is continuing. By classifying trees by diameter and determining the numbers in each class, these workers hope to predict the current and future reserves of mulga for drought feeding.

Feeding value of browse

Chemical analyses of several leguminous browse species are presented in Appendix 1. Generally, they contain a high percentage of crude protein in leaves, twigs and pods, are well supplied with vitamin A (Anson and Gartner, 1967), but are higher in fibre than other normal food material (Wilson, 1969). They constitute an extremely valuable source of feed for livestock, especially during drought periods (Everist, 1969) when grass is low in quantity and of poor quality. Grass declines in feeding value following maturity, whereas browse can maintain its nutrients for long periods. Browse also comes into leaf early and before the first rains (French, 1949).

Bonsma (1942) has shown that in southern Africa pastoralists converge

493

with their cattle on haakdoring veld containing haakdoring (*Acacia litakunensis*) for the winter or dry season. The pods contain 18.83 percent crude protein, 2.44 percent fat, 46.28 percent carbohydrates, 5.1 percent ash and 20.1 percent crude fibre. The pods are loosely attached to the branches and are easily detached by browsing animals. Bonsma estimated that the pods supply the equivalent of 0.7 kg of protein supplement per animal per day, which is adequate to maintain the cattle.

In many browse species it is only the foliage which is eaten. However, with leguminous species, the pod is eagerly sought and herders muster the animals to fallen pods or shake them from the trees for the waiting animals. Chemical analyses of the pods recorded in Appendix 1 show their high feeding value.

Utilization and management of browse

Browse can be used for day-to-day mixed grazing, for seasonal reserves or for drought feeding (Gray, 1970).

Daily feeding. Most browse is used on a day-to-day basis in the wet or dry tropics because lack of fencing or herding permits the animals to have free range. Under natural conditions a good deal of browse will be out of reach of the normal grazing animals and so utilization can be manipulated by mechanical treatment to make it available as needed as is done with mulga (*Acacia aneura*) for Merino sheep feeding in Australia (Everist, 1969).

Natural leaf through senescence is also an important day-to-day component of the diet of small animals (sheep make good use of such leaf fall in mulga country). In Africa, goats thrive on the leaf fall of *Acacia mellifera* (Dougall and Bogdan, 1958).

Browse plants are often cut by hand for feeding green to dairy cows in small village communities. "Living fences" of leguminous trees such as *Erythrina beteroana, Gliricidia maculata* and *Milettia thonningii* are used in this way (Paterson, 1949). Such leguminous toppings are also added to rice fields at "mudding up" to improve fertility.

In Colombia, the green foliage of *Cajanus cajan* is lopped for green feed for poultry or dried and hammer milled or ground into a meal for livestock feeding (Lotero, personal communication).

Seasonal reserves. Where browse is to be used for seasonal reserves, some fencing is required unless the material is out of reach of the animals. Fenced hedges or paddocks of *Leucaena leucocephala* have been utilized in Queensland, *Leucaena* and *Desmanthus virgatus* in Hawaii, and *Cajanus cajan* planted in contoured double rows in pangola grass in Hawaii and Brazil (von Schaaffhausen, 1966).

Drought feeding. In times of drought many of the gramineous and herbaceous species either die or are so heavily grazed that there is little dry matter available. It is then that ranchers turn to browse for animal nutrition.

In times of stress, animals will browse relatively unpalatable plants which they would not normally touch, and as stress becomes greater, selectivity decreases and even the bark of trees will be stripped to satisfy the animals' roughage requirements.

A completely new concept of browse has arisen with the introduction of urea supplementation of roughage in animal nutrition. With a change in the microbial flora of the rumen, animals have quite readily turned to previously untouched woody plants. This new tool in browse use and bush control is worthy of serious investigation and implementation.

Weller (1969) has shown that the introduction of molasses-urea drum lick feeders during a drought into the scrub and forest lands of Queensland has helped cattle to eliminate such previously uneaten species as sandalwood or buddah (*Eremophila mitchelli*) and young cypress pine (*Callistris glauca*).

In Australia, mulga (*Acacia aneura*) is used widely for drought-feeding sheep and cattle. Everist (1949, 1958, 1969) has recorded its use. This species is often pushed over with a bulldozer for feeding sheep.

Sheep eat the accessible lower leaves and cattle browse the higher foliage. Cattle can break down the lower branches of mulga trees, more often with their necks than with their horns. Total pushing over of mulga scrub on a face usually kills most of the trees, and in thick mulga scrubs new seedlings usually replace those pushed down when continuous wet weather for a few days germinates the fallen seeds. However, in thinner stands care must be taken to preserve the trees. Everist (1969) recommends lopping the upper branches with a chain saw or axe or breaking them selectively with a mechanical pusher arm to preserve the lower lateral branches for regeneration of the tree. It is also essential to leave some mature trees untouched to act as seed bearers.

Each browse species requires its own treatment to ensure preservation, and studies of browse behaviour under various systems and intensities of defoliation are required to develop satisfactory techniques.

Bark, ground pods and seeds have been used together with foliage in feeding. Dougall and Bogdan (1958) reported that in *Acacia seyal* the thick, smooth bark is the most valuable fodder. In the dry months of February and

March in the Baringo district in Kenya, thick branches are cut and thrown to the ground where the animals browse the bark and eat the leaves (at this time, rather few). Cattle eat 5 to 6 kg/head/day on a bark diet, which is sufficient for maintenance and production of 4.5 litres of milk. The material has 10.6 percent crude protein and over 4 percent calcium in February.

Marion *et al.* (1957) found that chipped and ground wood from green stems of mesquite (*Prosopis juliflora*), which infests millions of hectares of grassland in the southwestern United States, makes a very satisfactory ingredient in winter feeding of beef cattle. Mesquite wood meal contains 6 percent crude protein, 35 percent nitrogen-free extract, 45 percent fibre and 30 ppm carotene. Steers fed rations containing 50 percent mesquite meal gained up to 1.14 kg/head daily.

Use of fire

Fire has long been used to manipulate vegetation for grazing. Used intelligently, with full knowledge of the reaction of the species to the passage of fire at different seasons of the year and at different stages in the growth cycle, fire can be a useful tool.

Fire is usually used as a temporary measure to promote an ephemeral growth of new grass for a green bite for the animals at a time when the grass is dry enough to burn. The long-term effects of fire are rarely considered and are largely unknown.

As many of the browse legumes in semi-arid regions are *Acacia* species, and germination of their seed is usually stimulated by fire, the passage of fire often gives rise to extensive and rapid seedling growth (e.g. *Acacia flavescens*) in Queensland. Fire also stimulates lignotubers to produce new growth, as in the case of brigalow (*Acacia harpophylla*).

Fire has also been very effective as an aid in the control of bush regrowth. Lamprey (personal communication) found that if fire could be excluded from grassland for two years on the Serengeti Game Reserve in northern Tanzania, the browse species *Acacia drepanolobium* could regenerate in that locality. By contrast, Ivens (1970) found that fire was insufficient in controlling *A. drepanolobium* in the Kenya rangelands.

Fire has been used at the Santa Rita Experimental Range in Arizona to control *Aplopappus tenuisectus* in mesquite (*Prosopis juliflora* var. *velutina*). Burning in late spring and summer was considered to be a cheap and effective treatment.

496

Problem of thicket formation

Many of the leguminous shrubs propagate rapidly and ultimately form thickets. Some species sucker extensively (e.g. *Dichrostachys cinerea*). Other species seed heavily and are unpalatable in normal times, but are eaten in serious droughts. The spiny nature of some species such as *Acacia polyacantha* deters stock. The seeds of many species pass undamaged, and are even stimulated in viability through the animal gut so that they germinate thickly in dung patches in cattle camps near water. Each of these situations can lead to thicket formation.

In general, it is unwise to encourage thicket-forming species, as there is usually a more suitable forage available for each ecological niche.

Problem of toxicity

Many browse plants are toxic to animals at certain times of the year or to animals under conditions of stress.

Hungry animals entering a new area will readily browse whatever species they find in their path and will usually eat more than usual because of their starved condition. A species which is not toxic to animals in free-range grazing may become toxic to hungry stock travelling along well-frequented stock routes.

Many of the major cattle losses from plant poisoning in Australia occur in this way. Experienced stockhandlers become acquainted with the distribution of such plants and push travelling stock through the area quickly, giving them little time to browse.

Fretes, Samudio and Gay (1970) have tested some toxic leguminous browse plants in Paraguay, some of which (e.g. *Parkinsonia aculeata* and *Acacia* spp.) contain prussic acid. *Acacia georginae* (Georgina gidgee) in the basin of the Georgina River in central western Queensland has been the cause of heavy stock losses (Everist, 1969). Death is sudden; Oelrichs and McEwan (1962) found that the toxic principle was fluoracetic acid. Although both pods and suckers were proved toxic, the plant is considered to be a valuable fodder in the area and eradication would be both undesirable and impractical. McEwan (1964) found the fluoracetate ion to be the cause of toxicity in the leguminous heart-leaf poison bush (*Gastrolobium grandiflorum*). *Leucaena leucocephala* contains mimosine, which is toxic to monogastric animals such as the horse and pig, but is fed widely to ruminants.

497

Donaldson *et al.* (1970) found it caused enlarged thyroid glands in calves at birth, but generally could be used as a supplement for cattle.

The matter of seasonal or other toxicity in browse plants is accepted by pastoralists as a calculated risk associated with animal production, and management is designed to minimize possible losses. Closure of areas heavily infested with toxic plants is desirable. Where they are not plentiful, as much free range as possible should be allowed, as animals can select more palatable species. Paddocks should be leniently grazed to prevent animals eating toxic material, and in droughts supplementary feeding will help reduce intake of doubtful browse species. A careful watch should be kept on the animals to identify early symptoms of toxicity, even though death is often sudden.

Establishment of browse tree groves

Little effort has been made to establish browse groves as a fodder reserve for the dry season or drought feeding. Rose-Innes (1966) felt that the correct approach to combating extreme climatic conditions lay not in grass-legume pastures but in establishing palatable, nutritious evergreen browse in existing tree savanna. West (1950) and Everist (1969) suggest that browse be treated as a crop and planted on prepared seed beds or established from seedlings or small trees after preliminary growth in a plant nursery.

As seed germination can be difficult because of the hard seed coat, scarification or hot-water treatment may be necessary. Mulga (*Acacia aneura*) seeds, for example, should be boiled for three minutes before planting; leucaena seeds heated to 80°C for three to four minutes.

Selection of species in relation to soil type is important, as many browse plants will not establish readily in heavy, cracking clay, probably because of root rupture when the soils dry out and crack.

Jurriaanse (1950) established groves of the honey locust (*Gleditsia triacanthos*) under irrigation with three aims in view — to control erosion on a susceptible area, to provide shade and to provide fodder reserves. The trees were planted on the contour.

Browse performance

Rose-Innes and Mabey (1966) studied the use of the native leguminous browse *Griffonia simplicifolia*, which grows naturally on the Accra plains in Ghana. On natural rangeland, West African Shorthorn cattle ingested 60 percent grass and 40 percent *Griffonia* in a 12-hour day under free choice and with ample *Griffonia* at a population of one tree to 2 hectares. The daily intake under natural browsing was 60 percent greater than when the animals were stall fed.

The animals took the *Griffonia* and grass equally at the rate of 0.77 kg/hour, and the total dry legume intake was 3.7 and 3.64 kg per beast or 3.31 percent of the body weight. These workers found that animals on a restricted 12-hour grazing regime spent more beast-hours feeding and fewer ruminating than those on a 24-hour grazing regime, and that the shorter grazing time gave greater live-weight gain and total intake.

In Queensland, the mulga tree is utilized for drought-feeding sheep (Everist, Harvey and Bell, 1958). Approximately 1.4 kg of leaf per day will maintain a sheep. The mulga is pulled or lopped to feed several hundred thousand sheep for several months in drought years. Anson and Gartner (1967) found the carotene content of mulga leaf (47.3 mg/kg of dry matter) to be adequate to meet the vitamin A requirements where mulga is the sole source of feed and the sheep are eating enough for survival.

In the arid Bikaner district of Rajasthan in northern India, where the vegetational cover is only 2.1 percent around the villages, *Prosopis spicigera* yields 512 kg of fodder per hectare (Shankarnarayan, Pandey and Dhuvanarayan, 1965).

Legislative control of browse

Only a few land tenure regulations throughout the world attempt to conserve the natural vegetation indirectly by imposing limitations on livestock numbers. This may be due to a lack of accurate data on carrying capacities and variable stocking according to seasonal conditions. Only slight regulation is imposed by land rentals and taxation.

However, in some countries in which there has been serious devastation of vegetation, some legislative control is exerted. In Nigeria, there are laws to control lopping. Charreau and Vidal (1965) reported that *Acacia albida* trees are protected by law in Senegal.

Several grazing regulations in force in India are discussed by Whyte (1964). Many are aimed at conservation for forest purposes. One case involving a fodder species concerns babul (*Acacia nilotica* syn. *arabica*). In Maharashtra State, closure of *A. nilotica* land for five years in a working cycle is imposed to allow for regeneration.

499

The browse species

It is difficult to decide what actually constitutes a browse species. French (1949) stated that at least 75 percent of the trees and shrubs in Africa are browsed to some extent by domestic and game animals. Some plants are browsed regularly, others at some particular growth phase or under certain seasonal conditions. Others are eaten when urea-molasses mixtures are available to the ruminant.

In the following catalogue a number of leguminous browse species are listed alphabetically according to their botanical names. Both humid and semi-arid species are included.

I

Le Tourneau tree-crusher
clearing tropical rain forest
at Pucallpa, Peru

II

Results of
heavy application
of nitrogen
to Kikuyu grass
in white clover field

III

Soybean leaves
show (left)
potash deficiency
and (right)
adequate potash

IV

*Aeschynomene
falcata* sward in
native grasslands,
southeast Queensland

V

Effective
nodulation of
Centrosema sp.,
Pucallpa, Peru

VI

*Calopogonium
mucunoides,* Palmira
Experimental Station,
Cauca Valley,
Colombia

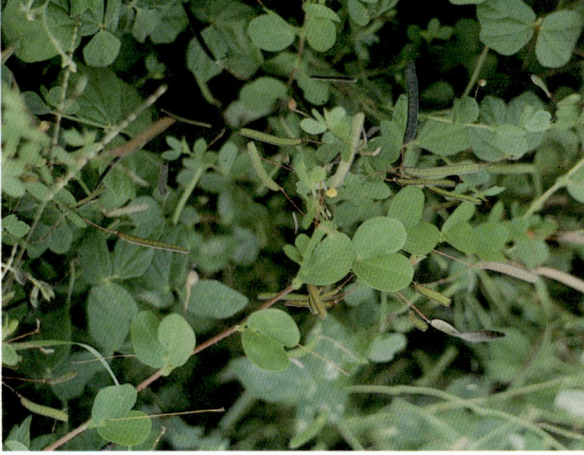

VII

Cassia rotundifolia
showing leaves,
flowers, and seed
pods in varying
stages of maturity

VIII

Centrosema pubescens Benth.

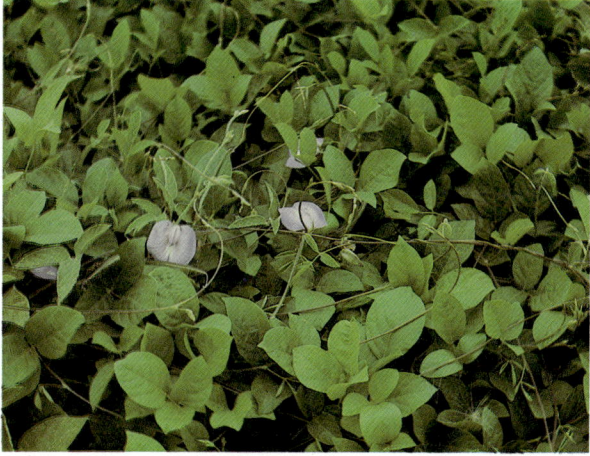

*Centrosema
virginianum*
foliage
and flowers

X

Clitoria ternatea
foliage and flowers

XI

Desmodium intortum
cv. Greenleaf
foliage, flowers
and seed pods
at Lansdown
Research Station,
Queensland, Australia

XII

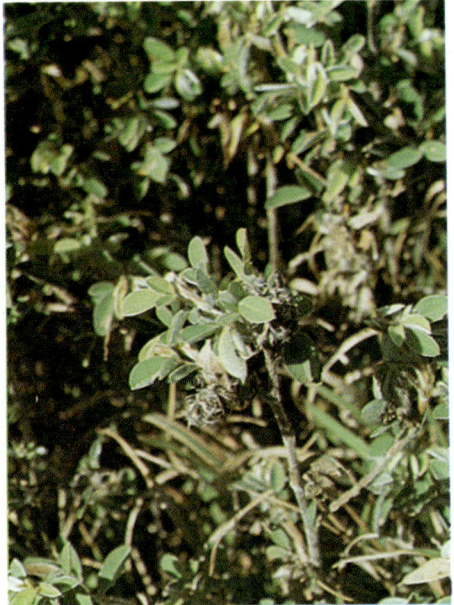

Desmodium barbatum,
Nova Odessa, Brazil

XIII

Desmodium barbatum
growing in
a greenhouse
late in the season

XIV

*Desmodium
"ovalifolium"*
growing at
David, Panama

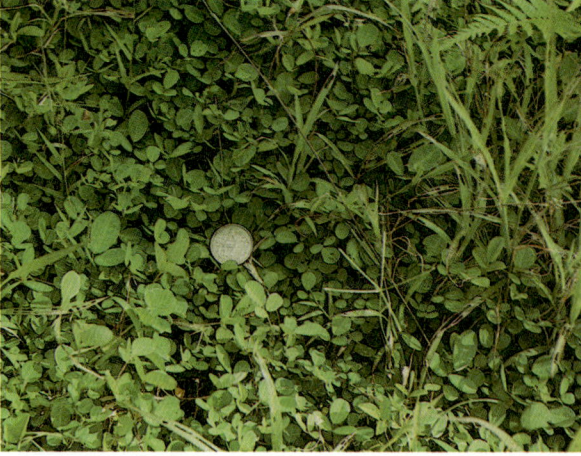

XV

Desmodium heterophyllum sward at Auki, Malaita

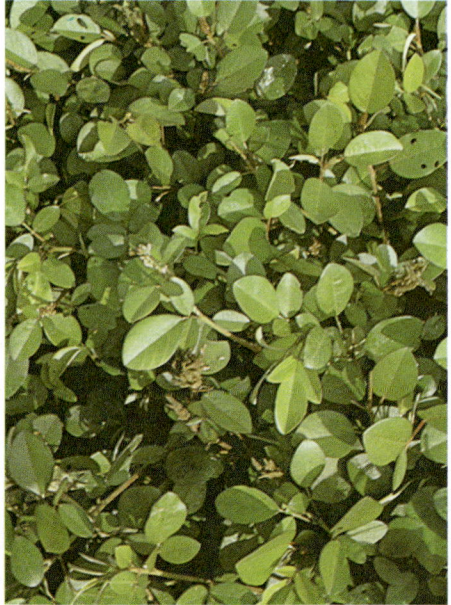

XVI

Desmodium "ovalifolium" growing at Palmira Experimental Station, Cauca Valley, Colombia

XVII

*Desmodium
uncinatum*
foliage and pods

XVIII

Galactia striata
foliage
in a nursery row

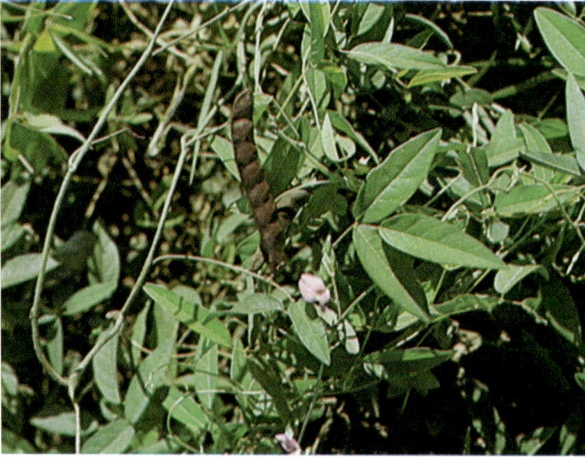

XIX

Galactia striata
foliage, flowers
and pods

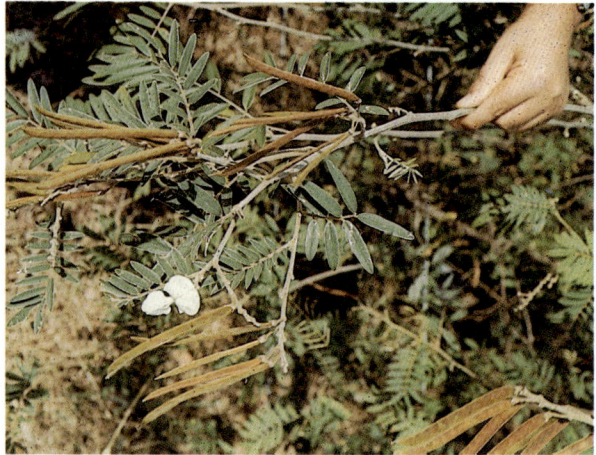

XX

Indigofera hirsuta
foliage, flowers
and seed pods

XXI

Lablab purpureus
cv. Rongai
foliage, flowers
and seed pods

XXII

A heavily flowering
and seeding
sward of
Lotononis bainesii

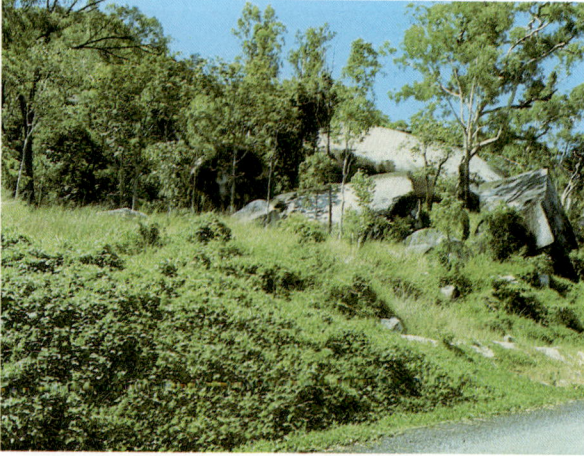

XXIII

Macroptilium atropurpureum cv. Siratro growing strongly beside a road in southeast Queensland

XXIV

Macroptilium atropurpureum foliage, flowers and seed pods

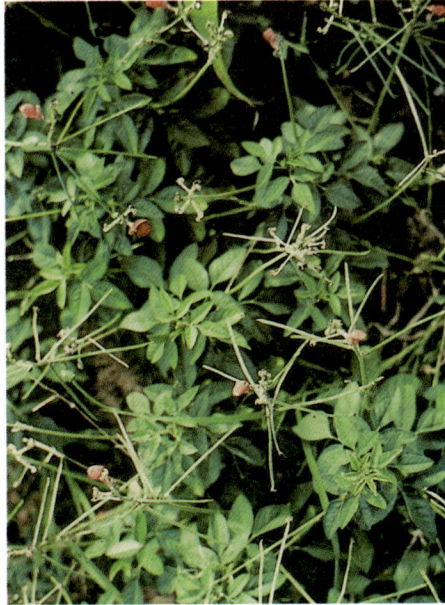

Macroptilium lathyroides
foliage, flowers
and seed pods

Macrotyloma axillare
foliage
in a nursery row
at David, Panama

XXVII

Macrotyloma axillare
foliage
and seed pods

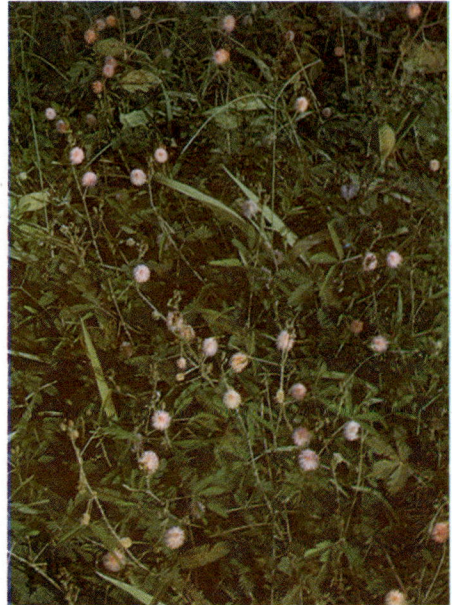

XXVIII

Mimosa pudica
foliage and flowers

XXIX

Neonotonia wightii
cv. Tinaroo
foliage and flowers

XXX

Strong root and stem
development of *Neonotonia wightii*

XXXI

Strong root development by glycine
at Fazenda da Prata, Brazil

XXXII

Pueraria phaseoloides
as a cover crop
in a sisal plantation,
Mlingano, Tanzania

XXXIII

Pueraria phaseoloides,
called puero or tropical kudzu,
showing foliage and flowers

XXXIV

Stylosanthes capitata
at CIAT, Quilichao,
Colombia

XXXV

Stylosanthes capitata
showing flowers and seed heads

XXXVI

Stylosanthes fruticosa
growing in
an enclosure
in native pasture
at Nahud, Kordofan,
the Sudan

XXXVII

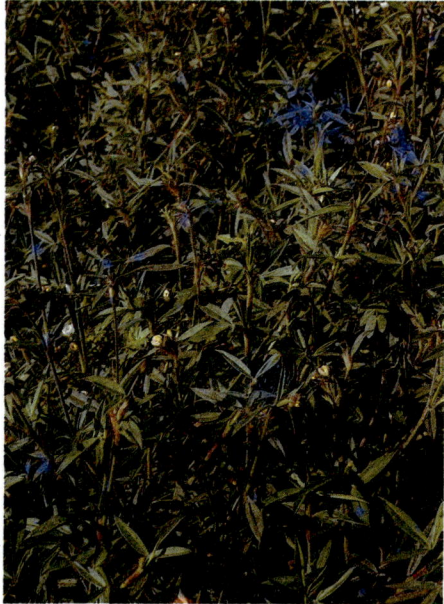

Stylosanthes guianensis
at Palmira Experimental Station,
Cauca Valley, Colombia

XXXVIII

Schofield stylo
growing with
Setaria sphacelata
cv. Kazungula,
Mount Cotton,
Queensland

Stylosanthes hamata cv. Verano
flowerhead

The diploid form of
Stylosanthes hamata
growing on a beach
in Antigua

XLI

Stylosanthes humilis
growing at Obello,
near Kupans, Timor

XLII

A sward of Townsville stylo
and *Heteropogon contortus*
at Rodd's Bay, central Queensland

XLIII

Stylosanthes humilis
showing termite
damage to a plant
at Khon Kaen,
Thailand

XLIV

Stylosanthes scabra
cv. Seca,
a single plant
midway through
the growing season

XLV

Ungrazed,
flowering plants of
Stylosanthes scabra
cv. Seca
at Katherine,
Northern Territory

XLVI

Ungrazed plants of
Stylosanthes scabra
cv. Seca
in north Queensland
early in the dry season

XLVII

Teramnus uncinatus
showing foliage and pods

XLVIII

Trifolium semipilosum
cv. Safari
under irrigation
at Samford Research
Station,
southeast Queensland

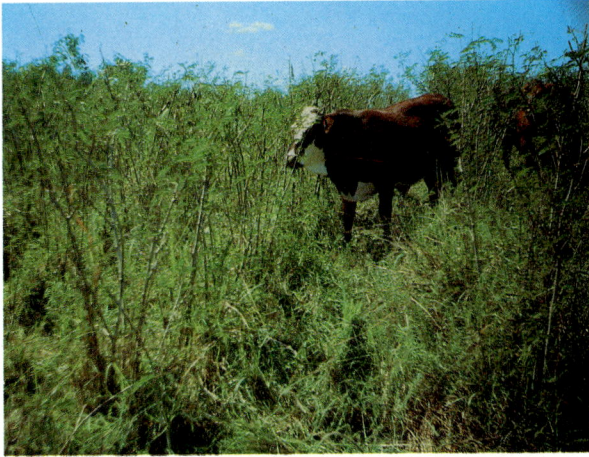

XLIX

Vigna parkeri sward
at Beerwah
Research Station,
southeast Queensland

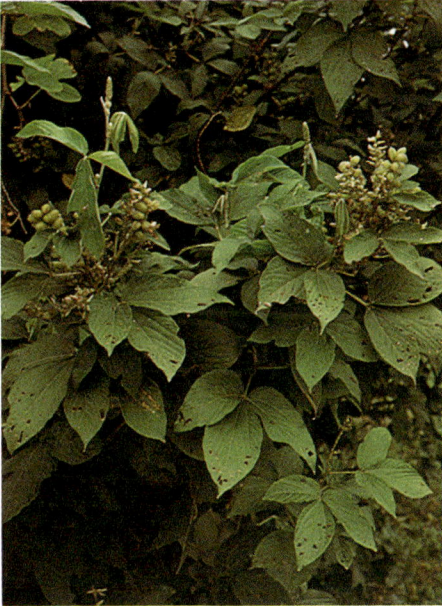

L

Flemingia macrophylla
showing foliage and flowers

LI

Pickup seed
harvester,
cutter-bar type

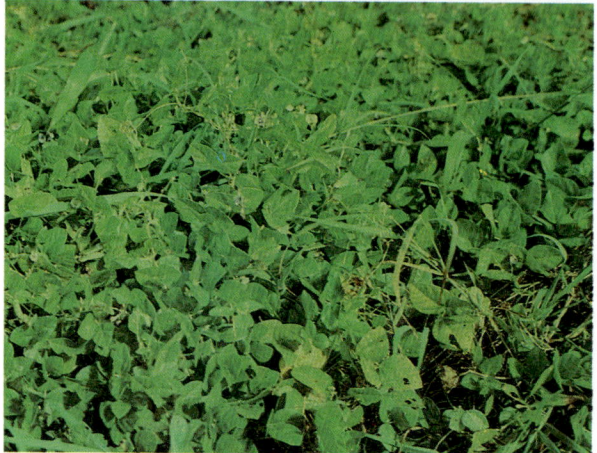

LII

*Leucaena
leucocephala*
interplanted with
Setaria sphacelata
being grazed
by beef cattle
at Samford
Research Station,
southeast Queensland

LIII

Leucaena leucocephala
foliage and flowers

LIV

Mesquite
(*Prosopis glandulosa*)
bushes in
population
and control studies
at Santa Rita
Experimental Range,
Arizona

the
browse species

CATALOGUE

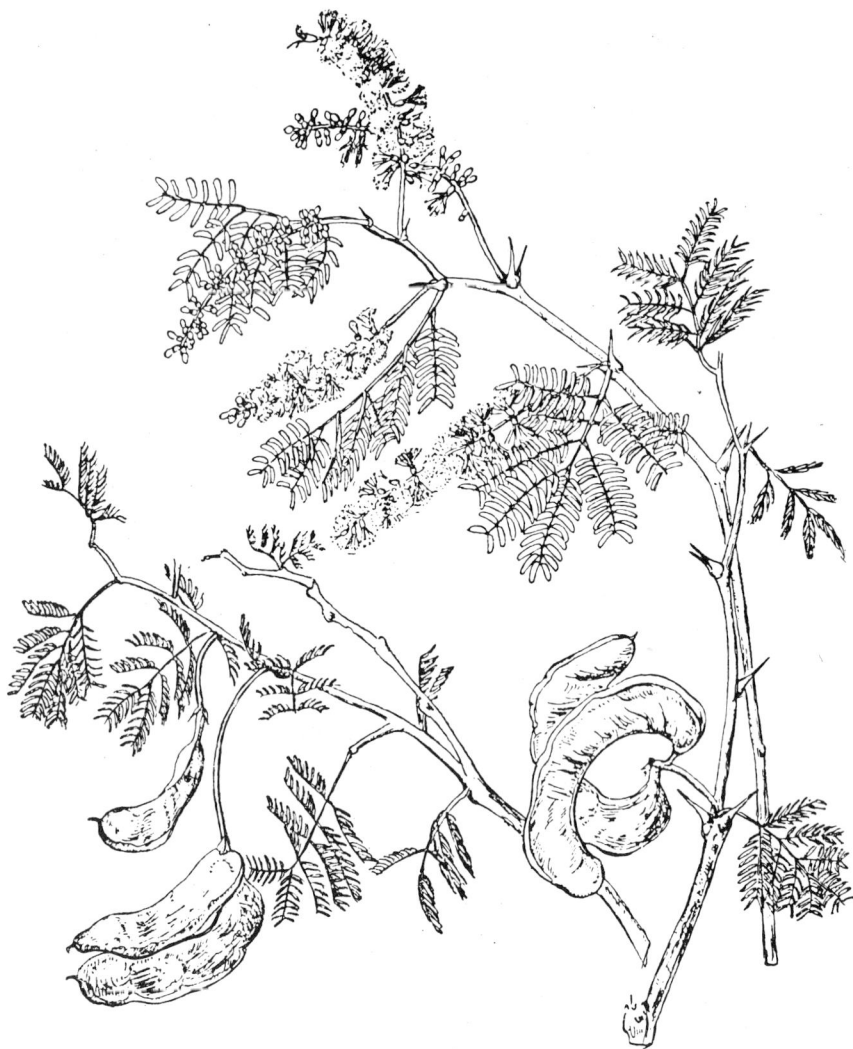

Figure 99. *Acacia albida*

Acacia spp.

Acacia albida Del.

Synonyms. *A. leucophloea; Faidherbia albida* (Del.) A. Chev.

Common names. Kad (Senegal), haraz (the Sudan), apple-ring acacia, winter thorn.

Description. One of the largest of the *Acacia* trees in Africa. Bark dull grey, fissured and scaly; slash pale brown, fibrous; pinnae in three to ten pairs; leaflets grey-green, 6-20. Flower spikes 7.5 to 10 cm long; pod orange-yellow, twisting into strange shapes as it ripens, frequently forming hoops and spirals (Andrews, 1952). As the elevation increases its size becomes smaller (Tothill, 1954).

Distribution. Throughout tropical Africa, from Egypt, Senegal and the Gambia south to the Transvaal and Natal; Syria; Palestine; India.

Characteristics. Commonly distributed on flood plains and banks of large rivers on alluvial soil, where it may occur in pure stands or as a constituent of riparian woodland (Wilson and Bredon, 1963). It produces a large number of twisted pods which are relished by camels, cattle and game animals. The pods are indehiscent, and Lamprey (1967) believes that passage through the animal is necessary to stimulate germination. Boudet (1970) suggested that it be planted as a windbreak in West Africa and lopped in the dry season for livestock feed.

Charreau and Vidal (1965) state that the fertility-building ability of *A. albida* has long been known in Senegal and that populations of the trees are protected and may reach a density of 40 to 50 trees/ha. From mineralization of leaf drop and nodulation there is a marked increase in fertility from the drip-ring of the tree to the trunk, particularly in nitrogen, phosphorus and exchangeable calcium. Millet (*Pennisetum typhoides*) yields are multiplied 2.5 times and protein content by a factor of 3 or 4 near the trees. Annual

503

accession of fertility per 150 m² has been calculated as 183 kg CaO, 39 kg MgO, 19 kg K_2O, 75 kg N, 27 kg P_2O_5 and 20 kg S.

It remains leafless during the rains and assumes new foliage and flowers after the commencement of the dry season (Dalzeil, 1955).

Analyses of the leaves and pods by Wilson and Bredon (1963) and Boudet (1970) are given in Table 15.1.

TABLE 15.1 **Compositional analysis of *Acacia albida* leaves and pods on a dry-matter basis**

Plant part	Crude protein	Crude fat	Crude fibre	Ash	N-free extract	Food units/ kg/DM	D.C.[a] Pro-tein (g/kg/DM)	Reference
			(%)					
Leaves	17.8	—	17.5	6.4	—	0.89	137	Boudet, 1970
Pods	11.7	—	21.4	4.1	—	0.88	76	Boudet, 1970
Pods (Dec.)	10.53	0.94	27.55	3.92	57.06	—	—	Wilson and Bredon, 1963
Pods (Aug.)	12.36	0.68	40.73	3.56	42.57	—	—	Wilson and Bredon, 1963
Green pods	10.19	0.49	30.36	4.02	54.94	—	—	Wilson and Bredon, 1963

[a] D.C. = Digestible crude.

Acacia aneura F. Muell. ex Benth.

Common name. Mulga (Australia).
Description. There are four forms:
● Low mulga — shrub with short, spurlike branches, kept trimmed on top by continual grazing by sheep and cattle. It is useful for grazing but a poor source of drought feed because of its low yield per hectare.
● Whipstick mulga — immature stands of thin trees, close together, with leaves only in the upper part. These are best utilized by pushing with a bulldozer or pulling with cable or chain.
● Umbrella mulga — mature trees, branched and very leafy. Vary consider-

Figure 100. Mulga (*Acacia aneura*) being bulldozed for drought feeding

ably in palatability; respond well to lopping and are most productive of leaf. About 175 to 200 trees per hectare give the best combination of mulga for drought feeding and grass for normal seasons.

● Tall mulga — old trees with bare trunks and leaves only at the top. They are usually most palatable, but are not amenable to lopping. Utilization without destruction is difficult or impossible. They usually produce large amounts of seed (Everist, 1969).

Distribution. Occurs throughout semi-arid Australia in the 300- to 450-mm rainfall zone with a summer dominance but with some useful winter rain, mostly on lateritic red earth soils acid in reaction and very low in available phosphorus, but also on sandy soils overlying such material. It does not grow on soils containing lime and is not tolerant of wet soils. The drainage lines in the mulga scrubs are usually occupied by *Eucalyptus* spp., notably *E. populnea* and *Acacia cambagei* (gidgee). Perry (1970) describes a grove-intergrove vegetational pattern in mulga communities.

It is tolerant of high manganese. *Macroptilium lathyroides* grown in lateritic red earth from the mulga scrubs at Charleville, western Queensland,

505

Australia, exhibited manganese toxicity symptoms (Cowie and Skerman, 1970). The hard-seedness of mulga can be broken by boiling the seed in water for three minutes. Seedlings usually germinate prolifically after about 7.5 to 10.0 cm rain spread over a few days in summer.

The seedlings are very attractive to sheep and to rabbits and should be protected in the regeneration period for five years or more.

For drought feeding, the mulga is pushed over with a bulldozer or lopped. Sheep will utilize 1.4 kg/head/day of leaves and so the area lopped is adjusted to this consumption.

Day-to-day checking of consumption will soon adjust lopping to the correct requirement. Supplements should be offered to the animals on mulga after three months of feeding since it takes flocks of Merino sheep about three months to adjust to and accept supplementary feed. The supplement recommended is bone flour at 28 g per head per day, salt (NaCl) at 10 to 15 percent of the supplement, and molasses at 168 g per head per day.

Umbrella mulga should be lopped to preserve the lateral branches (see Figure 101). Retain 250 trees per hectare as seed trees. Fire should be avoided (Everist, 1969).

Main reference. Everist (1969).

Acacia argyrodendron **Domin**

Common name. Black gidgee.
Characteristics. Small tree, dark grey, dry, furrowed and flaky bark; leaves narrow, greyish green; flowers yellow, in balls; pods narrow, produced in great profusion. This plant forms dense scrub on flood plains of the Belyando River and Torrens Creek, Queensland, where it has a good reputation as a sheep and cattle fodder. The leaves are eaten as windfalls and the pods are also readily eaten (Everist, 1969).

Acacia aulacocarpa **A. Cunn. ex Benth.**

Common name. Hickory wattle.
Characteristics. A small tree with grey bark broken up into small blocks, greenish-grey curved leaves with few prominent veins, slender spikes of yellow flowers and short, broad, flat, woody pods; fairly common in open forest country in coastal districts. Eaten in very dry seasons (Everist, 1969).

506

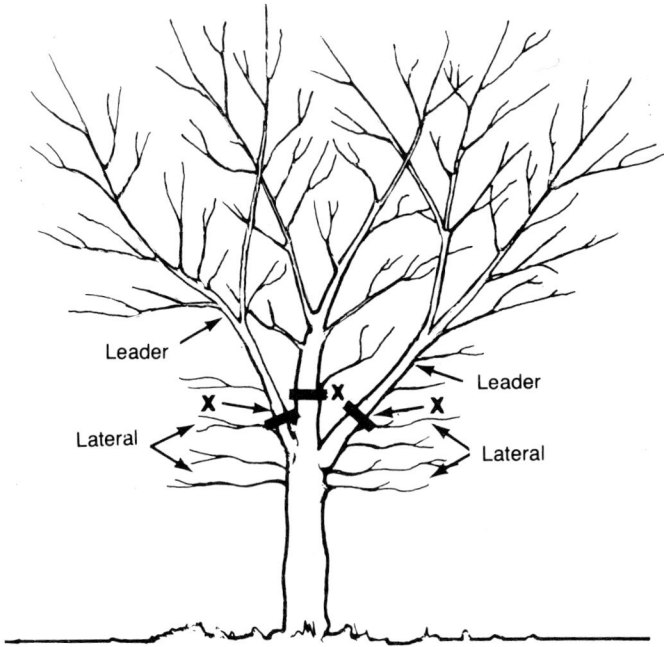

Figure 101. Drawing of an umbrella mulga tree (*Acacia aneura*), showing its characteristic branching system. When lopping for drought feeding, cut at point X, taking care to leave the lower lateral branches intact

Acacia bidwillii **Benth.**

Common name. Cordwood wattle.

Characteristics. A small tree with dark brown, somewhat corky, furrowed bark and green fernlike leaves; flowers in balls, pods fairly large and flat. It is somewhat similar in appearance to *A. sutherlandii*, but grows mostly in areas of higher rainfall south and east of that species, extending into the Burnett district and on to the coast about Bowen in Queensland. It favours clay and clay loam soils, particularly pebbly types (Everist, 1969).

507

Acacia sp.

Common name. Bowyakka or booyakka.
Characteristics. A shapely small tree with leafy, rounded head; leaves stiff, grey or slightly greyish green, narrow, straight or almost so; flowers in balls; pods narrow. In Queensland, it is most common in an area from south of Mitchell to about Talwood, forming scrubs of limited extent chiefly on pebbly brown clay loam soils. Small patches are known as far west as Cheepie. Cattle and sheep eat it readily (Everist, 1969).

Acacia brachystachya **Benth.**

Common name. Turpentine mulga.
Characteristics. A large shrub, usually as wide as it is high, the branches arching upward from the base; leaves very narrow, almost or quite straight; pods narrower and thicker than mulga. It occurs in the more arid parts of the mulga country, mostly west of the Paroo River in Queensland, and is generally eaten less readily than most forms of genuine mulga (Everist 1969).

Acacia brevispica **Harms.**

Description. A shrub 2 to 5 m high with long, thin, rambling branches covered with small scattered thorns. The leaves are fairly large and the white flowers are in globose heads. The pods are thin, flat and broad.
Distribution. It is widespread in Africa, being found in the Sudan, Ethiopia, Somalia, Kenya, Zaire, Angola, Natal and Cape Province. In Kenya, it occurs at 1 600 to 2 000 m in elevation, where it often forms dense thickets on well-drained slopes in deep and fertile red soil.
Characteristics. Goats browse on this bush to a considerable extent, eating the leaves, flowers and young pods. It is seldom eaten by cattle because of its thorny branches. The leaves and young pods have a protein content of 17 to 19 percent (Dougall and Bogdan, 1958).

Because of its thicket formation it is usually necessary to control this species. Ivens (1970) has found that digging out the trees is the only full control measure. The tree trunks when cut down sucker profusely. Repeated burning or repeated spraying of the foliage with 2,4-D mixtures exerts some effect. Spraying with this species should be done four to six months after cutting.

Acacia burrowii **Maid.**

Common name. Currawong.

Characteristics. Tree somewhat similar in appearance to *A. doratoxylon*, but usually not growing in dense scrubs. The tops are usually rounded and the leaves more curved. Occurs in somewhat stony soils in the 625 to 750-mm rainfall belt south of the tropics, in Queensland in the area from Tara to St. George. The leaves are eaten readily and in some districts the trees are cut for fodder in times of drought (Everist, 1969).

Acacia cambagei **R.T. Baker**

Common names. Gidgee, gidgea, gidyea, or gidya.

Characteristics. Small tree, up to about 8 mm; bark dark grey, dry, flaky; leaves narrow, curved, grey-green; flowers in balls, with an unpleasant onion-like odour. It forms dense scrubs in the 250- to 625-mm rainfall belt, mainly on clay loams and clays. It is most abundant in central-west Australia. Although the leaves are eaten as windfalls, stock will rarely eat them when trees are cut down or lopped. If the trees are burnt down, leaves are readily eaten by sheep. Gidgee is not normally regarded as an important drought reserve because of this unpalatability. If the scrub is cleared, the land produces a good growth of palatable grasses and herbage plants and it is usual to sacrifice the trees to get pasture (Everist, 1969).

Acacia cana **Maid.**

Common name. Boree.

Characteristics. A tree with a long trunk, rather sparse branches and a rounded top. The bark is furrowed and flaky, dry and grey. The leaves are long and narrow, silvery grey with very fine parallel veins. The flowers are yellow and wattle-like. It is common in central and northeastern Queensland and extends southward along the Grey Range, mainly on pebbly clay soils on the edges of the Mitchell grass country. It is eaten fairly readily by sheep, but is reported to cause impaction after about six weeks of continuous feeding (Everist, 1969).

509

Acacia catenulata C.T. White

Common name. Bendee.

Characteristics. Somewhat similar to mulga in general appearance, but with less regular branching and a darker bark. Pods are constricted between the seeds and are narrower than those of mulga. Grows mainly on stony outcrops or the rims of lateritic tablelands in the 375- to 625-mm rainfall belt south of about Charters Towers in Queensland. It is normally regarded as of only secondary value, since it is not particularly palatable (Everist, 1969).

Acacia coriacea DC

Common name. Desert oak.

Characteristics. A small tree with dark grey, dry, deeply furrowed bark, pale yellow inside; leaves very narrow, long and pendulous; pods grey, fairly large with constrictions between the seeds, somewhat resembling a necklace.

It is called desert oak because it resembles certain species of *Hakea* and *Grevillea*.

This species proliferates mainly in deep sandy soils from the Central Highlands, through northwest Queensland and into the Northern Territory, Australia.

Although *Acacia coriacea* is eaten freely by sheep and is fairly palatable, its leaf yield is small (Everist, 1969).

Acacia cunninghamii Hook.

Common name. Cunningham's wattle.

Characteristics. Known in the Brisbane district as black wattle, a name given to other species of *Acacia* in other districts. A small tree with furrowed bark; leaves green, sickle-shaped with prominent veins, the lowermost two of which are joined near the base; flowers yellow, in narrow spikes; pods narrow and thin, rather curly and borne in loose bunches. The species includes many different forms which are widespread in Queensland from the coast to more than 325 km inland, mostly on stony or gravelly soils. It is eaten in times of scarcity but is not of major importance as a drought fodder (Everist, 1969).

510

Acacia cyperophylla F. Muell. ex Benth.

Common name. Mineritchie or minnaritchee.
Characteristics. Small tree; bark peels in peculiar curly, red-brown flakes; leaves narrow, almost straight, grey-green. In Queensland it is confined to the far southwest, often in brown pebbly soils on broken gullies. It is eaten freely by cattle (Everist, 1969).

Acacia deanei (R.T. Bak.) Welch et al.

Common name. Green wattle.
Characteristics. The name green wattle is also applied to other species with greenish bark and fernlike leaves. Shrub or small tree with green bark; leaves fernlike, young leaves and stem tips golden yellow; yellow flowers in balls. It is common in patches in shallow soils on ridges in the western Darling Downs, Maranoa and Warrego districts in Queensland and is eaten readily by sheep in dry times (Everist, 1969).

Acacia doratoxylon A. Cunn.

Common name. Lancewood.
Characteristics. Tree with upright straight trunk; bark dark grey or grey-brown, finely furrowed; leaves grey-green, narrow, slightly curved, borne mainly on the topmost branches; flowers yellow, in spikes. Grows in dense scrubs on rather stony soils (such as the flat tops of mesas) in the 375- to 625-mm rainfall belt. The leaves are eaten fairly readily, though local differences in palatability are reported. The trunks are not suitable for fence posts but make excellent rails when used out of the ground (Everist, 1969).

Acacia excelsa Benth.

Common name. Ironwood or ironwood wattle.
Characteristics. Tree with grey, furrowed bark and very dark, hard wood; leaves green, small but variable in shape, prominently veined; flowers in balls, very pale yellow.
Widespread in western Queensland on soils ranging from sand to clay loam. Eaten freely, but is said to cause impaction after about six weeks unless

511

mixed with other species. Regeneration from seed is rapid if the young trees are protected from sheep (Everist, 1969).

Acacia farnesiana **(L.) Willd.**

Synonyms. *Mimosa farnesiana* L.; *M. acicularis* Poir.

Common names. Aromo (Panama); espino blanco, espino ruco (El Salvador), huisache (Mexico), opoponax (southern United States); mimosa bush, needle bush (Australia); klu (Hawaii); Ellington curse, vaivai vakavotona, ban baburi (Fiji).

Description. A many-branched, woody shrub, usually 2 to 3 m, but can reach 6 m in height; bipinnate, green leaves consisting of many small leaflets. The light coloured stems armed with large thorns; flowers globular, yellow and fragrant. Fruit (pods) borne in clusters, green, turning to dark brown at maturity, 5 to 8 cm long; each contains 12 to 14 seeds.

Distribution. Native of tropical America but now common throughout the tropics. Formerly a garden plant. Common in Viti Levu (Fiji), and northern Australia, southern Australia, southern France and Africa.

Characteristics. Considered one of the worst weeds in Fiji (Mune and Parham, 1967) and is also a problem in the Northern Territory (Hall, 1967). It is, however, one of the few woody plants which grows naturally on the heavy, cracking calcareous clay grasslands in northwest Queensland and along stream banks in northern Australia, and is of considerable value in supplementing the diet of roughage provided by the *Astrebla* and *Dichanthium* native grasses in the dry season (Everist, 1969). It is also one of the few species which can persist along highly saline artesian watering drains for sheep and cattle, where it also provides valuable shade. It has a crude protein content of 18.4 percent.

Control. In Fiji, it is controlled manually by digging out roots and seedlings; by building a fire around the stems at ground level and maintaining a fierce heat for several hours; or by spraying seedlings and young plants with 1.6 kg acid equivalent of 2,4,5-T ester diluted in 180 litres of water or diesel fuel oil applied as a fine spray to the leaves and stems. Spraying will most likely need to be repeated. Mature bushes should be sprayed with a basal application of 2,4,5-T ester at the rate of 1.6 kg acid equivalent diluted in 270 litres of diesel fuel oil. The mixture is applied as a fine spray or painted on the dry bark at the base of the stems from ground level to a height of 1 metre, to thoroughly wet the bark. Shrubs so treated take three to six months to die. Hall (1967)

512

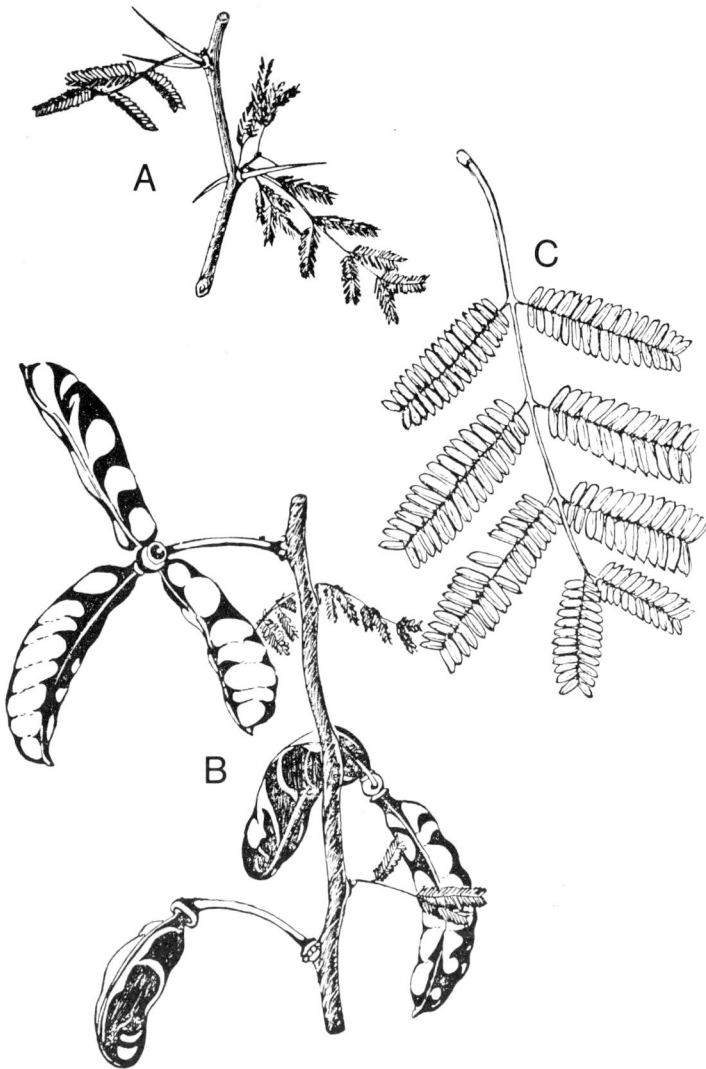

Figure 102. *Acacia farnesiana.* **A**-Stem, showing leaves and thorns **B**-Stem and pods **C**-Bipinnate leaf (**Source:** Weeds of Fiji, 1967)

513

recommends treating cut stumps with a mixture of 1 kg of 2,4,5-T in 45 litres of diesel oil.

Main reference. Hall, N.H. (1967).

Acacia georginae F.M. Bail

Common name. Georgina gidgee.
Description. A small, rather gnarled tree. Pods broad, flat, curled.
Distribution. The species is almost confined to the basin of the Georgina River and parts of the Northern Territory.
Characteristics. Leaves are grazed and the pods are eagerly sought by sheep and cattle. However, the seeds and pods are very poisonous, the leaves less so. Some trees appear to be nontoxic (Everist, 1969). Oelrichs and McEwan (1962) isolated and identified the toxic principle as fluoracetic acid. The plant is nevertheless considered a valuable fodder in the area.

Acacia gerrardii Benth.

Synonym. *A. hebecladoides* Harms.
Common names. Sebeldit, ol debbe (Kenya).
Description. Tree, usually 3 to 15 m high, with a flat, umbrella-shaped crown and rough grey bark, fissured. Young branches densely pubescent, the epidermis splitting to reveal a rusty red inner layer. Spines straight, short, about 1 cm long. Leaves rather large; pinnae five to ten pairs, leaflets 12 to 23 pairs, 3 to 7.5 mm long, 1 to 2 mm wide; flowers white, scented, occurring in globose heads. Pods large and sickle-shaped, dehiscent, 7 to 16 cm long, 0.6 to 1.1 cm wide. Seeds olive-brown, compressed, 9 to 12 mm long and 7 mm wide (Brennan 1959).
Distribution. Africa from Natal to the Sudan, westward to Nigeria.
Characteristics. Common in the *Acacia-Themeda* zone at altitudes from 1 230 to 2 000 m. Often occurs in places with impeded drainage. It is too tall for a browse plant, and branches are lopped to feed cattle and goats when other feed is not available. The leaves contain over 17 percent crude protein (Dougall and Bogdan, 1958).

Where control is needed, stands can be thinned by digging out the stumps or spraying the cut surface of the stumps of foliage with a picloram/2,4-D mixture (Ivens, 1970). A single burning or spraying of the cut stump will exert some control.

514

Acacia giraffae **Burch.**

Synonyms. A. *erioloba* E. Mey.; A. *giraffae* Willd.

Common name. Camel thorn.

Description. Medium to large tree, up to 12 m high; stipular spines, stout, brown and straight, up to 3 cm long; foliage green, leaves up to 6 cm long with one to three pairs of pinnae. Pinnules about 5 mm long and nearly 2 mm broad in 8 to 15 pairs. Flowers capitate and yellow.

Distribution. Occurs from the Cape Province, South Africa, to Zimbabwe on deep alluvium and sand.

Characteristics. Tolerant of frost but easily killed by fire. When felled it will coppice freely from the stems and will sucker from the roots. Flowers in September in Cape Province and ripe pods are shaken from the trees from July onward. The roughly falcate pods are thick and heavy and are covered with a dense grey felt. A yield of 275 kg of pods per tree has often been exceeded, and up to 500 kg has been reached. Peak production occurs at 20 years of age; four to eight trees per 0.5 ha are preferable (West, 1950).

Acacia harpophylla **F. Muell. ex Benth.**

Common name. Brigalow.

Characteristics. A tree of variable size and shape with black or dark grey, furrowed bark; leaves grey-green, grey or silvery grey, curved into a shallow sickle shape; flowers in balls, yellow. The species forms scrubs similar in structure to gidgee scrubs but somewhat taller and in a higher rainfall zone (500 to 875 mm). Along its eastern boundaries, brigalow scrub merges gradually into mixed softwood scrub or light rain forest. The leaves are to some extent eaten as windfalls but cut material is refused. Sheep will eat the very young suckers until they turn green and harden off (Everist, 1969). Brigalow is generally regarded as an undesirable species.

Acacia homalophylla **A. Cunn. ex Benth.**

Common name. Yarran.

Characteristics. A tree with clean trunk and leafy head; bark dark grey, rough, finely furrowed; leaves green, narrow, usually straight; flowers yel-

515

low, in balls. There are several forms of this species in Queensland, occurring mostly in the 500- to 625-mm rainfall belt on clay and clay loams. Sometimes they form dense clumps. The leaves are eaten readily and are regarded as about equal to *Acacia pendula* (myall) in fodder value (Everist, 1969).

Acacia koa Gray

Common name. Koa (Hawaii).
Description. An erect, somewhat spreading tree, 15 to 30 m tall with rather smooth branches. Leaves bipinnate with 12 to 15 pairs of leaflets, 0.6 cm long and 0.3 cm wide; phyllode simple, linear, slightly curved, 10 to 15 cm long and 1 to 1.5 cm wide, leathery. Flowers yellow in a globose head, 1 cm diameter at the end of a 1.25-cm stalk. Pod straight, flat, 7.5 to 12.5 cm long and 1.25 cm broad, smooth, brown when ripe. Seed oblong, flat, 1 cm long, dark, shiny (Hosaka and Ripperton, 1944).
Distribution. Occurs at elevations of 230 to 850 m in Hawaii.
Characteristics. Reproduces mainly by suckers; heavy grazing and trampling tend to eliminate it. In six years it reaches 3 m in height with a diameter of 1 m at breast height.

Acacia litakunensis Burch.

Synonyms. *A. heteracantha* Burch.; *A. spirocarpoides* Engl.; *A. spirocarpa* Hochst. ex A. Rich.
Common names. Haakdoring, umbrella thorn (South Africa), haak-ensteek (Transvaal).
Description. Small to medium-sized tree, with stipular spines of two types, alternating along the branches; strong, dark coloured, hook thorn and a pair of long white thorns about 2.5 cm further along the branch. The hook thorn becomes the white one next year. White thorns long and straight, 2.5 to 6.0 cm; leaves about 5 cm long with ten pair of pinnae. Pinnules very small and narrow, in 7 to 15 pairs. Flowers capitate, white. Pods 0.6 cm broad and 7.5 to 10 cm long, curled to form a spiral (Bonsma, 1942).
Distribution. Occurs from northern Cape Province, South Africa, to Tanzania.
Characteristics. It seems to have two growth forms: a young shrubby, dome-shaped type and a tall, often flat-topped tree. It flowers in November in Cape

516

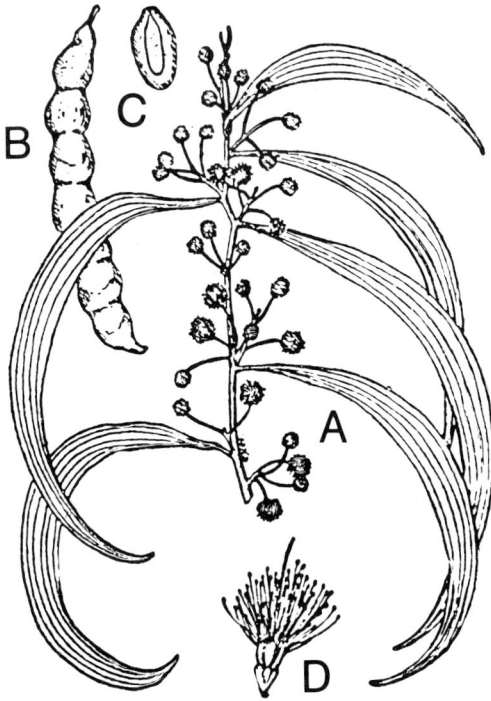

Figure 103. *Acacia koa.* **A**-Habit **B**-Pod **C**-Seed **D**-Flower

Province; the pods ripen early and drop before the end of July. Pods are very loosely attached to the branches; when cattle touch the twigs with their tongues, the pods fall into their mouths. West (1950) says that they are so eagerly sought after by cattle and game that it is difficult to collect any quantity of pods. This species often pioneers the invasion of grassland by bush and may form thickets.

Bonsma states that the seeds pass undigested through the animal and germinate in the dung around drinking places.

The pods contain 18.83 percent crude protein, 2.44 percent fat, 46.25 percent nitrogen-free extract, 5.1 percent ash, 20.1 percent crude fibre, and Bonsma assesses that the daily intake is equal to 0.7 kg of protein supplement. Cattle are kept in good condition during the winter on this diet, and the ranchers refer to the area where the trees grow as Haakdoring veldt.

Acacia macrothyrsa **Harms.**

Synonyms. A. *buchanaii* Harms.; A. *dalzielii* Craib.

Description. A tree 3 to 10 m high with large, finely pinnate, very shiny, drooping leaves and striking golden to yellow, scented flowers, in panicles up to 45 cm long. Pod dark red-brown, flat, oblong, 7.5 to 12.5 cm long, 1 to 2 cm broad, persisting on the tree for long periods, 6- to 12-seeded (Andrews, 1952).

Distribution. It is found in Ghana, Nigeria, Zaire, Angola, Swaziland, Zambia, Zimbabwe, and in the southern Sudan (Brennan, 1959). Van Rensburg (1968) reported it locally and frequently in *Combretum* mixed woodland in Zambia on shallow, gravelly soils. It has relatively few rather short, unobtrusive thorns. Young shoots, which develop in October/November in Zambia, are palatable and readily browsed by stock.

Acacia mellifera **(Vahl) Benth.**

Common name. Kitr (the Sudan).

Description. Shrub or small tree with ball-shaped crown. Leaves small, with only a few leaflets. Branches covered with very sharp recurved thorns. Flowers white and gathered in short, dense spikes: pods thin and flat.

Distribution. Widespread in Africa, occurring in Egypt, the Sudan, Somalia, Ethiopia, Angola, Kenya, Uganda and Tanzania, usually on clay soils. In Kenya, it occurs at altitudes between 1 000 and 1 400 m.

Characteristics. Goats are very fond of the leaves, which are either browsed from the standing shrubs or are picked up when they fall to the ground (normally during the dry season). The young leaves, which usually appear in February (middle of the dry season), are very much liked by goats, and where it is dominant, *A. mellifera* contributes very considerably to their diet. The tree is too spiny for cattle and the fallen leaves too small. Young leaves contain up to 42 percent protein (Dougall and Bogdan, 1958). Ivens (1970) showed that *A. mellifera* is easily killed by stumping and cutting, and repeated burning exerts a good deal of control.

Acacia nigrescens Oliver

Synonyms. *A. nigrescens* Oliv. var. *pallens* Benth.; *A. brosigii* Harms.
Description. Tree 4 to 25 m high with trunk usually with knobby prickles. Young branchlets smooth. Prickles on branchlets in pairs just below each node, hooked, blackish, persistent, 2.5 to 7.0 mm long. Pinnae two to four pair; leaflets one to two pair, 10 to 26 mm long and 7 to 22 mm wide, ovate to elliptical. Flowers cream or white in aggregate or solitary spikes. Pods darkish brown, dehiscent, smooth, oblong, straight, 7 to 14.5 cm long, 1.5 to 2.4 cm wide. Seeds sublenticular, 12 to 13 mm in diameter (Brennan, 1959).
Distribution. Tanzania, Botswana, the Transvaal and Lesotho.
Characteristics. Occurs in deciduous bushland and wooded grassland in black soil areas, at elevations of 240 to 1 160 m. McKay (1968) showed that goats browse this plant and are able to reduce the density of young plants 5 to 30 cm high.

Acacia nilotica (L.) Del. ssp. *subalata* (Vatke) Brennan

Synonyms. *A. arabica* (Lam.) Willd.; *A. subalata* Vatke.
Common names. Prickly acacia (Australia), mgunga (East Africa; ol'erbat in Masai), babul (Jhansi, India).
Description. A tree 8 m or more high, stems and branchlets usually dark coloured; spines straight, sharp, up to 7.5 cm long in young trees, mature trees commonly without thorns. Pinnae usually in 3 to 12 pairs; leaflets in 10 to 30 pairs, linear-oblong, about 0.4 cm long. Flowers yellow. Pods grey, thick, softly tomentose, straight or slightly curved, 10 to 15 cm on a pedicel, 0.5 to 1.2 cm long, with constrictions between the seeds giving a necklace appear-

519

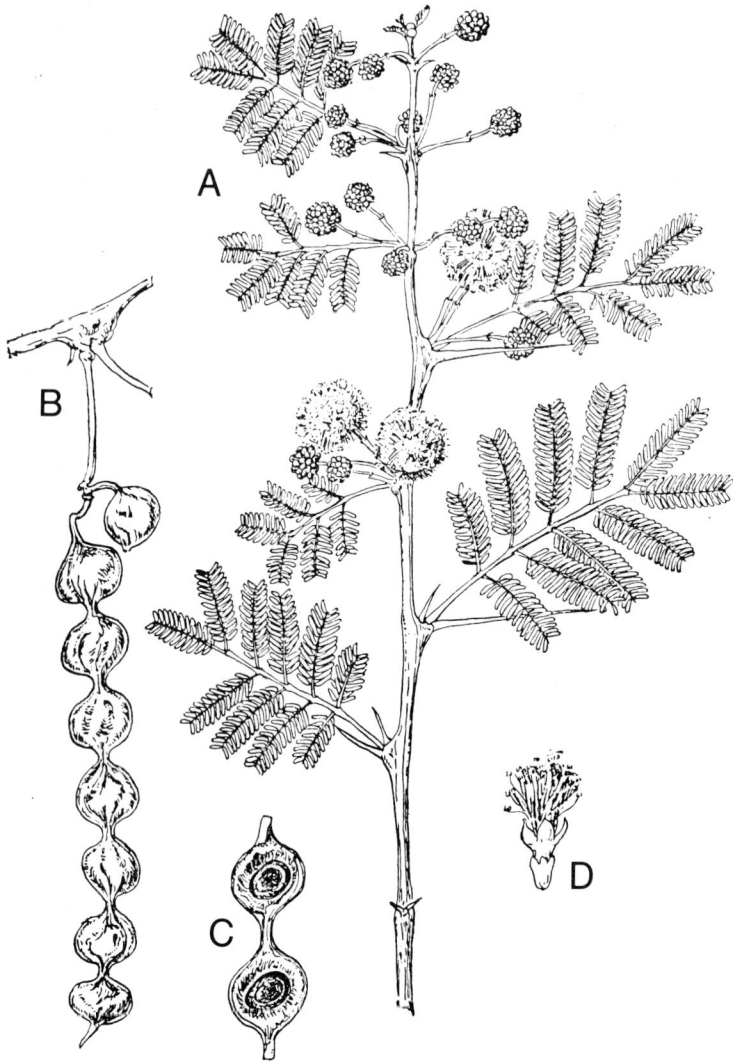

Figure 104. *Acacia nilotica.* **A**-Habit **B**-Pod **C**-Seeds **D**-Flower (**Source:** Andrews, 1952)

ance, fleshy when young, becoming black and hard at maturity (Andrews, 1952).

Distribution. Widespread in Africa and Asia, and occurs in Australia. In Kenya, it is mainly found at 900 to 2 000 m (Dougall and Bogdan, 1958). Found in the Gambia, the Sudan, Togo, Ghana, Nigeria, and on lateritic soil in the Himalayan foothills in India (Whyte, 1964).

Characteristics. Leaves used for feeding sheep and goats in the Hissar district in India. In Maharashtra State, areas of *A. nilotica* are closed by law for regeneration in five-year rotations (Whyte, 1964). The removal of leaves from this species should be restricted to prevent damage to flower development and a consequent loss of pods. In Kenya, the fleshy pods are readily eaten by goats, sheep and cattle, but some tribes believe they cause bloat. The leaves contain up to 12 percent crude protein and 21.35 percent crude fibre. They are occasionally browsed by goats (Dougall and Bogdan, 1958). In Queensland, cattle, sheep, and goats are fond of the plant, but because of the spiny nature of the young trees the species has been declared noxious throughout the state (Everist, 1969). Ivens (1970) has had some success in its control by hand digging, basal bark spraying with 2,4-D and with foliar spraying with a picloram/2,4-D mixture, but a number of treatments may be necessary.

Acacia nubica

Synonyms. *A. virchowiana* Vatke; *A. merkeri* Harms.; *A. orfota*.

Common name. Laot (Sudan).

Description. A somewhat obconical shrub, 1 to 5 m high, often with branches radiating from the base in all directions; branchlets grey-white, straight, stout; grey-white spines with brown tips, 0.5 to 1.5 cm long. Pinnae in 3 to 12 pairs, leaflets in 5 to 15 pairs, about 0.3 cm long. Flowers in globose heads, off-white, very fragant, though the bush often has an offensive smell. Pod pale yellow, longitudinally striate, distinctive, linear-elliptic, pointed at both ends, 5 to 10 cm long, 1.25 cm broad. Seeds olive-green, five to ten (Andrews, 1952).

Distribution. Found in northeast Africa from Egypt to Kenya, and in Iraq and Iran. It occurs on alluvial silts.

Characteristics. An important browse shrub in the drier parts of Kenya; the leaves and pods are high in calcium, but the leaves have more phosphorus (Dougall and Bogdan, 1958).

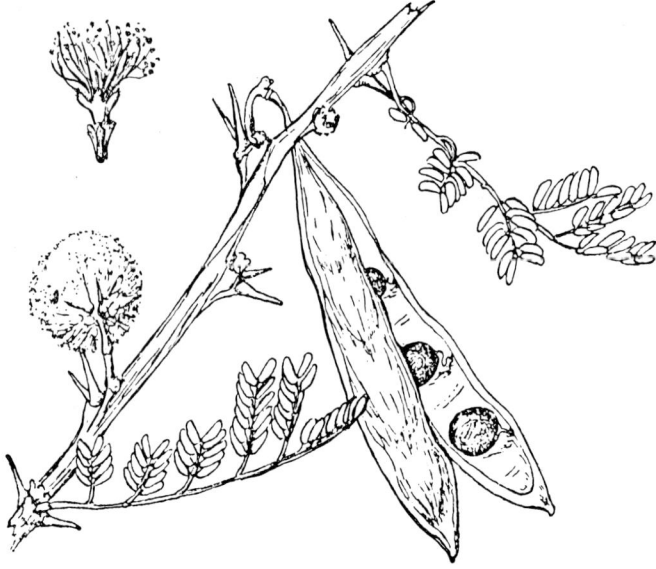

Figure 105. *Acacia nubica*

Acacia oswaldii **F. Muell.**

Common names. Nelia, midger, ram's-horn tree.

Characteristics. Densely branched shrub or small tree; limbs often with sharp "boomerang" bends; leaves short, stiff, ending in short sharp points; flowers in balls, pods curled into rough spiral. The plant is found mainly as scattered individuals among eucalypts or along edges of mulga country. It extends from the 375- to 625-mm rainfall zone and is eaten freely where other edible trees are scarce (Everist, 1969).

Acacia pendula **A. Cunn. ex G. Don**

Common name. Myall.

Description. Small tree with rounded crown and drooping branchlets; grows up to 6 m. Narrow silvery grey leaves, 1 to 1.5 cm long and 0.3 to 0.9 cm wide and somewhat curved. Flowers in small heads in short axillary racemes, yellow wattle type. Pods are winged, flat, and about 1.25 cm wide and 2.5 to 7.5 cm long.

Distribution. Occurs mainly south of the Tropic of Capricorn in Australia: on the western plains of Queensland and New South Wales on heavy brown and grey clays which are occasionally inundated, it is often pure stand. Its roots forage widely and crops cannot grow near the trunk. Regenerates from seed; seedling growth is rapid, young trees reaching 1.3 m in four years. The young plants are relished by sheep and cattle and should be protected until they are 45 cm high. The tree is attacked by gregarious bag caterpillars (family Liparidae) which often defoliate much of the tree (Stannard and Condon, 1958).

Acacia peuce **F. Muell.**

Common name. Waddy wood.

Characteristics. Small tree with scaly bark, very hard, dark wood and long pendulous leaves with very fine longitudinal ribs resembling those of beefwood (*Grevillea striata*).

Pods large, broad and flat, bluish-grey in colour.

Found in only three or four patches in the arid interior, two of them being known in Queensland. It occurs in a region where almost no other trees of suitable size are available and for that reason it has been widely used for fence posts and yard construction.

Leaves are eaten by cattle (Everist, 1969).

Acacia polyacantha **Willd.**

Synonyms. *A. campylacantha* Hochst. ex A. Rich.; *A. catechu* (L.) Willd.; *Mimosa suma* Roxb.

Description. Tree up to 21 m high, trunk with fissured bark and knobbly persistent prickles in pairs below each node, straw coloured to brown or black,

4 to 12 mm long. Leaf petiole glandular, pinnae 13 to 40 pairs, leaflets 6 mm long and 0.5 mm wide. Flowers cream or white in spikes 6 to 12.5 cm long. Pods brown, dehiscent, 7 to 18 cm long, 1 to 2 cm wide (Brennan, 1959).

Distribution. Widespread in tropical Africa from the Gambia to Eritrea, Ethiopia, in the north, to the Transvaal in the south. A subspecies, *polyacantha*, occurs in India.

In spite of its feeding quality, it is a most undesirable tree; has strong, pernicious, recurved spines, and spreads rapidly on fallow land in pasture, particularly on low-lying fertile alluvial soil along the drainage lines (in Zambia). It is expensive and difficult to eradicate. Large trees can be killed by ringbarking and the application of arboricides.

Acacia salicina Lindl.

Common names. Coolan, cooba, umung, black sally wattle, willow, black mattle.

Characteristics. Fairly large tree; grows mainly on river and creek channels and in levee soils; bark dark grey, almost black, furrowed; wood dark but fairly soft; leaves grey-green, often slightly succulent, with a prominent midrib; flowers pale yellow, in balls; pods thick and fleshy, froth if shaken in water; seeds black with a prominent red aril. It is abundant in the 375- to 625-mm rainfall belt, but extends almost to the coast. Leaves are eaten to some extent, but are not widely used. They contain large amounts of tannin and have been suspected of poisoning hungry cattle. The pods are rich in saponin (Everist, 1969).

Acacia senegal (L.) Willd.

Synonym. *A. verek* Guillem. and Perrott.

Common name. Gum-arabic tree, hashab.

Characteristics. Bush or tree, usually less than 5 m high, occasionally 10 m high, frequently forming thickets; stem short, usually low-branched; crown eventually flattened; bark pale brown to pale grey, scaly on the older parts; slash mottled red; prickles up to 0.5 cm long, the centre one sharply curved, the other two more or less straight and directed forward. Leaflets grey-green. Flowers creamy white (red in bud), usually appearing before the leaves in spikes 5 to 10 cm long either solitary or two to three together. Pod brown, flat,

524

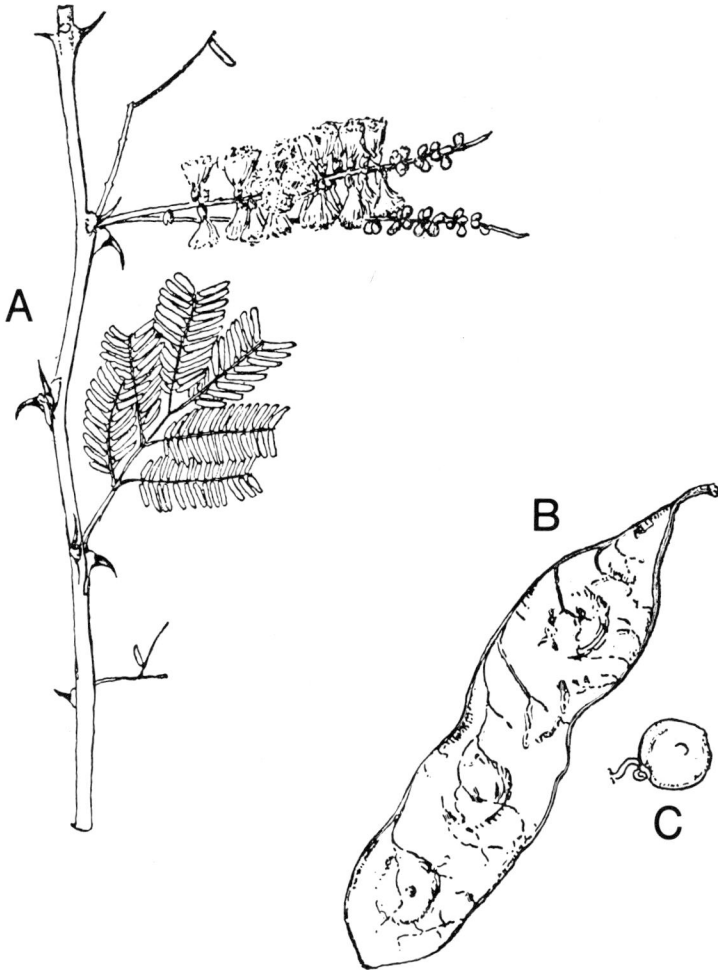

Figure 106. *Acacia senegal.* **A**-Flowering branch **B**-Pod **C**-Seed

525

papery, oblong, 2.5 to 12.5 cm long, sometimes constricted owing to the abortion of some of the seeds; seeds greenish-brown, one to five in number (Andrews, 1952).

Deciduous, drooping its leaves in November in the Sudan, the leaf fall being mineralized to build up the fertility of sandy soils for ensuing crops of groundnuts, sorghum, bulrush millet and sesame. Strips of bark are slashed from the branches in November, and in January the gum exudate is collected and sold as gum arabic. Camels and goats browse it, but are usually excluded from gum-bearing groves as they browse back the trees and reduce gum production (Skerman, 1966).

Acacia seyal **Del.**

Description. Slender tree, usually 6 to 11 m high, with a flat-topped, thin crown and rather scarce foliage. Bark smooth and greenish yellow in colour — when damaged it often becomes orange coloured. Yellow flowers in globose heads; pods sickle-shaped (Dougall and Bogdan, 1958).

Distribution. Widespread in northern tropical Africa, extending to Egypt.

Characteristics. The bark is the most valuable part of *Acacia seyal*. It is extensively used for feeding cattle, goats and sheep during the dry season. The bark is thick, smooth and relatively soft when fresh. In the dry months of February and March in Kenya, thick branches are thrown to the ground and animals browse the bark and eat the leaves, which, however, are relatively few at that time. Often the whole tree is cut to a height of 1.5 to 2 m and new growth arises from the previous cut. As animals often browse the bark of standing trees, deformation is common. They will eat up to 5.5 kg of bark a day, sufficient for maintenance and the production of 4.5 litres of milk. The crude protein content of the bark was 10.6 percent in February and 4.1 percent in August. The calcium values were very high, with over 4 percent in February, but the phosphorus content of under 0.1 percent was low (Dougall and Bogdan, 1958).

If some control is necessary, repeated burning or a basal bark spray with 2,4,5-T will help but will not give full control (Ivens, 1970).

Acacia shirleyi **Maid.**

Common name. Lancewood.

Characteristics. Tree with upright straight trunk; bark dark grey or grey-brown, finely furrowed; leaves grey-green, narrow, slightly curved, borne mainly on the topmost branches; flowers yellow, in spikes. Grows in dense scrubs on rather stony soils as the flat tops of mesas in the 375- to 625-mm rainfall belt. Fair palatability.

Acacia sieberiana DC

Synonyms. *A. verugera; A. woodii* (Burtt and Davy) Keay and Brennan.
Common name. White thorn.
Description. Tree, up to 15 m high with a flat, umbrella-shaped crown and yellow-brown scaly bark which exposes a powdery yellow surface. Widespread branches give a lot of shade. Spines are white, strong and straight, up to 10 cm long. Leaves up to 15 cm long, 10 to 30 pairs of pinnae; leaflets in 15 to 40 pairs. Flowers, creamy white, fragrant. Pod brown, 10 to 20 cm long, 2 to 2.5 cm broad, up to 1.5 cm thick, indehiscent, containing 12 to 15 brown seeds arranged in one row and embedded in spongy tissue. Flowering begins in November in the Transvaal, and the pods mature in August-September (West, 1950).

The subspecies are three: var. *sieberiana*, var. *vermoesenii*, and var. *woodii*.
Distribution. Widely, in mixed woodland in Africa from the central Sudan to the southern Sudan, Kenya, Zambia, Zimbabwe, Angola, Botswana and the Transvaal.
Characteristics. Bears heavy crops of edible pods which drop during the dry season. The pods are readily eaten and contain 8.3 percent crude protein. The seeds contain 27.4 percent crude protein, and the tips of the young shoots, also eaten toward the end of the dry season, contain 21.13 percent.

The branches are armed with numerous white thorns, making browsing hazardous. Limited numbers of trees could be retained to advantage in pasture. Encroachment should, however, be prevented as serious damage can be caused to the pasture. Eradication is a laborious and expensive process (van Rensburg, 1968).

Dougall and Bogdan (1958) state that the pods are liked by rhinoceros and elephants, and Lamprey (1967) found seedlings of both *Acacia albida* and *A. sieberiana*, the pods of which are indehiscent, growing in elephant dung in the Lake Manyarra Game Reserve in Tanzania.

527

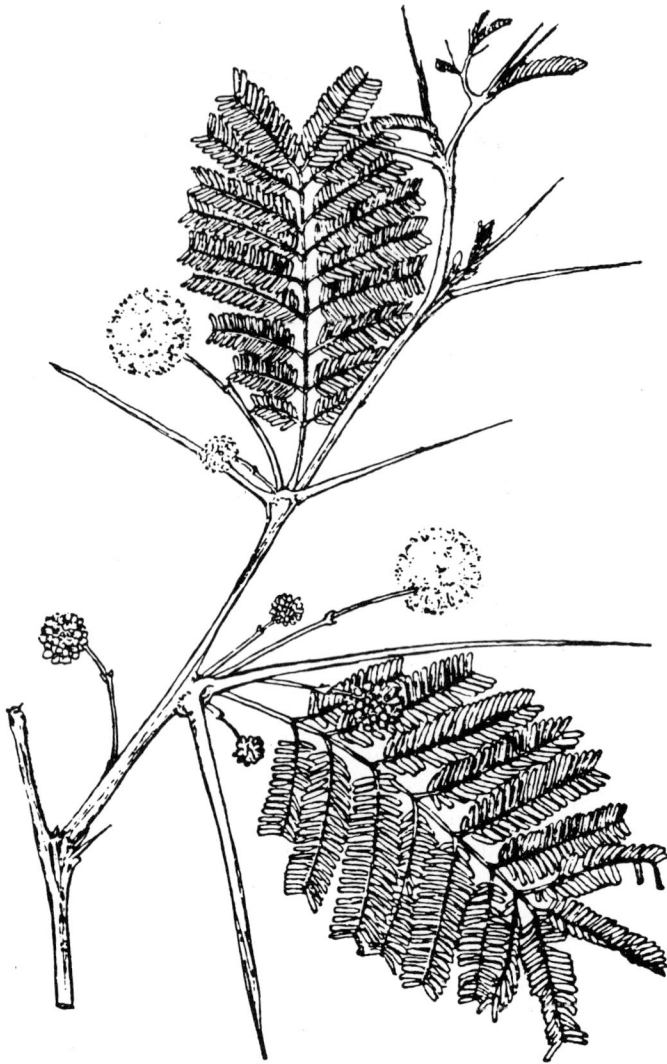

Figure 107. *Acacia sieberiana*

528

Acacia sparsiflora **Maid.**

Common name. Currawong.
Characteristics. Tree somewhat similar in appearance to *A. shirleyi*, but usually not growing in dense scrubs. The tops are usually rounded and the leaves more curved. They occur in somewhat stony soils in the 625- to 750-mm rainfall belt south of the tropics in Queensland, particularly in the area from Tara to St George. The leaves are eaten readily and in some districts the trees are cut for fodder in times of drought (Everist, 1969).

Acacia stenophylla **A. Cunn. ex Benth.**

Common names. Belalie, dunthy, river myall, black wattle.
Characteristics. Small tree with dark brown, finely furrowed bark; leaves long, usually dull green in colour, not curved; flowers pale yellow, in balls; pods grey or bluish grey, fairly long with constructions between the seeds similar to those of desert oak (*A. coriacea*). This plant is found mostly on river channels in Queensland in the area of 250- to 625-mm rainfall. It is eaten fairly readily but is not cut to any extent for drought feed (Everist, 1969).

Acacia sutherlandii **F. Muell.**

Common name. Corkwood wattle.
Characteristics. Tree with upright trunk and pendulous branches. Bark pale yellowish-grey, soft and corky; leaves bright green, fernlike, pendulous. Pods grey, 10 to 15 cm long and up to 2.5 cm wide. Grows on clay soils, mostly north of the 500-mm rainfall belt, and favours lightly timbered downs country. It is eaten fairly readily by sheep and cattle (Everist, 1969). It is native to Queensland.

Acacia tetragonophylla **F. Muell.**

Common name. Dead finish.
Characteristics. Shrub, usually with crooked irregular branches; leaves very narrow, bright green with sharp points, borne in tufts on short branchlets off the main branches, deciduous during the dry season; flowers bright yellow, in

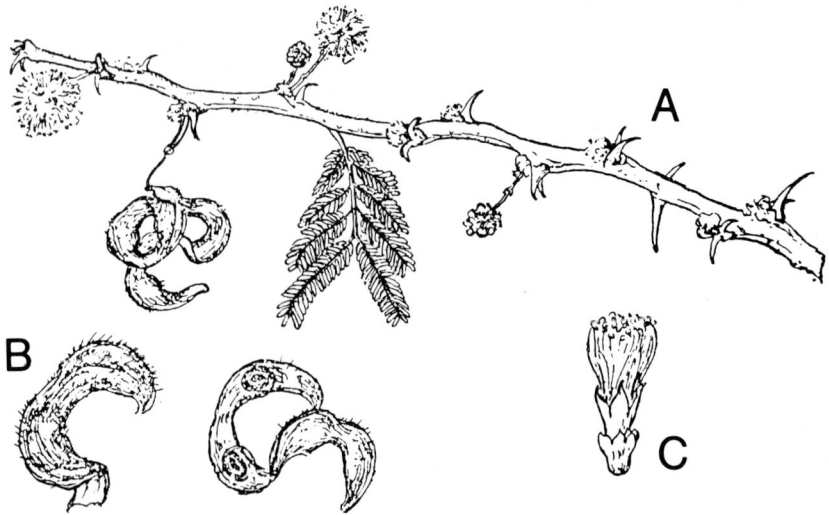

Figure 108. *Acacia tortilis.* **A**-Branch, showing flowers, pods, leaves, thorns **B**-Pods **C**-Flower (**Source:** Andrews, 1952)

balls; pods brown, very narrow and inclined to curl. In Queensland, this species occurs only west of the Parvo River, mainly on hard red-brown soils. It is eaten fairly readily but because the leaves are shed during drought, they are not available under real drought conditions.

Acacia tortilis (Forsk) Hayne

Synonym. *A. spirocarpa* Hochst. ex A. Rich.
Description. Gregarious, wide-spreading, flat-topped or umbrella-shaped tree, up to 4 m high; branchlets pubescent, red-brown; spines mixed, some white, straight, slender, and up to 7.5 cm long, others grey with black or brown tips, sharply curved, very small. Pinnae in three to ten pairs, leaflets in 7 to 15 pairs. Flower heads white to cream. Pod yellow-brown, pubescent, spirally twisted, slightly constricted between the seeds, circular in cross-section, 7.15 to 15 cm long, 0.6 to 0.8 cm thick (Andrews, 1952).

530

Distribution. Occurs in the drier areas of northern Africa, Senegal to Nigeria, in the Sudan, Kenya and Tanzania; often on flat alluvial areas. In Kenya it occupies the desert grass-bush zones at altitudes of 1 200 m.

Characteristics. Leaves of young trees are browsed by goats and sheep, but the main value of this species is in its pods, which can be very numerous and are picked up from the ground and eaten by all African livestock. At the time when pods are mature (usually in January-February in Kenya), they are often the main source of food for cattle, sheep and goats. Lamprey (1967) found in the Tarangere Game Reserve in Tanzania that seedlings of *Acacia tortilis* could only be found growing from the faecal pellets of the impala. He obtained no germination of seed from fallen pods, but 7 percent from seed collected from faecal pellets. Goat enclosures at Maswa, Tanzania, also have seedlings germinating from the dung, and Lamprey believes that such indehiscent pods must pass through an animal for regeneration to take place.

Acacia victoriae Benth.

Synonym. A. sentis.

Common names. Gundabluey or gundabluie.

Characteristics. Large shrub or small tree, usually much branched, the fine twigs often dead at the tips; leaves green or bluish green, narrow, straight, with a prominent midrib; flowers pale yellow, in balls. Grows mainly in open grassland in the 375- to 500-mm rainfall belt from the southern border to the Barkly Tableland in Queensland. Although it does not produce any great bulk of forage, it is eaten readily and is a valuable supplement in the dry season. During a succession of very wet years it can increase to such an extent that it becomes a nuisance, particularly around watering places (Everist, 1969).

Albizia spp.

Albizia adianthifolia (Schum.) W.F. Wight

Description. Tree 4 to 30 m high, crown flattened. Young branchlets rusty-pubescent; flowers white or greenish white. Pod oblong, flat, 9 to 19 cm long × 1.9 to 3.2 cm wide, pubescent. Seeds 7 to 9.5 mm × 6.5 to 8.5 mm, flattened (Brennan, 1959).
Distribution. Widespread in tropical Africa from the Gambia and Kenya to Angola and South Africa.
Characteristics. Found in lowland rain forest, deciduous woodland, wooded grassland and upland at altitudes of 30 to 1 680 m.

Cattle graze the coppice regrowth after the trees are fallen for cultivation. The leaves contain 20.13 percent crude protein, 0.20 percent Ca, 0.35 percent P, 0.20 percent Mg and 1.66 percent K (Lawton, 1968).

Albizia amara (Roxb.) Boiv.

Description. Tree 3 to 6 m high with wide, dense crown; white flowers in globose heads (often extremely numerous); at flowering time (January-February in Kenya) the crowns are white and resemble cherry trees in full blossom.
Distribution. Occurs in the Baringo district of Kenya at 1 775 metres.
Characteristics. Leaves eaten by cattle, goats and sheep (from lopped branches) but are not particularly well liked by these animals as the trees are lopped only when more palatable fodder is not available. Flowers are preferred and are either picked from the ground after they are shed or eaten from lopped branches. Both leaves and flowers appear to have a high feeding value, with a crude protein content of over 26 percent of the dry matter. Crude fibre figures are not very high, though they are less in the flowers than the leaves (Dougall and Bogdan, 1958).

Albizia basaltica **Benth.**

Common name. Dead finish.

Description. Shrub or small tree growing in some areas to a height of nearly 13 m. Bark dark grey, dry, deeply furrowed and broken up into rough blocks; leaves deep, dull green, fernlike; flowers pale yellow, in balls; pods about 7.5 cm long and 1 cm wide. It grows in the 500- to 750-mm rainfall belt and reaches its best development in the central highlands in Queensland, Australia. It is common on sandy or loamy, often gravelly, soils, or in shallow soils overlying basalt. Leaves are eaten readily by stock but are usually shed during the dry season, a characteristic that makes the tree very useful as a supplementary fodder during normal dry seasons but useless for feeding in a real drought (Everist, 1969).

Albizia harveyi **Fourn.**

Description. Tree about 10 to 15 m high, deciduous, crown flat, bark fissured and reticulate. Young branches have a grey pubescence. Leaves with a prominent gland on the upper side of the petiole; pinnae 6 to 20 pairs; leaflets 12 to 27 pairs, 2 to 6 mm long and 1 to 1.25 mm wide. Flowers white, bracteoles persistent. Pod oblong, 8 to 18 cm long, 2.5 to 3.5 cm wide, brown to purple. Seeds 8 to 12 mm long, 6 to 9 mm wide, flattened (Brennan, 1959).

Distribution. Occurs in Kenya, Tanzania, Zambia, and southward to Botswana and the Transvaal in deciduous *Combretum/Acacia* mixed woodland at elevations of 80 to 2 130 metres (Brennan, 1959). It is particularly prevalent around termitaria.

Characteristics. The pods are eaten and the leaves browsed during the dry season (van Rensburg, 1968).

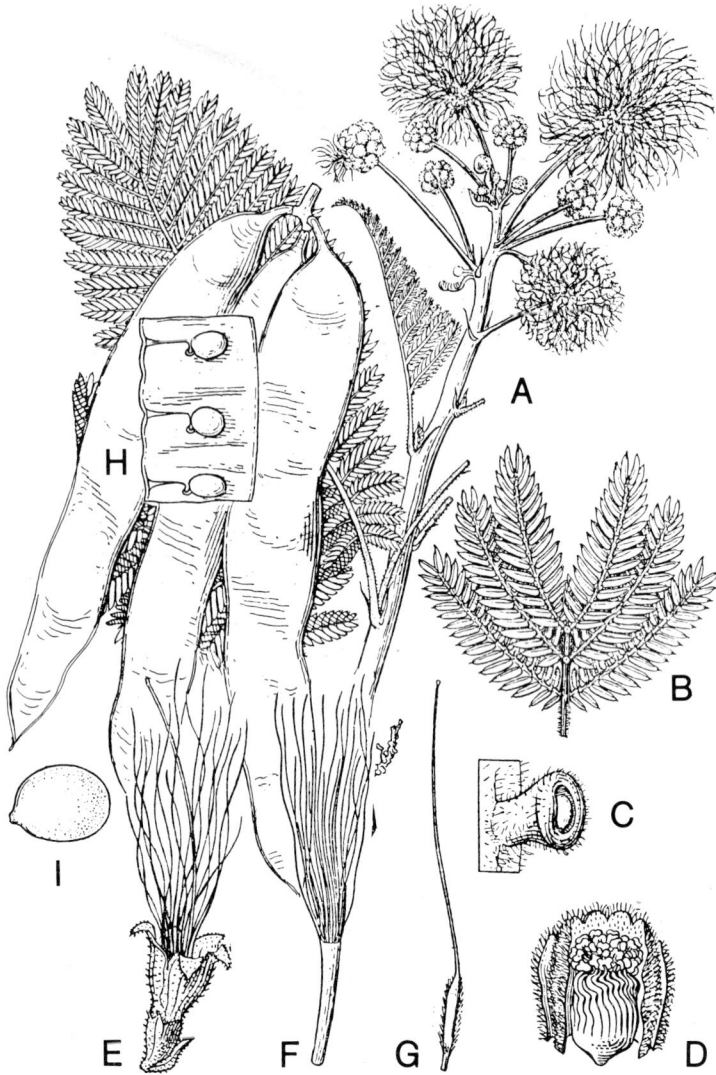

Figure 109. *Albizia harveyi*. **A**-Flowering branch **B**-Portion of leaf, showing glands on rachis **C**-Gland **D**-Flower-bud, opened **E**-Flower **F**-Stamen, filaments and tube **G**-Carpel **H**-Pods, one opened **I**-Seed

534

Figure 110. Merino sheep browsing a fallen tree

Albizia lebbek (**L.) Benth.**

Common names. Indian siris, acacia tree (Australia); mataratón (Colon, Panama); tibit tree, acacia (El Salvador).

Description. Large tree, up to about 20 m high and 28 m across. Bark pale grey or yellowish grey, flaky; leaves large, consisting of numerous rounded, bright green leaflets that are deciduous in the dry season; flowers pale, cream coloured, in large balls; pods large and flat, 20 to 30 cm long and 2.5 to 4 cm across, containing many flat, rounded seeds.

Characteristics. Commonly grown as a shade tree and often called "acacia tree" in western Queensland. It is possibly native in Cape York Peninsula. The leaves are eaten when they fall (Everist, 1969).

In the West Indies the tree is "pollarded" (the branches cut for fodder) and the pods are also used. In Barbados the deep root system makes it difficult to eradicate (Paterson, 1949).

535

Baphia spp.

Baphia nitida Lodd.

Common name. Camwood.

Description. Shrubby tree growing to about 3 m on the Accra Plains, Ghana, to about 10 m in the forests, with smooth branchlets and fragrant white flowers with a yellow centre, usually one to four together.

Distribution. Native to west central Africa; occurs in Sierra Leone, Liberia, Côte d'Ivoire, Togo, Benin, southern Nigeria, the Republic of Cameroon, Guinea, Gabon, Ghana and Zaire.

Characteristics. Favours moist seepage areas on the perimeters of termitaria on plains. Not always associated with thickets. A secondary species in forests; its frequency increases with rainfall. Grows easily from seeds or cuttings. High palatability (rated 4 out of 5); browsed at any time. Table 15.2 gives percentages of dry matter and its composition (Rose-Innes and Mabey, 1964a).

The earliest dye wood (Camwood) was brought from West Africa. The source of the dye, which is soluble in alkali, is the bark and heart of the tree.

TABLE 15.2 **Compositional and chemical analysis of *Baphia nitida* on a dry-matter basis**

	Dry matter	Crude protein	Crude fibre	Ca	P	Cl
Mean	36.5	21.2	26.8	0.85	0.11	0.74
Range	30.2-47.2	19.8-24.6	23.9-31.7	0.43-1.58	0.01-0.17	0.40-1.10
Date	March-July	August-March	June-November	March-July	January-August	July-March

Bauhinia spp.

Bauhinia spp.

Common names. Bauhinia or bohemia, *Bauhinia carronii, B. cunninghamii* and *B. hookeri* (sometimes known as pegunny, particularly *B. hookeri*).

Characteristics. Tree, usually with short trunk and branches of irregular shape; bark dark grey or almost black; dry, hard, coarsely flaky and somewhat furrowed; leaves deciduous, consisting of two equal rounded halves folded down the middle, usually with a notch between the two at the tip; flowers white in *B. hookeri*, pink in the other species; pods hard, dry, 7.5 to 10 cm long and 4 to 5 cm wide, flat, containing hard, brown, discoid seeds. Of these species, the first two are characteristic of the more westerly areas in Queensland, Australia, *B. carronii* being the most abundant. *B. hookeri* is common on drainage lines and banks of streams in some of the softwood scrubs of the 625- to 750-mm rainfall belt and also in the basalt country north of Hughenden, Australia. All grow on a fair range of soils. Leaves are somewhat hard but are eaten readily by stock; pods, too, are eaten. Bauhinia trees are leafless during late winter and spring and are usually leafless during major droughts (Everist, 1969).

Bauhinia rufescens Lam.

Common name. Kharroub (the Sudan).

Characteristics. Much-branched shrub or small tree up to 5 m high, bark white, smooth. Leaves bilobed almost to the base, rarely over 4 cm long, glaucous, glabrous. Racemes few-flowered. Calyx spathaceous. Petals cream-white, fragrant. Pod almost black, often twisted, 6.5 to 8.5 cm long, several-seeded. Occurs in the central and southern Sudan (Andrews, 1952).

Skerman (1966) reported this to be the most palatable browse for camels in the Sudan.

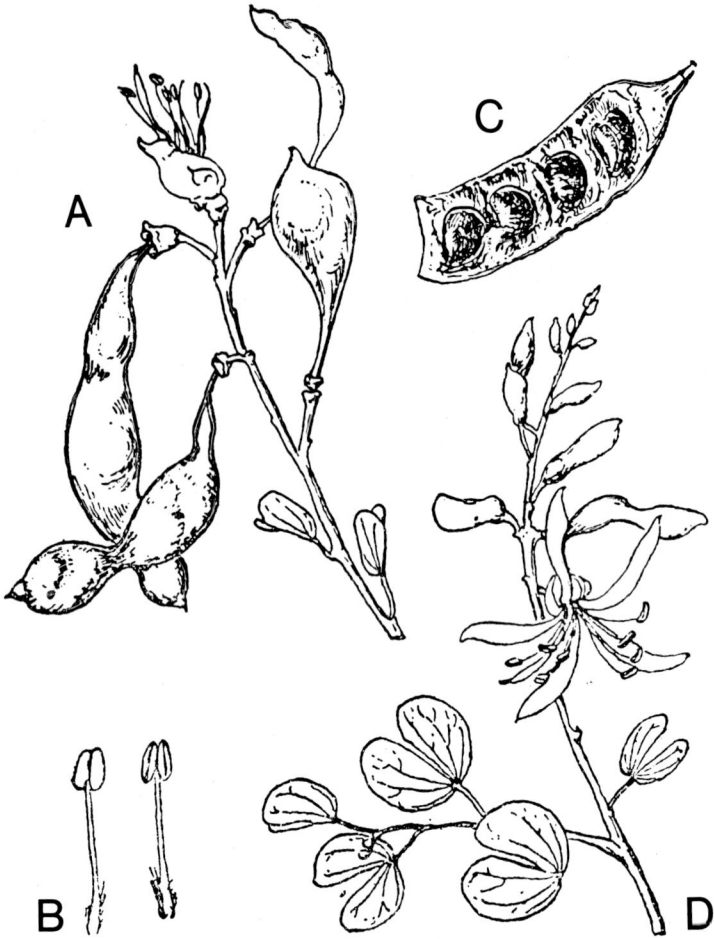

Figure 111. *Bauhinia rufescens.* **A**-Flowering, fruiting branch **B**-Seedlings **C**-Pod with seeds **D**-Habit

Brachystegia spp.

Brachystegia spiciformis Benth.

Description. Tree, about 10 m high; grows on sandy and gravelly soil.
Distribution. An important constituent of miombo woodland, which occurs extensively in Central and East Africa.
Characteristics. Leaves and thin twigs browsed by cattle during the dry season (van Rensburg, 1968).

Cajanus spp.

Cajanus cajan (L.) Millsp.

Synonyms. *C. indicus* Spreng.; *C. flavus* DC; *Cytisus cajanus* L.; *Cajanus cajan* (L.) Druce.
Common names. Guandul, poroto guandul, poroto paraguayo, sachacafé, falso café, arveja (Argentina), guando (Brazil), quinchoncho (Venezuela), frijol de árbol (Mexico), Cumandái (Paraguay), red gram, tur, arhar, dahl (India), catjang, kachang (Asia), pigeon pea (Australia), pigeon pea, angola pea (United Kingdom), pois d'angole (French-speaking West Africa), Puerto Rican bean, pigeon pea (Hawaii).

539

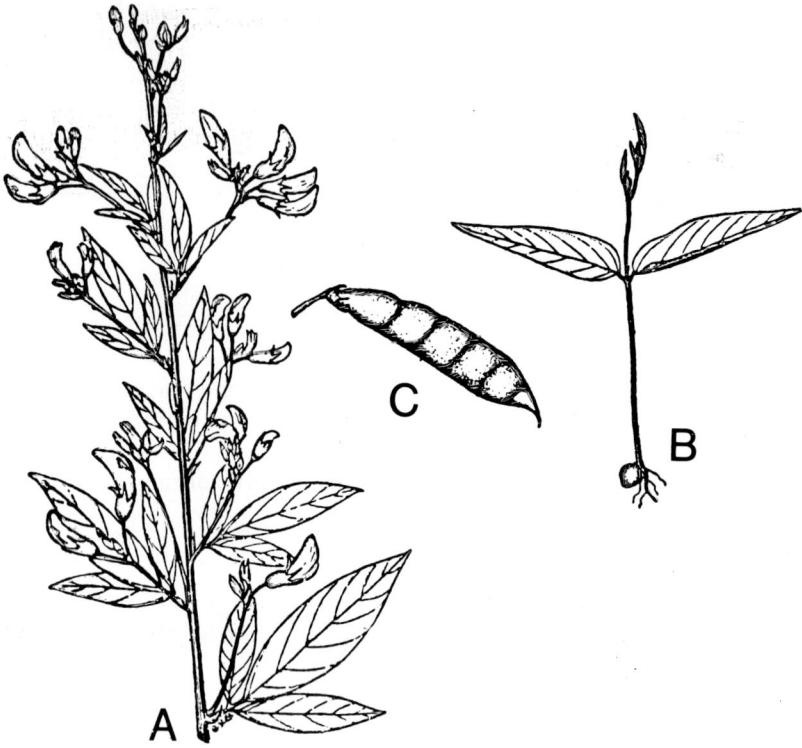

Figure 112. *Cajanus cajan.* **A**-Habit **B**-Seedling **C**-Pod

Description. Annual, or more usually short-term perennial, shrub growing to 4 m high, but usually 1 to 2 m, woody at the base. Leaves are narrow-lanceolate and hairy. Yellow, brown or purple flowers borne in terminal racemes. Pods are usually short (5 to 6 cm), though long-podded types are found in India. Pods contain four to six seeds in shades of brown, red or black. It has an extremely deep-rooting taproot. Persists up to five years in Colombia.

Distribution. Of Indian and African origin; now acclimatized in several tropical countries.

Temperature for growth. Prefers hot moist conditions. Under Hawaiian conditions grows between 18 and 30°C. Will grow at temperatures above 35°C under adequate soil conditions of moisture and fertility. Does not tolerate frost, but will grow in temperatures to just above frost level. Will seed as a perennial at 1 840 m down to a minimum night temperature of 10°C (Krause, 1932).

Subject to frost damage (Schofield, 1945). Tall plants may escape light frosts because of the height of the foliage.

Latitudinal limits. Extends to 30°S in South America (Burkart, 1952).

Altitude range. 1 250 m in Hawaii (Ripperton and Hosaka, 1942), at heights of 1 230 to 1 500 m it fails to set seed; 0 to 2 000 m in India (Krause, 1932) and Colombia (Crowder, 1960). Essentially a plant of the semi-dry lowlands but has wide adaptability. Generally should not be planted above 770 m.

Rainfall requirements. Wide range of rainfall tolerance, but prefers more than 625 mm and in elevated areas exceeding 2 000 m cold nights and cloudy weather interfere with fertilization of flowers. Flowers well where rainfall is 1 500 to 2 000 mm. On deep, well-structured soil will grow at a rainfall of 250 to 375 mm (Krause, 1932).

Tolerance of drought and flooding. Drought tolerance is good because of its deep-rooted habit. In the dry season there may be leaf fall. Damaged by waterlogging (Stanton, 1966).

Soil requirements. Tolerates a wide range of soils, from sands to heavy black clays.

Tolerates a wide range of pH, but the most favourable range is pH 5.0 to 7.0. It is sensitive to salt spray. Will grow in sand provided it does not contain more than 0.0005 g of sodium chloride per gram of soil (Krause, 1932).

Rhizobium relationships. Promiscuous and usually not inoculated. The cowpea type of inoculum can be used, the current Australian strain being CB756 and in South America, C95, based on the Australian material (Batthyany, 1970).

Oke (1967a) found that nodules appeared on the roots three weeks after seeding, and that young nodules of pigeon pea were the most active in fixing nitrogen.

Ability to spread naturally. Not easily, especially in pastures.

Land preparation for establishment. Thrives best in seed beds prepared by deep ploughing and cultivations to reduce weeds.

Sowing methods. Best sown in rows for interrow cultivation and mechanical harvesting. Can be seeded in holes about 2 m apart or in 35-cm rows for fine stems for silage and mechanical harvesting (van Rensburg, 1967). In India and Uganda, it is usually sown in alternate rows with sesame or *Eleusine coracana*; in Malawi, with maize; in Hawaii, some sow in double rows 3 m apart from centre to centre and intercrop with forage grasses. Planting is done by maize drill, by broadcasting or by hand. For oversowing into natural pastures, von Schaaffhausen (1966) recommends sowing *Cajanus* in contour furrows 2 m apart into pangola grass pastures, using 20 to 30 seeds per hole. A large percentage of mature seeds pass undigested into the dung and germinate and many seedlings are established in this way (Krause, 1932).

Seed is sown from 2.5 to 10 cm deep, the deeper figure being in hand dibbling.

Sow in spring to early summer at 0.5 to 4 kg/ha depending on required density, up to 5 to 7 kg/ha under irrigation.

Number of seeds per kg. 16 000. Percentage of hard seed is less than the 10 percent required for commercial sale.

Seed treatment before planting. To break dormancy: no treatment is required. Inoculation is usually not necessary. Pelleting is usually not necessary; if to protect *Rhizobium*, use rock phosphate. For insect and disease control, treat seed with insecticide and fungicide.

Nutrient requirements. Usually responds to phosphorus, and requires enough calcium, potash, magnesium. According to Nichols (1964) the following are deficiency symptoms observed in this plant for the major elements:

- On old leaves
- Leaf tip yellow or brown. Yellowing spreads from the tip onward along the leaf margin and may coalesce with similar areas at extremities of the lateral veins. Leaf tip becomes scorched as symptoms become more severe. The scorching may spread around the leaf margin but typically there is a yellow band between the scorched area and the healthy green tissue in the early stages. The affected leaves not showing symptoms are generally dark green. Plants stunted.

Deficiency: *potassium*.

542

● Mild interveinal chlorosis, veins remain dark green. Interveinal areas become rusty brown or bronzed and may become necrotic so that narrow elongated streaks of dying tissue appear between the veins. The margins of the young leaves frequently inrolled.
Deficiency: *magnesium*.

● On youngest leaves
● Leaves uniformly pale green, greenish-yellow, or pale yellow. Oldest leaves show similar symptoms.
Deficiency: *nitrogen*.
● Interveinal areas light green, veins and adjacent tissues remaining dark green.
Deficiency: *iron*.
● Light green patches, irregular in outline, appear randomly in interveinal areas around the leaf margin. Patches may turn brown, particularly in the youngest and most severely affected leaves. Severely affected young leaves absciss. Stems weak and prostrate.
Deficiency: *calcium*.

● No clearly defined symptoms
● Plants small and stunted, leaves dark green; no obvious leaf symptoms.
Deficiency: *phosphorus*.

Toxicity levels and symptoms. Manganese — the plant becomes chlorotic when growing in the high manganese soils of Hawaii (Krause, 1932).
Response to photoperiod and light. It is a short-day plant. Gooding (1962) reported two groups of plants in Trinidad — one group spans 60 to 106 days in the time taken from sowing to podding, the other up to 237 days. It flowers over about two months as days become shorter, and has flowers and ripe pods on the plant at the same time. It flowers in 98 days in the Kimberley district in northern Australia (lat. 15°S) with pods maturing in 178 days. It flowers in eight months in the Bombay district in India (lat. 20°N). Too much shading causes the plant to make spindly growth and to bear thin, pale green foliage and few pods (Krause, 1932).
Compatibility with grasses and other legumes. Grows well interplanted with Rhodes grass, *Cynodon dactylon* and molasses grass in Hawaii (Hosaka and Ripperton, 1944), and with pangola grass in Brazil (von Schaaffhausen, 1966).
Ability to compete with weeds. Not high in the seedling stage, but thereafter

543

quickly shades out weeds. Dry season leaf fall also helps suppress weeds (Stanton, 1966).

Tolerance of herbicides. It is fairly tolerant of herbicides. (Getner and Danielson, 1965; Kasasion, 1968).

Vigour of seedling, growth and growth rhythm. Seedling is fairly slow to start and weed control for the first two months of growth considerably improves its performance. When established, it makes vigorous growth.

Nitrogen-fixing ability. It is regarded as a good plant for restoration of fertility and is used in a rotation such as maize-groundnut-tobacco-pigeon pea for three to four years in Mauritius and Uganda (Stanton, 1966). Hosaka and Ripperton (1944) state that one of the advantages of pigeon pea is the increased growth and palatability of the grass interplanted with it. In Hawaii, a profitable rotation with pineapples — five years for each crop — was developed, to the great benefit of the pineapple crop (Krause, 1932). Oke (1967a) found maximum fixation in pigeon pea of 14.5 mg/day, compared with 10.3 mg for *Centrosema pubescens* and 4.6 mg for *Stylosanthes guianensis*. Eighty-eight percent of the nitrogen fixed was transferred to the tops of the plants.

Response to defoliation. It does not persist if heavily grazed (Ripperton and Hosaka, 1942). For cut forage, it should be topped at a height of about 0.8 m. Under grazing conditions, it makes vigorous growth during the first year and then declines and ordinarily requires replanting every four to five years.

Grazing management. Allow the plants to develop well before grazing and then lightly graze to a hedge at the desired height. The plant stems are brittle and easily broken, so grazing should not be continuous. In Colombia, where the plants are used for poultry rations, cutting height is 0.5 m in the first year and 1 m in the second year, and cattle are allowed to browse the regrowth. In São Paulo, Brazil, it can be grazed for up to five years if carefully handled.

Response to fire. Easily destroyed by fire (Downes, 1966).

Breeding system. Usually self-fertile, but some natural crossing (up to 40 percent) occurs (Khan, 1973). Chromosome number $2n = 22$.

Dry- and green-matter yields. Parbery (1967a) harvested 25.45 tonnes of dry matter in 372 days (unfertilized) and 37.96 tonnes/ha when fertilized with 100 kg N/ha on Cunnunurra clay but only 1 071 kg/ha, unfertilized, on Cockatoo sand in the Kimberley district of northern Australia. One hundred kg N/ha depressed its yield on Cockatoo sand. The yield of 37.96 tonnes when fertilized with 100 kg/ha represents 7 704 kg/ha of protein.

In Colombia, Herrera and Crowder (1963) obtained 14 000 kg DM/ha and more than 2 000 kg/ha of crude protein from cutting pigeon pea at ground level or at 15 cm when the plants were 150 cm tall. No regrowth occurred when the plants were cut to ground level.

Suitability for hay and silage. Pigeon pea is harvested for hay and for milling for meal in Hawaii when a large percentage of pods are mature — probably two-thirds to three-quarters of the pods in sight — because a large part of the nutritive value of the plant is contained in the seed. Harvest not more than the upper third of the plant to avoid the woody base unless the plant is spindly. Cure on the cut surface of the plant which has just been harvested for six to eight days and then mill into meal.

Pigeon-pea hay is an effective substitute for more expensive industrial concentrates.

Otero (1952) recorded that in Brazil it made good silage and was very palatable to the animals.

Value as standover or deferred feed. If *Cajanus* is not frosted, it will stand over very well for winter and dry-season feeding.

The plant has a high feeding value for beef and dairy cattle, swine, sheep and goats.

Feeding value.

● Chemical analysis and digestibility. The well-leaved and fully podded branches, cut at 0.8 metres, have 40 to 50 percent dry matter, and protein up to 16 percent of the dry matter (Takahashi and Ripperton, 1949).

The plant is also a good source of vitamin A. Otero (1952) recorded 11.46 percent crude protein and 22.6 percent crude fibre in the leaves, and 18.36 crude protein and 5.43 percent crude fibre in the seeds. From silage made from the foliage he recorded 15.09 percent crude protein, 26.05 percent crude fibre, and 32.8 percent nitrogen-free extract in the dry matter. The silage had 66.7 percent crude fibre, and 32.8 percent nitrogen-free extract in the dry matter. The silage had 66.7 percent moisture.

Krause's (1932) analyses are given in Table 15.3.

● Palatability. The forage is not relished by cattle in the immature stage. Grazing should be deferred to the early green-pod stage (Hosaka and Ripperton, 1944).

Toxicity. Mature plants may cause irritation of the rumen of cattle (Stanton, 1966).

Seed harvesting methods. Most of the seed pods are picked by hand in the tropics, although mechanical harvesting is done in Hawaii (Krause, 1932). The seed pods ripen irregularly and do not shatter easily. The final drying of the pods may be done after harvest. Threshing the seed from the stem

545

TABLE 15.3 **Compositional analysis of *Cajanus cajan* on a dry-matter basis**

Form	Moisture	Ash	Crude protein	Crude fibre	N-free extract	Nitrogen	Fat
				(%)			
Fresh green forage	70.00	2.64	7.11	10.72	7.88	1.13	1.65
Meal from whole plant	11.19	3.53	14.83	28.87	39.89	2.37	1.72
Seed and pod meal	11.45	3.85	17.65	30.73	34.53	2.82	1.49
Seed meal	12.26	3.55	22.34	6.44	53.94	3.57	1.46
Pod hull meal	11.30	2.66	8.75	35.44	39.22	1.40	1.03

requires a strongly constructed machine such as double-cylinder pea and bean threshers. The seed is then winnowed, graded and stored in airtight containers to prevent insect damage.

Seed yields. 900 to 1 150 kg/ha. Akinola and Whiteman (1972) recorded a yield of 7.5 tonnes/ha of shelled seed in experimental plots in Queensland, Australia.

Minimum germination and quality for commercial sale. Seventy percent minimum germination, with not more than 10 percent hard seed, and a purity of at least 98.8 percent in Queensland. The seed is germinated at 25°C under a cover.

Cultivars. There are no registered commercial cultivars in Australia, but in India there are numerous varieties — Whyte (1964) reported 86 different types. Commercially there are the late, tall, long-podded, many-grained "arhar" type of northwest and eastern India, and the early, small, few-seeded "tur" type of Madhya Pradesh, western and peninsular India.

Diseases. Whyte (1964) records that it is subject to wilt (*Fusarium udum*) and leaf spot. In Puerto Rico it has been affected by spotting of the leaves and pods and seed destruction by *Colletotrichum cajani* (Krause, 1932). Gooding (1962) reports collar and stem canker caused by *Physalaspora* and possibly *Diplodia*, and Hammerton (1975), rust (*Uredo cajani*).

Pests. Whyte (1964) records that it is attacked by caterpillars, pod borers and gall fly.

Colbran (1963) found that the roots were attacked by the nematode *Helicotylenchus dihystera*. In Hawaii it is attacked by the scale insect, *Coccus elongatus*, a stem borer, termites, a pod borer (*Lycaena boetica*) and leaf-

546

eating caterpillars (Krause, 1932). A 4 percent DDT spray controlled pod borers in Mauritius (Gooding, 1962).

Main attributes. It is more adaptable to higher elevations than *Leucaena leucocephala*, though not to wet conditions; it is palatable forage and there is a high proportion of seed in the forage; it is adaptable to a wide range of soil types.

Main deficiencies. Its short life, frost susceptibility and brittleness under grazing.

Performance. The grain is used for human and stock feed. Straw, husks and screenings are useful roughages. Nearly 2 350 000 hectares are grown in India yearly. In one single harvest in Hawaii, the fully podded tops gave 11.2 tonnes of green forage, 4.97 tonnes of dry matter and 400 kg protein per hectare (Takahashi and Ripperton, 1949).

Cattle fed wholly on pigeon-pea pasture have gained in weight from 0.7 to 1.25 kg per head per day at a carrying capacity of 1 to 3.75 beasts/ha, and liveweight gains of 200 to 500 kg per hectare per annum have been recorded (Krause, 1932).

In Brazil, von Schaaffhausen (1966) found that Zebu bulls grazing a pangola grass/pigeon pea pasture gained an average of 35 kg in 90 days during a severe drought, while animals on a control pasture lost 6 kg. On rotation pastures of *Lablab purpureus*, pigeon pea and grasses, 47 bulls gained 40 kg in 63 days.

Miscellaneous. The crop has long been used as a windbreak and shade for young coffee trees, forest seedling nurseries and vegetable beds, and is an important honey-producing plant. The canning of green pigeon peas is a major industry in Puerto Rico and Trinidad.

Main references. Gooding (1962); Krause (1932); Nichols (1964); Akinola, Whiteman and Wallis (1975).

Cassia spp.

Cassia eremophila **A. Cunn. ex Vog.**

Description. Small, attractive shrub, rarely more than 2 m high, more or less rounded in shape. Leaves finely cylindrical or narrow and flattened, 2.5 to 5.0 cm long, occurring in pairs at the end of a short, thin, cylindrical stalk. Flowers numerous, with bright yellow petals in short clusters. Pod thin, 7.5 to 10 cm long and 1.25 cm wide.
Distribution. Australian semi-arid subtropics.
Characteristics. Occurs on deep sands or sandy soils overlying clay, often in windswept areas. Grows quickly from seed to 1 m in height in three years. It is edible and can be regenerated by excluding stock (Stannard and Condon, 1958).

Desmanthus spp.

Desmanthus virgatus **(L.) Willd.**

Synonyms. *Mimosa virgata* L.; *Desmanthus depressus* H. et B. ex Willd.
Common names. Dwarf koa, desmanthus (Hawaii).

548

Description. Small shrub, 2 to 3 m tall, nearly erect or (more commonly) diffuse or decumbent; branchlets glabrous. Leaves moderately small, bipinnate; 10 to 20 leaflet pairs per pinna, petiole usually no more than 5 mm long. Inflorescence of axillary, pedunculate heads toward the tips of the twigs; head small, dense, few-flowered, the flowers all erect, whitish, sessile; pod linear, 4 to 6 cm long and 3 to 4 m wide, flat, glabrous, shortbeaked, dehiscent on both valves, seeds oblique.

Distribution. Florida and Texas, United States, to Argentina; West Indies; Galapagos Islands; Hawaii. It is restricted to roadsides and unused blocks of land, and does not occur in pastures in Hawaii.

Characteristics. A perennial browse similar to *Leucaena leucocephala* but has slender, angular, pithy stems, smaller leaflets and narrow pods.

Grows in sandy and other open-textured soils under a rainfall regime of 1 000 to 1 500 mm at elevations from sea level up to 300 m. In Paraguay it occurs on clay soils. Its pH preference is from 5.0 to 6.5 (Fretes, Samudio and Gray, 1970). Grows best in hot weather; its frost tolerance is unknown. It is quite drought tolerant. Grows well with tall grasses, but stoloniferous grasses compete with it. It is sown on a well-prepared seed bed at 2 kg/ha and at a depth of 1 to 1.5 cm. The seed should be treated with concentrated sulphuric acid for 8 minutes to break dormancy. It appears to fix ample nitrogen. Desmanthus is a plant of high palatability which can be harvested four times a year in Hawaii, cutting it at the early pod stage. It flowers 45 to 50 days after cutting. In Hawaii, the plants were cut 5 to 7.5 cm above ground with a mower. Cutting at 91-day (four cuts per year) intervals gave the highest yield, 23.68 tonnes/ha/year over three years, *Leucaena leucocephala* giving a significantly higher yield than desmanthus. The crude-protein content of the whole plant cut at 61-, 91- and 122-day intervals was 10.55 percent, 12.27 percent and 15.52 percent respectively and the average protein content of the leaves was 22.4 percent, of the stems 7.10 percent. Leucaena surpassed desmanthus in protein percentage by one-third. Under the cutting treatments, desmanthus developed a crown similar in many respects to that of alfalfa (lucerne). The size of the crown gradually increased with each successive crop until it was up to 15 cm across after three years. This well-developed, vigorous crown produced as many as 50 slender, erect stems, seldom branched, pithy in the centre, brittle and fairly soft. The yield of oven-dry forage decreased in the second and third years, but there was no mortality up to the fourth year. In every way, *Leucaena leucocephala* performed better than desmanthus in Hawaii, but desmanthus is non-toxic (Takahashi and Ripperton, 1949). At Sigatoka in Fiji, desmanthus yielded an average of 7 590 kg DM/ha, of which

64 percent was produced in the wet season and 26 percent in the dry season in a 2 000-mm rainfall regime (Payne *et al.*, 1955). At the Kimberley Research Station in northern Australia, Parbery (1967a) obtained dry-matter yields of 35.08 tonnes/ha/year without added nitrogen on Cunnunurra clay and 26.05 tonnes/ha with 100 kg/ha of added nitrogen, cutting the whole plant. When the foliage above 45 cm only was taken, he harvested 63.29 tonnes/ha unfertilized and 70.33 tonnes/ha with 100 kg added N per hectare, each from four cuts per year.

It is regarded as a useful forage plant in Argentina (Burkart, 1943) and Colombia (Bermudez *et al.*, 1968).

Desmodium spp.

Desmodium capitatum DC

Description. A woody plant, silvery when young. Petioles 1.25 to 2.5 cm long, leaflets green, fairly smooth above but densely pubescent with white silky hairs below. Terminal leaflets 2.5 to 4 cm long. Racemes copious, terminal, dense, 2.5 to 5 cm long; pod 1.25 to 2 cm long, smooth, 4- to 6-jointed and 0.3 cm broad (Hooker, 1879).

Distribution. India, Sri Lanka, Malaysia, Philippines.

Characteristics. Farinas (1966) records this as a drought-resistant species which also tolerates stagnant water up to at least two months.

Desmodium discolor Vog.

Synonym. *Meiboma discolor* Vog.

Common names. Horse marmalade (South Africa), marmelada de cavalo (Zambia, Brazil), discoloured clover (Fiji).

Description. Upright perennial plant growing to a height of 2.5 to 3 m with stems which become woody when mature. Has a well-developed, deep root system. Trifoliate leaves are hairy and dark green, although they often have a reddish tinge when the plant is growing vigorously. Flowers are pink or white and borne in large open panicles. Seed is fine, about the size of lucerne (alfalfa).

Distribution. Found in the subtropics of Brazil (Saõ Paulo, Mato Grosso), Argentina (Misiones), Paraguay. It has been introduced to Colombia, Cuba, Fiji, South Africa and Zambia.

Temperature for growth. It prefers warmer climates (Boelcke, 1964). It starts growing early in the season as soon as temperatures rise. Young plants are susceptible to frost, but established plants are little affected.

Rainfall requirements. It has a wide range, from relatively dry to subhumid (Otero, 1952).

Tolerance of flooding. Does not tolerate "wet feet." Waterlogged sites should be avoided.

Soil requirements. Does best in well-drained, deep fertile soils but will grow on a wide range from acid sandy loams to loams (Whyte *et al.*, 1969), but if the soil is too acid, liming will benefit it (Otero, 1952).

Rhizobium *requirement*. No inoculation has been found necessary at the Henderson Research Station, Zimbabwe (Boultwood, 1964) but the "Desmodium"-type *Rhizobium* may be beneficial.

Land preparation for establishment. Performs best on a well-prepared seed bed with cultivation early in the planting season to destroy young weed growth. Subsoiling to a depth of 40 cm is beneficial in hard soils (Otero, 1952).

Sowing methods. On a prepared seed bed it can be drilled in rows 0.40 m apart; for seed purposes sowing on a square 0.60 × 0.60 m allows interrow cultivation. It can also be sown by cuttings. In Zimbabwe, (Boultwood, 1964) it is often sown in maize, which acts as a nurse crop providing a protective canopy, reducing soil crusting from raindrop impact and preventing rapid drying. Undersowing is best done after at least one cultivation of the maize to control the early flush of weeds, and the seed is broadcast between the maize rows onto the loose soil.

551

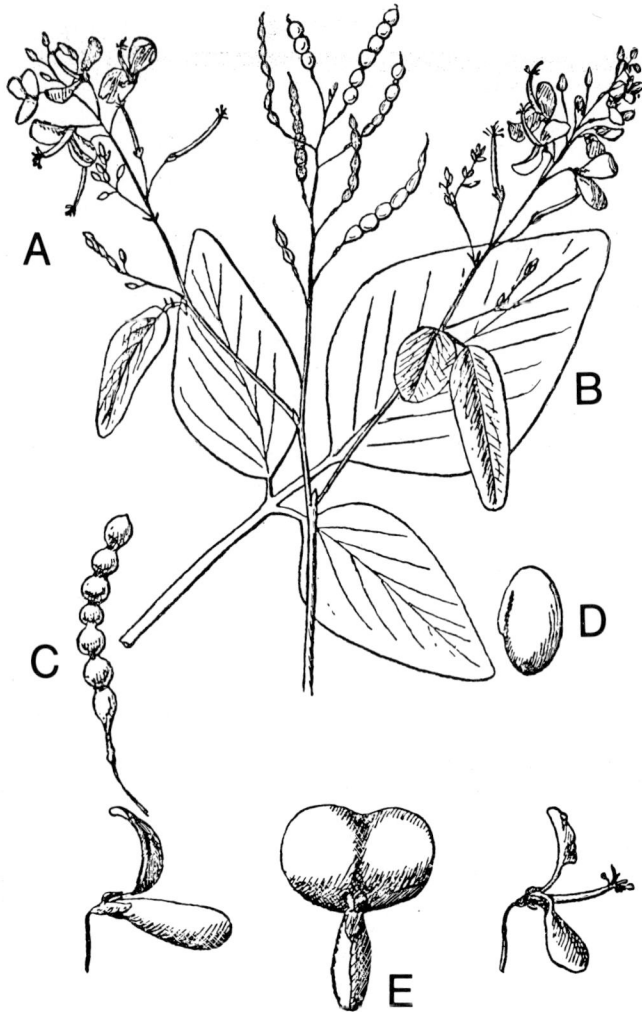

Figure 113. *Desmodium discolor.* **A**-Flowering, fruiting branch **B**-Trifoliate leaf **C**-Pod **D**-Seed **E**-Flowers

Seed is broadcast on the surface or drilled at 1 to 2 cm and rolled after being lightly covered.

It is planted just before the rains or early in the growing season (October to November in Brazil). Sowing rate is 5 to 8 kg/ha in 0.60 × 0.60-m rows; 30 to 35 kg/ha broadcast or seeded in 0.40-m rows. One hundred litres of seed weigh 40 kg.

Trials at Henderson Research Station, Zimbabwe (Boultwood, 1964), indicate that *D. discolor* may be successfully introduced into rough veld grazing. The land is disc-harrowed in strips through the veld, after burning just before the rains. This reduces the competition from the grasses and provides an adequate seed bed.

Percentage of hard seed. An average sample of seed contains 12 percent rotten seed, 27 percent viable seed which will germinate immediately, and 61 percent hard seed (Otero, 1952).

Seed treatment before planting. Dormancy can be broken by treatment with concentrated sulphuric acid. Nicking the seed will also accomplish this (Otero, 1952).

Nutrient requirements. No trials have been undertaken to determine requirements. In Zimbabwe, 200 kg/ha single superphosphate are applied for establishment and a similar quantity is used for yearly maintenance.

Vigour of growth. Grows slowly in the first year and is of little use until the second year (Boultwood, 1964). At Katapola Farm Institute in Zambia, it survived competition and produced vigorous growth (van Rensburg, 1967).

Response to light and day length. A short-day plant, it flowers in April and seeds in May in Brazil. It is able to tolerate deep shading in the early stages of growth.

Compatibility with grasses. Grows well in natural pastures and in maize.

Ability to compete with weeds. In the young stages, weeding is recommended. It rapidly grows taller when it becomes established and dominates weeds.

Nitrogen-fixing ability. No measurements have been undertaken. It nodulates freely and grass growing in association with it assumes a darker green colour.

Grazing management. Stands should be allowed to become well established before being used intensively, and it is advisable to leave first-year stands to grow for seed production and then graze them in the dry season. Although cattle will graze it at all stages of growth, they will not take the stems if they are mature and woody. Boelcke (1964) recommends cutting and feeding the cut material to cattle.

Response to fire. Resists fire very well.

553

Breeding system. Self-fertile, chromosome number 2n = 22 (Rotar and Ukio, 1967).

Yield. Twenty-four to 30 tonnes of green material per hectare have been harvested in Zimbabwe, and in Brazil 3 430 kg/ha, which produced 1 000 kg of hay (Otero, 1952).

Suitability for hay and silage. Makes excellent hay in Brazil. Can be cut at a height above 10 cm and will give five cuts a year. The hay should be handled carefully to prevent leaf shedding. Also makes excellent silage if cut at a height of 1.5 m. It is readily harvested with a flail-type forage harvester. It is advisable to add molasses at the rate of 2 percent by weight of the green material (Boultwood, 1964).

Feeding value.

● Chemical analysis and digestibility. Otero (1952) recorded the figures shown in Table 15.4.

TABLE 15.4 **Chemical analysis and digestibility of *Desmodium discolor***

	Moisture	Crude protein	Fat	N-free extract	Fibre	Ash
	(%)					
Green material	81.50	3.15	0.45	6.48	6.68	1.74
Dry-matter	—	17.04	2.44	35.00	36.10	9.42
Hay	10.48	16.60	1.63	35.07	23.09	13.13
Digestibility	—	59.00	46.5	57.00	46.00	—

Source: Otero, 1952

In Zimbabwe, the protein content of the dry matter of the green material was recorded as 15.5 percent and of silage made from material cut in mid-season as 11 to 13 percent.

● Palatability. All reports indicate that the plant is extremely palatable.

Value as standover feed. It is especially useful for supplementary protein-rich grazing in late summer and autumn when the feed value of grasses is declining.

Seed harvesting methods. So far only hand harvesting has been practised. The seed heads are cut, allowed to dry thoroughly and are fed through a slow-running hammer mill. A combination of winnowing and sieving then cleans the seed.

Seed yield. Up to 200 kg/ha of clean seed can be expected in Zimbabwe and Otero (1952) records 150 kg/ha in Brazil, but up to 300 kg from hybrids.

Diseases. The plants differ in their resistance to nematodes, and selection for nematode-resistant plants is proceeding at the Henderson Research Station in Zimbabwe. A rustlike fungus attacks the plant in the winter in Brazil (Otero, 1952).

Pests. At Mt. Makulu in Zambia, severe damage occurs early in the season from attacks of a "flea beetle" (*Monolepta gossypiperda* Br.) (van Rensburg, 1967).

Main references. Boultwood (1964); Otero (1952).

Desmodium distortum (Aubl.) Macbride

Description. Upright herbaceous perennial plant growing 2 m high, becoming woody at the base; well branched; leaves oval to oblong, pointed; inflorescence broomlike, hairy. Pods of one to six ovate segments (Pittier, 1944).

Distribution. Occurs in Venezuela, but is not common.

Characteristics. In Zambia during the rainy season it produced a massive crop of fodder of 7 355 kg dry matter per hectare. It produced a very useful secondary growth between March and June and continued to grow during the latter part of the dry season.

Palatability. In spite of its stemmy nature and coarse appearance, it is relished by cattle and even much of the stem is eaten.

Owing to its stemmy growth habit, this species is not very suitable for inclusion in grass-legume pastures, but it is a very promising fodder plant for specialized use (van Rensburg, 1967). Horrell (1963) reported that it seeded well at Serere (Uganda) and was a possible fodder plant for that area. In Tanzania, it remained green well into the dry season and was relished by cattle (Tanzania, 1955). Early growth was slow.

Desmodium gyroides DC

Synonym. Codariocalyx gyroides.

Description. Erect, often much-branched shrub growing from 1 to 2.5 m. Rachises of the inflorescence and top of stem clothed with long hairs; termi-

555

Figure 114. *Desmodium distortum* growing at Mt Makalu Research Station, Zambia (**Photo:** H.J. van Rensburg)

nal leaflet oval to oblong obovate, 3.5 to 7.5 cm long × 1.75 to 5 cm broad; corolla light pink at first, turning to violet pink, and afterwards light lilac or bluish lilac. Pod densely clothed with long hairs, 2.5 to 4.75 cm long, 5- to 13-seeded (Backer and van den Brink, 1963). Chromosome number 2n = 22.

Distribution. Indonesia.

Characteristics. Bryan (1966) reported that the plant retained its leaves in winter in Queensland, Australia. The lower half of the plant is deciduous if ungrazed, but under a system of grazing that maintains the whole plant within reach of the grazing animal, it is deciduous. Plants have persisted under grazing for five years.

It is unaffected by high water-tables and poor drainage. Roberts (personal communication) says that it is better suited to wet conditions than leucaena in Fiji; cattle browse the leaf down the stem without taking much stem.

Bryan stated that it did not have a high requirement for Ca or Mg. He envisaged its chief use as a high-protein feed saved for autumn and winter in Queensland and browsed to control it at other times of the year. In midwinter 1965, a typical 18-month-old stand, 165 cm high, yielded 1 980 kg of edible dry matter per hectare containing 3.1 percent N, equivalent to 385 kg/ha of protein.

556

Desmodium lasiocarpum (Beauv.) DC

Description. Half-woody, erect, hairy under-shrub growing to 1 to 2 m high; branches densely brown, pubescent. Leaves large, ovate or rhomboid, rounded to acute at the apex: up to 15 cm long and 8 cm broad, tomentose on both surfaces. Flowers pink or purplish, small in dense sessile racemes. Pod about 1.8 cm long, densely hairy, deeply indented on one side (Andrews, 1952).

Distribution. Occurs in the southern Sudan and is being tested in Maracay, Venezuela.

Characteristics. Adapted to high rainfall areas (up to 3 000 mm annually). Regarded as a browse plant.

Desmodium tortuosum (Sw.) DC

Synonyms. *D. purpureum* Fawe and Rendle; *Hedysarum tortuosum* Sw.; *D. spirale* (Sw.) DC.

Common name. Florida beggarweed (United States).

Description. Erect, short-lived annual or perennial herb, 0.6 to 3 m tall, somewhat woody at base, cylindrical, smooth in lower parts of stem and clothed with short, fine hairs in upper part. Leaves trifoliate, leaflets lanceolate or elliptic, sparsely covered with short hairs, the central leaflet larger and on a stalk four times longer than those of the laterals. Flowers purple or pale blue, about 4 mm long, borne on slender stalk about 1.25 cm long. Seed pod is indented about equally on both sutures. It is chainlike, with 3 to 7 joints, and covered with short hairs (Hosaka and Ripperton, 1944). Somatic chromosome number $2n = 22$.

Distribution. Native to the West Indies; widespread in Florida, Central and South America and also in tropical Africa. Occurs in Hawaii below 900 m elevation.

Characteristics. Cultivated in Florida for forage. Generally used for green chop or hay in the tropics. Very palatable; a good plant for grazing or hay. Vigorous, recovers rapidly from cutting or grazing and persists well under favourable conditions up to four years. Adapted to a wide range of soils and often volunteers on stubble. Because it is so palatable it tends to disappear under grazing.

557

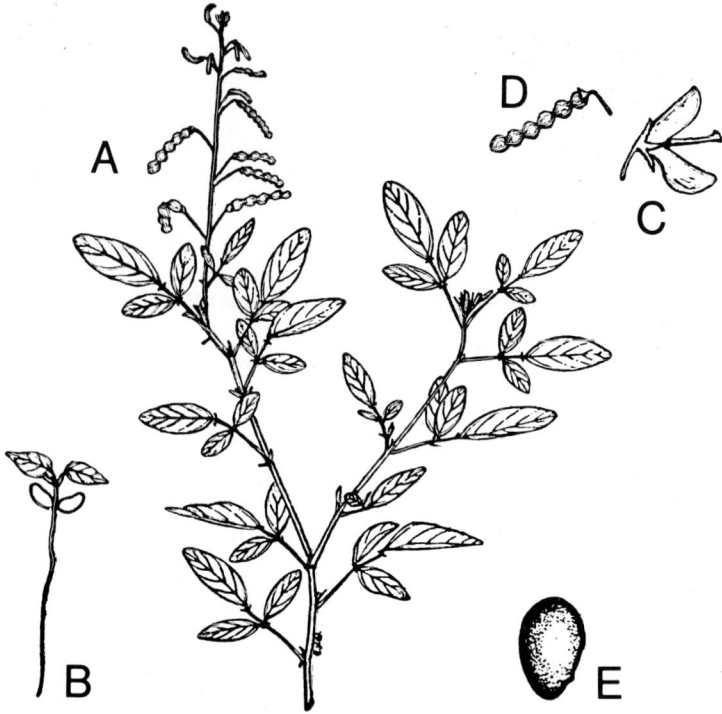

Figure 115. *Desmodium tortuosum.* **A**-Habit **B**-Seedling **C**-Flower **D**-Pod **E**-Seed

Dichrostachys spp.

***Dichrostachys cinerea* (L.) Wight and Arn.**

Synonyms. *D. glomerata* (Forsk.) Chiov.; *D. nutans* (Pers.) Benth.; *Mimosa cinerea* L.

Common name. Kakada (the Sudan).

Description. Shrub or tree, usually 3 to 5 m high, branchlets armed with sharp woody spines which terminate the lateral branchlets and often bear leaves. Leaf-rachis pubescent; pinnae in 7 to 15 pairs with a rodlike gland between each pair; leaflets in 15 to 30 pairs, variable in size, usually less than 6 mm long. Flowers spikes pendulous, 1.25 to 7.5 cm long; distal (functional) flowers composed of ten long pink or mauve staminodes. Pod dark brown, about 10 cm long, indehiscent, glabrous, twisting to form strangely shaped bundles that remain on the tree for a long period, usually four-seeded (Andrews, 1952).

Distribution. Widespread in tropical Africa, from Cape Verde, the Gambia and the Sudan south to Natal and Angola; also in Iran, tropical Asia and Australia. Introduced in Florida, the United States, and in Cuba.

Occurs in frost-free localities in medium rainfall areas (West, 1950).

Characteristics. French (1934) stated that the pods form a valuable source of protein in central Tanzania, when they ripen and fall in August, a time when grazing is scarce. Camels eat the leaves and cattle eat the pods in the Sudan (Tothill, 1954), and McKay (1968) reported that cattle and goats browse the pods in Botswana — after three years of goat grazing the plants were diminished in size. Van Rensburg (1968) found that it had a high crude protein content of 15 percent in the leaves and shoots, with 21.6 percent crude fibre, 1.53 percent Ca and 0.18 percent P. He reported that the plant forms an almost impenetrable thicket (it produces suckers), which is fire-proof, and although it is a good fodder plant, it should be eradicated in Zambia.

559

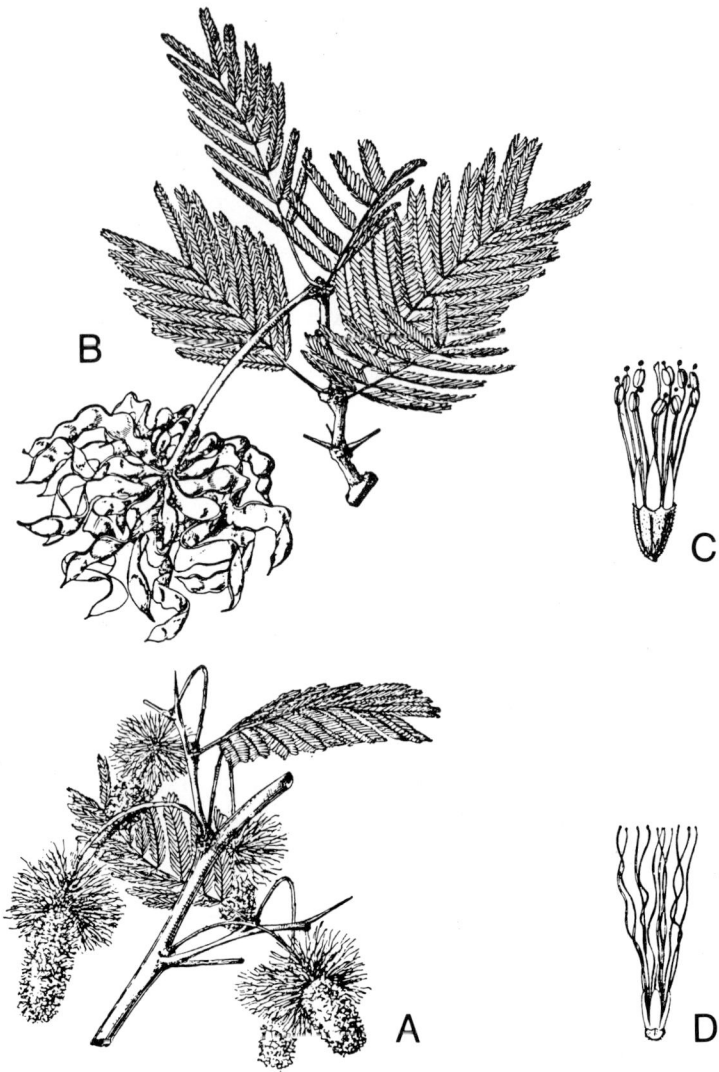

Figure 116. *Dichrostachys cinerea.* **A**-Flowering branch **B**-Fruiting branch **C**-Bisexual flower **D**-Neuter flower

560

Eriosema spp.

Eriosema psoraloides (Lam.) Don

Synonym. *E. cajancides* (Guillem and Perrott) Benth. and Stock.
Description. Leguminous shrub about 1.5 m high with yellow flowers and hairy two-seeded pods.
Distribution. It is widely distributed in *Combretum* mixed woodland and *Hyparrhenia filipendula* grassland on loam and sandy loam soils in Zambia, and in the southern Sudan.
Characteristics. It appears to be unpalatable and is avoided by stock except at the end of the dry season (van Rensburg, 1968).

Flemingia spp.

Flemingia macrophylla (Willd.) Merrill (see Colour plate L)

Synonyms. *Flemingia congesta* Roxb.; *F. latifolia* Benth.; *Moghania macrophylla* Willd.
Description. Erect shrub, up to 2 m high; leaves digitately trifoliate. Leaflets

often longer than 10 cm. Flowers in dense racemes with greenish standard with red blotches or stripes. Pod 11 to 15 mm long.

Distribution. Occurs in southeast Asia in brushwood, waterways and shaded locations.

Characteristics. Used as a green manure in plantation crops. Kannegieter (1966), after testing numerous legumes in the forests (1 125- to 2 000-mm rainfall) of Ghana, selected *Flemingia macrophylla* with centro as the most promising for mixing with grasses for temporary leys. In Malaysia it is a useful bush to plant with creeping legumes as it provides support for them to climb and is deep rooting. It fixed a useful amount of nitrogen and tolerates light shade (Establishment and maintenance of legume covers, 1958).

Gleditsia spp.

Gleditsia triacanthos L.

Common names. Honey locust (United States, Australia), driedoring, sprinkaan boom (South Africa), acacia negra (Argentina).

Description. Adult trees have an erect trunk and spreading leafy branches, but the young growth forms very thorny thickets. Leaves are bright green, fernlike; pods dark brown, large, with sweet pulp and numerous hard brown seeds. Stems of the younger plants bear very large, thick, sharp thorns up to 12.5 cm long and 0.6 cm diameter, dark glossy brown in colour (Everist, 1969).

Characteristics. Although stock relish its foliage, the honey locust is cultivated for its large pods (30 to 45 cm long and 2.5 to 5.0 cm broad). In the United States, some trees bear pods containing 40 percent sugar, in other cases sugar content is very low. The pods contain 12 to 14 percent crude protein, and 230 kg per tree are obtained from the fourth or fifth year. *Gleditsia* is dioecious and is unreliable as a pod bearer. It is best grown on alluvial soils, (e.g. along the Ohio and Mississippi rivers, United States) under irrigation with interrow cultivation. Seeds are scarified in hot water to break dormancy. It can stand severe droughts.

Budding and grafting are difficult, but root cuttings can be planted. Male trees are required for adequate pollination (Jurriaanse, 1950). In Argentina, the variety *inermis* has been selected for planting (Whyte *et al.*, 1969). Everist does not recommend it to be planted in Queensland, Australia, because of its habit of forming thickets and becoming a serious pest.

Analyses show that the dry-matter content of the leaves varies from 49.4 percent in May to 28.1 percent in April with a mean of 40.0 percent. Crude protein content ranges from 17.3 percent in August to 14.3 percent in February with a mean of 15.7 percent; crude fibre, from 33.4 percent in November to 25.9 percent in May. Calcium percentage of the dry matter ranges from 2.85 percent in January-February to 1.75 percent in October, with a mean of 2.20 percent. Phosphorus ranges from 0.16 percent in May to 0.06 percent in February with a mean of 0.12 percent, and chlorine has a mean of 0.02 percent. The digestibility of the dry matter for cattle was 68.7 percent, of the organic matter 69.8 percent, crude protein 81.4 percent and the crude fibre 59 percent (Rose-Innes and Mabey, 1964).

Gliricidia spp.

Gliricidia sepium (Jacq.) Walp.

Synonym. *Gliricidia maculata* Kunth.
Common names. Gliricidia (Sri Lanka, India, Australia), mata-ratón (Venezuela).
Description. Tree, growing up to 5 to 15 m. Leaves pinnate, glabrous; pinnae

563

in 3 to 11 pairs, leaflets wedge-shaped at the base, acute at the tip, 2.25 to 6.0 cm long; light green on the surface, pale underneath with bronze spots; racemes numerous, erect, 10 to 12 cm long. Flowers pedicellate, isolated, glabrous, 1.8 cm long. Pod glabrous, 22 cm long and 1.5 cm broad (Pittier, 1944).

Distribution. Native to continental Mexico and the West Indies, now widespread in tropical plantation agriculture.

Characteristics. Gliricidia has long been used as a shade tree in tropical plantations of tea, coffee and cocoa. It is also used as a living fence, the yearly increment of new growth being lopped for fodder or for adding to rice fields at "mudding up" to build up organic matter and add some nitrogen. The roots are toxic to rodents and the leaves toxic to horses, but cattle and goats are not affected (Pittier, 1944). Payne (1955) found it a useful browse at Sigatoka, Fiji.

Griffonia spp.

Griffonia simplicifolia (Vahl ex DC) Baill.

Synonym. *Schotia simplicifolia* (Vahl ex DC) Baill.

Description. Stout, woody, climbing shrub growing to about 3 m with greenish flowers and inflated black pods.

Distribution. Occurs mainly in West-Central Africa.

Characteristics. Occurs in thickets, usually associated with mounds of the termite *Macrotermes* on plains, in forests, in secondary vegetation and on old farms. It is evergreen, vigorous and has wide adaptability. Produces fair quantities of highly palatable herbage of good nutritive value, and fair quantities of large, viable seeds.

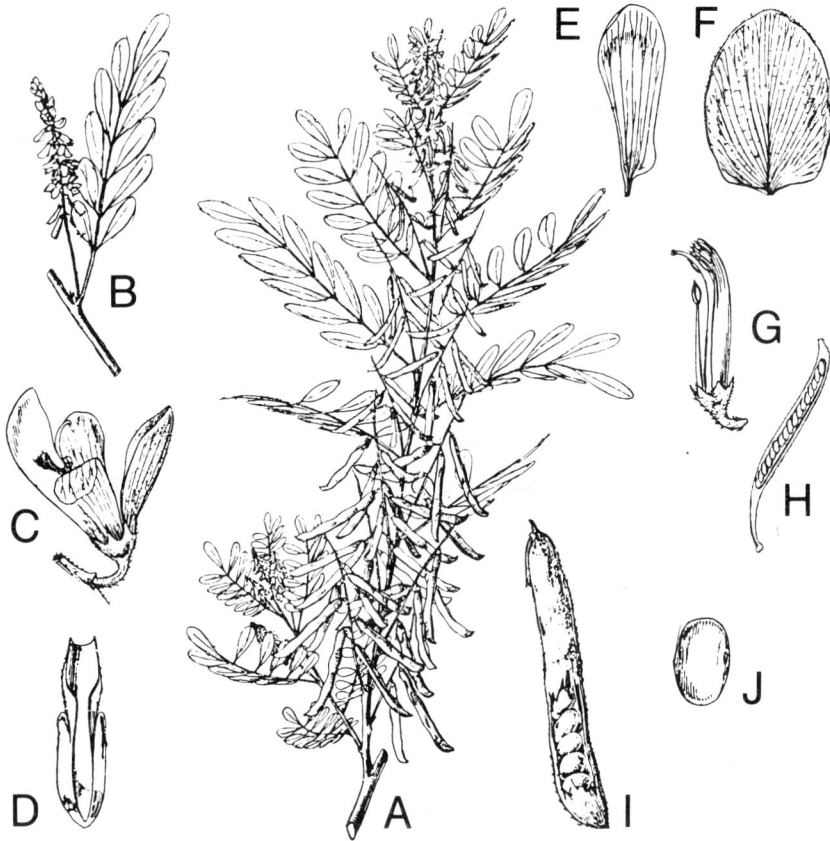

Figure 117. *Indigofera arrecta.* **A**-Fruiting branch **B**-Inflorescence and leaf **C**-Flower **D**-Keel **E**-Wing **F**-Standard **G**-Flower, corolla removed **H**-Carpel **I**-Pod **J**-Seed

Indigofera spp.

Indigofera arrecta Hochst. ex A. Rich.

Description. A shrub about 2 m high, erect, branched: leaflets in five to eight pairs, oblanceolate, about 1 to 1.5 cm long, smooth above and hairy beneath. Flowers yellow in 6- to 12-flowered racemes. Pods chestnut-brown when mature, 1.5 to 4 cm long, polished, six- to eight-seeded (Andrews, 1952).

Distribution. Occurs in the central and southern Sudan and is locally abundant in Zambia in *Combretum/Acacia* mixed woodland, particularly in disturbed areas.

Characteristics. It has low palatability during the rainy season, but is well browsed toward the end of the dry season when young subsidiary shoots are also eaten readily.

Leucaena spp.

Leucaena leucocephala (Lam.) de Wit. (see Colour plates LII, LIII)

Synonyms. *Leucaena glauca* (L.) Benth.; *Mimosa glauca* L.; *Acacia glauca* (L.) Moench; *Mimosa leucocephala* Lam.

Common names. Koa haole (Hawaii), leucaena (Australia), vaivai (Fiji), ipil-ipil (Philippines), lead tree (Caribbean), tan-tan (Virgin Islands), jumbie bean (Bahamas), acacia bella rosa (Colombia), aroma blanco (Cuba), hediondilla (Puerto Rico), wild tamarind (West Indies).

Description. Small tree; leaves bipinnate: pinnae in four to nine pairs on a rachis 15 to 20 cm long, pinnae up to 10 cm long; leaflets 11 to 17 pairs, oblong-lanceolate 2.0 to 3.5 mm wide, 7 to 10 mm long. Inflorescence capitate or globose, solitary, axillary, with peduncle up to 5 cm long and densely many-flowered. Flowers white. Pods thin and flat, up to 20 cm long and 2 cm wide, acuminate. Seeds elliptic compressed, brown (Barnard, 1967). It has a very deep root system, with laterals growing downwards at a sharp angle to the taproot. Roots reached 2 m in one year and 5 m at five years (Dijkman, 1950) (see Fig. 118).

Distribution. Originated in Mexico, has spread throughout the tropics and become naturalized in most tropical countries.

Season of growth. Perennial summer-growing browse plant which is subject to frost. It defoliates during the dry season in Hawaii.

Temperature for growth. Optimum, 22 to 30°C. Minimum 10°C (Hutton and Gray, 1959). It rarely does well with temperatures below 15.5°C. At Rodd's Bay in the coastal area of central Queensland, Australia (lat. 23°15'S), it goes into the winter in leaf and gives early spring growth. Has low frost tolerance, but escapes some frost because it is tall. At Brian Pastures, Gayndah, Queensland (lat. 25°38'S), it has persisted in temperatures down to −10°C in winter and up to 42.5°C in summer.

Latitudinal limits. About 30°N and S.

Altitude response. Thrives in naturalized stands in Hawaii from sea level up to 305 m, in Indonesia up to 490 m, with planted stands up to 1 600 m (Takahashi and Ripperton, 1949; Dijkman, 1950). At higher elevations, leucaena growth becomes slower and the plant tends to remain a shrub. It grows poorly above 220 m in Hawaii. In Colombia it grows from sea level to 1 800 m (Crowder, 1960).

Rainfall requirements. It requires 750 mm or more annual rainfall. Areas where leucaena grows wild in Hawaii have rainfall ranging from 600 to 1 550 mm, and in Indonesia from 675 to 1 650 mm, while in Indonesia it is used in areas with rainfall up to 3 900 mm. In Queensland, it has been used in areas with rainfall over 800 mm (Gray, 1968). Takahashi and Ripperton (1949) recommend one irrigation per crop if the rainfall is less than 1 000 mm in Hawaii.

Tolerance of drought and flooding. Persists well in a drought but may

567

Figure 118. Leucaena leucocephala

defoliate (e.g. in Hawaii it can persist in a 375-mm rainfall area). Tolerance of flooding is very poor; requires good drainage, but thrives where roots can reach the water-table (Oakes, 1968).

Soil requirements. Leucaena is not very specific in its soil requirements but it flourishes in deep red latosolic soils. In Queensland it grows on podzolic and black earth soils; in the Philippines on calcareous and coral soils with high pH (Gantt, 1953); and in Hawaii on ferruginous humid latosols, low humid latosols, regosols, alluvium, hydromorphic and grey hydromorphic soils (Takahashi and Ripperton, 1949). It tolerates a wide range in pH from acid to alkaline (pH 5.0 to 8.0). On acid soils, growth is slow and the plant tends to remain a shrub; therefore it does not do well in tea plantations at pH 4 to 6.5 (Dijkman, 1950). It is only moderately tolerant of salinity.

Rhizobium *requirement.* It has a highly specific *Rhizobium* requirement and

568

needs inoculation with a special *Leucaena* inoculant based on NGR 8 (Date, 1969). In Hawaii it does not need to be inoculated.

Ability to spread naturally. Excellent in Hawaii and in cultivated deep latosolic soils at Redland Bay, Queensland, but does not spread in pastureland.

Land preparation for establishment. It is slow in its early growth and no preparation of land as for a crop is preferable; however, on some latosolic soils it grows well with little cultivation. Takahashi and Ripperton (1949) recommend two deep ploughings, each followed by disc harrowing to kill weed growth.

Sowing methods. Drill the seed into a well-prepared seed bed in rows about 1 m apart (Risopoulos, 1966; Takahashi and Ripperton, 1949). The plants should eventually be 2.5 to 7.5 cm apart in the rows. It is suggested for Queensland subcoastal frost-free areas that one quarter of the area be sown to leucaena and the remainder to fertilized Townsville stylo for a productive pasture for beef cattle. Leucaena can be established by transplanting stumps — pieces of trunk up to 10 cm in diameter and 4 m long can be rooted in a nursery and after trimming roots to a length of 20 cm transplanted to the field (Dijkman, 1950). For erosion control, rows spaced at a quarter-degree slope are used in plantations. During the first three months, interrow cultivation should be given to control weeds. It is too slow in its early growth to survive pasture competition, at least without the use of herbicides to weaken the pasture; therefore oversowing is not successful. Some strip preparation of land with a rotary hoe before planting leucaena would be successful.

Seed should be drilled from 2.5 to 5.0 cm deep just before the main rainy season, at 2 to 10 kg/ha, according to seed supply. At Brian Pastures in Queensland (rainfall 710 mm) it has been drilled in 3-m rows at 2 kg/ha.

Number of seeds per kg. 20 000. Akamine (1942) reports up to 90 percent hard seed; Oakes (1968) up to 95 percent. Dormancy normally disappears in two to three months, but complete loss of hardseededness takes several years (Davies and Hutton, 1970).

Seed treatment before planting. To break dormancy: place seed in hot water at 76.5°C, allow temperature to drop to 37.5°C (Akamine, 1942) — the seed deteriorates rapidly; or immerse in concentrated sulphuric acid (60° Baumé) for 13 minutes, wash and dry (Takahashi and Ripperton, 1949); or treat with water at 80°C for two minutes, followed by rapid drying (Gray, 1962). The seed remained viable for 15 months after treatment; or immerse in boiling water for half a minute (Roberts, personal communication); or scarify mechanically (Dijkman, 1950).

569

Fifty-eight percent of the seeds pass undigested through the animals (Takahashi and Ripperton, 1969).

● Inoculation. *Leucaena leucocephala* is highly strain-specific with respect to *Rhizobium* (Galli, 1958; Trinick, 1968). Norris (1965) found a wide range in effectiveness of several strains. The current recommendation for Australia is to inoculate with strain NGR 8 (1970). Edwards (1963) found inoculation with *Rhizobium* with a 10 percent sucrose solution to be beneficial in Costa Rica.

● Pelleting. The *Rhizobium* from leucaena is a fast-growing acid-producing type. Norris (1965) suggests that this is evidence that leucaena is adapted to alkaline soils. In fact he recommends that the seed be lime-pelleted for planting.

Nutrient requirements. Leucaena requires adequate calcium, phosphorus and sulphur for growth. Phosphorus stimulates nodulation. At Rodd's Bay, Queensland, 125 kg superphosphate/ha/year are applied to solodic soils to establish and maintain leucaena. Boron may be necessary in some soils (Kinch and Ripperton, 1962). Molybdenum is necessary for the *Rhizobium* but not for the legume in Taiwan (Wu, 1964).

● Calcium. Andrew and Hegarty (1969) found that the calcium uptake of leucaena was 1.4 to 1.7 percent of the dry matter.

● Nitrogen. Parbery (1967a) lifted protein percentages in leucaena by about 2 percent by adding 100 kg N/ha to Cockatoo sand, at the Kimberley Research Station, Western Australia.

● Phosphorus. Andrew and Robins (1969a) found in soil culture that leucaena gave 81 percent of its maximum yield at the equivalent of 1 350 kg/ha of superphosphate when fertilized with only 250 kg/ha. It was also found to be the most efficient of the legumes tested in accumulating nitrogen in its foliage as a result of applied phosphorus.

Toxicity levels and symptoms. Manganese — the toxicity threshold value for manganese in soil culture is 550 ppm (Andrew and Hegarty, 1969). Aluminium — leucaena is tolerant to high levels of aluminium (Oakes, 1968) and to low levels of iron (Oakes and Skov, 1967). Has very low seeding vigour and requires weed-free conditions in the seedling stage.

Vigour of seedlings, growth and growth rhythm. It is a vigorous grower in the wet season and produces a large body of foliage.

Response to light. Flowers in February (El Salvador), in March-April (Peru) and at latitude 27°22'S in Queensland. *Leucaena leucocephala* cannot tolerate deep shade (Dijkman, 1950).

Compatibility with grasses and other legumes. It is usually used as hedges,

570

either in cultivated rows or uncultivated, but Davies and Hutton (1970) suggest that much greater productivity would ensue from leucaena row-planted in pangola grass (*Digitaria decumbens*). It does very well with paspalum (*Paspalum dilatatum*), guinea grass (*Panicum maximum*) and napier grass (*Pennisetum purpureum*). Pendleton (1934) reports its value in reclaiming waste grassland of *Imperata cylindrica* in the Philippines.

Ability to compete with weeds. In the young stages it is a poor competitor, but at maturity it is little affected.

Tolerance of herbicides. Kinch and Ripperton (1962) showed that leucaena is susceptible to herbicides in the seedling stage. In the Northern Territory (anon., 1964a), monuron was found effective in controlling weeds on the crop, especially during establishment.

Nitrogen-fixing ability. Very high. Dijkman (1950) reported that from 1 000 trees per hectare, pruned every two months, nitrogen equivalent to 100 kg of sulphate of ammonia and phosphoric acid equivalent to 100 kg/ha of double superphosphate were added to the soil yearly. Gates (1970) showed that growth and nitrogen fixation in leucaena were a continuous process under favourable conditions of light, moisture and mineral balance between phosphorus, nitrogen and sulphur, with no rhythmic pattern of nodulation. He subjected seedlings to 16 harvests in pots over a period of 84 days' growth. Nitrogen production, nodule dry weight and dry-matter production in the whole plant were almost linear after six weeks' growth.

Response to defoliation. It appears to stand a good deal of defoliation when it is established, much more so than *Cajanus cajan* (Davies and Hutton, 1970). Payne (1955) recorded leucaena as slow to establish in Fiji, but once established it is very persistent under grazing. Takahashi and Ripperton (1949) cut it at 5 to 7.5 cm above ground level with a motor-scythe up to six times per year in Hawaii, when the crop was 1 to 2 m tall. Herrera (1967) found in Colombia that cutting the plants at 10 cm developed a crown which gave rise to new shoots and increased plant persistence.

Grazing management. The plant should be browsed lightly in the first year and then can be grazed rather frequently. Crowder (1960) began grazing at 0.8- to 1-m height and grazed it back to 10 cm in height. He grazed it first five to eight months after planting and every three to four months thereafter. Risopoulos (1966) suggests grazing to a hedge of 30 cm in height. Grazing should cease when the leaves and twigs up to 0.6 cm in diameter have been removed. Care should be taken to ration grazing to prevent undue toxic effects on the grazing animal.

Response to fire. Firing of grassland containing leucaena in the Philippines twice a year suppresses it, but if roots are well established they will shoot again after fire (Gantt, 1953). In Hawaii it appears to recover from the effects of fire. Brown (1921) reported that it could not survive repeated damage by fire and is vulnerable to grass fires up to three years of age.

Breeding system. It is self-fertilized, although the flower structure would permit open pollination. Crosses between varieties can be made quite successfully (Gonzalez, Brewbaker and Hamill, 1967; Hutton and Gray, 1959; Gray, 1967a, b, c). Erect habit is dominant to bushy habit and strong basal branching is recessive. The somatic chromosome number is 2n = 104.

Dry- and green-matter yields. Takahashi and Ripperton (1949) obtained a mean annual yield of 20 to 25 tonnes dry matter/ha containing 2 730 to 3 450 kg protein. This was obtained by cutting plots 5 to 7.5 cm above ground level with a motor-scythe, taking three, four or six cuts a year. Over 12 months in Mauritius under an annual rainfall of 1 540 mm Anslow (1957) obtained 7.5 tonnes dry matter/ha containing 1 900 kg protein. Hutton and Bonner (1960) at Samford, southeastern Queensland, by a plucking method, obtained a yield of edible dry matter from 'Peru' cultivar of 12 360 kg/ha over nine months containing 3 540 kg protein. It also yielded 16 500 kg/ha of green herbage in ten weeks, after being first cut back to a height of 30 cm in early summer. Kinch and Ripperton (1962) have obtained 20 to 25 tonnes/ha dry matter in Hawaii from high-yielding strains under continuous cropping with four or five harvests a year. In the Virgin Islands, Oakes and Skov (1967) reported yields of 7.5 to 20 tonnes of dry forage per hectare per year. At Tengeru, Tanzania, leucaena yielded 10 875 kg of leaves and young shoots per hectare (van Rensburg, 1968). Parbery (1967a) obtained up to 21 450 kg of dry matter per hectare with 'El Salvador' cultivar on fertilized Cunnunurra clay after 53 days' growth.

Suitability for hay and silage. It can be chopped and made into dehydrated meals. In Hawaii a forage harvester is used and a conventional alfalfa (lucerne) dehydrator used for drying the material. In Mauritius, Anslow (1957) both artificially dried and sun-dried the crop and subsequently prepared meal from the dry matter. Sun-dried material from the last cut in September/October contained 31 percent crude protein and 14 percent fibre.

Pure leucaena makes satisfactory silage, but mixing it with napier or elephant grass (*Pennisetum purpureum*), sugar cane or *Panicum maximum* in the proportions of 1:1 or 1:2 improves the quality of the silage. Ripperton (1942) recommended that leucaena should be chopped, mixed with 5 percent molasses and well packed. He reported that this gave satisfactory silage but should not be fed to horses. The material was harvested mechanically with a mower or maize ensilage harvester. Anslow (1957) in Mauritius made silage that was palatable and pleasant smelling, containing 29 percent dry matter, 25 percent crude protein in the dry matter and 24 percent crude fibre. The high fibre content reduced its feeding value.

Value as standover or deferred feed. It is a good source of feed during early drought periods in Hawaii before it defoliates. In Queensland, it is most valuable in frost-free areas.

Feeding value. The value of fresh green leaves for poultry-feeding has been demonstrated by Palafox (1948) in Hawaii, and Dingayan and Fronda (1950) and Molina (1953) in the Philippines. However, Sadoval (1955) and Springhall and Ross (1965) found no advantage in a laying ration containing leucaena, and the latter actually recorded later sexual maturity and lower body weight with such a ration. In Colombia, it is added to the ration to improve yolk colour in eggs. Iwanaga, Otagaki and Wayman (1957) found that dehydrated leucaena meal was useful for growing and fattening pigs; and leucaena has been used extensively in Hawaii for feeding dairy and beef cattle (Kinch and Ripperton, 1962). The leaves contain three times the protein percentage of the stems, and the plant contains more potash than lucerne (Takahashi and Ripperton, 1949).

● Chemical analysis and digestibility. Parbery (1967a) recorded 11.6 percent crude protein in the foliage of leucaena grown on unfertilized Cockatoo sand, and 13.6 percent when the plants were fertilized with 100 kg N/ha. On both fertilized and unfertilized Cunnunurra clay, the protein content was 20 percent of the dry matter. The digestibility of the dry matter is 65 to 87 percent. Vitamin A content is superior to that of alfalfa or lucerne (Oakes, 1968). Dingayan and Fronda (1950) gave analyses of young green leaves and shoots for green feed for chickens as 74.89 percent moisture, 0.36 percent crude protein, 2.77 percent crude fat, 17.69 percent carbohydrates, 2.25 percent crude fibre and 2.04 percent ash, with a calorific value of 97.00 compared with 80.00 for *Centrosema pubescens*. Gantt (1953) gave analyses of leucaena meal in the Philippines as 24.4 percent protein, 6.4 percent fat, 9.6 percent fibre and 10.6 percent ash, with total digestible nutrients at 67.9, compared with 53.6 for alfalfa. Takahashi and Ripperton (1949) found that the crude-protein contents of the leaves from plants cut three, four and six times a year were 24.13 percent, 26.05 percent and 30.07 percent of the dry matter respectively, and of the stems 8.06 percent, 9.14 percent and 10.04 percent.

● Palatability. Its high palatability has been reported in Colombia, Guatemala, Fiji (Payne *et al.*, 1955), Hawaii (Takahashi and Ripperton,

1949), Australia, (Davies and Hutton, 1970), the Philippines (Farinas, 1966) and Papua New Guinea (anon., 1966).

Toxicity. Wayman and Iwanaga (1957) showed that at 15 percent level in the ration of pigs, dried *Leucaena leucocephala* feed reduced the ability of gilts to conceive and reduced average litter size and weight. In the early work on feeding *Leucaena leucocephala*, it was shown that where it comprised a high proportion of the diet of nonruminants (monogastric animals), especially horses, it could cause depilation, growth reduction and general ill health (Owen, 1958; Hutton and Gray, 1959).

It was originally thought that toxicity was associated with selenium accumulation from soils (Gray, 1968) but Yoshida (1944), working with rats, showed that toxicity was due to the amino-acid mimosine, which occurs in the leaves and seeds at a rate of up to 10 percent in the growing points. Several workers have since investigated mimosine (review by Gray, 1968). Springhall and Ross (1965) reported that the addition of ferrous sulphate solution to leucaena material just before mixing poultry rations reduced toxicity. Toxicity to ruminants is of major concern. Damseaux (1956) reported that Suffolk sheep had been affected by feeding the plant. Compère (1959) found toxic effects in ruminants and nonruminants fed large quantities of leucaena in Zaire. Letts (1963) described the loss of hair of a buffalo calf from eating leucaena, but the hair grew again when the calf's diet was changed.

Hegarty, Schinkel and Court (1964) found that sheep fed a sole diet of leucaena shed their fleece, owing to intake of mimosine. They suggested that sheep could be conditioned to a diet, as they can detoxify mimosine in the rumen. Subsequently, however, Donaldson *et al.* (1970) did not recommend it as a diet for sheep. Reporting results with leucaena feeding at Samford, southeastern Queensland, they found that it did not affect oestrous cycle length, conception rate or gestation length in dairy cows, and did not affect heifers prior to or after calving. Some mild incoordination occurred. The birth weight of calves was lower from leucaena-fed cows (19.8 kg) than from control cows (28.5 kg), and the calves had enlarged thyroid glands at birth. They concluded that leucaena could be used safely as a grazing supplement for breeding cattle, but not as a sole diet. Under commercial conditions it is unlikely that it would be used as a sole diet.

Leucaena imparts a flavour to the milk of dairy cows fed the plant within two hours of milking (Oakes, 1968). Cattle browsing the plant for long periods develop yellow-coloured fat (Takahashi and Ripperton, 1949).

Seed harvesting methods. The pods are easy to collect by hand and are usually harvested in this way. Dry pods are gathered and threshed by placing in a sack and beating with a flail. Seeds are separated from empty pods by shaking them through a 6.35-mm mesh screen (Takahashi and Ripperton, 1949).

Seed yields. Seed yields do not appear to have been recorded in the literature. Seeds should be stored in sealed drums in the humid tropics, as they lose up to 50 percent of their viability after four to five months under natural conditions (Dijkman, 1950).

Minimum germination percentage and quality required for sale. Sixty percent germination, with a maximum of 10 percent hard seed and a minimum purity of 97.5 percent in Queensland. The seed is germinated at 25°C after cutting the seed coat (Prodonoff, 1968).

Cultivars. There are four cultivars and several ecotypes belonging to three well-defined types (Gray, 1968).

● Cv. Peru — slightly taller than cultivar Hawaii, with larger leaves; flower heads are larger but contain fewer flowers; pods are longer and broader, seeds larger and flatter and the stem tips and young pods glabrous. 'Peru' grows into a symmetrical, regularly branched tree which under good conditions may attain a height of more than 2 m during the season with a lateral spread of about the same distance. The main stem is erect but relatively short compared with 'El Salvador'; basal branching is also well developed, the main branches originating from low down on the main stem. The seed was introduced into Australia as CPI 18614 from Argentina, but it originated in Peru. It was released for commercial use in Australia in 1962 (Barnard, 1967).

● Cv. El Salvador — an erect, tall plant with very little basal branching which in southeast Queensland may attain a height of up to 4 m, but without much lateral spread. Leaves are longer than those of 'Hawaii', while heads, pods and seeds are similar. Stem tips and young pods may be either glabrous or pubescent. Its disadvantage compared with 'Peru' is its tall erect growth and paucity of low branches. It was introduced as CPI 18623 to Queensland.

● Cv. Hawaii — a small, bushy type common in Hawaii. This cultivar is being crossed with 'Peru' (Davies and Hutton, 1970) to incorporate high leaf yield and dense branching habit. It is a heavy flowering and seeding type which flowers throughout the year, but is low yielding (9 000 to 10 000 kg dry matter/ha). The Australian introduction is number CPI 18624.

● Cv. Guatemala — a tall, sparsely branched type. It was introduced into Australia as CPI 18228.

Diseases. Leucaena is very free of diseases. In new areas it may act as an alternate host of *Fomes lignosus*, *Rosellina bunodes*, *R. arcuata* and *Ganoderma pseudo-ferreum*, all active on tea (Dijkman, 1950).

575

Pests. It was heavily invaded by leaf-harvesting ants (*Zompopos*) at El Prado, Guatemala, and had to be sprayed with aldrin (Johnston, personal communication). *Pseudococcus citri* (above 600 m elevation) and *Ferrisia virgata* may attack pods in Indonesia (Dijkman, 1950).

Main attributes. Outstanding vegetative vigour and yield of forage of high palatability and high protein content; persistence.

Main deficiencies. Slowness of establishment; prolific seeding habit in many varieties leading to thicket production; rapid erect growth habit in some varieties; presence of the toxic substance mimosine (Gray, 1968).

Performance. It was early used as a green manure and shade crop in Indonesia and other countries. Dijkman (1950) estimated that a stand of 1 000 leucaena plants per hectare, pruned every two months, added 35 tonnes of green leaves and twigs per hectare per year. Henke, Work and Burt (1940) reported live-weight gains of up to 0.7 kg/day in steers grazing almost pure stands in Hawaii. On weathered lava, 80 hectares of leucaena/guinea grass pastures gave 660 to 770 kg live-weight gain/ha.

Lonchocarpus spp.

Lonchocarpus capassa Rolfe

Description. Tree, 10 to 12 m high.

Distribution. Occurs frequently in *Combretum/Acacia* woodland.

Characteristics. Not generally eaten by cattle during the rainy season, but fairly extensively browsed toward the end of the dry season (van Rensburg, 1968).

Millettia spp.

Millettia thonningii (Schum. et Thonn.) Bak.

Synonym. *Robinia thonningii* Schum. et Thonn.

Description. Deciduous tree growing to about 10 m. Purple flowers appear with young leaves.

Distribution. Native to West and Central Africa, occurring in Côte d'Ivoire, Togo, northern and southern Nigeria, Ghana and Zaire. It was introduced in Angola.

Characteristics. Fairly palatable (rated 1 to 2 out of 5). Has a tendency to shed its leaves in the dry season. Occurs as clumps or thickets on the Accra Plains in Ghana, usually associated with termitaria belonging to the genus *Macrotermes*.

Analyses by Rose-Innes and Mabey (1964) are presented in Table 15.5.

TABLE 15.5 **Chemical analysis of *Millettia thonningii***

	Dry matter	Crude protein	Crude fibre	Ca	P	Cl
Mean	40.6	19.0	29.5	1.42	0.10	0.18
Range	29.1-58.8	16.9-22.6	24.7-32.2	1.0-1.8	0.04-0.18	0.12-0.23
Date	May-August	January-March	February-April	March-November	August-November	April-January

Parkinsonia spp.

Parkinsonia aculeata L.

Common names. Parkinsonia or Jerusalem thorn; *cina-cina* (Paraguay).
Characteristics. Small tree of irregular habit; leaves consisting of a long, flat central portion with numerous small, rounded green leaflets along each edge, pods long, pencil shaped, with numerous hard seeds. Extremely hardy and will grow in inland areas where no other trees do well. Can become a pest around watering places. Leaves are eaten freely by sheep and cattle (Everist, 1969). Has been reported as toxic in Paraguay, at times containing hydrocyanic acid (Fretes, Samudio and Gay, 1970).

Piliostigma spp.

Piliostigma thonningii (Schumach.) Milne-Redh

Synonym. *Bauhinia reticulata* (non-DC) Broun and Massey.
Description. Shrub or tree, to 6 m high, often of crooked growth, with dark brown to black fissured bark. Leaves digitately 11- to 12-nerved, the central nerve prolonged as a point into the notch between the lobes of the leaf.

578

Leaves 7.5 to 15 cm long, leathery, very strongly reticulate. Flowers white, fragrant, drooping, about 2.5 cm long, in racemes alternately leaf-opposed and axillary along each branch and borne somewhat horizontally. Pod shortly pedicellate, oblong, 15 to 20 cm long 2.5 to 5 cm broad, at first densely red-brown, hairy, ultimately smooth (Andrews, 1952).

Distribution. Occurs in the central and southern Sudan and is widely distributed in Zambia in woodland over a wide range of soil conditions.

Characteristics. Yields heavy crops of pods which are eaten by livestock, game, monkeys, and also by humans. Pods collected in August in Zambia had a crude protein content of 5.1 percent.

The leaves are browsed by cattle during the dry season. Young shoots and leaves collected in November contained 9.88 percent crude protein, 0.70 percent calcium and 2.45 percent potash (van Rensburg, 1968).

Prosopis spp.

Prosopis chilensis (Mol.) Stuntz

Synonym. *Ceratonia chilensis* Molina.

Common names. Algaroba, kiawe (Hawaii), tamarugo, algarrobo blanco, algarrobo de Chile (Latin America).

Description. Tree of variable height, reaching 15 m in Hawaii. Shallow, spreading root system; branches freely, wood hard and reddish. Leaves bipinnate, 4 to 7.5 cm long, with small round glands, leaflets linear-oblong, smooth or slightly hairy. Flowers greenish yellow, about 5 mm long, borne in a raceme 5 to 10 cm long. Seed pod slender, slightly curved or straight, flat at maturity, 10 to 20 cm long, smooth, yellow when ripe, borne in drooping clusters. Seed angular-oblong, about 0.6 cm long and 0.3 wide, light brown, enclosed in a horny structure (Hosaka and Ripperton, 1944).

Distribution. Native to Central America, it is common along the Pacific coast of Peru, northern Chile and northern Argentina. It is naturalized in Hawaii and is common as thickets on the Pacific beaches.

Characteristics. In Hawaii it is regarded as a useful shade and fodder plant in the dry and hot zones. The pods or beans drop between June and November and are an excellent feed containing much sugar. The pods are ground into a meal for use in concentrate rations. The leaves are grazed to some extent, but from a forage point of view they are not as important as the beans. A mature tree may yield 80 kg annually. In India, it is recommended for planting as a fodder tree at altitudes ranging from 340 to 1 230 m (Gowda, Range and Ramaswamy, 1960).

Prosopis glandulosa **Torr.** (see Colour plate LIV)

Synonym. *P. juliflora* (Sw.) DC.

The "mesquite" of United States literature is generally *P. juliflora*. Taxonomy has been confused, and Schuster (1969) states that this latter species does not extend north of Mexico.

Description. Woody tree growing 4 to 8 m, very variable. For var. *glandulosa*, leaflets are 30 to 45 mm long and 8 to 15 times as long as broad, with 6 to 13 pairs per pinna. Var. *torreyana* has leaflets 15 to 25 mm long, 5 to 8 times as long as broad, with 10 to 15 pairs of pinnae.

Distribution. The typical variety occurs in most of Texas and northeastern Mexico; var. *torreyana* is found in western Texas, New Mexico, extreme southern Arizona, and California (United States), and in Coahuila, Chihuahua, Baja California, and parts of Sonora (Mexico).

Characteristics. The mesquites provide food and shelter for livestock. Pods are relished by cattle, horses, sheep, goats and pigs. They are useful when ground into a meal (Marion *et al.*, 1957). Sometimes jaw and tongue trouble occur in cattle from feeding mesquite beans (Dollahite and Anthony, 1957). Leaves are browsed to some extent by cattle, sheep and goats; they are eaten in the spring when young and tender. Generally, however, thickets of mesquite develop and become uncontrolled. Efforts are being made to eradicate or at least thin it out. Spraying with 2, 4, 5-T ester at 250 g/ha in a carrier of 1 part diesel oil: 7 parts water to make a total of 45 litres per hectare (Tschirley, 1962) is the best chemical control treatment.

Prosopis pallida H.B.K.

Common name. Algaroba.

Description. Tree, somewhat resembling *Acacia nilotica* in appearance. It is also thorny in the young stages, but appears to have little or no tendency to grow in thickets. Flowers are in spikes and the pods are yellowish in colour, thick, with a sweet pulp between the seeds.

Characteristics. Both leaves and pods are eaten by all classes of stock (Everist, 1969).

Prosopis spicigera L.

Description. Low tree with slender grey branches. Prickles few, none, or copious, 0.3 to 0.6 cm long, nearly straight, at first straw-coloured. Pinnae 2.5 to 5.0 cm long: leaflets 16 to 24. Flowers in axillary spikes, 5.0 to 7.5 cm long, and in terminal panicles. Pod straight, edible, smooth, 10 to 15 cm long, 0.6 cm thick, narrowed gradually to a short stalk, the mesocarp starchy. Seeds 10 to 15, dull brown, oblong (Hooker, 1879).

Distribution. Punjab, western India, Afghanistan, Iran.

Characteristics. Ganguli, Kaul and Nambiar (1964) stated that *Prosopis spicigera* was considered to be the best browse plant for cattle, sheep and camels in Jodhpur (India) on the basis of availability, palatability and nutritive value. The highest yield of dry fodder (125 kg/ha) was achieved with a density of 14 percent in the natural grazing lands. It is lopped in winter and can stand recurrent and severe loppings without detriment either to its growth or the subsequent leaf yield. A moderately sized tree produces more than 45 kg of dry leaf forage, locally known as "loong". They recorded the palatability as *Acacia nilotica* > *Prosopis spicigera* > *Acacia senegal* > *Albizia lebbek*.

Robinia spp.

Robinia pseudoacacia L.

Common name. Black locust (United States, Australia).
Description. Tree which can grow to a height of 12 to 15 m but often forms dense thorny thickets from root suckers. Leaves consist of several rounded, rather thin, dull green leaflets; flowers white, pea-shaped; pods fairly large, dark brown or almost black (Everist, 1969).
Distribution. United States, South Africa, Australia.
Characteristics. Leaves and pods are edible and can be toxic if eaten in excess (Jurriaanse, 1950; Everist, 1969). Everist does not recommend planting this species.

Samanea spp.

Samanea saman (Jacq.) Merrill

Synonyms. *Mimosa saman* Jacq.; *Pithecellobium saman* (Jacq.) Benth.; *Enterolobium saman* (Jacq.) Prain.; *Inga saman* Willd.; *Calliandra saman* Greseb.
Common names. Rain tree, saman, samaan, algarrobo, monkey pod (Hawaii); French tamarind, guango (Jamaica).

582

Figure 119. A single plant of *Sesbania cannabina* growing in a roadside drain

583

Description. A tree up to 25 m high with a stout trunk, to 1.5 m in diameter, and large spreading canopy. Low, spreading branches, more or less deciduous. Leaves with two to four pairs of pinnae. Leaflets obtuse, ovate-oblong or roundish, oblique, pubescent, up to 4 cm long. Inflorescence one or two together on peduncles 5 to 9 cm long; flowers in heads on short pedicels. Calyx 6 mm long, greenish. Corolla yellow or red, 10 to 12 mm long, stamens silky, pale crimson. Pods fleshy, 10 to 25 × 15 to 18 mm, straight, more or less flat, black when ripe. Seeds 16 to 20, rounded, truncate at one end, pointed at the other, dark reddish brown with a paler ring at each side, 10 to 11 × 5 to 6 mm (Adams, 1972).

Distribution. Native to Latin America from Nicaragua to Brazil, but particularly Venezuela. *S. saman* was introduced to the West Indies, where it is now widely cultivated and has escaped and is often naturalized. It has been widely introduced elsewhere, for example Sri Lanka, where it is planted as an avenue tree.

General features. *S. saman* has a wide range of useful products, the pod in particular producing an edible pulp. When ripe, the pulp is sweet and sugary. It can also be dried and ground into a meal for animal feed. The timber is strong and hard, with a rich, dark colour, and makes good furniture. It is also valuable as a shade tree in pastures, stimulating grass growth. It is an admirable street tree. The leaves fold together on the approach of rain.

Main reference. Adams (1972).

Sesbania spp.

Sesbania cannabina (Retz.) Poir.

Synonyms. *S. aculeata* auct., non Poir., *Aeschynomene cannabina* Retzius; *S. australia* F. Mueller; *S. serica* DC. var. *glabra* Domin.; *S. serica* DC. var. *subsingulifolia* Domin.

Description. An erect, annual legume, subshrub, to about 3.5 m high, stems with spreading hairs when young, more or less glabrous later, terete or slightly angular. Leaves with 12 to 30 pairs of leaflets. Leaflets narrowly-oblong, apex obtuse or truncate-apiculate, 7 to 8 × 1.5 to 3 mm glabrous or with a few appressed hairs. Inflorescence a raceme of one to three (five) flowers. Pedicels more or less equal to the calyx. Calyx 3 to 4 mm long, the lobes up to about half as long as the tube. Corolla yellow. Standard conspicuously streaked on the back, 0.8 to 1.1 cm long. Pods slender, 12 to 20 × 0.25 to 0.3 cm, with a slender terminal point. Seeds dark brown, about 4 mm long (Stanley and Ross, 1983; Bailey, 1900) (see Fig. 119).

Distribution. Long confused with the Indian species *S. aculeata*, *S. cannabina* is essentially an Australian species, although it does occur northwards into Malaysia. In Australia, it is particularly common in subcoastal and adjacent areas of the northern states, but extends inland in some areas, especially Queensland, where it is widespread. It extends into South Australia along the Diamantina River, and into New South Wales along the Darling River. It is the most widespread of the *Sesbania* spp. and commonly grows in areas subject to inundation, where it is able to withstand waterlogged conditions. It can, at times, be a weed of irrigated rice fields and of low-lying wet cultivation paddocks (Burbidge, 1965).

General features. It nodulates readily and heavily but is strain specific, although there are affinities with the garden bean, lupin and cowpea cross-inoculation groups (Bowen, 1956). Its palatability is low to moderate at best, as it rapidly becomes fibrous and hard. Usually only hungry stock will eat it. There are no known toxicities.

Sesbania sesban (L.) Merr.

Synonyms. *S. aegyptiaca* (Poir) Pers.; *Aeschynomene sesban* L.; *S. punctata* DC.

Description. Tall shrub or small tree, up to 6 m high, copiously branched; leaflets oblong, up to 3.5 cm long, but usually smaller; flowers yellow with red spots; pods long-acuminate at the apex, transversely divided, up to 15 cm long, glabrous, twisted, distinctly torulose (Andrews, 1952).

Distribution. Widespread in tropical Africa, and through tropical Asia and Australia. It is common beside streams and sandbanks.

Characteristics. Dougall and Bogdan (1958) say that it is a useful browse in Kenya. Leaves collected in October contained 25 to 97 percent protein, 2.64

585

percent ether extract, 14.39 percent crude fibre, 49.43 percent nitrogen-free extract, 7.5 percent ash, 1.11 percent calcium and 0.27 percent phosphorus.

Sesbania grandiflora (L.) Pers.

Synonym. Robinia grandiflora L.
Common names. Pico de flamenco (Yucatan Peninsula, Mexico), gallito, cresta de gallo (Puerto Rico).
Description. Small tree or shrub with rather stout branchlets, pubescent when young. Flowers very large, few, pink or red or cream-white (Hutchinson and Dalziel, 1954).
Distribution. Widely cultivated in the tropics. It is common in the wetter parts of West Africa and in Puerto Rico.
Characteristics. Payne (1955) reported it to be a useful browse at Sigatoka, Fiji, where annual rainfall is about 2 000 mm. Kareem and Sudararaj (1967) found that the leaves contained 10 percent moisture, 10.46 percent ash, 42.30 percent carbohydrates, 30.13 percent crude protein, 5.1 percent crude fibre and 2.01 percent fat.

Tamarindus spp.

Tamarindus indica L.

Common name. Tamarind.
Description. Large, unarmed tree. Leaves abruptly pinnate with 20 to 40 oblong opposite leaflets. Flowers few, together in copious lax racemes at the ends of the branchlets, yellow with red-striped petals. Pod 7.5 to 15 × 2.5 cm, three- to ten-seeded (Hooker, 1879).

586

Distribution. Probably indigenous to Africa and now well spread throughout the tropics.

Characteristics. Patel and Patel (1957) found that of the fodder trees analyzed from the Kaira district in India, *Tamarindus indica* had one of the best nutrient contents. The green pods are used for flavouring drinks.

Bibliography

ABRUNA, F. & FIGARELLA, J. Some effects of calcium and phosphorus fertilization on the
1957   yield and composition of a tropical kudzu-grass pasture. *J. Agric. Univ. P.R.*, 41:
231-235.

ABRUNA, F., VICENTE-CHANDLER, J. & PEARSON, R.M. Effects of liming on yields and
1964   composition of heavily fertilized grasses and on soil properties under humid tropical conditions. *Soil Sci. Soc. Am., Proc.*, 28: 657-661.

ADAMS, C.D. *Flowering plants of Jamaica*. Mona, Jamaica, Univ. West Indies.
1972

ADAMS, W.E., PEARSON, R.W., JACKSON, W.A. & McCREERY, R.A. Influence of lime-
1967   stone and nitrogen on soil pH and coastal Bermuda grass yield. *Agron. J.*, 59: 450-
453.

ADDISON, K.B. The effect of fertilising, espacement and date of planting on the yield of jack
1957   bean (*Canavalia ensiformis*). *Rhod. Agric. J.*, 54: 521-532.

ADDISON, K.B. The jack bean. *Rhod. Farmer*, 28: 17-18.
1958

AFFLECK, H. Jack bean poisoning in cattle. *Rhod. Agric. J.*, 58: 21.
1961

AGLIBUT, F.B. & CASTILLO, L.S. Estrogen content of legumes in the Philippines. *Philipp.*
1963   *Agric.*, 46: 673-680.

AGRICULTURAL RESEARCH COUNCIL. *The nutrient requirements of farm livestock*. Technical
1965   Reviews and Summaries. London, Agric. Res. Council.

AKAMINE, E.K. *Methods of increasing the germination of Koa haole seed*. Honolulu, Hawaii
1942   Agric. Exp. St. Circ. No. 21.

ALI, S.I. *Flora of West Pakistan No. 100, Papilionaceae*. Nasir, E. & Ali, S.I., eds. Karachi,
1977   Ferozsons.

ALLARD, R.E. *Principles of plant breeding*. New York, Wiley.
1960

ALLEN, G.H. Glycine — a pasture legume for Queensland. *Queensl. Agric. J.*, 86: 273-275.
1960

ALLEN, G.H. Glycine proves its value. *Queensl. Agric. J.*, 87: 96-97.
1961a

ALLEN, G.H. Tinaroo glycine. *Queensl. Agric. J.*, 87: 768.
1961b

ALLEN, G.H. & COWDRY, W.A.R. Beef gains from irrigated pastures. *Queensl. Agric. J.*,
1961a   87: 175-179.

ALLEN, G.H. & COWDRY, W.A.R. Yields from irrigated pastures in the Burdekin. *Queensl.*
1961b   *Agric. J.*, 87: 207-213.

ALLEN, O.N. & BALDWIN, I.L. Rhizobia-legume relationships. *Soil Sci.*, 78: 415-427.
1954

ANDERSON, A.J. Responses of subterranean clover and lucerne to boron. *J. Aust. Inst.*
1952   *Agric. Sci.*, 18: 159-163.

ANDERSON, G.D. Increasing coconut yields and income on the sandy soil of the Tanganyika
1967   coast. *E. Afr. Agric. For. J.*, 32: 310-314.

ANDERSON, G.D. Effects of fertilisers on botanical composition and productivity of pasture
1968a   on the sandy soils of the Tanganyika coast. *E. Afr. Agric. For. J.*, 34: 207-216.

ANDERSON, G.D. Pasture research at the Livestock Breeding Station, Tanga. Dar-es-
1968b   Salaam, Tanzania, Dept. Agric. Co-op.

ANDERSON, G.D. & NAVEH, Z. Promising pasture plants for Northern Tanzania. V. Overall
1968   comparisons of promising plants. *E. Afr. Agric. For. J.*, 34: 84-105.

ANDRADE, I.F. Production and chemical composition of tropical forage legumes grown on
1981   the Cerrado. *Rev. Soc. Braz. Zootec.*, 10(1): 103-122.

DE ANDRADE, R.P., THOMAS, D. & FERGUSON, J.E. Seed production of pasture species in
1983   a tropical savanna region of Brazil. I. Legumes. *Trop. Grassl.*, 17: 54-59.

ANDREW, C.S. Copper deficiency symptoms of some tropical and temperate pasture
1963   legumes. *Aust. J. Agric. Res.*, 14: 654-659.

ANDREW, C.S. Mineral nutrition of plants. In *Some concepts and methods in subtropical pas-*
1964   *ture research*. Hurley, Berkshire, UK, Com. Bur. of Pastures and Field Crops.
Bull. No. 47.

ANDREW, C.S. A kinetic study of phosphorus absorption by excised roots of *Stylosanthes*
1966   *humilis, Phaseolus lathyroides, Desmodium uncinatum, Medicago sativa* and *Hordeum vulgare. Aust. J. Agric. Sci.*, 17: 611-624.

ANDREW, C.S. & BRYAN, W.W. Pasture studies on the coastal lowlands of subtropical
1965   Queensland. I. Introduction and initial plant nutrient studies. *Aust. J. Agric. Res.*,
6: 265-290.

ANDREW, C.S. & BRYAN, W.W. Pasture studies on the coastal lowlands of subtropical
1958   Queensland. III. The nutrient requirements and potentialities of *Desmodium uncinatum* and white clover on a lateritic podzolic soil. *Aust. J. Agric. Res.*, 9: 267-285.

ANDREW, C.S. & HEGARTY, M.P. Comparative responses to manganese excess of eight
1969   tropical and four temperate pasture legume species. *Aust. J. Agric. Res.*, 20: 687-696.

ANDREW, C.S. & NORRIS, D.O. Comparative response to calcium of five tropical and four
1961   temperate pasture species. *Aust. J. Agric. Res.*, 12: 40-55.

ANDREW, C.S. & PIETERS, W.H.J. *Effect of potassium on the growth and chemical compos-*
1970a   *ition of some pasture legumes. III. Deficiency symptoms of 10 tropical pasture legumes.* Melbourne, CSIRO Aust. Div. Trop. Pastures. Tech. Paper No. 5.

ANDREW, C.S. & PIETERS, W.H.J. *Manganese toxicity symptoms of one temperate and seven*
1970b   *tropical pasture legumes.* Melbourne, CSIRO Aust. Div. Trop. Pastures. Tech. Paper No. 4.

ANDREW, C.S. & ROBINS, M.F. The effect of phosphorus on the growth and chemical com-
1969a   position of some tropical pasture legumes. I. Growth and critical percentage of phosphorus. *Aust. J. Agric. Res.*, 20: 665-674.

ANDREW, C.S. & ROBINS, M.F. The effect of phosphorus on the growth and chemical com-
1969b   position of some tropical pasture legumes. II. Nitrogen, calcium, magnesium, potassium and sodium contents. *Aust. J. Agric. Res.*, 20: 675-685.

ANDREW, C.S. & ROBINS, M.F. The effect of potassium on the growth and chemical com-
1969c   position of some pasture legumes. I. Growth and critical potassium percentages. *Aust. J. Agric. Res.*, 20: 999-1007.

ANDREW, C.S. & ROBINS, M.F. The effect of potassium on the growth and chemical com-
1969d   position of some pasture legumes. II. Potassium, calcium, magnesium, sodium, nitrogen, phosphorus, and chloride. *Aust. J. Agric. Res.*, 20: 1009-1021.

ANDREW, C.S. & THORNE, P.M. Comparative responses to copper of some tropical and
1962   temperate pasture legumes. *Aust. J. Agric. Res.*, 13: 821-835.

591

ANDREWS, F.W. *The flowering plants of the Anglo-Egyptian Sudan.* UK, J. Bungel.
1952

ANSLOW, R.C. Investigation into the potential productivity of "Acacia" (*Leucaena glauca*)
1957   in Mauritius. *Rev. Agric. Ile Maurice*, 36: 39-49.

ANSON, R.J. & GARTNER, R.J.W. Mulga supplies ample vitamin A. *Queensl. Agric. J.*, 93:
1967   463-465.

ARONOVICH, S., SERPA, A. & RIBEIRO, H. Effect of nitrogen fertilizer and legumes upon
1970   beef production of pangola grass pastures. *Proc. 11th Int. Grassl. Congr.,* Surfers
        Paradise, Australia, 786-800.

ARVIER, A.C. *Storage of seed in warm climates.* Queensl. Dep. Prim. Ind. Misc. Publ.
1983

ASHLEY, D.A., BENNETT, O.L., DOSS, B.D. & SCARSBROOK, C.E. Effect of nitrogen rate
1965   and irrigation on yield and residual nitrogen recovery by warm-season grasses.
        Agron. J., 57: 370-372.

ATKINSON, W.T. High altitude-low latitude forage plants from Mexico and Latin America.
1970   *Proc. 11th Int. Grassl. Congr.,* Surfers Paradise, Australia, 181-184.

AXTMAYER, J.H. HERNANDEZ, G.R. & COOK, D.H. The nutritive value of some forage
1938   crops of Puerto Rico. *J. Agric. Univ. P.R.*, 22: 455-481.

BACKER, C.A. & VAN DEN BRINK, R.C. *Flora of Java.* Vol. 1. Groningen, the Netherlands,
1963   Noordhoff.

BAILEY, D.R. Stylo tolerates 2,4-D in trial. *Queensl. Agric. J.*, 90: 236-238.
1964

BAILEY, D.R. Weed control during pasture establishment in the wet tropics. *Queensl.*
1965a   *Agric. J.*, 91: 2-11.

BAILEY, D.R. Effect of 2,4-D preplant soil spraying in the establishment of *Centrosema*
1965b   *pubescens. Queensl. J. Agric. Anim. Sci.*, 21.

BAILEY, D.R. Effect of 2,4,5-T preplant soil spraying on the establishment of *Stylosanthes*
1965c   *gracilis. Queensl. J. Agric. Anim. Sci.*, 22: 215-219.

BAILEY, D.R. Observations on the use of preplant herbicides in pasture establishment in the
1967a   wet tropics. *Queensl. J. Agric. Anim. Sci.*, 24: 31-40.

BAILEY, D.R. Effect of some phenoxy herbicides on *Vigna marina* and *Glycine javanica.*
1967b   *Queensl. J. Agric. Anim. Sci.*, 24: 121-124.

592

BAILEY, F.M. *The Queensland Flora. Part II.* Brisbane, H.J. Diddams & Co.
1900

BAILEY, F.M. *Weeds and suspected poisonous plants of Queensland.* Brisbane, H. Pole.
1906

BAKER, R.D. & BAKER, H.K. The output of swards on commercial farms in relation to fer-
1965   tiliser and other management practices. *J. Brit. Grassl. Soc.*, 20: 182-187.

BALDWIN, P.H. & FAGERLUND, G.O. The effect of cattle grazing on Koa reproduction in
1943   Hawaii National Park. *Ecology*, 24: 118-122.

BARKER, S.J. & KYNEUR, G.W. Leguminous pasture silage as a production ration for dairy
1962   cattle during the northern dry season. *Proc. Nth. Queensl. Agros. Conf.*, 12/4.

BARNARD, C. *Australian herbage plant register.* Canberra, A.C.T., CSIRO Div. Plant
1967   Industry.

BARNARD, C. *Herbage plant species.* Canberra, A.C.T., CSIRO Div. Plant Industry.
1969

BARNES, D.L. Some studies on nitrogen fertilising of dryland ley grasses on sandveld. *Rhod.*
1960   *Agric. J.*, 57: 311-317.

BARRAU, J. *Research in Queensland on tropical plant and animal industries.* South Pacific
1953   Comm. Tech. Paper No. 43.

BARROW, N.J. & JENKINSON, D.S. The effect of waterlogging on fixation of nitrogen by soil
1962   incubated with straw. *Plant Soil*, 16: 258-262.

BATTHYANY, C. Rhizobium - *nómina de técnicos, cepas e inoculantes de América Latina.*
1970   Montevideo, Uruguay, Laboratorio de Microbiología de Suelos y Control de
         Inoculantes, Ministerio de Ganadería y Agricultura.

BEAR, F.E. *Soils and fertilizers.* New York, Wiley.
1949

BECKETT, P.H.T. Method and scale of land resource surveys in relation to precision and
1968   cost. *In* Stewart, G.A., ed. *Land evaluation.* South Melbourne, Macmillan of
         Australia, pp. 53-63.

BECKING, J.H. Studies on nitrogen-fixing bacteria of the genus *Beijerinckia*. I. Geographical
1961   and ecological distribution in soils. *Plant Soil*, 14: 49-81.

BEESTON, G.R. Plant species lists. *In* Western arid region - land use study. Part IV. *QDPI*
1978   *Div. Land Util. Tech. Bull.*, 23: App. III.

593

BEGG, J.E. Comparative responses of indigenous, naturalized and commercial legumes to
1963   phosphorus and sulphur. *Aust. J. Exp. Agric. Anim. Husb.*, 3: 17-19.

BENNETT, H.H. *Elements of soil conservation.* 2nd ed. New York, McGraw-Hill.
1955

BERGERSEN, F.J. The growth of *Rhizobium* in synthetic medium. *Aust. J. Biol. Sci.*, 14: 349-
1961   360.

BERMÚDEZ, L.A., CEBALLOS, E. & CHAVERRA, H. Las leguminosas espontáneas en el
1968   Valle del Sinu. *Agric. Trop.*, 24: 589-603.

BINDON, B.M. & LAMOND, D.R. Examination of tropical legumes for deleterious effects on
1966   animal production. *Proc. Aust. Soc. Anim. Prod.*, 6: 109-116.

BIRCH, H.F. & DOUGALL, H.W. Effect of a legume on soil nitrogen mineralisation and per-
1967   centage nitrogen in grasses. *Plant Soil*, 27: 292-296.

BISSET, W.J. Fine-stem stylo in spear grass country. *Queensl. Agric. J.*, 94: 23.
1968

BLASCO, L.M. & BOHÓRQUEZ, A.N. Pasture species in the Amazon region (of Colombia).
1968   I. Analysis of some chemical components. *Agric. Trop.*, 24: 175-177.

BLOUARD, R. & THURIAUX, L. *Stylosanthes gracilis*, son comportement et son utilisation au
1962   Congo. *Bull. Inf. INEAC*, 11 (4-6): 339-355.

BLUE, W.G., GAMMON, N. JNR. & LUNDY, H.W. *Proc. Soil Crop. Sci. Soc. Fla*, 21: 56.
1961

BLUNT, C.G. & HUMPHREYS, L.R. Phosphate response of mixed swards at Mount Cotton,
1970   southeastern Queensland. *Aust. J. Exp. Agric. Anim. Husb.*, 10: 431-443.

BLYDENSTEIN, J., LOUIS, S., TOLEDO, J. & CAMARGO, A. Productivity of tropical pastures.
1969   I. Pangola grass. *J. Brit. Grassl. Soc.*, 24: 71-75.

BOELCKE, O. Forrajeras. *Enciclopedia Argentina de agricultura y jardinería.* Buenos Aires,
1964   Acme.

BOGDAN, A.V. Observations on palatability of some leguminous plants of Kenya. *E. Afr.*
1949   *Agric. J.*, 15: 38-41.

BOGDAN, A.V. Indigenous clovers of Kenya. *E. Afr. Agric. J.*, 22: 40-45.
1956

BOGDAN, A.V. Three interesting introductions from the local grass flora of Kitale, Kenya.
1963   *J. Brit. Grassl. Soc.*, 18: 247-248.

594

BOGDAN, A.V. Cultivated varieties of tropical and subtropical herbage plants in Kenya. *E.*
1965   *Afr. Agric. For. J.*, 30: 330-338.

BOGDAN, A.V. *Glycine javanica* under experimental cultivation in Kenya. *Trop. Agric.*
1966   *(Trin.)*, 43: 99-105.

BOGDAN, A.V. *Tropical pasture and fodder plants.* London, Longman Group Ltd.
1977

BONSMA, J.C. Useful bushveld trees and shrubs. *Fmg S. Afr.*, 226-239.
1942

BOTHA, A.D.P. Die invloed van grass (*Eragrostis curvula*) op die ammonium en nitraat-
1963   stikstof in neen swart kleigrond. *S. Afr. J. Agric. Sci.*, 6: 3-20.

BOTTON, H. *Les plantes de couverture en Côte-d'Ivoire.* Le Mans, Imprimerie Monnoy.
1958

BOUDET, G. Management of savanna woodland in west Africa. *Proc. 11th Int. Grassl.*
1970   *Congr.,* Surfers Paradise, Australia, 1-3.

BOULTWOOD, J.N. Two valuable perennial legumes - horse marmalade (*Desmodium dis-*
1964   *color*) and kuru vine (*D. intortum*). *Rhod. Agric. J.*, 61: 70-72.

BOWEN, G.D. Nodulation of legumes indigenous to Queensland. *Queensl. J. Agric. Sci.*, 13:
1956   47-60.

BOWEN, G.D. Field studies on nodulation and growth of *Centrosema pubescens* Benth.
1959a   *Queensl. J. Agric. Anim. Sci.*, 16: 253-265.

BOWEN, G.D. Specificity and nitrogen fixation in the *Rhizobium* symbiosis of *Centrosema*
1959b   *pubescens* Benth. *Queensl. J. Agric. Sci.*, 16: 267-282.

BOWEN, G.D. & KENNEDY, M. Effect of high soil temperatures of *Rhizobium* spp. *Queensl.*
1959c   *J. Agric. Sci.*, 16: 177-197.

BOWEN, G.D. & KENNEDY, MARGARET M. Heritable variation in nodulation of *Centrosema*
1961   *pubescens* Benth. *Queensl. J. Agric. Sci.*, 18: 161-170.

BOWYER, J.W., ATHERTON, J.G., TEAKLE, D.S. & AHERN, GABRIELLE A. Mycoplasma-
1969   like bodies in plants affected by legume little leaf, tomato big bud, and lucerne
         witches' broom diseases. *Aust. J. Biol. Sci.*, 22: 271-274.

BOYLAND, D.E. *Plant species list. In* "Western arid region - land use study. Part I". *QDPI*
1974   *Div. Land Util. Tech. Bull.* 12: v-7.

BRAITHWAITE, B.M. Tropical legume pest survey and clover root weevil studies. *Ann. Rep.*
1967   *Wollongbar Agric. Res. Sta.*, N.S.W., 1966/67, pp. 31-33.

BRAITHWAITE, B.M., JANE, A. & SWAIN, F.G. *Amnemus quadrituberculatus* (Bob.), a
1958   weevil pest of clover pastures on the north coast of New South Wales. *J. Aust. Inst. Agric. Sci.*, 24: 146-154.

BRANNON, C.C., KING, W.A. & COOK, W.C. Green chop feeding of coastal Bermuda grass.
1966   *J. Dairy Sci.*, 49(6): 717.

BRAY, R.A. The effect of creeping-rootedness on survival in lucerne. *J. Aust. Inst. Agric.*
1967   *Sci.*, 33: 46-47.

BRAY, R.A. Variation in and correlations between yield and creeping-rootedness in
1969   lucerne. *Aust. J. Agric. Res.,* 20: 47-55.

BREAKWELL, E.J. & JENKINS, H.V. A pasture overseeding implement. *J. Aust. Inst. Agric.*
1953   *Sci.*, 19: 109-110.

BRENCHLEY, WINIFRED E. & THORNTON, H.G. The relation between the development,
1925   structure, and functioning of the nodules in *Vicia faba*, as influenced by the presence or absence of boron in the nutrient medium. *Proc. Roy. Soc. London*, 98: 373-399.

BRENNAN, J.P.M. *Flora of tropical east Africa*. Vol. 24, Mimosoideae. Millbank, London,
1959   Crown Agents for Overseas Governments.

BRENNAN, J.P.M. Leguminosae, Sub family Caesalpinioideae. *In* Milne-Redhead, E. and
1967   Polhill, R.M., eds. *Flora of tropical East Africa*. London, Crown Agents for Overseas Govts and Admin.

BRITISH VIRGIN ISLANDS. DEPARTMENT OF AGRICULTURE. *Ann. Rept 1963*.
1963

BRITTEN, E.J. The influence of genotype and temperature on flowering in *Trifolium repens*.
1961   *Agron. J.*, 53: 11-14.

BRITTEN, E.J. & DE LACY, I.M. Assessment of the genetic potential for pasture purposes of
1979   the *Psoralea eriantha - patens* complex, a native legume of the semi-arid zone. *Aust. J. Exp. Agric. Anim. Husb.*, 19: 53-58.

BROCKINGTON, N.R. Studies on the growth of a *Hyparrhenia*-dominant grassland in Northern Rhodesia. I. Growth and reaction to cutting. *J. Brit. Grassl. Soc.*, 15: 323-338.
1960

BROCKWELL, J. & HELY, F.W. Symbiotic characteristics of *Rhizobium meliloti*: an appraisal
1966   of the systematic treatment of nodulation and nitrogen fixation interactions between hosts and rhizobia of diverse origins. *Aust. J. Agric. Res.*, 17: 885-899.

BROLMANN, J.B. Flood tolerance in *Stylosanthes*, a tropical legume. *Proc. Soil. Crop Sci.*
1978   *Soc. Fla.*, 37: 37-39.

596

BROWN, W.H. *Minor products of Philippine forests*. Quezon City, Bureau of Forest
1921   Development. Philippine Bur. For. Dev. Bull. No. 22.

BRUCE, R.C. Effect of *Centrosema pubescens* Benth. on soil fertility in the humid tropics.
1965   *Queensl. J. Agric. Anim. Sci.*, 22: 221-226.

BRUCE, R.C. Tropical pastures lift soil nitrogen. *Queensl. Agric. J.*, 93: 562-564.
1967

BRYAN, W.W. *Lotononis bainesii* Baker, a legume for subtropical pasture. *Aust. J. Exp.*
1961   *Agric. Anim. Husb.*, 1: 4-10.

BRYAN, W.W. The role of the legume in legume/grass pastures. *Commonw. Bur. Pastures*
1962   *Field Crops Bull.*, 46.

BRYAN, W.W. The pasture value of species of *Desmodium*. *Proc. 10th Int. Grassl. Congr.*,
1966   Helsinki, 311-315.

BRYAN, W.W. Grazing trials on the Wallum of southeastern Queensland. Comparison of
1968a   four pastures. *Aust. J. Exp. Agric. Anim. Husb.*, 8: 512-520.

BRYAN, W.W. Grazing trials on the Wallum of southeastern Queensland. II. Complex mix-
1968b   tures under common grazing. *Aust. J. Exp. Agric. Anim. Husb.*, 8: 683-690.

BRYAN, W.W. *Desmodium intortum* and *Desmodium uncinatum*. *Herb. Abstr.*, 39: 187-
1969   191.

BRYAN, W.W. & ANDREW, C.S. Pasture studies on the coastal lowlands of subtropical
1955   Queensland. II. The interaction of legumes, *Rhizobium* and calcium. *Aust. J.*
        Agric. Res., 6: 291-298.

BRYAN, W.W. & EVANS, T.R. *Legume-based pastures*. Ann. Rep. CSIRO Aust. Divn.
1968   Trop. Pastures, p. 21-22.

BRYAN, W.W., SHARPE, J.P. & HAYDOCK, K.P. Some factors affecting the growth of
1971   lotononis (*Lotononis bainesii*). *Aust. J. Exp. Agric. Anim. Husb.*, 11 (48): 29-34.

BRYAN, W.W. & SHAW, N.H. *Paspalum plicatulum* Michx. — two useful varieties for pas-
1964   ture in regions of summer rainfall. *Aust. J. Exp. Agric. Anim. Husb.*, 4: 17-21.

BRYANT, H.T., HAMMER, R.C., BLASER, R.E. JNR & FONTENOT, J.P. Effect of stocking
1965   pressure on animal and acre output. *Agron. J.*, 57: 273-276.

BULLER, R.E., ARONOVICH, S., QUINN, L.R. & BISSCHOFF, W.V.A. Performance of trop-
1970   ical legumes in the upland savanna of central Brazil. *Proc. 11th Int. Grassl. Congr.*,
        Surfers Paradise, Australia, 143-146.

BUNTING, B. Coconut cultivation in Ceylon. *Malaya Agric. J.*, 18: 378-390.
1930

BURBIDGE, N.T. Australian species of *Sesbania* Scopoli (Leguminosae). *Aust. J. Bot.*, 13:
1965   103-141.

BURKART, A. *Las leguminosas argentinas*. Buenos Aires, Acme.
1952

BURKART, A. *Report 4th Meeting on Forage Plants*. Argentina, Dept. Agric. 145 p.
1954

BURKILL, I.H. *et al. A dictionary of the economic products of the Malay Peninsula*. Vol. 1.
1935   London, Crown Agents for the Colonies.

BURROWS, W.H. & BEALE, I.F. Dimensions and production relations of mulga (*Acacia*
1970   *aneura* F. Muell.) trees in semiarid Queensland. *Proc. 11th Int. Grassl. Congr.*,
        Surfers Paradise, Australia, 33-35.

BURT, R.L., CAMERON, D.G., CAMERON, D.F., 'T MANNETJE, L. & LENNE, J. Stylosanthes.
1983   In *The role of* Centrosema, Desmodium *and* Stylosanthes *in improving tropical pas-
        tures*, p. 141-181. Eds. R.L. Burt, P.P. Rotar, J.R. Walker and M.W. Silvey. Boul-
        der, Colorado, Westview Press.

BURTON, G.W. *Bull. Ga. Agric. Exp. Stn*, 2.
1954

BURTON, G.W. & JACKSON, J.R. Effect of rate and frequency of applying six nitrogen
1962   sources on coastal Bermuda grass. *Agron. J.*, 54: 40-43.

BUTTERWORTH, M.H. Digestibility trials on forages in Trinidad and their use in the predic-
1963   tion of nutritive value. *J. Agric. Sci.*, 60: 341.

BUTTERWORTH, M.H. The digestibility of tropical grasses. *Nutr. Abst. Rev.*, 37: 349-368.
1967

BYTH, D.E. Breeding system and chromosome number in *Lotononis bainesii* Baker.
1964   *Nature*, 202: 830-831.

CABRERA, J.L. & RIVERA-BRENTES, L. The value of grass silage for feeding dairy cows in
1953   Puerto Rico. *J. Agric. Univ. P.R.*, 37: 59-73.

CALMA, V.C., VALERA, F.O. & SANTOS, F.I. Effect of spacing upon the yield of tick clover.
1959   *Philipp. Agric.*, 42: 404-413.

CAMERON, D.F. Variation in flowering time and in some growth characteristics of
1965   Townsville lucerne (*Stylosanthes humilis*). *Aust. J. Exp. Agric. Anim. Husb.* 5: 49-
        51.

CAMERON, D.F. Flowering in Townsville lucerne (*Stylosanthes humilis*). II. The effect of
1967a   latitude and time of sowing on the flowering time of single plants. *Aust. J. Exp.
Agric. Anim. Husb.*, 7: 495-500.

CAMERON, D.F. Flowering time and the natural distribution and dry-matter production of
1967b   Townsville lucerne (*Stylosanthes humilis*) populations. *Aust. J. Exp. Agric. Agric.
Anim. Husb.*, 7: 501-508.

CAMERON, D.F. Chromosome number and morphology of some introduced *Stylosanthes*
1967c   species. *Aust. J. Agric. Res.*, 18: 375-379.

CAMERON, D.F. *Stylosanthes* species. *CSIRO Div. Trop. Past. Ann. Rep. 1966-67*, pp. 98-
1967d   101.

CAMERON, D.F. Hard seededness and seed dormancy of Townsville lucerne (*Stylosanthes
1967e   humilis*) selections. *Aust. J. Exp. Agric. Anim. Husb.*, 7: 237-240.

CAMERON, D.F. Studies of the ecology and genetics of Townsville lucerne (*Stylosanthes
1968   humilis* H.B.K.). University of Queensland. (PhD. thesis).

CAMERON, D.F. & 'T MANNETJE, L. Effect of photoperiod and temperature of flowering of
1977   twelve *Stylosanthes* spp. *Aust. J. Exp. Agric. Anim. Husb.*, 17: 417-424.

CAMERON, D.G. & MULLALY, J.D. The preliminary evaluation of leguminous plants for
1969   pasture and forage in sub-coastal central Queensland 1962-69. *Aust. Plant. Introd.
Rev.*, 6: 29-54.

CARO-COSTAS, R. & VICENTE- CHANDLER, J. Effect of fertilisation on carrying capacity and
1961a   beef produced by napier grass pastures. *Agron. J.*, 53: 204-205.

CARO-COSTAS, R., VICENTE-CHANDLER, J. & BURLEIGH, C. Beef production and carrying
1961b   capacity of heavily fertilised, irrigated guinea, napier and pangola grass pastures on
the semi-arid south coast of Puerto Rico. *J. Agric. Univ. P.R.*, 45: 32-36.

CARO-COSTAS, R. & VICENTE-CHANDLER, J. Milk production with all-grass rations from
1969   steep, intensely managed tropical pastures. *J. Agric. Univ. P.R.*, 53: 251-258.

CARO-COSTAS, R., VICENTE-CHANDLER, J. & FIGARELLA, J. Productivity of intensely man-
1965   aged pastures of five grasses on steep slopes in the humid mountains of Puerto Rico.
J. Agric. Univ. P.R., 49:99-111.

CARTER, E.D. Some relationship between superphosphate use and consequent animal pro-
1965   duction from pasture in South Australia. *Proc. 9th Int. Grassl. Congr.*, 1027-1032.

CASS-SMITH, W.P. & PITTMAN, H.A.J. The influence of methods of planting on the effective
1939   inoculation and establishment of subterranean clover. *J. Dept. Agric. W. Aust.*, 16:
61-73.

CATCHPOOLE, V.R. Laboratory ensilage of three tropical pasture legumes — *Phaseolus*
1970   *atropurpureus, Desmodium intortum*, and *Lotononis bainesii. Aust. J. Exp. Agric.
Anim. Husb.*, 10: 568-576.

CATCHPOOLE, V.R. & HENZELL, E.F. Silage and silage-making from tropical herbage
1971   species. *Herb. Abstr.*, 41(3): 213-221.

CHARREAU, C. & VIDAL, P. Influence de *l'Acacia albida* Del. sur le sol, la nutrition
1965   minérale et les rendements de milo *Pennisetum* au Sénégal. *Agron. Trop.* (Paris),
20: 600-626.

CHATTERJEE, B.N. & SINGH, R.D. Some promising forage legumes. *Allahabad Fmr*, 40:
1966   225-226.

CHRISTIAN, C.S. & SHAW, N.H. *Proc. Spec. Conf. Agric., Brit. Commonw. Sci. Off.,*
1951   *Australia, 1949*, p. 225. London, HMSO.

CHRISTIAN, C.S. & SHAW, N.H. A study of two strains of rhodes grass (*Chloris gayana*
1952   Kunth.) and of lucerne (*Medicago sativa* L.) as component of a mixed pasture at
Lawes in southeast Queensland. *Aust. J. Agric. Res.*, 3: 277-299.

CHRISTIAN, C.S. & STEWART, C.A. Survey of Katherine-Darwin region, 1946. Melbourne,
1952   CSIRO Aust. Land. Res. Serv.

CIAT. *Ann. Rept.* Cali, Colombia.
1978

CIAT. Promising germplasm for the major ecosystems. 1979 Trop. Past. Programme. *Ann.
1979   Rept.* Cali, Colombia.

CLATWORTHY, J.N. The response of giant rhodes grass (*Chloris gayana* Kunth.) to heavy
1967   dressings of nitrogenous fertiliser. *Rhod. J. Agric. Res.*, 5: 87-89.

CLATWORTHY, J.N. Legumes and fertilisers as sources of pasture nitrogen. *Proc. 11th Int.
1970   Grassl. Congr.*, Surfers Paradise, Australia, 408-411.

CLATWORTHY, J.N. Introduction and preliminary screening of pasture legumes at Maran-
1975   dellas, Rhodesia, 1967-73. *Proc. Grassl. Soc. Sth. Afr.*, 10: 57-63.

CLEMENTS, R.J. *Preliminary evaluation of accessions of* Centrosema virginianum *in sub-
1983   humid south-east Queensland*. Melbourne CSIRO Div. Trop. Crops Pastures.
Gen. Res. Comm. No. 4.

CLEMENTS, R.J. & WILLIAMS, R.J. Genetic diversity in *Centrosema. In* Summerfield, R.J.
1983   & Bunting, A.H., eds. *Advances in legume science.* Kew, UK, Royal Botanical
Garden, pp. 559-567.

600

CLEMENTS, R.J., WILLIAMS, R.J., GROF, B. & HACKER, J.B. *Centrosema. In* Burt, R.L.,
1983   Rotar, P.P., Walker, J.L. & Silvey, M.W., eds. *The role of* Centrosema,
Desmodium *and* Stylosanthes *in improving tropical pastures.* Boulder, Colorado,
Westview Press, pp. 69-96.

CLEMENTS, R.J., WINTER, W.H. & REID, R. Evaluation of some *Centrosema* species in
1984   small plots in Northern Australia. *Trop. Grassl.*, 18: 83-91.

CLOONAN, M.J. Black nodules in *Dolichos. Aust. J. Sci.*, 26:121.
1963

CLOONAN, M.J. The root nodule bacteria as factors in the establishment of tropical legumes.
1966   *J. Aust. Inst. Agric. Sci.*, 29: 149-153.

CLOONAN, M.J. & VINCENT, J.M. The nodulation of annual summer legumes sown on the
1967   far north coast of New South Wales. *Aust. J. Exp. Agric. Anim. Husb.*, 7: 181-189.

COBLEY, L.S. *An introduction to the botany of tropical crops.* London, Longmans.
1956

COLBRAN, R.C. Studies of plant and soil nematodes. VII. Queensland records of the order
1963   *Tylenchida* and the genera *Trichodorus* and *Xiphinema. Queensl. J. Agric. Sci.*, 21:
77-123.

COLBRAN, R.C. Cover crops for nematode control in old banana land. *Queensl. J. Agric.*
1964   *Sci.*, 21: 233-236.

COLEMAN, R.G. *Frosts and low night temperatures as limitations to pasture development in*
1964   *subtropical eastern Australia.* Melbourne, CSIRO Aust. Div. Trop. Past. Tech.
Paper No. 3.

COLMAN, R.L., HOLDER, J.M. & SWAIN, F.G. Production from dairy cattle on improved
1966   pasture in a subtropical environment. *Proc. 10th Grassl. Congr.*, Helsinki, 499-503.

COMPÈRE, R. Etude toxicologique de *Leucaena glauca* chez les bovins. *Bull. agric. Congo*
1959   *belg.*, 50: 1311-1320.

COMPÈRE, R. Cinq ans d'exploitation de la culture fourragère de *Desmodium intortum*
1961   (Mill.) Urb. à la Station de Mulungu (Kivu). *Bull. agric. Congo*, 3(3).

CORRELL, D.S. & JOHNSTON, M.C. *Manual of the vascular plants of Texas.* Renner, Texas,
1970   Texas Res. Found.

COWDRY, W.A.R. Growing glycine for seed in the lower Burdekin. *Queensl. Agric. J.*, 86:
1960   677-680.

COWDRY, W.A.R. & VERHOEVEN, G. Stylo seed harvesting. *Queensl. Agric. J.,* 87: 179.
1961

601

COWIE, A.M. & SKERMAN, P.J. Fertility status of a red earth soil at Charleville, southwest
1970   Queensland. *Proc. Aust. Arid Zone Res. Conf.*

CRAUFURD, R.Q. & PRINS, W.H. Munkolo (*Rhynchosia sublobata*) a promising pasture
1979   legume for Zambia. *Trop. Grassl.*, 13: 45-52.

CROFTS, F.C., GEDDES, H.J. & CARTER, O.G. Water harvesting and planned pasture pro-
1963   duction at Badgery's Creek. Univ. Sydney School Agric. Rep. No. 6.

CROWDER, L.V. Gramíneas y leguminosas forrajeras en Colombia. Bogota. DIA Boletín
1960   Técnico. No. 8.

CROWDER, L.V., MICHELIN, A. & BASTIDAS, A. The response of pangola grass (*Digitaria*
1964   *decumbens* Stent.) to rate and time of nitrogen application in Colombia. *Trop.
        Agric. (Trin.)*, 41: 21-29.

DALZEIL, J.A. *The useful plants of west tropical Africa*. London, Crown Agents for Over-
1955   seas Governments and Administrations.

DAMSEAUX, J. Etude de trois légumineuses fourragères introduites au Congo belge en vue
1956   de l'alimentation du betail. *Bull. agric. Congo Belg.*, 47: 83-111.

DART, P.J. & MERCER, F.V. The effect of growth temperature, level of ammonium nitrate
1965   and light intensity on the growth and nodulation of cowpea. *Aust. J. Agric. Res.*,
        16: 321-345.

DATE, R.A. A decade of legume inoculant quality control in Australia. *J. Aust. Inst. Agric.*
1969   *Sci.*, 35: 27-37.

DAVIDSON, D.E. Five pasture plants for Queensland. *Queensl. Agric. J.*, 92: 460-466.
1966

DAVIES, J.G. Pasture and forage legumes for the dry subtropics of Australia. *Proc. 8th Int.*
1960   *Grassl. Congr.*, Reading, UK, 381-385.

DAVIES, J.G. Pasture improvement in the tropics. *Proc. 9th Int. Grassl. Congr.*, São Paulo,
1965   Brazil, 217-220.

DAVIES, J.G. & HUTTON, E.M. Tropical and subtropical pasture species. *In* Moore, R.M.,
1970   ed. *Australian grasslands*. Canberra, ANU Press.

DAYTON, W.A. *Important Western browse plants*. USDA Misc. Pub. No. 101.
1931

DE LACY, I.H. & BRITTEN, E.J. Cytogenetics of the species of *Psoralea* native to the
1970   semiarid Australian tropics. *Proc. 11th Int. Grassl. Congr.*, Surfers Paradise,
        Australia, 218-284.

DELHAYE, R.E. Some consideration of agrostological problems in Madagascar and in par-
1964   ticular in the region of Lake Alaotra. *Agron. Trop.* (Paris), 19: 137-160.

DENARIE, J., ANDREAMANANTENA, S. & RAMONJY, J. *Legume inoculation in Madagascar:*
1968   *results from trials in 1966-67.*

DIATLOFF, A. Poor legume nodulation lowers pasture yields. *Queensl. Agric. J.*, 93: 682-
1967a   684.

DIATLOFF, A. Effect of soil moisture fluctuation on legume nodulation and nitrogen fixation
1967b   in a black earth soil. *Queensl. J. Agric. Anim. Sci.*, 24: 315-321.

DIATLOFF, A. Nodulation and nitrogen fixation in some *Desmodium* species. *Queensl.*
1968   *Agric. Anim. Sci.*, 25: 165-167.

DIJKMAN, M.J. *Leucaena glauca* — a promising soil erosion control plant. *Econ. Bot.*, 4:
1950   337-349.

DIJKSTRA, N.D. & DIRVEN, J.G.P. Digestibility and feeding value of some tropical grasses
1962   and Kudzu. *Neth. J. Agric. Sci.*, 10: 275-285.

DILLON, J.L. & BURLEY, H.T. A note on the economics of grazing and its experimental
1961   investigation. *Aust. J. Agric. Econ.*, 5: 123-132.

DINGAYAN, A.B. & FRONDA, F.M. A comparative study of the influence of the leaves and
1950   young shoots of *Centrosema*, ipil-ipil, and sweet potato as green feed on the growth
        of chicks. *Philipp. Agric.*, 34: 110-113.

DIRVEN, J.G.P. The protein content in Surinam roughages. *Landbouwproefstn. Suriname,*
1965   *Bull.*, 82.

DIRVEN, J.G.P. & EHRENCRON, V.K.R. Deficiency symptoms in *Pueraria phaseoloides*
1969   (Roxb.) Benth. *Fertil. Orthogenie*, 1a(34): 12.

DIRVEN, J.G.P. & 'T HART, M.L. The feeding value of Kudzu. Jaarversal. *Landbouwproef-*
1959   *stn. Suriname, Bull.*, 65-66.

DÖBEREINER, J. & ARONOVICH, S. Efecto da calagem e da temperatura do solo na fixaçao
1965   de nitrogênio de *Centrosema pubescens* Benth. em solo com toxidez de manganes.
        Proc. 9th Int. Grassl. Congr., São Paulo, Brazil, 2: 1121-1124.

DOHERTY, N.W. Mung bean success. *Queensl. Agric. J.*, 89: 176-177.
1963a

DOHERTY, N.W. Sod-seeded summer legumes give autumn feed. *Queensl. Agric. J.*, 89:
1963b   200-204.

603

DOLLAHITE, J.W. & ANTHONY, W.V. Malnutrition in cattle on an unbalanced diet of mes-
1957   quite beans. *Texas Agric. Exp. Stn. Prog. Rep. 1931.*

DONALD, C.M. Competition among crop and pasture plants. *Adv. Agron.* 15: 1-118.
1963

DONALDSON, L.E., HAMILTON, R.I., LAMBOURNE, L.J. & LITTLE, D.A. Assessing
1970   *Leucaena leucocephala* for deleterious effects in cattle and sheep. *Proc. 11th Int.
Grassl. Congr.*, Surfers Paradise, Australia, 780-782.

DOUGALL, H.W. & BODGAN, A.V. Browse plants of Kenya — with special reference to
1958   those occurring in South Baringo. *E. Afr. Agric. J.*, 23: 236-245.

DOUGALL, H.W. & BODGAN, A.V. The chemical composition of some leguminous plants
1966   grown in the herbage nursery at Kitale, Kenya. *E. Afr. Agr. For. J.*, 32: 45-49.

DOUGLAS, N.J. Tropical legumes on a blady grass burn. *Queensl. Agric. J.*, 91: 36-39.
1965

DOWNES, R.W. The introduction and preliminary evaluation of species for improving cattle
1966   feed in the dry tropics of Queensland. Brisbane, Queensl. Dept Prim. Ind., Agric.
Branch. Tech. Rep. No. 1.

DOWNES, R.W. Establishment of legumes in pasture of savanna woodland in north Queens-
1967   land. *Queensl. J. Agric. Anim. Sci.*, 24: 23-29.

DOWNES, R.W., STAPLES, I.B., COLMAN, P.H. & PEDLEY, L. The effect of day-length and
1967   temperature on the growth and reproduction of six strains of Townsville lucerne
(*Stylosanthes humilis*). *Aust. J. Exp. Agric. Anim. Husb.*, 7: 351-356.

DUNSMORE, J.R. & ONG, C.B. Preliminary work on pasture species and beef production in
1969   Sarawak, Malaysia. *Trop. Grassl.*, 3: 117-121.

EAVIS, B.W., CUMBERBATCH, E.R. ST. J. & MEDFORD, D.L. Factors influencing regener-
1974   ation of natural vegetation on reformed Scotland District soils of Barbados. *Trop.
Agric. (Trin.)*, 51: 293-303.

EBERSOHN, J.E. Grazing management of dairy pastures. *Queensl. Agric. J.*, 95: 574-582.
1969

EDGLEY, W.H.R. Milk yield lifted by glycine. *Queensl. Agric. J.*, 88: 6-7.
1962

EDWARDS, D.C. & BOGDAN, A.V. *Important grassland plants of Kenya.* Nairobi, Pitman.
1951

EDYE, L.A. Yield comparisons of thirty-eight introductions of *Glycine javanica* in swards in
1967   three environments. *Aust. J. Exp. Agric. Anim. Husb.*, 7: 342-350.

EDYE, L.A. & HAYDOCK, K.P. Breeding creeping-rooted lucerne for the subtropics. *Aust.*
1967   *J. Agric. Res.*, 18: 891-901.

EDYE, L.A. & KIERS, H.J. Variation in maturity, stolon development, and frost resistance
1966   of *Glycine javanica. Aust. J. Exp. Agric. Anim. Husb.*, 6: 380-387.

EDYE, L.A., GROF, B. & WALKER, B. Agronomic variation and potential utilization of
1984   *Stylosanthes. In* Stace, H.M. & Edye, L.A., eds. *The biology and agronomy of*
         Stylosanthes. North Ryde, Australia, Academic Press, pp. 547-570.

ELLIOTT, R.C. & CROFT, A.G. Digestion trials on Rhodesian feedstuffs. III. *Rhod. Agric.*
1958   *J.*, 55: 40-49.

ELLIOTT, R.C. & FOKKEMA, K. Digestion trials on Rhodesian feedstuffs. III. *Rhod. Agric.*
1960   *J.*, 57: 252-256.

ENGLISH, B.H. & HOPKINSON, J.M. Verano stylo seed production. *Queensl. Agric. J.*,
1985   111(1): 59-63.

Establishment and maintenance of legume covers. *Planter's Bull.*, 39: 129-133.
1958

EVANS, J. Cattle fattening — Parada research station. Brisbane, Queensl. Dept Prim. Ind.,
1969   Agrost. Tech. Rept.

EVANS, T.R. Preliminary evaluation of grasses and legumes for the northern Wallum of
1967   southeast Queensland. *Trop. Grassl.*, 1: 143-152.

EVANS, T.R. Sources of nitrogen for beef production in the Wallum. *Trop. Grassl.*, 2: 192-
1968   194.

EVANS, T.R. Beef production from nitrogen-fertilised pangola grass (*Digitaria decumbens*)
1969   on the coastal lowlands of southern Queensland. *Aust. J. Exp. Agric. Anim. Husb.*,
         9: 282-286.

EVANS, T.R. Some factors affecting beef production from subtropical pastures in the coastal
1970   lowlands of southeast . Queensland. *Proc. 11th Int. Grassl. Congr.*, Surfers
         Paradise, Australia, 803-807.

EVERIST, S.L. Mulga (*A. aneura* F. Muell. ex Benth.) in Queensland. *Queensl. J. Agric.*
1949   *Sci.*, 6: 87-139.

EVERIST, S.L. *Use of fodder trees and shrubs.* Brisbane, Queensl. Dept Prim. Ind., Div. Plt.
1969   Ind. Advisory Leaflet No. 1024.

EVERIST, S.L., HARVEY, J.M. & BELL, A.T. *Feeding sheep on mulga.* Queensl. Dept Agr.
1958   and Stock. Reprint 7/4/58.

EZEDINMA, F.O.C. The influence of seed size and fertiliser on the development and yield of
1965 cowpea (*Vigna sinensis* Endl.). *Nigeria Agric. J.*, 2: 75-79.

FAO. *Review of national milk and milk product policies 1969/70.* Rome.
1970

FARINAS, E.C. Ipil-ipil, the "alfalfa" of the tropics: its establishment, culture and utilisation
1951 as a fodder and pasture crop. *Philipp. J. Anim. Ind.*, 12: 65-85.

FARINAS, E.C. Production and distribution of forage seed and vegetative propagation mate-
1965 rials in the Philippines. *Proc. 9th Int. Grassl. Congr.*, São Paulo, Brazil, 551-558.

FISHER, M.J. The growth and development of Townsville lucerne (*Stylosanthes humilis*) in
1969 ungrazed swards at Katherine, Northern Territory. *Aust. J. Exp. Agric. Anim.
Husb.*, 9: 196-208.

FISHER, M.J. The effects of phosphorus and water stress on Townsville lucerne
1970 (*Stylosanthes humilis* H.B.K.) *Proc. 11th Int. Grassl. Congr.*, Surfers Paradise,
Australia, 481-483.

FITZPATRICK, E.A. & NIX, H.A. The climatic factor in Australian grassland ecology. *In*
1970 Moore, R. Milton, ed. *Australian grasslands.* Canberra, Aust. Nat. Univ. Press.

FLAY, A. Report to the Honorary Commission of Plan Agropecuario.
1968

Fodder crops and fallows in Dahomey. *Soils Afr.*, 10: 225-240.
1965b

FOSTER, W.H. Note on the establishment of a legume in Northern Nigeria. *Emp. J. Agric.*,
1961 29: 319-312.

FOSTER, W.H. & MUNDY, E.J. Forage species in Northern Nigeria. *Trop. Agric. (Trin.)*, 38:
1961 311-318.

FRANKEL, O.H. & BENNETT, E. *Genetic resources in plants: their extraction and conserva-
1970 tion.* Oxford, Blackwell. IBP Handbook No. 11.

FRENCH, M.H. The nutritive value of the pods of *Acacia arabica* and *Dichrostachys
1934 glomerata. Ann. Rept. Dept. Vet. Sci. Anim. Husb., Tanganyika*, 6: 79-83.

FRENCH, M.H. The nutritive value of legume hays. *Ann. Rept. Dept. Vet. Sci. Anim. Husb.,
1935 Tanganyika*, 7: 104-106.

FRENCH, M.H. Comparative nutritive value of legume hays. *Ann. Rept. Dept. Vet. Sci.
1937 Anim. Husb., Tanganyika*, 9: 86-89.

FRENCH, M.H. Some use and misuse of shrubs and trees as fodder. *E. Afr. Agric. J.*, 14: 157-
1949   165.

FRENCH, M.H. Nutritional value of tropical grasses and fodders. *Herb. Abstr.*, 27: 1-9.
1957

FRETES, R., SAMUDIO, R. & GAY, C. Las praderas naturales del Paraguay. I. Clasificación
1970   y descripción. Min. Agric. Gan., Pub. Misc., 5.

GALLI, F. Inoculacoes cruzadas com bacterias dos nodulas de leguminosas tropicales. *Rev.*
1958   *Agrícola (Mozambique)*, 33: 139-150.

GANGULI, B.N., KAUL, R.N. & NAMBIAR, K.T.N. Preliminary studies on a few topfeed
1964   species. *Ann. Arid Zone*, 3: 33-37.

GANTT, P.A. Utilization of *Leucaena glauca* as a feed in the Philippines. *Proc. 8th Pacific*
1953   *Sci. Congr.*, IVB, pp. 601-603.

GARDEN, D.L. A comparison of African clovers and temperate legumes on the North Coast
1977   of New South Wales. *Trop. Grassl.*, 11: 125-132.

GARDENER, C.J. Tolerance of perennating *Stylosanthes* plants to fire. *Aust. J. Exp. Agric.*
1980   *Anim. Husb.*, 20: 587-593.

GARTNER, J.A. *Queensl. Dept. Prim. Ind., Agrost. Tech. Ann. Rep.*, 1: 1.
1968

GARTNER, J.A. Effect of fertiliser nitrogen on a dense sward of kikuyu, paspalum and car-
1969   pet grass. *Queensl. J. Agric. Anim. Sci.*, 28: 21-30.

GARTNER, J.A. & FISHER, A.E. Improving pastures on the Atherton Tableland. *Queensl.*
1966   *Agric. J.*, 92: 356-361.

GATES, C.T. Physiological aspects of the rhizobial symbiosis in *Stylosanthes humilis,*
1970   *Leucaena leucocephala* and *Phaseolus atropurpureus*. Proc. 11th Int. Grassl.
        Cong., Surfers Paradise, Australia, 442-446.

GATES, C.T., HAYDOCK, K.P. & CLARINGBOLD, P.J. Response to salinity in Glycine. II.
1966   Varietal differences in *Glycine javanica*. *Aust. J. Exp. Agric. Anim. Husb.*, 6: 374-
        379.

GATES, C.T., HAYDOCK, K.P. & LITTLE, I.P. Response to salinity in Glycine. I. *Glycine*
1966   *javanica. Aust. J. Exp. Agric. Anim. Husb.*, 6: 261-265.

GATES, C.T., WILSON, J.R. & SHAW, N.H. Growth and chemical composition of Townsville
1966   lucerne (*Stylosanthes humilis*). *Aust. J. Exp. Agric. Anim. Husb.*, 6: 266-276.

607

GETHIN-JONES, G.H. The effect of a leguminous cover crop in building up soil fertility. *E.*
1942   *Afr. Agric. J.*, 8: 48-52.

GILCHRIST, E.C. A place for stylo in north Queensland pastures. *Queensl. Agric. J.*, 93: 344-
1967   349.

GILL, G.S. & BATRA, P.C. Influence of weather on seed setting of fodder crops. *Indian J.*
1968   *Agron.*, 13: 26-34.

GILLARD, P. Pasture development in the dry tropics of north Queensland. *Proc. 11th Int.*
1970   *Grassl. Congr.*, Surfers Paradise, Australia, 807-810.

GILLETT, J.B. The genus *Trifolium* L. in southern Arabia and in Africa, south of the Sahara.
1952   *Kew Bull.*, 3: 367-404.

GILLETT, J.B. III Trifolieae. *Flora Congo Belg. Ruanda-Urundi*, 4: 289-300.
1953

GILLETT, J.B., POLHILL, R.M. & VERDCOURT, B. *Leguminosae* (Part 4) Sub family
1971   Papilionoideae (2). *In* Milne-Redhead, E. & Polhill, R.M., eds. *Flora of tropical*
         East Africa. London, Crown Agents for Overseas Govt. and Admin.

GLOVER, P.E., STEWART, J. & GWYNNE, M.D. Masai and Kipsigis notes on east African
1966   plants. Part I. Grazing, browse, animal associated and poisonous plants. *E. Afr.*
         Agric. For. J., 32: 184-191.

GONZALES, V., BREWBAKER, J.L. & HAMILL, D.E. Leucaena cytogenetics in relation to the
1967   breeding of low mimosine lines. *Crop Sci.*, 7: 140-143.

GOODING, E.G.B., LOVELESS, A.R. & PROCTOR, G.R. *Flora of Barbados*. London,
1965   H.M.S.O.

GOODING, H.J. The agronomic aspects of pigeon peas. *Field Crop Abstr.*, 15: 1-15.
1962

GOWDA, D. RANGE & RAMASWAMY, M.N. The utilisation of *Prosopis juliflora*. *Indian For.*,
1960   86: 432-434.

GRAHAM, E.H. II Flora of the Kartabo Region, British Guiana. *Ann. Carnegie Mus.*, 22:
1933   161.

GRAHAM, N. McC. The net energy value of three subtropical forages. *Aust. J. Agric. Res.*,
1967   18: 137-147.

GRAHAM, T.G. Stepping up supply of Townsville lucerne seed. *Queensl Agric. J.*, 89: 90-93.
1963a

GRAHAM, T.G. Strip-planting Townsville lucerne in spear grass. *Queensl. Agric. J.*, 89: 670-
1963b   674.

GRAINER, P. Note on the management of low-lying ground in Madagascar for the produc-
1965   tion of forage. *Proc. 9th Int. Grassl. Congr.*, São Paulo, Brazil, 1005-1007.

GRAY, S.G. Hot water seed treatment for *Leucaena glauca* (L.) Benth. *Aust. J. Exp. Agric.*
1962   *Anim. Husb* , 2: 178-180.

GRAY, S.G. Inheritance of growth habit and quantitative characters in intervarietal crosses
1967a   in *Leucaena leucocephala* (Lam.) De Wit. *Aust. J. Agric. Res.*, 18: 63-70.

GRAY, S.G. General and specific combining ability in varieties of *Leucaena leucocephala*
1967b   (Lam.) De Wit. *Aust. J. Agric. Res.*, 18: 71-76.

GRAY, S.G. The components of variation in an intervarietal cross in *Leucaena leucocephala*
1967c   (Lam.) De Wit. *Aust. J. Agric. Res.*, 18:77-83.

GRAY, S.G. A review of research on *Leucaena leucocephala. Trop. Grassl.*, 2: 19-30.
1968

GRAY, S.G. The place of trees and shrubs as sources of forage in tropical and subtropical
1970   pastures. *Trop. Grassl.*, 4: 57-62.

GREENLAND, D.J. Nitrate fluctuations in tropical soils. *J. Agric. Sci.*, 50: 82-92.
1958

GRISEBACH, A.H.R. *Flora of the British West Indian islands.* New York, Hafner.
1963

GROF, B. Establishment of legumes in the humid tropics of northeastern Australia. *Proc.*
1965   *9th Int. Grassl. Congr.*, São Paulo, Brazil, 1137-1142.

GROF, B. & HARDING, W. *Stylosanthes guyanensis* frequency of cutting trial. *Queensl. Dept*
1968   *Prim. Ind., Agrost. Tech. Ann. Rep.*, 1: 14-15.

GROF, B., SCHULTZE-KRAFT, R. & MULLER, F. *Stylosanthes capitata* Vog., some agronomic
1979   attributes, and resistance to anthracnose (*Colletotrichum gloeosporioides* Penz).
    Trop. Grassl., 13: 28-37.

GROF, B. Performance of *Desmodium ovalifolium* Wall. in legume-grass association. *Trop.*
1982   *Agric. (Trin.)*, 59 (1): 33-37.

GRUNDY, G.H.F. *Leguminous cover crops.* Tanganyika Sisal Grower's Assoc. Sisal Res.
1959   Stn. Bull. No. 20.

609

GUAYADEEN, K.D. A note on some promising legume forages. *Trop. Agric. (Trin.)*, 28:
1951   231-232.

GUDE, J.B. Growing our own centro seed. *Queensl. Agric. J.*, 85: 139-140.
1959

GUKOVA, M.M. *Trudy S-kh. Akad. Timiryazeva*, 30: 33-42.
1945

GUNN, LAI TEIK. Kuala Lumpur, Fed. Malaya Dept. Agric. Scientific Series No. 24.
1951

GURGEL, M.A. & FERNANDES, A.A.O. Programme for improvement and management of
1980   pasture in the northeast. In *Relat. Ann. Pesq. Zoui. Brazil*, pp. 7-26.

GUTTERIDGE, R.C. & WHITEMAN, P.C. Effect of defoliation frequency on growth and sur-
1975   vival of four accessions of *Psoralea eriantha*. *Aust. J. Exp. Agric. Anim. Husb.*, 15:
493-497.

HACKLAND, N.G.E., BOOYSEN, P. DE V. & SOMNER, M.E. The response of *Trifolium*
1976   *africanum* (Ser) var. *glabellum* and *Trifolium repens* (L) cv. Ladino to lime and fertilizer. *Proc. Grassl. Soc. Sth. Afr.*, 11: 103-108.

HALL, N.H. Weeds in the N.T. *Acacia farnesiana* (needle bush). Darwin, Aust., Primary
1967   Ind. Branch, N.T. Admin. (*Herb. Abstr.*, 38: 129). Pamphlet No. 11.

HALL, T.D., MEREDITH, D. & ALTONA, R.E. Production from grassland in South Africa
1950   (Fertilizer treatments and livestock gains on veld). *Emp. J. Exp. Agric.*, 18: 8-18.

HALL, T.J. Attack on the legume *Stylosanthes scabra* (Vog.) by *Platyomposis pedicornia* (F)
1980   (Coleopterce: Cerambycidae). *J. Aust. Ent. Soc.*, 19: 277-279.

HALLIDAY, J. Field responses by tropical forage legumes to inoculation with *Rhizobium. In*
1979   Sanchez, P.A. & Tergas, L.E., eds. *Pasture production in acid soils of the tropics*.
Cali, Colombia, CIAT, pp. 123-137.

HALLSWORTH, E.G., GREENWOOD, E.A.N. & YATES, M.G. Studies on the nutrition of for-
1964   age legumes. III. The effect of copper on nodulation of *Trifolium subterraneum* L.
and *Trifolium repens*. *Plant Soil*, 20: 17-33.

HALLSWORTH, E.G., WILSON, S.B. & & GREENWOOD, E.A.N. Copper and cobalt in nitro-
1960   gen fixation. *Nature (London) Phys.*, 187: 79-80.

HAMILTON, C.P. & MCCARTHY, W.O. Increased beef production through improved pas-
1970   tures in Queensland. *Proc. Aust. Soc. Anim. Prod.*, 6: 263-267.

HAMILTON, R.A. & PILLAY, K.A. The manuring of *Centrosema pubescens*. *J. Rubber Res.*
1941   *Inst. Malays*. II. Communication No. 255.

610

HAMILTON, R.I. Measuring pasture quality in terms of milk production. *Trop. Grassl.*, 3:
1969   88-90.

HAMILTON, R.I., DONALDSON, L.E. & LAMBOURNE, L.J. *Leucaena leucocephala* as a feed
1971   for dairy cows: direct effect on reproduction and residual effect on the calf and lactation. *Aust. J. Agric. Res.*, 22: 681-692.

HAMILTON, R.I., FRASER, J. & ARMITT, J.D. Preliminary assessment of tropical pasture
1969   species for taint in milk. *Aust. J Dairy Tech.*, 24: 62-65.

HAMILTON, R.I., LAMBOURNE, L.J., ROE, R. & MINSON, D.J. Quality of tropical grasses for milk
1970   production. *Proc. 11th Int. Grassl. Congr.*, Surfers Paradise, Australia, 860-864.

HAMILTON, R.I. & RUTH, G. Bloat in *Dolichos lab-lab. Trop. Grassl.*, 2: 135-136.
1968

HARDISON, W.A. Tech. Bull. Dairy Training Res. Inst., Univ. of Philippines, 1.
1966

HARDISON, W.A., REID, J.T., MARTIN, C.M. & WOOLFOLK, P.G. Degree of herbage selec-
1954   tion by grazing cattle. *J. Dairy Sci.*, 37: 89-102.

HARDY, F. Seasonal fluctuations of soil moisture and nitrate in a humid tropical climate
1946   (Trinidad, B.W.I.). *Trop. Agric. (Trin.)*, 23: 40-49.

HARRISON, E. Digestibility trials on green fodders. *Trop. Agric. (Trin.)*, 19: 147.
1942

HARTLEY, W. Plant collecting expedition to sub-tropical South America, 1947-48. In
1949   *CSIRO (Aust.) Div. Plant Ind. Rep.*, 7. Canberra.

HARTY, R.L. Effect of superphosphate on the germination of Townsville lucerne
1967   (*Stylosanthes humilis* H.B.K.). *Queensl. J. Agric. Anim. Sci.*, 24: 235-236.

HASSELL, O.L. Native pasture legumes on the Central Coast. *Queensl. Agric. J.*, 60: 5-13.
1945

HAUCK, R.D. & BREMNER, J.M. *Proc. Biology and Ecology of Nitrogen Conf.*, Univ. of
1969   Calif., Davis, 1967, p. 31.

HAUMAN, L. in *Flore du Congo belge et du Ruanda-Urundi*. Vol. 6. Brussels, INEAC.
1954

HAYDOCK, K.P. & NORRIS, D.O. Opposed curves for nitrogen per cent dry weight given by
1967   *Rhizobium*-dependent and nitrate-dependent legumes. *Aust. J. Sci.*, 29: 426-427.

HEADY, H.F. *Range management in east Africa*. Nairobi, Govt Printer.
1960

611

HEGARTY, M.P. & POUND, A.W. Indospicine, a new hepatotoxic amino acid from *Indigo-*
1968   *fera spicata. Nature*, 217: 354-355.

HEGARTY, M.P., SCHINKEL, P.G. & COURT, R.D. Reaction of sheep to the consumption of
1964   *Leucaena glauca* Benth. and to its toxic principle, mimosine. Appendix I - A simple
        method for the isolation of mimosine from the seed of *Leucaena glauca* Benth.
        Aust. J. Agric. Res., 15: 153-167.

HEINRICHS, D.H. Developing creeping-rooted alfalfa for pasture. *Can. J. Agric. Sci.*, 34:
1954   269-280.

HELY, F. Symbiotic variation in *Trifolium ambiguum* M. Bieb. with special reference to the
1957   nature of resistance. *Aust. J. Biol. Sci.*, 10: 1-16.

HELY, F. Relation between effective nodulation and time to initial nodulation in a diploid
1963   line of *Trifolium ambiguum* M. Bieb. *Aust. J. Biol. Sci.*, 16: 43-45.

HEMSLEY, W.B. *Flora Cent. Am.*, 1: 294.
1888

HENDRICKSEN, R.E. Effect of stage and intensity of defoliation on the growth of cowpea.
1965   *Queensl. J. Agric. Anim. Sci.*, 22: 343-345.

HENDRICKSEN, R.E. How to manage grazing cowpeas. *Queensl. Agric. J.*, 95: 189-190.
1969

HENKE, L.A., WORK, S.H. & BURT, A.W. Beef cattle feeding trials in Hawaii. *Hawaii*
1940   *Agric. Exp. Sta. Bull.*, 85.

HENKE, L.A., WORK, S.H. & MARUYAMA, C. Koa-haole roughage as a substitute for
1942   imported protein supplements fed to dairy cows. Honolulu, Hawaii Agric. Exp.
        Sta. Progress Note 34 (*Herb. Abstr.*, 17: 310). (Mimeo)

HENKE, L.A. *et al.* Koa haole (*Leucaena glauca*) as the sole roughage fed to milking cows.
1951   *Hawaii Agric. Exp. Sta., Bienn. Rep. 1948-50*, 56-58.

HENRY, T. Rep. 4th Meeting on forage plants. Buenos Aires, Dept of Agric., Argentina,
1954   p. 206.

HENZELL, E.F. Nitrogen fixation and transfer by some tropical and temperate pasture
1962a   legumes in sand culture. *Aust. J. Exp. Agric. Anim. Husb.*, 2: 132-40.

HENZELL, E.F. The use of nitrogenous fertilisers on pastures in the subtropics and tropics.
1962b   *Bull. Commonw. Bur. Past. Fld Crops*, 46: 161-172.

HENZELL, E.F. Nitrogen fertiliser responses of pasture grasses in southeastern Queensland.
1963   *Aust. J. Exp. Agric. Anim. Husb.*, 3: 290-299.

612

HENZELL, E.F. Tropical pasture legumes in northern Australia. *Proc. Soil Crop. Sci. Soc.*
1967   *Fla*, 27: 322-338.

HENZELL, E.F. Sources of nitrogen for Queensland pastures. *Trop. Grassl.*, 2: 1-17.
1968

HENZELL, E.F. Problems in comparing the nitrogen economies of legume-based and nitro-
1970a   gen-fertilised pasture systems. *Proc. 11th. Int. Grass. Congr.*, Surfers Paradise,
Australia, A112-A120.

HENZELL, E.F. Use of nitrogenous fertilisers on Subtropical Pastures in Queensland. *J.*
1970b   *Aust. Inst. Agric. Sci.*, 36: 206-213.

HENZELL, E.F., FERGUS, I.F. & MARTIN, A.E. Accumulation of soil nitrogen and carbon
1966   under a *Desmodium uncinatum* pasture. *Aust. J. Exp. Agric. Anim. Husb.*, 6: 157-
160.

HENZELL, E.F., MARTIN, A.E., ROSS, P.J. & HAYDOCK, K.P. Isotopic studies on the
1968   uptake of nitrogen by pasture plants. IV. Uptake of nitrogen from labelled plant
material by rhodes grass and siratro. *Aust. J. Agric. Res.*, 19: 65-77.

HENZELL, E.F. & NORRIS, D.O. Processes by which nitrogen is added to the soil/plant sys-
1962a   tem. *Bull. Commonw. Bur. Past. Fld Crops*, 46: 1-18.

HENZELL, E.F. & NORRIS, D.O. A review of nitrogen in the tropics with particular refer-
1962b   ence to pastures. *Bull. Commonw. Bur. Past. Fld Crops*, 46: 1-18.

HENZELL, E.F. & OXENHAM, D.J. Seasonal changes in the nitrogen content of three warm-
1964   climate pasture grasses. *Aust. J. Exp. Agric. Anim. Husb.*, 4: 336-344.

Herb. Abstr., 38: 52.
1964b

HERMANN, F.J. *A revision of the genus* Glycine *and its immediate allies*. Washington, D.C.,
1962   USDA Tech. Bull. No. 1268.

HERMANN, F.J. *A synopsis of genus* Arachis. Washington, D.C., USDA.
1954

HERRERA, G. & CROWDER, L.V. Influencia del corte en el rendimiento del guandul
1963   (*Cajanus cajan* (L.) Mills). *Agric. Trop.*, 19: 521-531.

HERRERA, P.G. Effect of height of cutting on pigeon pea and koa haole. *Agric. Trop.*, 23:
1967   34-42.

HIGGINS, B.B. Origin and early history of the peanut. In *The peanut - the unpredictable*
1951   *legume*. Washington, D.C., Nat. Fert. Assoc., pp. 18-27.

613

HILDER, E.J. & MOTTERSHEAD, B.E. The distribution of plant nutrients through freegraz-
1963   ing sheep. *Aust. J. Sci.*, 26: 88-89.

HODGES, E.M., KILLINGER, G.B., McCALEB, J.E., RUELKE, O.C., ALLEN R.J., JNR,
1967   SCHANK, S.C. & KRETSCHMER, A.E., JR. Gainsville., Inst. Food. Agric. Sci.,
Univ. Fla, Bull. No. 718.

HODGES, E.M., KRETSCHMER, A.E., MISLEVY, P., ROUSH, R.D., RUELKE, O.C. &
1982   SNYDER, G.H. Production and utilization of the tropical legume *Aeschynomene*.
Fla. Agric. Exp. Stn. Circ., S-290.

HOLDER, J.M. Milk production from tropical pastures. *Trop. Grassl.*, 1: 135-141.
1967

HOLDER, J.M., SWAIN, F.G. & COLMAN, R.L. The use of sod-sown vetch (*Vicia sativa*) as
1963   a supplement by dairy cows on the far north coast of New South Wales. *Aust. J.
Exp. Agric. Anim. Husb.*, 3: 153-160.

HOOKER, J.D. *The flora of British India*. Vol. II. Ashford, UK, Reese.
1879

HOPKINSON, D. Leguminous cover crops for maintaining soil fertility in sisal in Tanzania.
1969   *Exp. Agric.*, 5: 283-294.

HOPKINSON, D. & BREITENSTEIN, J. Tropical kudzu as a cover crop in sisal. *Wld Crops*, 21:
1969   256-258.

HOPKINSON, J.M. & REID, R. Significance of climate in tropical pasture legume seed pro-
1979   duction. *In* Sanchez, P.A. & Tergas, L.E., eds. *Pasture production in acid soils of
the tropics*. Cali, Colombia, CIAT, pp. 343-360.

HORRELL, C.R. Herbage plants at Serere Experiment Station, Uganda, 1954-57. 2.
1958   Legumes. *E. Afr. Agric. J.*, 24: 133-138.

HORRELL, C.R. Herbage plants at Serere, Uganda, 1957-61. *E. Afr. Agric. For. J.*, 28: 174-
1963   180.

HORRELL, C.R. & COURT, M.N. Effect of the legume *Stylosanthes-gracilis* on pasture yields
1965   at Serere, Uganda. *J. Br. Grassl. Soc.*, 20: 72-76.

HORRELL, C.R. & NEWHOUSE, P.W. Yields of sown pastures in Uganda as influenced by
1965   legumes and fertilizers. *Proc. 9th Int. Grassl. Congr.*, São Paulo, Brazil, 1133-1136.

HOSAKA, E.Y. & RIPPERTON, J.C. Legumes of the Hawaiian ranges. *Hawaii Agric. Exp. St.
1944   Bull.*, 93.

HUANG, C. A brief on several major forage legumes in Taiwan. Vol. II. Forage operations
1967   under small farm system in Taiwan. *Chin-Am. Joint Comm. Rural Reconstr.*,
          Taipei (quoted by Bryan, 1969).

HUDSON, W.J.C., SWAIN, F.G., HOLDER, J.M., COLMAN, R.L. & BIRD, J.C. Development
1965   and evaluation of pastures for dairy cattle in a subtropical environment. *Proc. 9th
          Int. Grassl. Congr.*, São Paulo, Brazil, 923-927.

HUGHES, J. DE R. & SEARLE, P.G.E. Observations on the residual value of accumulated
1964   phosphorus in a red loam. *Aust. J. Agric. Res*, 15: 377-383.

HUMPHREYS, L.R. Townsville lucerne: history and prospect. *J. Aust. Inst. Agric. Sci.*, 33:
1967   3-13.

HUTCHINSON, J. & DALZIEL, J.M. *Flora of west tropical Africa*. London, Crown Agents for
1954   Overseas Governments and Administrations.

HUTTON, E.M. Flowering and pollination in *Indigofera spicata, Phaseolus lathyroides,
1960   Desmodium uncinatum* and some other tropical pasture legumes. *Emp. J. Exp.
          Agric.*, 28: 235-243.

HUTTON, E.M. Siratro - a tropical pasture legume bred from *Phaseolus atropurpureus*. *Aust.
1962   J. Exp. Agric. Anim. Husb.*, 2: 117-25.

HUTTON, E.M. Pasture species and beef production on the north coast of New South Wales.
1968   *Trop. Grassl.*, 2: 74-79.

HUTTON, E.M. The contribution of research in plant introduction to the development of
1970a   Australian pastures. *Proc. 11th Int. Grassl. Congr.*, Surfers Paradise, Australia,
          pp. A1 - A22.

HUTTON, E.M. Legume "little leaf" resistance and susceptibility in pasture legumes adapted
1970b   to the Australian tropics. *Sabrao Newslett*, 2(2): 151-153.

HUTTON, E.M. Tropical Pastures. *Adv. Agron.*, 22: 2-66.
1970c

HUTTON, E.M. & BEALL, L.B. Root-knot nematode resistance in two pasture species of
1957   *Phaseolus. J. Aust. Inst. Agric. Sci.*, 23: 158.

HUTTON, E.M. & BONNER, I. Dry matter and protein yields of four strains of *Leucaena
1960   glauca* Benth. *J. Aust. Inst. Agric. Sci.*, 26: 276-277.

HUTTON, E.M. & COOTE, J.N. Tannin content of some tropical legumes. *J. Aust. Inst.
1966   Agric. Sci.*, 32: 139-140.

HUTTON, E.M. & GRAY, S.G. Problems in adapting *Leucaena glauca* as a forage for the
1959   Australian tropics. *Emp. J. Exp. Agric.*, 27: 187-196.

615

HUTTON, E.M. & GRAY, S.G. Hybridisation between the legumes *Desmodium intortum*,
1967   *D. uncinatum* and *D. sandwicense*. *J. Aust. Inst. Agric. Sci.*, 33: 122-123.

HUTTON, E.M. & GRYLLS, N.E. Legume "little leaf": a virus disease of subtropical pasture
1956   species. *Aust. J. Agric. Res.*, 7: 85-97.

HUTTON, E.M. & GUERASIMOFF, J. Problems in breeding the legume *Indigofera spicata* for
1966   tropical pastures. *Euphytica*, 15: 353-361.

HUTTON, E.M., WINDRUM, G.M. & KRATZING, C.C. Studies on the toxicity of *Indigofera*
1958   *endecaphilla*. I. Toxicity for rabbits. *J. Nutr.*, 64: 321-328.

HYMOWITZ, T., STEENMEIJER, H.P., CARDOSO, A. & NUTI, P. Preliminary information on
1967   *Stylosanthes gracilis* CV IRI 1022, the lucerne of the northeast. *Zootecnia (São
        Paulo)*, 5: 39-41.

IBACH, D.B. & ADAMS, J.R. Fertilizer use in the United States. Washington, D.C., USDA.
1967   Statist. Bull. No. 408.

IMRIE, B.C., JONES, R.M. & KERRIDGE, P.C. *Desmodium. In* Burt, R.L., Rotar, P.P.,
1983   Walker, J.L. & Silvey, M.W., eds. *The role of* Centrosema, Desmodium *and*
        Stylosanthes *in improving tropical pastures*. Westview Press, Boulder, Colorado,
        pp. 97-140.

I'ONS, J.H. The development of tropical pastures for Swaziland middle veld. *Proc. Grassl.*
1968   *Soc. S. Afr.*, 3: 67-73.

ISON, R.L. & HUMPHREYS, L.R. Reproductive physiology of *Stylosanthes. In* Stace, H.M.
1984   & Edye, L.A., eds. *The biology and agronomy of* Stylosanthes. North Ryde,
        Australia, Academic Press, pp. 257-277.

IVENS, G.W. Report to FAO on brush control in Kenya.
1970

IWANAGA, I.I., OTAGAKI, K.K. & WAYMAN, O. Dehydrated Koa haole (*Leucaena glauca*)
1957   in rations for growing and fattening swine. *Proc. Western Sect., Amer. Soc. Anim.
        Prod.*, 8: 1-14.

JAMIESON, G.I. Effect of superphosphate application rate on pasture establishment in
1969   Queensland's wet tropical coast. *Queensl. J. Agric. Anim. Sci.*, 26: 529-536.

JENSEN, R.C. Farm development plans including tropical pastures for dairy farms in the
1968   Cooroy area of Queensland. *Rev. Marketing Agric. Econ.*, 36: 139-147.

JOHNSON, D.T. The cowpea in the African areas of Rhodesia. *Rhod. Agric. J.*, 67: 61-64.
1970

616

JOHNSON, R.W. Clearing the scrub. *Queensl. Agric. J.*, 88: 736-750.
1965

JONES, M.J. & BROMFIELD, A.R. Nitrogen in the rainfall at Samaru, Nigeria. *Nature (Lon-*
1970   *don)*, 227: 86.

JONES, R.J. The use of cyclodiene insecticides as liquid seed dressing to control bean fly
1965   (*Melanagromyza phaseoli*) in species of *Phaseolus* and *Vigna marina* in southeastern Queensland. *Aust. J. Exp. Agric. Anim. Husb.*, 5: 458-465.

JONES, R.J. Nutrient requirements of improved pasture on podsolic soils developed on phyl-
1966   lite at North Deep Creek. *Proc. Trop. Grassl. Soc. Aust.*, 6: 23-34.

JONES, R.J. Effects of close cutting and nitrogen fertiliser on growth of Siratro (*Phaseolus*
1967a   *atropurpureus*) pasture at Samford, southeastern Queensland. *Aust. J. Exp. Agric.*
Anim. Husb., 7: 157-161.

JONES, R.J. Losses of dry matter and nitrogen from autumn-saved improved pastures during
1967b   the winter at Samford, southeastern Queensland. *Aust. J. Exp. Agric. Anim.*
Husb., 7: 72-77.

JONES, R.J. A note on the *in vitro* digestibility of two tropical legumes - *Phaseolus atropur-*
1969   *pureus* and *Desmodium intortum. J. Aust. Int. Agric. Sci.*, 35: 62.

JONES, R.J. & ANDREW, C.S. Effects of seed and phosphate rate on growth of Siratro.
1967   *CSIRO, Aust., Divn Trop. Past., Ann. Rept.* 1966-67, p. 46.

JONES, R.J., DAVIES, J.G. & WAITE, R.B. The contribution of some tropical legumes to
1967   pasture yields of dry matter and nitrogen at Samford, southeastern Queensland.
Aust. J. Exp. Agric. Anim. Husb., 7: 56-65.

JONES, R.J., DAVIES, J.G., WAITE, R.B. & FERGUS, I.F. The production and persistance of
1968   grazed irrigated pasture mixtures in southeastern Queensland. *Aust. J. Exp. Agric.*
Anim. Husb., 8: 177-189.

JONES, R.K. Initial and residual effects of superphosphate on a Townsville lucerne pasture
1968   in northeast Queensland. *Aust. J. Exp. Agric. Anim. Husb.*, 8: 521-527.

JONES, R.K. A study of phosphorus responses of a wide range of accessions from genus
1974   *Stylosanthes. Aust. J. Agric. Res.*, 25: 847-862.

JONES, R.M. Mortality of some tropical grasses and legumes following frosting in the first
1969   winter after sowing. *Trop. Grassl.*, 3: 57-63.

JONES, R.M., ALCORN, J.L. & REES, M.C. Death of siratro due to Violet Root Rot. *Trop.*
1969   *Grassl.*, 3: 137-139.

JONES, R.M., STRIJDOM, B.W. & THERON, E.P. The indigenous South African clovers (*T.*
1974   *africanum* Ser. and *T. burchellianum* Ser.) and their potential as pasture legumes.
Trop. Grassl., 8: 7-16.

JONES, R.M. & COOK, B.G. Agronomy of Kenya White clover - cultivar Safari. Melbourne,
1981   CSIRO Div. Trop. Crops Pastures. Inf. Serv. Sheet No. 41-3.

JUKO, C.D. & BREDON, R.M. The chemical composition of leaves and whole plants as an
1961   indicator of the range of available nutrients for selective grazing by cattle. *Trop.
Agric. (Trin.)*, 38: 179-187.

JURRIAANSE, A. Fodder trees. *Farming S. Afr.*, 25.
1950

JUTZI, S.C. & NOSBERGER, J. Seed production and growth of the tropical pasture legume
1984   *Zornia latifolia* Accession CIAT 728. *Trop. Grassl.*, 18: 138-148.

KALMBACHER, R.S., HODGES, E.M. & MARTIN, F.G. Effect of cutting height on yield and
1980   quality of *Indigofera hirsuta. Trop. Grassl.*, 14: 14-18.

KANNEGIETER, A. The cultivation of grasses and legumes in the forest zone of Ghana. *Proc.
1965   9th Int. Grassl. Congr.*, São Paulo, Brazil, 313-318.

KAREEM, S. & SUDARARAJ, E.D. Why sesbania makes nutritious cattle feed. *Indian Fmg*,
1967   17: 20.

KARLOVSKY, J. Pasture improvement in Uruguay. Report to FAO.
1969

KAWATAKE, M., SHIMURA, K., ISHIDA, R. & NISHIMURA, G. Studies on the relation of sow-
1959   ing and cutting time on yield of forage crops in paddy fields. 2. On the forage crops
suitable for before-crop to the late season cultivation of rice plants. *Bull. Tokai-
Kinki Agric. Exp. Sta.*, 1: 16-32.

KELLY, T.K. Results from replanting pasture. *Queensl. Agric. J.*, 88: 9-120.
1962

KELLY, T.K. Ten tons of glycine seed from Atherton. *Queensl. Agric. J.*, 89: 227-228.
1963

KELLY, T.K. Desmodium at Topaz. *Queensl. Agric. J.*, 90: 272-275.
1964

KENNEDY, M.M. Notes on the symbiosis of selected strains of rhizobia and *Glycine javanica*.
1962   *Queensl. J. Agric. Sci.*, 19: 425-428.

KERRIDGE, P.C. & SKERMAN, P.J. The distribution and growth characteristics of the native
1968   legume *Psoralea eriantha* in western Queensland. *Trop. Grassl.*, 2: 41-50.

618

KINCH, D.M. & RIPPERTON, J.C. *Koa haole: production and processing. Hawaii Agric. Exp.*
1962   *Sta. Bull.*, 129.

KING, W.A. Clemson College. S. Carol. Agric. Exp. Stn. Dairy Sci. Res. Ser. No. 34.
1964

KING, N.J., MUNGOMERY, R.W. & HUGHES, C.G. *Manual of cane growing.* Sydney, Angus
1965   and Robertson.

KLAGES, K.H.W. *Economic crop geography.* New York, Macmillan.
1942

KOPPEN, W. *Grundriss der Klimakunde.* Berlin, Walter de Gruyter.
1931

KRAPOVICKAS, A. & RIGONI, V.A. *Darwinana*, 11: 433-455.
1959

KRAUSE, F.G. Leguminous crops for Hawaii. *Hawaii Agric. Exp. Sta. Bull.*, 23.
1911

KRAUSE, F.G. The pigeon pea (*Cajanus indicus*), its improvements, culture and utilization
1932   in Hawaii. Honolulu. *Hawaii Agric. Exp. Sta. Bull.*, 64.

KRETSCHMER, A.E. JUN. *Proc. Soil Crop. Sci. Soc. Fla*, 24: 167.
1964

KRETSCHMER, A.E. JUN. The effect of nitrogen fertilisation of mature pangola grass just
1965   prior to utilisation in the winter on yields, dry matter and crude protein contents.
         Agron. J., 57: 529-534.

KRETSCHMER, A.E. JUN. Four years' results with Siratro (*Phaseolus atropurpureus* DC) in
1966   south Florida. *Proc. Soil Crop. Sci. Soc. Fla*, 26: 238-245.

KRETSCHMER, A.E. JUN. *Stylosanthes humilis.* Florida Agric. Exp. Sta. Circ. S-184.
1968

KRETSCHMER, A.E. JUN. Production of annual and perennial tropical legumes in mixtures
1970   with pangola grass and other grasses in Florida. *Proc. 11th Int. Grassl. Cong.*, Sur-
         fers Paradise, Australia, 149-153.

KRETSCHMER, A.E. JUN., BROLMANN, J.B., SNYDER, G.H. & COLEMAN, S.W. 'Florida'
1976   *Carpon desmodium*, a perennial legume for use in South Florida. *Proc. Soil Crop
         Sci. Soc. Fla*, 35: 25-31.

KRETSCHMER, A.E., BROLMANN, J.B., SNYDER, G.H. & COLEMAN, S.W. Florida *Carpon
1970   desmodium* - a perennial tropical forage legume for use in South Florida.
         Fla Agric. Exp. Stn. Circ., S-260.

619

KYNEUR, G.W. Glycine on the Atherton Tableland. *Queensl. Agric. J.*, 86: 507-513.
1960

KYNEUR, G.W. The role of fertilisers in establishing *Glycine javanica*. L. on latosol soil.
1962   *Proc. N. Queensl. Agros. Conf.*, 10-15.

KYNEUR, G.W. *Proc. 5th Aust. Agrost. Conf.*, Melbourne.
1963

LAMPREY, H.F. Notes on the dispersal and germination of some tree seeds through the
1967   agency of mammals and birds. *E. Afr. Wildlife J.*, 5: 179-180.

LANDRAU, P., SAMUELS, G. & RODRIGUEZ, P. Influence of fertilizers, minor elements and
1953   soil pH on the growth and protein content of tropical kudzu. *J. Agric. Univ. P.R.*,
        37: 81-95.

LANSBURY, T.J. The composition and digestibility of some conserved fodder crops for dry
1959   season feeding in Ghana..3. Fodders for irrigation. *Trop. Agric. (Trin.)*, 36: 305-
        308.

LANSBURY, T.J., ROSE-INNES, R. & MABEY, G.L. Studies on Ghana grasslands: yield and
1965   composition on the Accra Plains. *Trop. Agric. (Trin.)*, 42: 1-29.

LAWTON, R.M. The value of browse in the dry tropics. *E. Afr. Agric. For. J.*, 33: 227-230.
1968

LEE, ALMA T. The *Psoralea patens* complex. *Telopea*, 2: 129-141.
1980

LEE, L.C. *Proc. Pan Indian Ocean Sci. Congr., Pt. II*. Perth, W. Australia.
1954

LENNE, J.M. & CALDERON, M.A. Disease and pest problems of *Stylosanthes*. *In* Stace,
1984   H.M. & Edye, L.A., eds. *The biology and agronomy of* Stylosanthes. North Ryde,
        Australia, Academic Press, pp. 279-293.

LESLIE, J.K. Factors responsible for failure in the establishment of summer grasses on the
1965   black earths of the Darling Downs, Queensland. *Queensl. J. Agric. Anim. Sci.*, 22:
        17-38.

LETTS, G.A. *Leucaena glauca* and ruminants. *Aust. Vet. J.*, 39: 287-288.
1963

LITTLE, D.A. The examination of Townsville lucerne (*Stylosanthes humilis*) for oestrogenic
1969   activity. *Aust. Vet. J.*, 45: 24-26.

LITTLE, D.A., McIVOR, J.G. & McLEAN, R.W. The chemical composition and nutritive
1984   value of *Stylosanthes*. *In* Stace, H.M. & Edye, L.A., eds. *The biology and
        agronomy of* Stylosanthes. North Ryde, Australia, Academic Press, pp. 381-403.

LOOSLI, J.K., VILLEGAS, V.E. & YNALVEZ, L.A. Preliminary report on the composition
1954   and digestibility of some Philippine forages. *Phil. J. Anim. Ind.*, 15: 265.

LORIMER, M.S. Forage selection studies. 1. The botanical composition of forage selected by
1978   sheep grazing *Astrebla* spp. pasture in north-west Queensland. *Trop. Grassl.* 12:
        97-108.

LOTERO, J., WOODHOUSE, W.W. JUN & PETERSEN, R.G. Local effect on fertility of urine
1966   voided by grazing cattle. *Agron. J.*, 58: 262-265.

LOUSTALOT, A.J. & TELFORD, E.A. Physiological experiments with tropical kudzu. *J.*
1948   *Amer. Soc. Agron.*, 40: 503-511.

LOVADINI, L.A.C. & MIYASAKA, SHIRO. *Cultura de Soja Perene.* São Paulo, Sec. Agric.
1968   Est., Inst. Agron. Bol. No. 186.

LUCK, P.E. *Dolichos axillaris* - Prospects in Queensland. *Aust. Plant Introd. Rev.*, 1: 50a-
1965a   51a.

LUCK, P.E. *Dolichos lab-lab* - a valuable grazing crop. *Queensl. Agric. J.*, 91: 308-309.
1965b

LUCK, P.E. & DOUGLAS, N.J. Dairy pasture research and development in the near north
1966   coast centred at Cooroy, Queensland. *Proc. Trop. Grassl. Soc., Aust.*, 6: 35-53.

LUDLOW, M.M. & WILSON, G.L. Studies on the productivity of tropical pasture plants. 1.
1967   Growth analysis, photosynthesis, and respiration of Hamil grass and Siratro in a
        controlled environment. *Aust. J. Agric. Res.*, 19: 35-45.

LUDLOW, M.M. & WILSON, G.L. Growth of some tropical grasses and legumes at two tem-
1970   peratures. *J. Aust. Inst. Agric. Sci.*, 36: 43-45.

LYCHATCHYNSKY, D. & STEENMEYER, H.P. Teste de palatabilidade de 12 leguminosas trop-
1968   icals selecionades. Belem, Para, Brazil. IRI.

MABEY, G.L. & ROSE-INNES, R. Studies on browse plants in Ghana. II. Digestibility. *Emp.*
1964   *J. Exp. Agric.*, 32: 125-130.

MAJUMDAR, B.R., SEN, S. & ROY, S.R. Raise rice bean for rich nutritious fodder. *Indian*
1968   *Fmg*, 18: 29-30.

'T MANNETJE, L. The effect of photoperiod on flowering, growth habit, and dry matter pro-
1965   duction in four species of the genus *Stylosanthes* Sw. *Aust. J. Agric. Res.*, 16: 767-
        771.

'T MANNETJE, L. A punch card key to species of *Trifolium* L. in Africa south of the Sahara
1966   excluding Ethiopia. *E. Afr. Agric. For. J.*, 31: 261-270.

'T MANNETJE, L. A revision of varieties of *Stylosanthes guianensis* (Aubl.) Sw. *Aust. J. Bot.*,
1977   25: 347-362.

'T MANNETJE, L. Considerations on the taxonomy of the genus *Stylosanthes. In* Stace, H.M.
1984   & Edye, L.A., eds. *The biology and agronomy of* Stylosanthes. North Ryde,
        Australia, Academic Press, pp. 1-21.

'T MANNETJE, L., SHAW, N.H. & ELICH, T.W. The residual effect of molybdenum fertiliser
1963   on improved pastures on a prairie-like soil in subtropical Queensland. *Aust. J. Exp.
        Agric. Anim. Husb.*, 3: 20-25.

MARION, P.T., FISHER, C.E. & ROBINSON, E.D. Ground mesquite wood as a roughage for
1957   yearling steers. *Texas Agr. Exp. Sta. Progr. Rep.*, 1972.

MARTIN, A.E. & COX, J.E. Nitrogen studies on black soils from the Darling Downs,
1956   Queensland. I. Seasonal variations in moisture and mineral nitrogen fractions.
        Aust. J. Agric. Res., 7: 169-193.

DE MATTOS, H.B. & ALCANTARA, P.B. *Galactia striata,* a promising legume for Central
1976   Brazil. *Zootecnia*, 14: 51-57.

DE MATTOS, H.B. & WERNER, J.C. Effect of mineral nitrogen and legumes on Coloniao
1979   grass (*Panicum maximum* Jacq.) dry matter production. *Bol. Ind. Anim.*, 36: 147-
        156.

MAWSON, W.F. Suspension fence saves time, money. *Queensl. Agric. J.*, 89: 752-755.
1963

MAYLAND, H.F., McINTOSH, T.H. & FULLER, W.H. Fixation of isotopic nitrogen on a
1966   semiarid soil by algal crust organisms. *Proc. Soil. Sci. Soc. Amer.*, 30: 56-60.

McADAM, A.C. & SWAIN, F.G. Summer growing legumes for the far north coast of N.S.W.
1969   *Agric. Gaz., N.S.W.*, 70: 242-244.

McCARTHY, W.O. Factors influencing the adoption of improved pastures in Queensland.
1969   *Trop. Grassl.* 3: 162-164.

McCLUNG, A.C. & QUINN, L.R. *Sulfur and phosphorus responses of Batatais grass* (Pas-
1959   palum notatum) IBEC Res. Inst. Bull. No. 18.

McEWAN, T. Isolation and identification of the toxic principle of *Gastrolobium grand-*
1964   *iflorum. Queensl. J. Agric. Sci.*, 21: 1-13.

McGARITY, J.W. The influence of sod-seeded legumes on the nitrogen economy of grass-
1959   land soils at Lismore, N.S.W. *J. Aust. Inst. Agric. Sci.*, 25: 287-293.

McILVANE, T.C. & POHLMAN, W.E. Crop growth and soil reaction. *Bull. W. Va. Agric.
1949   Exp. Sta.*, 337.

622

McKAY, A.D. Rangeland productivity in Botswana, *E. Afr. Agric. For. J.*, 34: 178-192.
1968

McKEAGUE, P.J., MILLER, C.P. & ANNING, P. Verano — a new stylo for the dry trop-
1978   ics. *Queensl. Agric. J.*, 104: 31-35.

McKEE, R. Other legumes—hairy indigo. *In* Hughes, H.D., Heath, M.E. & Metcalfe,
1952   D.S., eds. *Forages - science of grassland agriculture.* Ames, Iowa, Iowa State Univ.
        Press, p. 252.

McKEON, G.M. & MOTT, J.J. Seed biology of *Stylosanthes. In* Stace, II.M. & Edye, L.A.,
1984   eds. *The biology and agronomy of* Stylosanthes. North Ryde, Australia, Academic
        Press, pp. 311-332.

McLEOD, M.N. & MINSON, D.J. Sources of variation in the *in vitro* digestibility of tropical
1969   grasses. *J. Brit. Grassl. Soc.*, 24: 244-49.

McMEEKAN, C.P. Grazing management and animal production. *Proc. 7th Int. Grassl.*
1956   *Congr.*, Palmerston North, NZ, 146-156.

McMEEKAN C.P. Grazing management. *Proc. 8th Int. Grassl. Congr.*, Reading, UK, 21-26.
1960

McWHIRTER, K.S. Cytoplasmic male sterility in *Desmodium. Aust. J. Agric. Res.*, 20: 227-
1969   241.

McWILLIAM, J.R. & DOWLING, P.M. Factors influencing the germination and establish-
1970   ment of pasture seed on the soil surface. *Proc. 11th Int. Grassl. Congr.*, Surfers
        Paradise, Australia, 578-583.

MEARS, P.T. Tropical pastures in the Richmond-Tweed region of New South Wales —
1967   recent experiences and future development. *Trop. Grassl.*, 1: 98-105.

MEARS, P.T. & BARKUS, B. Response of *Glycine wightii* to molybdenised superphosphate
1970   on a krasnozem. *Aust. J. Exp. Agric. Anim. Husb.*, 10: 415-425.

MEARS, P.T., MURTAGH, G.J. & WILSON, G.P.M. Silver leaf desmodium - a perennial sub-
1964   tropical legume for the north coast of New South Wales. *Agri. Gaz, N.S.W.*, 75:
        1332-1335.

MENEGARIO, A. *Soja perenne em pastagens.* Secretaria da Agricultura do Est., São Paulo,
1964   Seccao de Leguminosas.

MENENDEZ, J. *Teramnus* Swartz. *Pastos Forrajes*, 5: 251-263.
1982

MES, M.G. Influence of temperature on the symbiotic nitrogen fixation of legumes. *Nature*,
1959   184: 2032-2033.

623

MIDDLETON, C.H. Some effects of grass-legume sowing rates on tropical species establish-
1970   ment and production. *Proc. 11th Int. Grassl. Congr.*, Surfers Paradise, Australia,
         119-123.

MIDDLETON, C.H. & MELLOR, W. Grazing assessment of the tropical legume
1982   *Calopogonium caeruleum. Trop. Grassl.*, 16: 213-216.

MILES, J.W. Natural out crossing in *Stylosanthes capitata. Trop. Grassl.*, 17: 114-117.
1983

MILFORD, R. The importance of age, leafiness and chemical composition in determining
1963   feeding values of subtropical pasture plants. *3rd Aust. Grassl. Conf.*, Melbourne,
         16-20.

MILFORD, R. Nutritive values and chemical composition of seven tropical legumes and
1967   lucerne grown in subtropical southeastern Queensland. *Aust. J. Exp. Agric. Anim.
         Husb.*, 7: 540-545.

MILFORD, R. & HAYDOCK K.P. The nutritive value of protein in subtropical pasture species
1965   grown in southeast Queensland. *Aust. J. Exp. Agric. Anim. Husb.*, 5: 13-17.

MILFORD, R. & MINSON, D.J. Intake of tropical pasture species. *Proc. 9th Int. Grassl.
1965   Congr.*, São Paulo, Brazil, pp. 815-822.

MILFORD, R. & MINSON, D.J. The feeding value of tropical pastures. *In* Davis, W. & Skid-
1966   more, C.L., eds. *Tropical Pastures.* London, Faber and Faber, pp. 106-114.

MILFORD, R. & MINSON, D.J. The effect of age and method of haymaking on the digestibil-
1968   ity and voluntary intake of the forage legumes *Dolichos lab-lab* and *Vigna sinensis.
         Aust. J. Exp. Agric. Anim. Husb.*, 8: 409-412.

MILLER, H.P. The effect of pretreatment of native pasture and sowing rate on establishment
1967   of Townsville lucerne on Tippera clay loam at Katherine, N.T. *Aust. J. Exp. Agric.
         Anim. Husb.*, 7: 515-517.

MILLER, T.B. & RAINS, A.B. The nutritive value and agronomic aspects of some fodders in
1963   E. Nigeria. *J. Brit. Grassl. Soc.*, 18: 158-167.

MILLER, T.B., RAINS, A.B. & THORPE, R.J. The nutritive value and agronomic aspects of
1964   some fodders in Northern Nigeria. 3. Hays and dried crop residues. *J. Brit. Grassl.
         Soc.*, 19: 77-80.

MILLS, W.R. Nitrate accumulation in Uganda soils. *East Afr. Agric. J.*, 19: 53-54.
1953

MINSON, D.J. The digestibility and voluntary intake of six varieties of *Panicum. Aust. J.
1971   Exp. Agric. Anim. Husb.*, 11: 18-25.

MINSON, D.J. & McLEOD, M.N. The digestibility of temperate and tropical grasses. *Proc.
1970   11th Int. Grassl. Congr.*, Surfers Paradise, Australia, 719-722.

MINSON, D.J. & MILFORD, R. The energy values and nutritive value indices of *Digitaria*
1966   *decumbens, Sorghum almum,* and *Phaseolus atropurpureus. Aust. J. Agric. Res.*
17: 411-423.

MISHUTIN, E.N. & SHILNIKOVA, V.K. The biological fixation of atmospheric nitrogen by
1969   free-living bacteria. *Biol. Sol. (Paris),* 65: 150.

MISRA, D.K., BHAN, S. & PRASAD, R. Guar, multipurpose summer legume. *Allahabad*
1968   *Fmr,* 42: 239-244.

MOHLENBROCK, R.H. A revision of the genus *Stylosanthes. Ann. Mo. Bot. Gard.,* 44: 299-
1957   355.

MOHLENBROCK, R.H. A monograph of the leguminous genus *Zornia. Webbia. Racc. Scr.*
1961   *Bot.,* 16: 1-141.

MOHLENBROCK, R.H. Further considerations on *Stylosanthes* (Leguminosae). *Rhodora,* 65:
1963   248-258.

MOLINA, D.G. The influence of 5 and 10 per cent levels of ipil-ipil (*Leucaena glauca*) leaf
1953   meal in the College all-mash ration for chicks. *Philipp. Agric.,* 37: 142-145.

MONTEIRO, M. DO C. DA C. & ARONOVICH, S. Anatomia do sistema vegetativo da *Cen-*
1965   *trosema pubescens* Benth. *Proc. 9th Int. Grassl. Congr.,* São Paulo, Brazil, 1729-
1732.

MOOMAW, J.C. & TAKAHASHI, M. Forage and beef production from Pangola grass —
1962   *Desmodium intortum* in Hawaii. *Agron. Abstr.,* 90.

MOORE, A.W. The influence of a legume on soil fertility under a grazed tropical pasture.
1962   *Emp. J. Exp. Agric.,* 30: 239-248.

MOORE, A.W. The influence of fertilisation and cutting on a tropical grass/legume pasture.
1965   *Exp. Agric.,* 1: 193-200.

MOORE, A.W. Nonsymbiotic nitrogen fixation in soil and soil-plant systems. *Soils Fertil.,*
1966   29: 113-128.

MOORE, A.W. & JAIYEBO, E.O. The influence of cover on nitrate and nitrifiable nitrogen
1963   content of the soil in a tropical rain forest environment. *Emp. J. Exp. Agric.,* 31:
189-198.

MOORE, R.M. *Australian grasslands.* Canberra, Aust. Nat. Univ. Press.
1970

MORRIS, J.G. & LEVITT, M.S. The intake and digestibility of silages made from *Dolichos*
1968   *lab-lab* L. alone and with sorghum cv. sugar drip. *Proc. Aust. Soc. Anim. Prod., 7th*
Bienn. Conf., 7: 78-85.

625

MOTT, G.O. Grazing pressure and the measurement of pasture production. *Proc. 8th Int.*
1960   *Grassl. Congr.*, Reading, UK, 606-611.

MOTT, G.O., QUINN, L.R. & BISSCHOFF, W.V.A. The retention of nitrogen in a soil-plant-
1970   animal system in Guinea grass (*Panicum maximum*) pastures in Brazil. *Proc. 11th
        Int. Grassl. Congr.*, Surfers Paradise, Australia, 414-416.

MOTT, J.J. High-temperature contact treatment of hard seed in *Stylosanthes. Aust. J. Agric.*
1979   *Res.*, 30: 847-54.

MOTT, J.J. Fire survival of *Stylosanthes* spp. in the dry savanna woodlands of the Northern
1982   Territory. *Aust. J. Agric. Res.*, 33: 203-211.

MOUAT, M.C.H. & WALKER, T.W. Competition for nutrients between grasses and white
1959   clover. I. Effect of grass species and nitrogen supply. *Plant Soil*, 11: 30-40.

DE MOURA, M.P., WERNER, J.C., MOTEIRO, F.A. & BOIN, C. Haymaking rate, leaf/stem
1975   ratio and protein percentage in the blades and stems of some perennial tropical
        legumes and molasses grass. *Bol. Ind. Anim.*, 32: 363-370.

MULDER, E.G., LIE, T.A. & WOLDENDORP, J.W. Biology and soil fertility. In *Soil biology:*
1969   *reviews and research.* Paris, Unesco, pp. 163-201.

MUNE, T.L. & PARHAM, J.W. The declared noxious weeds of Fiji and their control. *Fiji*
1967   *Dep. Agric. Bull.*, 48: 14-16.

MUNNS, D.N. Soil acidity and growth of a legume. II. Reactions of aluminium and phos-
1965   phate in solution and effects of aluminium phosphate, calcium, and pH in solution
        culture. *Aust. J. Agric. Res.*, 16: 743-755.

MUNNS, D.N. Nodulation of *Medicago sativa* in solution culture. I. Acid-sensitive steps.
1968   *Plant Soil*, 28: 129-146.

MURTAGH, G.J. Herbicides as an aid to the establishment of sod-sown *Glycine javanica.*
1963   *Aust. J. Exp. Agric. Anim. Husb.*, 3: 175-176.

MURTAGH, G.J. Effect of temperature on the germination of *Glycine javanica. Proc. 11th*
1970   *Int. Grassl. Congr.,* Surfers Paradise, Australia, 514-518.

MURTAGH, G.J. & DOUGHERTY, A.B. Relative yields of lab-lab and velvet bean. *Trop.*
1968   *Grassl.*, 2: 57-63.

MURTAGH, G.J. & WILSON, G.P.M. Glycine, a summer growing legume. *Agric. Gaz.*
1962   *N.S.W.*, 73: 634-637.

MWAKHA, M. *Aust. Plant Introd. Rev.*, 7: 24.
1970

NAVEH, Z. Selection of promising pasture plants for northern Tanzania. Part III. The early
1966   performance of drought-resistant *Glycine javanica* ecotypes. *E. Afr. Agric. For. J.*,
        32: 103-107.

NAVEH, Z. & ANDERSON, G.D. The introduction and selection of promising pasture plants
1966   for the Arusha and Kilimanjaro regions of northern Tanzania. 1. Introduction
        problems of pasture research and development. *E. Afr. Agric. J.*, 32: 41-44.

NEGI, S.S. Soil conservation in Uttar Pradesh under the Uttar Pradesh-Rajasthan Border
1956   Afforestation Scheme. *Indian For.*, 82: 114-115.

NEME, N.A. Phosphated fertilisers and lime in the production of 'Soja perenne' (*Glycine
1965   javanica*) forage in "Latosol Roxo". *Proc. 9th Int. Grassl. Congr.*, São Paulo,
        Brazil, 677-681.

NEME, N.A. Scarifier "I.A.C." for seeds of perennial soybean (*Glycine javanica*). *Bragan-
1966   tia*, 25: 19-21 (*Herb. Abstr.*, 38: 128).

NEME, N.A. & LOVADINI, L.A.E. Efecto de adubos fosfatados e calcário na produção de
1967   forragem de soja perene (*Glycine javanica*) em terra de cerrado. *Bragantia*, 26: 365-
        371.

NEME, N.A. & NERY, J.P. Influence of mineral fertilizers and lime on the yield and chemical
1965   composition of perennial forage legumes. *Proc. 9th Int. Grassl. Congr.*, São Paulo,
        Brazil, 665-670.

NESTEL, B.L. & CREEK, M.J. Animal production studies in Jamaica. I. Live weight produc-
1964   tion from pangola grass pastures used for rearing and fattening beef cattle and the
        economic implications of the yield. *J. Agric. Sci.*, 62: 187-198.

NESTEL, B.L. & CREEK, M.J. An economic investigation of pasture improvement using
1965   pangola grass (*Digitaria decumbens* Stent.) on beef farms in Jamaica. *Proc. 9th Int.
        Grassl. Congr.*, São Paulo, Brazil, 2: 1571-1574.

NEWMAN, D.M.R. A comparison of the nutritional value of *Stylosanthes humilis*,
1968   *Stylosanthes guyanensis* and introduced grasses in northern Australia. *Aust. J. Exp.
        Agric. Anim. Husb.*, 8: 172-176.

NICHOLAS, D.B. Genotypic variation in growth and nodulation in *Glycine wightii*. *J. Aust.
1970   Inst. Agri. Sci.*, 37 (1): 69-70.

NICHOLS, R. Studies on the major element deficiencies of the pigeon pea (*Cajanus cajan*) in
1964   sand culture. 2. The effects of major element deficiencies on nodulation growth,
        and mineral composition. *Plant Soil*, 22: 112-116.

NJOKU, E. Photoperiodic response of some Nigerian plants. *J.W. Afr. Sci. Assoc.*, 4: 99-111.
1958

627

NORMAN, M.J.T. *Influence of fertilizers on the yield and nodulation of Townsville lucerne*
1959   (Stylosanthes sundaica *Taub.) at Katherine, N.T.* Melbourne, CSIRO Div. Land
Use Res. Tech. Pap. No. 5.

NORMAN, M.J.T. *The establishment of pasture species on arable land at Katherine, N.T. (Au-*
1960   *stralia).* Melbourne, CSIRO Div. Land Use Res. Tech. Pap. No. 8.

NORMAN, M.J.T. *The establishment of pasture species with minimum cultivation at*
1961   *Katherine, N.T. (Australia).* Melbourne, CSIRO Div. Land Use Res. Tech. Pap.
No. 14.

NORMAN, M.J.T. Response of native pastures to nitrogen and phosphate fertiliser at
1962   Katherine, N.T. *Aust. J. Exp. Agric. Anim. Husb.*, 2: 29-34.

NORMAN, M.J.T. The pattern of dry matter and nutrient content changes in native pastures
1963   at Katherine, N.T. *Aust. J. Exp. Agric. Anim. Husb.*, 3: 119-124.

NORMAN, M.J.T. Postestablishment grazing management of Townsville lucerne on
1965a   uncleared land at Katherine, N.T. *J. Aust. Inst. Agric. Sci.*, 31: 311-313.

NORMAN, M.J.T. The response of a birdwood grass/Townsville lucerne pasture to phos-
1965b   phate fertilizers at Katherine, N.T. (Australia). *Aust. J. Exp. Agric. Anim. Husb.*,
5: 120-124.

NORMAN, M.J.T. Seasonal performance of beef cattle on native pasture at Katherine, N. T.
1965c   *Aust. J. Exp. Agric. Anim. Husb.*, 5: 227-231.

NORMAN, M.J.T. Companion grasses for Townsville lucerne at Katherine. *J. Aust. Inst.*
1967a   *Agric. Sci.*, 33: 14-22.

NORMAN, M.J.T. The "critical period" for beef cattle grazing standing forage at Katherine,
1967b   N.T. *J. Aust. Inst. Agric. Sci.*, 33: 130-132.

NORMAN, M.J.T. The performance of beef cattle on different sequences of Townsville
1968,   lucerne and native pasture at Katherine, N.T. *Aust. J. Exp. Agric. Anim. Husb.*,
8: 21-25.

NORMAN, M.J.T. Relationships between liveweight gain of grazing beef steers and availabil-
1970   ity of Townsville lucerne. *Proc. 11th Int. Grassl. Congr.*, Surfers Paradise,
Australia, 829-832.

NORMAN, M.J.T. & STEWART, G.A. Investigations on the feeding of beef cattle in the
1964   Katherine region. *J. Aust. Inst. Agric. Sci.*, 30: 39-46.

NORMAN, M.J.T. & WETSELAAR, R. Losses of nitrogen on burning native pasture at
1960   Katherine, N.T. *J. Aust. Inst. Agric. Sci.*, 26: 272-273.

628

NORRIS, D.O. Legumes and the *Rhizobium* symbiosis. *Emp. J. Exp. Agric.*, 24: 247-270.
1956

NORRIS, D.O. A red strain of *Rhizobium* from *Lotononis bainesii* Baker. *Aust. J. Agric.*
1958a   *Res.*, 9: 629-632.

NORRIS, D.O. Lime in relation to the nodulation of tropical legumes. *In* Hallsworth, E.G.,
1958b   ed. *Nutrition of the legumes*. London, Butterworths, pp. 164-182.

NORRIS, D.O. *Rhizobium* affinities of African species of *Trifolium. Emp. J. Exp. Agric.*, 27:
1959a   87-97.

NORRIS, D.O. The role of calcium and magnesium in the nutrition of *Rhizobium. Aust. J.*
1959b   *Agric. Res.*, 10: 651-698.

NORRIS, D.O. Acid production by *Rhizobium* - a unifying concept. *Plant Soil*. 22: 143-166.
1965

NORRIS, D.O. *Rhizobium* relationships in legumes. *Proc. 9th Int. Grassl. Congr.*, São
1965   Paulo, Brazil, 1088-1092.

NORRIS, D.O. The intelligent use of inoculants and lime pelleting for tropical legumes.
1967   *Trop. Grassl.*, 1: 107-121.

NORRIS, D.O. The contribution of research in legume bacteriology to the development of
1970   Australian pastures. *Proc. 11th Int. Grassl. Congr.*, Surfers Paradise, Australia,
A22-A30.

NORRIS, D.O. & 'T MANNETJE, L. The symbiotic specialization of African *Trifolium* spp. in
1964   relation to their taxonomy and their agronomic use. *E. Afr. Agric. For. J.*, 24: 214-
235.

NORTHERN TERRITORY AGRICULTURAL BRANCH. Pastures and fodder crops. Rept. 1960-
1961   61.

NORTHERN TERRITORY. *Administration Ann. Rept, 1963/64.*
1964

Notes on Kenya Agriculture. VI. Grass leys and grassland plants. *E. Afr. Agric. J.*, 24: 223-
1959   236.

NUTMAN, P.S. The physiology of nodule formation. *In* Hallsworth, E.G., ed. *Nutrition of*
1958   *the legumes*. London, Butterworth's, pp. 87-107.

NWOSU, N.A. Conservation and utilisation of *Stylosanthes gracilis. Trop. Agric. (Trin.)*, 37:
1960   61-66.

NYE, P.H. & GREENLAND, D.J. The soil under shifting cultivation. *Commonw. Bur. Soils*
1960   *Tech. Commun*, 51.

OAKES, A.J. Pangola grass (*Digitaria decumbens* Stent) on the Caribbean. *Proc. 8th Int.*
1960   *Grassl. Cong.*, Reading, UK, 386-389.

OAKES, A.J. *Leucaena leucocephala:* description, culture, utilisation. *Adv. Frontiers Plant*
1968   *Sci.*, 20: 1-114.

OAKES, A.J. & SKOV, O. Response of four pasture grasses to nitrogen in the dry tropics.
1962a   *Agron. J.*, 54: 176-178.

OAKES, A.J. & SKOV, O. Some woody legumes as forage crops for the dry tropics. *Trop.*
1962b   *Agric. (Trin.)*, 39: 281-287.

OAKES, A.J. & SKOV, O. Yield trials of *Leucaena* in the U.S. Virgin Islands. *J. Agric. Univ.*
1967   *P.R.*, 51: 176-181.

ODEJAR, D.Y. Chemical analysis of some commonly and some uncommonly used Philip-
1949   pine feeds and feedstuffs. *Philipp. J. Anim. Ind.*, 10: 371.

ODEJAR, D.Y. & MASANKAY-ARENAS, L.R. Chemical analysis of some commonly and
1951   some uncommonly used Philippine feeds and feedstuffs. *Philipp. J. Anim. Ind.*, 12:
        29.

OELRICHS, P.B. & MCEWAN, T. The toxic principle of *Acacia georginae*. *Queensl. J. Agric.*
1962   *Sci.*, 19: 1-16.

OHASHI, H. Asiatic species of *Desmodium* and its allied genera (Leguminosae) Ginkgoana
1973   - Cont. to the Flora of Asia and the Pacific Region. Tokyo, Academia Sci. Books.

OKE, O.L. Nitrogen fixation capacity of some Nigerian legumes. *Exp. Agric.*, 3: 315-321.
1967a

OKE, O.L. Nitrogen fixing capacity of *Calopogonium* and *Pueraria*. *Trop. Sci.*, 9: 90-99.
1967b

ONLEY, J.A. & SILLAR, D.I. Pasture-crop sequence for beef all the year. *Queensl. Agric. J.*,
1965   91: 470-473.

OSTROWSKI, H. Tropical pastures for the Brisbane district. *Queensl. Agric. J.*, 92: 106-116.
1966

OTERO, J. DE R. *Informacoes sobre algunes plantas forrageires*. Rio de Janeiro, Directora
1952   de Estatistica da Produccao.

OWEN, N.L. Hair loss and other toxic effects of *Leucaena glauca* (Jumbay). *Vet. Rec.*, 70:
1958   454-457.

630

OYENUGA, V.A. The composition and agricultural value of some grass species in Nigeria.
1957   *Emp. J. Exp. Agric.*, 25: 237-255.

OZANNE, P.G., KIRTON, D.J. & SHAW, T.C. The loss of phosphorus from sandy soils. *Aust.*
1961   *J. Agric. Res.*, 12: 409-423.

PALAFOX, H.L. The use of fresh green koa haole leaves in chick rations. *Poul. Sci.* 27: 558-
1948   561.

PALTRIDGE, T.B. Sown pastures for southeast Queensland. *CSIRO Aust. Bull.*, 274.
1955

PALTRIDGE, T.B. & SANTHIRASEGARAM, K. Studies on the nutrient status of some coconut
1957   soils in Ceylon. I. The lateritic soils of Bandirippuwa Estate. II. The Cinnamon
        sand in Horrekelly Estate. *Coconut Res. Inst. Bull.*, 12.

PAPUA NEW GUINEA, Dept. Agric. and Stock and Fisheries. *Ann. Rep. 1959-60.*
1961

PARBERY, D.P. Pasture and fodder crop plant introduction at Kimberley Research Station,
1967a   W.A. 1963-64. Part I. Perennial legumes. Melbourne, CSIRO Aust. Div. Land
        Use Res. Tech. Mem. 67/6.

PARBERY, D.P. Pasture and fodder crop plant introduction at Kimberley Research Station,
1967b   W.A. 1963-64. Part II. Annual legumes. Melbourne, CSIRO Aust. Div. Land Use
        Res. Tech. Mem. 67/10.

PARBERY, N.H. University of Sydney. (D. Sc. Agr. thesis)
1940

PATEL, B.M. & PATEL, P.S. Fodder value of tree and vegetable leaves in Kaira District.
1957   *Indian J. Agric. Sci.*, 27: 307-315.

PATERSON, D.D. The provision of animal fodder in tropical and subtropical countries. Part
1949   I. *Commonw. Bur. Pastures Field Crops Bull.*, 31.

PATIL, G.D. Anthesis and pollination of field Wal. *Poona Agric. Coll. Mag.*, 49: 95-102.
1958

PAUL, W.R.C. Notes on legumes. *Trop. Agriculturalist (Colombo)*, 107: 15-20.
1951

PAYNE, W.J.A. Vaivai as a fodder. *Fiji Agric. J.*, 25: 8.
1954

PAYNE, W.J.A. Sigatoka Agricultural Station. *SPC Quart. Bull.*, (Apr.-July).
1955

631

PAYNE, W.J.A., LAING, W.I., MILES, N.S. & MASON, R.R. Fodder and pasture investiga-
1955   tional work at Sigatoka, 1949-53. *Fiji Agric. J.*, 26: 38-60.

PEIXOTO, A.M., DE MORAES, S. & PRÓSPERO, A.O. Contribution to the study of the chem-
1965   ical composition and the digestibility of the hay of *Glycine javanica*. *Proc. 9th Int.
        Grassl. Congr.*, São Paulo, Brazil, 791-796.

PENDLETON, R.L. Philippine experience in reforestation with ipil-ipil and its application to
1934   conditions in Kwantung Province, China. *Lingan Sci. J.*, 13: 215-224.

PENTNEY, C.J., WHITEMAN, P.C. & SIVASUPIRAMIAM, S. Studies on the germination,
1984   phenology and rhizobium requirements of *Vigna parkeri*. *Trop. Grassl.*, 18: 66-74.

PEREIRA, H.C., OHENERY, E.M. & MILLS, W.R. The transient effects of grasses on the
1954   structure of tropical soils. *Emp. J. Exp. Agric.*, 22: 148-160.

PEREZ-GUERREROZ, J. Advantages and limitations in the utilization of native legumes from
1980   tropical America in animal production. Case study: *Galactia striata* var. Palenque.
        Turrialba, Costa Rica, Centro Agron. Trop. de Inv. y Enseñ.

PERRY, R.A. The effects in grass and browse production of various treatments on a mulga
1970   community in central Australia. *Proc. 11th Int. Grassl. Congr.*, Surfers Paradise,
        Australia, 63-69.

PETERSEN, R.G., LUCAS, H.L. & MOTT, G.O. Relationship between rate of stocking and
1965   per animal and per acre performance on pasture. *Agron. J.*, 57: 27-30.

PETERSON, R.A. *Management of natural and sown pastures.* Fifth Inter-American Meeting
1962   on Animal Production and Health, FAO.

PHILPOTTS, H. The effect of soil temperature on nodulation of cowpea (*Vigna sinensis*).
1967   *Aust. J. Exp. Agric. Anim. Husb.*, 7: 372-376.

PITTIER, H. *Leguminosas de Venezuela.* Caracas, Min. Agric. y Cria. Boletin Tecnico No. 5.
1944

PLAYNE, M.J. The effect of dicalcium phosphate supplements on the intake and digestibility
1969   of Townsville lucerne and spear grass by sheep. *Aust. J. Exp. Agric. Anim. Husb.*,
        9: 192-195.

PLAYNE, M.J. The sodium concentration in some tropical pasture species with reference to
1970   animal requirement. *Aust. J. Exp. Agric. Anim. Husb.*, 10: 32-35.

PLUCKNETT, D.L. Productivity of tropical pastures in Hawaii. *Proc. 11th Int. Grassl.
1970   Congr.*, Surfers Paradise, Australia, A38-A49.

632

PLUCKNETT, D.L. & FOX, R.L. Effects of P fertilisation on yields and composition of Pan-
1965   gola grass and *Desmodium intortum. Proc. 9th Int. Grassl. Congr.*, São Paulo,
         Brazil, 1525-1529.

PORTER, L.K. & GRABBLE, A.R. Fixation of atmosphere nitrogen by nonlegumes in wet
1969   mountain meadows. *Agron. J.*, 61: 521-523.

POULSEN, D.E. Seed production. *Proc. Trop. Grassl. Soc. Aust.*, 6: 55-60.
1966

PRESCOTT, J.A. Single value climatic factors. *Trans. Roy. Soc. S. Aust.*, 58: 48-61.
1934

PRITCHARD, A.J. & WUTOH, J.G. Chromosome numbers in the genus *Glycine* L. *Nature*,
1964   202: 322.

PRODONOFF, E.T. *Seed testing in Queensland*. Brisbane, Dept Prim. Ind.
1968

PULLE, A. *Flora of Suriname*. Lonjouw, J. & Stoffer, A.L., eds. Vol. 2, part 2. Leiden, the
1976   Netherlands, E.J. Brill.

PURSEGLOVE, J.M. *Tropical crops. 1. Dicotyledons*. London, Longmans.
1968

PURSS, G.S. Stem rot: a disease of cowpeas caused by an undescribed species of
1957   *Phytophthora. Queensl. J. Agric. Sci.*, 14: 125-154.

QUARTERMAN, J. Carbohydrate studies on herbage. *Rept E. Afr. Vet. Res. Org., 1955-56.*
1956   Nairobi, pp. 98-100.

QUEENSLAND DEPARTMENT OF PRIMARY INDUSTRIES. *Kairi Research Station Bulletin.*
1965a

QUEENSLAND DEPARTMENT OF PRIMARY INDUSTRIES. *Ann. Rept 1966-67*, p. 18.
1967

QUINN, L.R., MOTT, G.O., BISSCHOFF, W.V.A. & DE FREITAS, L.M.M. Production of beef
1970   from winter vs. summer nitrogen-fertilised colonial Guinea grass (*Panicum
         maximum*) pastures in Brazil. *Proc. 11th Int. Grassl. Congr.*, Surfers Paradise,
         Australia, 832-835.

QUINN, L.R., MOTT, G.O., BISSCHOFF, W.V.A., JONES, M.B. & DA ROCHA, G.L. Beef
1965   production of six tropical grasses in central Brazil. *Proc. 9th Int. Grassl. Congr.*,
         São Paulo, Brazil, 2: 1015-1020.

RANJHAN, S.K., TALAPATRA, S.K. & KALA, A.C. Yield and nutritive value of dual purpose
1967   cowpea (*Vigna catjang*)-cowpea hay as a growth and production ration. *Ind. J.
         Dairy Sci.*, 20: 146.

633

REDRUP, J. An approach to the commercial production of the seeds of tropical pasture
1965   plant. *Proc. 9th Int. Grassl. Congr.*, São Paulo, Brazil, 521-526.

VAN RENSBURG, H.J. *Pasture legumes and grasses in Zambia*. Lusaka, Govt Printer.
1967

VAN RENSBURG, H.J. *Palatability of plants in Zambia*. Lusaka, Govt Printer.
1968

VAN RENSBURG, H.J. & HORNBY, H.E. The place of goats in Tanganyika farming systems.
1948   *E. Afr. Agric. J.*, 14: 94-98.

REYES, B. The digestibility of *Centrosema pubescens* and *Pueraria javanica*. *Philipp. J.*
1955   *Agric.*, 39: 27-29.

REYNOLDS, S.T. & PEDLEY, L. A revision of *Atylosia* (Leguminosae) in Australia.
1981   *Austrobaileya*, 1: 420-8.

RICHARDS, B.N. & BEVEGE, D.I. The productivity and nitrogen economy of artificial
1967   ecosystems comprising various combinations of perennial legumes and coniferous
        tree species. *Aust. J. Bot.*, 15: 467-480.

RICHARDS, J.A. Effects of fertilizers and management on three promising tropical grasses
1965   in Jamaica. *Exp. Agr.*, 1: 281-288.

RICHARDS, J.A. Productivity of Guinea grass (*Panicum maximum*). *Proc. 9th Int. Grassl.*
1965   *Congr.*, São Paulo, Brazil, 2: 1033-1035.

RICHARDS, J.A. Productivity of tropical pastures in the Caribbean. *Proc. 11th Int. Grassl.*
1970   *Congr.*, Surfers, Paradise, Australia, A49-A56.

RICKERT, K.G. Some influences of straw mulch, nitrogen fertilizer and oat companion crops
1970   on establishment of Sabi panic. *Trop. Grassl.*, 4: 71-75.

RIEPMA, P. Neburon, a selective weedicide for use in tropical legumes. *Weed Res.*, 5: 52-60.
1965

RIEWE, M.E. Use of the relationship of stocking rate to gain of cattle in an experimental
1961   design for grazing trials. *Agron. J.*, 53: 309-313.

RIJKEBUSCH, P.A.H. Notes on leguminous cover crops in sisal. Tanganyika Sisal Growers
1967   Assocn. *Sisal Res. Sta. Res. Bull.*, 44.

RIPPERTON, J.C. Vaivai (*Leucaena glauca*) as a feed. *Fiji Agric. J.*, 13: 124.
1942

RIPPERTON, J.C. & HOSAKA, E.Y. Vegetation zones of Hawaii. *Hawaii Agric. Exp. Sta.*
1942   *Bull.*, 89.

634

RISOPOULOS, S.A. *Management and use of grasslands.* Democratic Republic of Congo.
1966   FAO Pastures and Fodder Crop Studies No. 1.

RIVERA-BRENES, L., COLON-TORRES, E.N., GELPI, F. & TORRES-MAS, J. Influence of ni-
1958   trogenous fertilisers on guinea grass yield and carrying capacity in Lajas Valley. *J.
Agric. Univ. P.R.,* 42: 239-247.

RIVEROS, F. *Productivity of* Desmodium intortum *and* Setaria anceps *in relation to defolia-
1969   tion.* Brisbane, University of Queensland. (Ph. D. thesis)

RIVEROS, F. & WILSON, G.L. Response of a *Setaria-Desmodium intortum* mixture to height
1970   and frequency of cutting. *Proc. 11th Int. Grassl. Congr.,* Surfers Paradise,
Australia, 666-668.

ROBERTS, F.J. & CARBON, B.A. Growth of tropical and temperate grasses and legumes
1969   under irrigation in southwest Australia. *Trop. Grassl.,* 3: 109-116.

ROBERTSON, W.G. Final report of the agronomist. Pastures and Fodder Crops, FAO/
1971   UNDP/SF Project No. 170, Uruguay, S.A.

ROBINSON, D.W. & STEWART, G.A. Protein digestibility in sheep and cattle in north-west-
1968   ern Australia. *Aust. J. Exp. Agric. Anim. Husb.,* 8: 419-424.

ROBINSON, J.B.D. The feeding of dairy cattle in Barbados. 1. The feeding value of some
1950   tropical fodders. *Trop. Agric. (Trin.),* 27: 56-62.

RODEL, M.G.W. Herbage yields of five grasses and their ability to withstand intensive graz-
1970   ing. *Proc. 11th Int. Grassl. Congr.,* Surfers Paradise, Australia, 618-621.

ROE, R. & JONES, R.J. Soil fertility and pasture species investigations on soils derived from
1966   phyllites in the North Deep Creek and Kin Kin areas, Gympie, Queensland. *Proc.
Trop. Grassl. Soc. Aust.,* 6: 13-22.

ROSE-INNES, R. The concept of the 'woody pasture' in low altitude tree savanna environ-
1965   ments. *Proc. 9th Int. Grassl. Congr.,* São Paulo, Brazil, 1419-1423.

ROSE-INNES, R. & MABEY, G.L. Studies on browse plants in Ghana. 1. Chemical composi-
1964a   tion. *Emp. J. Exp. Agric.,* 32: 114-124.

ROSE-INNES, R. & MABEY, G.L. Studies on browse plants in Ghana. III. Browse/grass
1964b   ingestion rations. *Emp. J. Exp. Agric.,* 32: 180-190.

ROTAR, P.P. Tannins and crude proteins of tick clovers (*Desmodium* spp.). *Trop. Agric.*
1965   *(Trin.),* 42: 333-337.

ROTAR, P.P. Variation in agronomic characteristics of *Desmodium intortum* (Mill.) Urb.
1970   and a related species. *Proc. 11th Int. Grassl. Congr.,* Surfers Paradise, Australia,
296-299.

635

ROTAR, P.P., BROWN, T.A. & LYMAN, C. *Kiami clover*. Univ. of Hawaii Coop. Ext. Serv.
1967   Misc. Pub. No. 114.

ROTAR, P.P. & UKIO, URATA. Cytological studies on the genus *Desmodium* - some chromo-
1967   some counts. *Amer. J. Bot.*, 54: 1-4.

ROUX, E. *Grass: a story of Frankenwald*. Cape Town, Oxford.
1969

RUDD, V.E. American species of *Aeschynomene*. *Contrib. US Natl Herb.*, 32(1):
1955   1:131.

RUDD, V.E. The genus *Aeschynomene* in Malaysia (Leguminosae - Papilionatae).
1959   *Reinwardtia*, 5: 23-36.

RUSSELL, J.S. Plant growth on a low calcium status solodic soil in a subtropical environment.
1966   1. Legume species, calcium carbonate, zinc, and other minor element interactions.
        Aust. J. Agric. Res., 17: 673-686.

RUSSELL, M.J. & COALDRAKE, J.E. The effects of some chlorinated hydrocarbon insec-
1966   ticides on nodulation of *Medicago sativa* and *Glycine javanica*. *J. Aust. Inst. Agric.*
        Sci., 32: 214-216.

RYLEY, J.W. Nitrogenous fertilisers and animal production. *Proc Trop. Grassl. Soc. Aust.*,
1966   7: 10-12.

SADOVAL, J.R. The influence of 5 and 10 parts of ipil-ipil leaf meal in the College laying
1955   ration. *Philipp. Agric.*, 38: 574-582.

SAINT-SMITH, J.H. Pasture weeds of the wet tropics. *Queensl. Agric. J.*, 90: 289-300.
1964

SALDANHA, C.J. & NICOLSON, D.H. *Flora of the Hassan District, Karnataka, India.*
1976   Washington, D.C., Smithsonian Inst.

SALETTE, J.E. Nitrogen use and intensive management of grasses in the wet tropics. *Proc.*
1970   *11th Int. Grassl. Congr.*, Surfers Paradise, Australia, 404-407.

SAMPSON, H.C. Cultivated crop plants of the British Empire and the Anglo-Egyptian Sudan
1936   (tropical and subtropical). *Kew Bull. Addit. Ser.*, 12: 138.

SANOGHO, S. Contribution à l'étude des *Rhizobium* de quelques espèces de légumineuses
1977   spontanées de la région de Bamako (Mali). *Cah. ORSTOM Ser. Biol.*, 12(2): 145-
        165.

636

SANTHIRASEGARAM, K. Studies on the nutrient status of some coconut soils in Ceylon. *Trop.*
1967   *Agric. (Trin.)*, 44: 193-207.

SANTHIRASEGARAM, K., COALDRAKE, J.E. & SALIH, M.H.M. Yield of a mixed subtropical
1966   pasture in relation to frequency and height of cutting and leaf area. *Proc. 10th Int.*
         Grassl. Congr., 125-129.

SATYANARAYAN, Y. & GAUR, Y.D. Preliminary studies on the nodulation of arid zone
1965   legumes. *Curr. Sci.*, 34: 21-2.

SAVAGE G.F. Brisbane, University of Queensland. (Hon. Thesis)
1970

VON SCHAAFFHAUSEN, R. Weight increase of zebu cattle grazing on the legumes *Dolichos*
1965   *lab- lab* and *Cajanus indicus*. *Proc. 9th Int. Grassl. Congr.*, São Paolo, Brazil, 965-
         968.

SCHMIDT, P.E. Pasture research at Tanga Livestock Breeding Station. Dar-es-Salaam, Tan-
1967   zania, Dept Agric. Coop.

SCHOFIELD, J.L. Introduced legumes in north Queensland. *Queensl. Agric. J.*, 56: 378-388.
1941

SCHOFIELD, J.L. A comparison of soil nitrate nitrogen values under bare fallow and after
1945   ploughing in various perennial tropical legumes and cowpeas. *Queensl. J. Agric.*
         Sci., 2: 170-174.

VAN SCHREVEN, D.A. Some factors affecting the uptake of nitrogen by legumes. *In*
1958   Hallsworth, E.G., ed. *Nutrition of the legumes*. London, Butterworth's, pp. 137-
         163.

SCHULTZE-KRAFT, R. & GIACOMETTI, D. Genetic resources of forage legumes for the acid,
1979   infertile savannas of tropical America. *In* Sanchez, P.A. & Tergas, E. eds. *Pasture*
         production in acid soils of the tropics. Cali, Colombia, CIAT, pp. 55-64.

SCHUSTER, J.L. Estimating browse from twig and stem measurements. *J. Range Manage*,
1965   18: 220-222.

SCHUSTER, J.L. Literature on the Mesquite (*Prosopis* L.) of North America. An annotated
1969   bibliography. Texas Tech. Univ. Special Rep. No. 26.

SEARS, P.D. Pasture growth and soil fertility. I. The influence of red and white clovers,
1953   superphosphate, lime and sheep grazing on pasture yields and botanical composi-
         tion. N.Z. *J. Sci. Tech.*, 35A (suppl. 1): 1-29.

SEKIZUKA, S. Studies on the native wild grasses for fodder. 5. Crop- scientific studies of *Cas-*
1960   *sia mimosoides* L. as forage. *J. Kanto-Tosan Agric. Exp. Sta.*, 15: 74-90.

SEN, A. & PAUL, N.B. Changes in nitrogen and organic matter contents of soil associated
1959   with the growth of some summer wild legumes. *Indian J. Agric. Sci.*, 29: 104-146.

SERPA, A. Melhoramento da *Centrosema pubescens* Benth. 1. Obtenčâo de variedades de
1965   sementes premeãveis. *Proc. 9th Int. Grassl. Congr.*, São Paulo, Brazil, 171-174.

SHANKAR, V., VELAYUDHAN, K.C. & TRIVEDI, B.K. Pattern of interspecific association in
1975   *Sehima-Heteropogon* grasslands under three different systems of management.
        Ann. Arid Zone, 14: 358-366.

SHANKARNARAYAN, K.A., DABADGHAO, P.M., RAI, P. & KUMAR, R. Techniques for
1975   establishment of pasture legumes in grasslands of *Heteropogon contortus* (Roem
        and Schult.) Beauv. at Jhansi. *Ind. J. Agric. Sci.*, 45: 194-198.

SHAW, K.A. & QUINLAN, T.J. Safari - a new pasture legume for the sub-tropics. *Queensl.*
1976   *Agric. J.*, 102: 251-255.

SHAW, N.H. Increased beef production from Townsville lucerne (*Stylosanthes sundaica*
1961   Taub.) in the spear grass pastures of central coastal Queensland. *Aust. J. Exp.*
        Agric. Anim. Husb., 1: 73-80.

SHAW, N.H., GATES, C.T. & WILSON, J.R. Growth and chemical composition of Townsville
1966   lucerne (*Stylosanthes humilis*). 1. Dry matter yield and nitrogen content in
        response to superphosphate. *Aust. J. Exp. Agric. Anim. Husb.*, 6: 149-156.

SHAW, N.H. & 'T MANNETJE, L. Studies on a spear grass pasture in central coastal Queens-
1970   land - The effect of fertiliser, stocking rate, and oversowing with *Stylosanthes*
        humilis on beef production and botanical composition. *Trop. Grassl.*, 4: 43-56.

SHAW, W.J.D. A beef producer's attitude to the use of sown pastures. *Trop. Grassl.*, 3: 160-
1969   162.

SHONE, D.K. Toxicity of jack bean. *Rhodesia Agric. J.*, 58: 18-20.
1961

SHONE, D.K. & DRUMMOND, R.B. Poisonous plants of Rhodesia. *Rhodesia Agric. J.*,
1965   62: 1-6.

SHREVE, F. *Vegetation and flora of the Sonoran Desert*. Vol. 2. Stanford, California, Stan-
1964   ford Univ. Press.

SHUKLA, K.S., RANJHAN, S.K. & KATIYAR, R.C. *Rhynchosia minima* as a feed for sheep.
1970   *Indian J. Dairy Sci.*, 23: 82-84.

SILLAR, D.I. Effect of shade on growth of Townsville lucerne (*Stylosanthes humilis*
1967   H.B.K.). *Queensl. J. Agric. Anim. Sci.*, 24: 237-240.

638

SILLAR, D.I. Townsville lucerne in Queensland. *Queensl. Agric. J.*, 95: 2-11.
1969a

SILLAR, D.I. Review of technical resources available to the farmer and progress in the use
1969b   of sown pastures. *Trop. Grassl.*, 3: 154-157.

SILVEY, M.W. & CARLISLE, V.W. Influence of Zn, P and Ca on yield and chemical compo-
1972   sition of hairy indigo in eastern Panama. *Proc. Soil Crop Sci. Soc. Fla.*, 31: 26-31.

SINGH, G.S. Studies on an indigenous pasture legume - Wild kulthi (*Atylosia*
1962   *scarabaeoides*). *Sci. Cult.*, 28: 21-23.

SKERMAN, P.J. The vegetation of the Cooper Country. In *The Channel Country of South-*
1947   *west Queensland - with special reference to Cooper's Creek*. Brisbane, Bur. of Investigation. Queensl. Tech. Bull. No. 1.

SKERMAN, P.J. The brigalow country and its importance to Queensland. *J. Aust Inst. Agric.*
1953   *Sci.*, 19: 167-176.

SKERMAN, P.J. Bullamon lucerne - *Psoralea eriantha* Benth. - a plant worth watching. *J.*
1957   *Aust. Inst. Agric. Sci.*, 23:337.

SKERMAN, P.J. Heat waves and their significance in Queensland primary industries. *Arid*
1958a   *Zone Res. Unesco*, 11: 195-198.

SKERMAN, P.J. Cropping for fodder conservation and pasture production in the woolgrow-
1958b   ing areas of western Queensland. *Univ. Queensl. Papers, Dept Agric.*, 1: 89-146.

SKERMAN, P.J. *Report on pastures and livestock in the project area.* UNDP/SF/FAO land and
1966   water use survey in Kordofan Province, Republic of the Sudan. Doxiadis
Associates, Athens.

SKERMAN, P.J. *Stylosanthes mucronata* Willd., an important natural perennial legume in
1970   eastern Africa. *Proc. 11th Int. Grassl. Congr.*, Surfers Paradise, Australia, 196-
198.

SLAYTER, R.O. Measurements of precipitation interception by an arid zone plant commun-
1965   ity (*Acacia aneura*, F. Muell.). *Arid Zone*, 25: 81-92.

SMITH, C.A. Studies on the Northern Rhodesia *Hyparrhenia* veld. IV. The effects of nitro-
1964   gen fertiliser and defoliation. *J. Agric. Sci.*, 62: 299-306.

SMITH, C.A. Studies on the *Hyparrhenia* veld. VI. The fertiliser value of cattle excreta. *J.*
1965   *Agric. Sci.*, 64: 403-406.

SMITH, J.H. Irrigated pasture gives high yield at Cooroy. *Queensl. Agric. J.*, 87: 89-91.
1961

639

SMITH, J.H. & CHANDLER, J.V. Tropical Kudzu moves into Puerto Rico. *Crops Soils*, 3: 12-
1951   14.

SOUTO, S.M. Estabelecimiento de leguminosas forrageiras tropicais numa regiao de Baix-
1969   ada Fluminense. I. Encontro de Técnicos de região centro-sul para discussão dos
problemas relacionados ás leguminosas "forrageiras" 1969. (Unpublished)

SOUTO, S.M. & DÖBEREINER, J. Problems in the establishment of perennial soybean
1970   (*Glycine javanica* L.) in a tropical region. *Proc. 11th Int. Grassl. Congr.*, Surfers
Paradise, Australia, 127-131.

SPRINGHALL, J.A. & ROSS, E. Preliminary studies with poultry rations for the Territory of
1965   Papua and New Guinea. I. Grower rations with copra, sago, and *Leucaena
leucocephala. Papua New Guinea Agric. J.*, 17: 117-121.

SPURWAY, C.H. Soil reaction (pH) preference for plants. *Bull. Mich. Agric. Exp. Sta.*, No.
1941   306.

SQUIBB, R.L., FALLA, A., FUENTES, J.A. & LOVE, H.T. Value of *Desmodium*, pigeon pea
1950   forages and Guatemalan and United States alfalfa meals in rations for baby chicks.
Poult. Sci., 29: 482-485.

SQUIBB, R.L., GUZMAN, M. & SCRIMSHAN, N.S. Dehydrated *Desmodium*, Kikuyu grass,
1953   ramie and banana leaf forages as supplements of protein, riboflavin and
carotenoids in chick rations. *Poult. Sci.*, 32: 1078-1083.

STACE, H.C.T., HUBBLE, G.D., BREWER, R., NORTHCOTE, K.H., SLEEMAN, J.R., MUL-
1968   CAHY, M.J. & HALLSWORTH, E.G. *A handbook of Australian soils*. Glenside, S.
Aust., Rellim.

STACE, H.M. Breeding systems in *Stylosanthes*. 1. Observations of outcrossing in *S. scabra*
1982   at an alcohol dehydrogenase locus. *Aust. J. Agric. Res.*, 33: 87-96.

STACE, H.M. & CAMERON, D.F. Cytogenetics and evolution of *Stylosanthes. In* Stace, H.M.
1984   & Edye, L.A., eds. *The biology and agronomy of* Stylosanthes. North Ryde,
Australia, Academic Press, pp. 49-72.

STANDLEY, P.C. *Flora of the Panama Canal Zone*. Washington, D.C., Smithsonian Insti-
1928   tute.

STANLEY, T.D. & ROSS, E.M. *Flora of south-eastern Queensland* Vol. 1. Queensl. Dept
1983   Prim. Ind. Misc. Publ. No. 81 020.

STANNARD, M.E. & CONDON, R.W. Further fodder trees and shrubs of western New South
1958   Wales. *J. Soil Consv. N.S.W.*, 14: 73-81.

STANTON, W.R. *Grain legumes in Africa*. Rome, FAO.
1966

640

STAPLES, I.B. A promising new legume. *Queensl. Agric. J.*, 92: 388-392.
1966

STAPLES, R.R. Grazing tests at Mpwapwa. *Ann. Rept Dept Vet. Serv. Anim. Husb.*, 120-
1938   125.

STEWART, G.A. *Land evaluation.* South Melbourne, Macmillan of Australia.
1968

STOBBS, T.H. The influence of inorganic fertilisers upon the adaptation, persistency and
1969a   production of grass and grass/legume swards in eastern Uganda. *E. Afr. Agric. For.
J.*, 35: 112-117.

STOBBS, T.H. The value of *Centrosema pubescens* (Benth.) for increasing animal production
1969b   and improving soil fertility in northern Uganda. *E. Afr. Agric. For. J.*, 35: 197-203.

STOBBS, T.H. Beef production from pasture leys in Uganda. *J. Brit. Grassl. Soc.*, 24: 81-86.
1969c

STOBBS, T.H. The use of liveweight-gain trials for pasture evaluation in the tropics. III. The
1969d   measurement of large pasture differences. *J. Brit. Grassl. Soc.*, 24: 177-183.

STOBBS, T.H. The use of liveweight-gain trials for pasture evaluation in the tropics. IV. Ani-
1969e   mal replication. *J. Brit. Grassl. Soc.*, 24: 258-263.

STOBBS, T.H. The use of liveweight-gain trials for pasture evaluation in the tropics. V. Type
1969f   of stock. *J. Brit. Grassl. Soc.*, 24: 345-348.

STOBBS, T.H. The use of liveweight-gain trials for pasture evaluation in the tropics. VI. A
1969g   fixed stocking rate design. *J. Brit. Grassl. Soc.*, 25: 73-77.

STOBBS, T.H. The effect of grazing management upon pasture productivity in Uganda. I.
1969h   Stocking rate. *Trop. Agric. (Trin.)*, 46: 187-194.

STOBBS, T.H. The effect of grazing management upon pasture productivity in Uganda. II.
1969i   Grazing frequency. *Trop. Agric. (Trin.)*, 46: 195-200.

STOBBS, T.H. The effect of grazing management upon pasture productivity in Uganda. III.
1969j   Rotational and continuous grazing. *Trop. Agric. (Trin.)*, 46: 293-301.

STOBBS, T.H. The effect of grazing management upon pasture productivity in Uganda. IV.
1969k   Selective grazing. *Trop. Agric. (Trin.)*, 46: 303-309.

STOBBS, T.H. & JOBLIN, A.D.H. The use of liveweight-gain trials for pasture evaluation in
1966a   the tropics. I. An animal latin-square design. *J. Brit. Grassl. Soc.*, 21: 49-55.

STOBBS, T.H. & JOBLIN, A.D.H. The use of liveweight-gain trials for pasture evaluation in
1966b   the tropics. II. Variable stocking rate designs. *J. Brit. Grassl. Soc.*, 21: 181-185.

641

STOCKER, G.C. & STURTZ, J.D. The use of fire to establish Townsville lucerne in the North-
1966   ern Territory. *Aust. J. Exp. Agric. Anim. Husb.*, 6: 277-279.

STONARD, P. Fine-stem stylo, a legume of promise. *Queensl. Agric. J.*, 94: 478-484.
1968

STONARD, P. & BISSET, W.J. Fine-stem stylo: a perennial legume for the improvement of
1970   subtropical pasture in Queensland. *Proc. 11th Int. Grassl. Congr.*, Surfers
        Paradise, Australia, 153-158.

STRACHAN, R.T., LAMBERT, F.C. & FINLAY, M.C. A way to establish Townsville lucerne.
1967   *Queensl. Agric. J.*, 93: 110-112.

STRANGE, R. Preliminary trials of grasses and legumes under grazing. *E. Afr. Agric. J.*, 24:
1958   92-102.

STURGESS, O.W. & EGAN, B.T. A wilt disease of velvet beans. Indooroopilly, Queensl.,
1960   Bur. Sugar Exp. Sta. Tech. Com. No. 2.

SUTHERST, R.W., JONES, R.J. & SCHNITZERLING, H.J. Tropical legumes of the genus
1982   *Stylosanthes* immobilize and kill cattle ticks. *Nature (London)*, 295: 320-321.

SUTTIE, J.M. Pasture legume research in Kenya. *E. Afr. Agric. For. J.*, 33: 281-285.
1968

SUTTIE, J.M. & MOORE, C.E.M. *Desmodium uncinatum. Kenya Farmer*, 116: 18.
1966

SUTTIE, J.M. & OGADA, J. The production of *Desmodium uncinatum* seed with special
1967   reference to mechanised harvesting. *Kenya Farmer*, 131: 22-36.

SWAIN, F.G., MEARS, P.T., DRANE, F.H., MURTAGH, G.J., BIRD, J.G., COLMAN, R.L. &
1970   YABSLEY, G.M. Commercial evaluation of a new farming system. *Proc. 11th Int.
        Grassl. Congr.*, Surfers Paradise, Australia, p. 925-929.

TAKAHASHI, M. & RIPPERTON, J.C. Koa haole (*Leucaena glauca*): its establishment, cul-
1949   ture, and utilization as a forage crop. *Hawaii Agric. Exp. Sta. Bull.*, 100.

TALBOT, L.M. & TALBOT, M.H. Food preferences of some east African wild ungulates. *E.
1962   Afr. Agric. J.*, 27: 131-138.

TANZANIA. DEPARTMENT OF AGRICULTURE. *Ann. Rept 1954*. Dar-es-Salaam, p. 103.
1955

TEITZEL, J.K. Responses to phosphorus, copper, and potassium on a granite loam of the wet
1969a   tropical coast of Queensland. *Trop. Grassl.*, 3: 43-48.

642

TEITZEL, J.K. Pastures for the wet tropical coast. *Queensl. Agric. J.*, 95: 304-314; 380-385;
1969b   464-471; 532-537.

TEITZEL, J.K., ABBOT, R.A. & MELLOR, W. Beef cattle pastures in the wet tropics 3.
1974   *Queensl. Agric. J.*, 100: 185-189.

THERON, J.J. The influence of plants on the mineralisation of nitrogen and the maintenance
1951   of organic matter in the soil. *J. Agric. Sci.*, 41: 289-296.

THOMAS, R. & HUMPHREYS, L.R. Pasture improvement at Na Pheng, central Laos. *Trop.*
1970   *Grassl.*, 4: 229-236.

THOMPSON, D.P. & DE MEDEIROS, R.B. Monitoring of seed production in *Stylosanthes*
1981   *scabra* cv. Seca. *Trop. Grassl.*, 15: 112-114.

THORNTHWAITE, C.W. An approach toward a rational classification of climate. *Geog. Rev.*,
1948   38: 35-94.

THURBON, P., BYFORD, I. & WINKS, L. Evaluation of hays of *Dolichos lab-lab* cv. Rongai,
1970   a sorghum/Sudan grass hybrid cv. Zulu, and Townsville lucerne (*Stylosanthes*
         humilis H.B.K.) on the basis of organic matter and crude protein digestibility.
         Proc. 11th Int. Grassl. Congr., Surfers Paradise, Australia, 743-747.

TORSSELL, B.W.R., BEGG, J.E., ROSE, C.W. & BYRNE, G.F. Stand morphology of
1968   Townsville stylo (*Stylosanthes humilis*): seasonal growth and root development.
         Aust. J. Exp. Agric. Anim. Husb., 8: 533-543.

TOSI, H., DE FARIA, V.P., SILVEIRA, A.C. & PERIERA, R.L. Evaluation of tropical forage
1975   legumes as forage plants. *Pesqui Agropecu. Brasil. Ser. Zootec.* 10: 19-22.

TOSI, H., MAKAGAWA, J., SILVEIRA, A.C. & KRONKA, S. DO N. Comparison of forage
1979   legumes with four rates of phosphate fertilizer. *Rev. Soc. Bras. Zootec.*, 8: 376-385.

TOTHILL, J.D. *Agriculture in the Sudan.* Oxford Univ. Press.
1954

TOW, P.G. Some ways of establishing glycine. *Queensl. Agric. J.*, 86: 425-427.
1960

TOW, P.G. New research findings on tropical pastures. *Queensl. Agric. J.*, 93: 602-606.
1967

TRINICK, M.J. Nodulation of tropical legumes. I. Specificity in the *Rhizobium* symbiosis of
1968   *Leucaena leucocephala. Exp. Agric.*, 4: 243-253.

TRUMBLE, H.C. & SHAPTER, R.E. The influence of nitrogen and phosphorus treatment on
1937   the yield and chemical composition of Wimmera ryegrass and subterranean clover,
         grown separately and in association. *Bull. CSIRO Aust.*, 105: 25-36.

643

TRUONG, N.V., ANDREW, C.S. & SKERMAN, P.J. Response by siratro (*Phaseolus atropur-*
1967   *pureus*) and white clover (*Trifolium repens*) to nutrients on solodic soils at
Beaudesert, Queensland. *Aust. J. Exp. Agric. Anim. Husb.*, 7: 232-236.

TSCHAN, Y.T. Studies on N-fixing bacteria. V. Presence of *Beijerinckia* in northern
1953   Australia and geographical distribution of nonsymbiotic N-fixing organisms. *Proc.
Linn. Soc. NSW*, 78: 171-178.

TSCHIRLEY, F.H. Controlling mesquite with 2,4,5-T. Arizona Agr. Exp. Sta. and Co-op.
1962   Ext. Service, Folder 981.

TULEY, P. *Stylosanthes gracilis. Herb. Abstr.*, 38: 87-94.
1968

TUTT, H.S. & LUCK, P.E. Mist-flower control with pasture. *Queensl. Agric. J.*, 95: 387-391.
1969

UNIVERSITY OF THE WEST INDIES. *Herbicide Research Unit Report.*
1963

USDA. Cattle feeding in the United States. Agric. Econ. Rep. No. 186.
1970

VALENZA, J. Notes about stocking rate trials on grazing lands in the Republic of Senegal.
1965   *Proc. 9th Int. Grassl. Congr.*, São Paulo, Brazil, 1381-1385.

VALLIS, I. Measurement of nitrogen changes in pastures. Brisbane, University of Queens-
1969   land. (Ph. D. thesis)

VALLIS, I., HAYDOCK, K.P., ROSS, P.J. & HENZELL, E.F. Isotopic studies on the uptake of
1967   nitrogen by pasture plants. 3. The uptake of small additions of 15N-labelled fertiliser by rhodes grass and Townsville lucerne. *Aust. J. Agric. Res.*, 18: 865-877.

VARGAS, M.A. & SUHET, A.R. Efficiency of commercial inoculants and native *Rhizobium*
1981   strains on six forage legumes on a cerrado soil. *Pesqui Agropecu. Brasil. Ser.
Zootec.*, 16: 357-362.

VAVILOV, N.I. The origin, variation, immunity and breeding of cultivated plants. *Chron.
1951   Bot.*, 13: 1-364.

VERA, R.R., PIZARRO, E.A., MARTINS, M. & VIANA, J.A.C. Yield and quality of tropical
1983   legumes during the dry season: *Galactia striata* (Jacq) Urb. *Proc. 14th Int. Grassl.
Cong.*, Lexington, Kentucky, USA, 786-791.

VERBOOM, W.C. Legumes in Zambia. *Trop. Agric. (Trin.)*, 42: 229-242.
1965

VERGARA, J. Minerals for beef cattle. *Agric. Trop.*, 23: 373-383.
1967

VERHOEVEN, G. Tropical legume seed can be harvested commercially. *Queensl. Agric. J.*,
1958   84: 77-82.

VICENTE-CHANDLER, J. Role of fertilizers in hot, humid climates. *Proc. Soil Crop Sci. Soc.*
1966   *Fla*, 26: 328-360.

VICENTE-CHANDLER, J., CARO-COSTAS, R. & FIGARELLA, J. The effect of two heights of
1953   cutting and three fertility levels on the yield, protein content, and species compo-
        sition of a tropical Kudzu and molasses grass pasture. *Agron. J.*, 45: 397-400.

VICENTE-CHANDLER, J., SILVA, S. & FIGARELLA, J. The effect of nitrogen fertilisation and
1959   frequency of cutting on the yield and composition of three tropical grasses. *Agron.*
        J., 51: 202-206.

VILLARREAL, O. Effects of legumes or fertiliser nitrogen on the productivity of grazed pan-
1967   gola grass (*Digitaria decumbens*) pastures. Brisbane, Univ. of Queensland. (M.
        Agr. Sci. thesis)

VINCENT, J.M. Influence of calcium and magnesium upon the growth of *Rhizobium*. *J. Gen.*
1962   *Microb.*, 28: 653-663.

VINCENT, J.M. *A manual for the practical study of root nodule bacteria.* Oxford and Edin-
1970   burgh, Blackwells. IBP Handbook No. 15.

VISSER, J.H. Bemesting van die veld. *Proc. Grassl. Soc. Sth Afr.*, 1: 41-48.
1966

VIVIAN, L.A. The leguminous fodder "stylo" or "tropical lucerne" in Kelantin (*Stylosanthes*
1959   *gracilis*). *Malay. Agric. J.*, 42: 183-198.

WALKER, B. Effects of nitrogen fertiliser on natural pastures in western Tanzania. *Exp.*
1969a   *Agric.*, 5: 215-222.

WALKER, B. Effects of nitrogen fertilisers and forage legumes on a *Cenchrus ciliaris* pasture
1969b   in western Tanzania. *East Afr. Agric. For. J.*, 35: 2-5.

WALLACE, A.T., KILLINGER, G.P., BLEDSOE, R.W. & DUNCAN, D.B. *Fla. Agric. Exp. Sta.*
1957   *Bull.*, 581.

WANG, C.C. Growth, flowering and forage production of some grasses and legumes in
1961   response to different photoperiods. *J. Agric. Assoc. China*, 36: 27-52.

WARMKE, H.E. & FREYRE, R.H. Legume palatability. Rep. P. R. Agric. Exp. Sta., p. 21.
1952

WARREN, M. University of Witwatersrand, South Africa. (Ph. D. thesis)
1956

WATSON, G.A. Cover plants in rubber cultivation. *J. Rubber Res. Inst. Malays.*, 15: 8.
1957a

WATSON, G.A. Nitrogen fixation by *Centrosema pubescens*. *J. Rubber Res. Inst. Malays.*,
1957b   15: 168-174. Communication 314.

WATSON, G.A. Interactions of lime and molybdate in the nutrition of *Centrosema pubescens*
1960   and *Pueraria phaseoloides*. *J. Rubber Res. Inst. Malays*, 16: Communication 331.

WAYMAN, O. & IWANAGA, I.I. The inhibiting effect of *Leucaena glauca* (Koa haole) on
1957   reproduction performance in swine. *Proc. Western Sec. Amer. Soc. Anim. Prod.*,
        8: 5.

WEIR, W.C. & TORRELL, D.T. Selective grazing by sheep as shown by a comparison of the
1959   chemical composition of range and pasture forage obtained by hard clipping and
        that collected by oesophageal-fistulated sheep. *J. Anim. Sci.*, 18: 141-649.

WELLER, M.C. Urea-molasses makes the difference. *Queensl. Agric. J.*, 95: 662-663.
1969

WELLS, H.D. & FORBES, I. Anthracnose of *Desmodium* in Georgia. *Pl. Dis. Rep.*, 47: 837-
1963   839.

WENDT, W.B., STOBBS, T.H., TILEY, G.E.D. & TUCKER, G.G. *Pasture handbook.*
1970   Entebbe, Min. of agric. for Uganda.

WEST, O. Indigenous tree crops for southern Rhodesia. *Rhod. Agric. J.*, 47: 204-217.
1950

WETSELAAR, R. Estimation of nitrogen fixation by four legumes in a dry monsoonal area of
1967   northwestern Australia. *Aust. J. Exp. Agric. Anim. Husb.*, 29: 518-522.

WETSELAAR, R. & HUTTON, J.T. The ionic composition of rainwater at Katherine, N.T.,
1963   and its part in the cycling of plant nutrients. *Aust. J. Agric. Res.*, 14: 319-329.

WHEELER, W.A. *Forage and pasture crops.* New Jersey, van Nostrand.
1950

WHITE, R.E. & HAYDOCK, K.P. Phosphate concentration in Siratro as a guide to its phos-
1970   phate status in the field. *Aust. J. Exp. Agric. Anim. Husb.*, 10: 426-430.

WHITEMAN, P.C. Effects of temperature on the vegetative growth of six tropical legume
1968   species. *Aust. J. Exp. Agric. Anim. Husb.*, 8: 528-532.

WHITEMAN, P.C. The effects of close grazing and cutting on the yield, persistence and nitro-
1969   gen content of four tropical legumes with Rhodes grass at Samford, southeastern
        Queensland. *Aust. J. Exp. Agric. Anim. Husb.*, 9: 287-294.

WHITEMAN, P.C. Seasonal changes in growth and nodulation of perennial tropical pasture
1970   legumes in the field. I. The influence of planting date and grazing and cutting on
        Desmodium uncinatum and *Phaseolus atropurpureus*. *Aust. J. Agric. Res.*, 21: 195-
        206. II. Effects of controlled defoliation levels on nodulation of *Desmodium intor-
        tum* and *Phaseolus atropurpureus*. *Aust. J. Agric. Res.*, 21: 207-214. III. Effects of
        flowering on nodulation of three *Desmodium* species. *Aust. J. Agric. Res.*, 21: 215-
        222.

WHITEMAN, P.C. & LULHAM, A. Seasonal changes on growth and nodulation of perennial
1970   tropical pasture legumes in the field. I. The influence of planting date and grazing
        and cutting on *Desmodium uncinatum* and *Phaseolus atropurpureus*. *Aus. J. Agric.
        Res.*, 21: 195-206.

WHITEMAN, P.C., SEITLHEKO, M., SIREGAR, M.E., CHUDASAMA, A.K. & JAVIER, R.R.
1984   Short-term flooding tolerance of seventeen commercial tropical pasture legumes.
        Trop. Grassl., 18: 91-96.

WHITNEY, A.S. Nitrogen fixation by three tropical forage legumes and the utilisation of
1966   legume-fixed nitrogen by their associated grasses. *Herb. Abstr.*, 38: 143.

WHITNEY, A.S. Effects of harvesting interval, height of cut, and nitrogen fertilisation on the
1970   performance of *Desmodium intortum* mixtures in Hawaii. *Proc. 11th Int. Grassl.
        Congr.*, Surfers Paradise, Australia, 632-636.

WHITNEY, A.S. & GREEN, R.E. Pangola grass performance under different levels of nitro-
1969a   gen fertilisation in Hawaii. *Agron. J.*, 61: 577-581.

WHITNEY, A.S. & GREEN, R.E. Legume contributions to yields and compositions of
1969b   *Desmodium* spp. - Pangola grass mixtures. *Agron. J.*, 61: 741-745.

WHITNEY, A.S. & KANEHIRO, Y. Pathways of nitrogen transfer in some tropical legume-
1967   grass associations. *Agron. J.*, 59: 585-588.

WHITNEY, A.S., KANEHIRO, Y. & SHERMAN, G.D. Nitrogen relationships of three tropical
1967   forage legumes in pure stands and in grass mixtures. *Agron. J.*, 59: 47-50.

WHYTE, R.O. *The grassland and fodder resources of India*. Indian Coun. Agric. Res., N.
1964   Delhi.

WHYTE, R.O., NILSSON-LEISSNER, G. & TRUMBLE, H.C. *Legumes in agriculture*. FAO,
1969   Rome.

647

WHYTE, R.O., NILSSON-LEISSNER, G. & TRUMBLE, H.C. *Legumes in agriculture.* Rome,
1953   FAO. FAO Agricultural Studies No. 21.

WILCZEK, R. In *Flore du Congo belge et du Ruanda-Urundi*, Vol. 6. INEAC, Brussels.
1954

WILDIN, J.H. & MIDDLETON, C.H. Fitzroy stylo - a hardy pasture legume. *Queensl. Agric.
J.* (In press)

WILLIAMS, J. & GARDENER, C.J. Environmental constraints to growth and survival of
1984   *Stylosanthes. In* Stace, H.M. & Edye, L.A., eds. *The biology and agronomy of*
Stylosanthes. North Ryde, Australia, Academic Press, pp. 181-201.

WILLIAMS, R.J., REID, R., SCHULTZE-KRAFT, R., SOUSA COSTA, N.M. & THOMAS, E.D.
1984   Natural distribution of *Stylosanthes. In* Stace, H.M. & Edye, L.A., eds. *The biology and agronomy of* Stylosanthes. North Ryde, Australia, Academic Press, pp.
73-101.

WILLOUGHBY, W.M. Limitations to animal production imposed by seasonal fluctuations in
1959   pasture and by pasture management. *Aust. J. Agric. Res.*, 10: 248-268.

WILSON, A.D. A review of browse in the nutrition of grazing animals. *J. Range Manage.*, 22:
1969   23-28.

WILSON, A.S.B. & LANSBURY, T.J. *Centrosema pubescens*: ground cover and forage crop in
1958   cleared rainforest of Ghana. *Emp. J. Exp. Agric.*, 26: 351-364.

WILSON, G.P. & MURTAGH, G.J. Lab-lab - new forage crop for the north coast. *Agric. Gaz.
1962   N.S.W.*, 73: 460-462.

WILSON, J.G. & BREDON, R.M. Nutritional value of some common cattle browse and fod-
1963   der plants of Karamoja, Northern Province, Uganda. *E. Afr. Agric. For. J.*, 31:
204-208.

WILSON, J.R., HAYDOCK, K.D. & ROBINS, M.F. Responses to salinity in glycine. 5.
1970   Changes in the chemical composition of three Australian species and *G. wightii (G.
javanica)* over a range of salinity stresses. *Aust. J. Exp. Agric. Anim. Husb.*, 10:
156-165.

WOODS, L.E. A survey of Townsville stylo (*Townsville lucerne*) pastures established in the
1969   Northern Territory up to 1969. *Trop. Grassl.*, 3: 91-98.

WOODS, L.E. & DANCE, R.A. Seed and nutritional aspects of grass-Townsville stylo com-
1970   petition. *J. Aust. Inst. Agric. Sci.*, 36: 45-47.

WOODSON, R.E. & SCHERY, R.W. Flora of Panama. Part 5, Fascicle 3 (Leguminosae, sec-
1951   ond part). *Ann. Mo. Bot. Gdn*, 35: 1-94.

WRIGHT, C.B. *Lotononis* in the Wallum. *Queensl. Agric. J.*, 90: 93-94, 96.
1964

WU, M.H. Effect of lime, molybdenum and inoculation on the growth of *Leucaena glauca*
1964   on acid soil. *J. Agric. Assn China*, 47: 57-60.

WUTOH, J.G., HUTTON, E.M. & PRITCHARD, A.J. The effects of photoperiod and temper-
1968a   ature on flowering in *Glycine javanica*. *Aust. J. Exp. Agric. Anim. Husb.*, 8: 544-
547.

WUTOH, J.G., HUTTON, E.M. & PRITCHARD, A.J. Combining ability in *Glycine javanica*.
1968b   *Aust. J. Agric. Res.*, 19: 411-418.

WUTOH, J.G., HUTTON, E.M. & PRITCHARD, A.J. Inheritance of flowering time, yield, and
1968c   stolon development in *Glycine javanica* L. *Aust. J. Exp. Agric. Anim. Husb.*, 8:
317-322.

WYCHERLEY, P.R. Seed germination of some tropical legumes. *J. Rubber Res. Inst.*
1960   *Malays.*, 16: 99-117.

van WYK, H.P.D., OCSTHUIZEN, S.A., MEYER, E.E., BREVIS, J.G. & GROBIER, J.H. The
1955   nutritive value of South African feeds. 3. Hay and pasture crops, silage, cereals,
tubers and pods. *Union S. Afr. Dept Agric. Sci. Bull.*, 354.

YOHO, J.G., DUTROW, G.F. & MOAK, J.E. What it costs to practise forestry. *Forest Farmer*,
1969   July.

YOSHIDA, R.K. A chemical and physiological study of the nature and properties of the toxic
1944   principle in *Leucaena glauca* (Koa haole) Minneapolis, Univ. of Minn. (Ph. D.
thesis)

YOUNG, J.G. & CHIPPENDALE, F. Beef cattle performance on pastures on heath plains in
1970   southeast Queensland. *Proc. 11th Int. Grassl. Congr.*, Surfers Paradise, Australia,
849-852.

YOUNG, N.D., FOX, N.F. & BURNS, M.A. A study of three important pasture mixtures in
1959   the Queensland subtropics. *Queensl. J. Agric. Sci.*, 16: 199-215.

YOUNGE, O.R. & PLUCKNETT, D.L. Beef production with heavy phosphorus fertilisation in
1965   infertile wetlands of Hawaii. *Proc. 9th Int. Grassl. Congr.*, São Paulo, Brazil, 959-
963.

YOUNGE, O.R. & PLUCKNETT, D.L. Quenching the high phosphorus fixation of Hawaiian
1966   latosol. *Proc. Soil. Sci. Am.*, 30: 653-655.

YOUNGE, O.R., PLUCKNETT, D.L. & ROTAR, P.P. Culture and yield performance of
1964   *Desmodium intortum* and *D. canum* in Hawaii. *Hawaii Agric. Exp. Sta. Tech.*
Bull., 59.

649

Appendix 1
Tables

TABLE 1. **Crude protein and crude fibre percentage of tropical legumes**

Legume	Crude protein		Crude fibre		Reference
	Mean (%)	Range (%)	Mean (%)	Range (%)	
Alysicarpus rugosus	16.4		33.8		Verboom, 1965
A. rugosus	15.9	9.9-21.4			van Rensburg, 1967
Atylosia scarabaeoides		8.6-12.9			Singh, 1962
Calopogonium brachycarpum	23.0		16.7		Otero, 1952
C. mucunoides	16.7				Bermudez, Ceballos and Chaverra, 1968
Canavalia ensiformis	13.2	11.1-19.2	28.8	19.30-31.8	Otero, 1952
Centrosema plumieri	22.4				Bermudez, Ceballos and Chaverra, 1968
C. pubescens	22.6		28.6		Odejar, 1949
C. pubescens	25.3	22.6-27.6	25.9	23.8-27.2	Guayadeen, 1951
C. pubescens	13.11		21.6		Otero, 1952
C.pubescens	20.1		34.7		Loosli, Villegas and Ynalves, 1954
C. pubescens	20.7		34.1		Reyes, 1955
C. pubescens	16.9	15.8-18.7	30.7	29.8-31.3	Oyenuga, 1957
C. pubescens		17.0-23.7	25.8	25.8-38.1	Wilson and Lansbury, 1958
C. pubescens	18.2		29.6		Lansbury, 1959
C. pubescens	18.9		35.8		Allen and Cowdry, 1961a

TABLE 1. **Crude protein and crude fibre percentage of tropical legumes** *(continued)*

Legume	Crude protein		Crude fibre		Reference
	Mean (%)	Range (%)	Mean (%)	Range (%)	
C. pubescens	18.8	11.6-22.1			Andrew and Norris, 1961
C. pubescens	14.4		38.0		Miller and Rains, 1963
C. pubescens	18.7	18.1-19.4			Neme and Nery, 1965
C. pubescens	14.7				Whitney, Kanehiro, and Sherman, 1967
C. pubescens	15.9				van Rensburg, 1967
C. pubescens	18.5				Bermudez, Ceballos and Chaverra, 1968
C. pubescens	17.8	13.3-21.1			Bryan, 1968
C. pubescens	19.6	17.3-22.6			Andrew and Robins, 1969b
C. pubescens	24.4	21.4-29.9			Andrew and Robins, 1969c
Crotalaria lanceolata	8.8		42.6		Milford, 1967
Desmodium adscendens	15.3		19.42		Otero, 1952
Desmodium canum	11.8				Whitney, Kanehiro and Sherman, 1967
D. canum	15.6				Bermudez, Ceballos and Chaverra, 1968
Desmodium intortum	22.1		22.2		Odejar and Masankay-Arenas, 1951
D. intortum leaves stems	17.8 23.63 8.92	12.9-23.3	30.7 24.52 43.60	28.1-34.1	Compère, 1961 Compère, 1961 Compère, 1961
D. intortum	19.1	18.2-20.0	31.0	30.8-31.3	Dougall and Bogdan, 1966
D. intortum	14.54	11.38-20.11			van Rensburg, 1967
D. intortum	12.6				Whitney, Kanehiro and Sherman, 1967
D. intortum		9.0-24.5			Bryan, 1969
D. intortum	19.9	15.6-24.1			Andrew and Robins, 1969b

652

Legume	Crude protein		Crude fibre		Reference
	Mean (%)	Range (%)	Mean (%)	Range (%)	
D. intortum	26.4	24.4-28.5			Andrew and Robins, 1969c
D. intortum	13.0	10.8-18.2			Blunt and Humphreys, 1970
D. intortum	16.6	15.6-17.5			Catchpoole, 1970
Desmodium repandum	15.11		27.72		Compère, 1961
Desmodium sandwicense	11.9	10.4-13.7			Jones, Davies and Waite, 1967
Desmodium scorpiurus	15.5		29.3		Miller and Rains, 1963
D. scorpiurus	19.3				Bermudez, Ceballos and Chaverra, 1968
Desmodium tortuosum	18.6		44.6		Compère, 1961
D. tortuosum	20.0				Bermudez, Ceballos and Chaverra, 1968
Desmodium triflorum	13.6		35.0		Otero, 1952
D. triflorum	14.0		34.2		Compère, 1961
D. triflorum	18.0				Bermudez, Ceballos and Chaverra, 1968
Desmodium uncinatum	14.4	10.0-16.9			Andrew and Norris, 1961
D. uncinatum	23.0		28.0		Dougall and Bogdan, 1966
D. uncinatum	12.0		38.0		Graham, 1967
D. uncinatum	13.0	10.9-18.2	38.8	32.5-43.4	Milford, 1967
D. uncinatum	18.4	14.7-21.5			Andrew and Robins, 1969b
D. uncinatum	23.3	22.1-24.6			Andrew and Robins, 1969c
D. uncinatum		10.7-21.5			Bryan, 1969
Desmodium varians	12.3		37.9		Compère, 1961

653

TABLE 1. **Crude protein and crude fibre percentage of tropical legumes** (continued)

Legume	Crude protein		Crude fibre		Reference
	Mean (%)	Range (%)	Mean (%)	Range (%)	
Dolichos sericeus	15.3	11.2-19.0			van Rensburg, 1967
Indigofera sp.	16.0				Quarterman, 1956
I. arrecta	26.2		25.5		Dougall and Bogdan, 1966
Indigofera campestris	14.1		8.6		Otero, 1952
Indogofera hirsuta	23.8		15.1		Dougall and Bogdan, 1966
I. hirsuta	9.14		11.23		Otero, 1952
Indigofera spicata	18.0	16.4-20.4			Andrew and Norris, 1961
I. spicata	16.4	15.5-17.7			Jones, Davies and Waite, 1967
Indigofera subulata	16.8		31.8		Robinson, 1950
I. subulata	28.2	25.6-32.9	14.3	13.3-15.7	Guayadeen, 1951
Indigofera suffruticosa	18.8				Bermudez, Ceballos and Chaverra, 1968
Indigofera vohemarensis	24.1		23.7		Dougall and Bogdan, 1966
Lablab purpureus	14.6	14.4-14.8	28.6	28.3-29.0	Elliot and Fokkema, 1960
L. purpureus	19.2		30.5		Dougall and Bogdan, 1966
L. purpureus	14.44	11.50-23.4			van Rensburg, 1967
L. purpureus	14.4	10.2-17.1			Milford and Minson, 1968
L. purpureus	22.1		27.4		Neme, 1970
Leucaena leucocephala	18.8		37.7		Henke, 1945
L. leucocephala	24.3		23.9		Work, 1946
L. leucocephala	12.5		12.4		Farinas, 1951
L. leucocephala	24.4		9.6		Gantt, 1953

654

TABLE 1. **Crude protein and crude fibre percentage of tropical legumes** *(continued)*

Legume	Crude protein		Crude fibre		Reference
	Mean (%)	Range (%)	Mean (%)	Range (%)	
L. leucocephala	18.5		27.4		Loosli, Villegas and Ynalvez, 1954
L. leucocephala	15.5	14.9-16.0			Oakes and Skov, 1967
L. leucocephala	28.8	20.2-35.8			Hutton and Bonner, 1960
L. leucocephala	27.5		16.4		van Rensburg, 1960
L. leucocephala	17.9	13.8-20.2			Andrew and Robins, 1969b
Lotononis bainesii	18.3	10.4-25.9			Bryan, 1961
L. bainesii	14.1	12.4-15.3			Jones, Davies and Waite, 1967
L. bainesii	20.0		27.0		Milford, 1967
L. bainesii	16.1	13.6-20.1			Bryan, 1968
L. bainesii	20.2	15.1-24.5			Andrew and Robins, 1969b
L. bainesii	23.6	21.7-25.1			Andrew and Robins, 1969b
L. bainesii	14.8	10.2-22.9			Blunt and Humphreys, 1970
L. bainesii	15.3	15.0-15.6			Catchpoole, 1970
Mimosa invisa	16.5		21.8		Dougall and Bogdan, 1966
Macroptilium atropurpureum	24.1	22.4-25.9			Dougall and Bogdan, 1966
M. atropurpureum	15.0	12.5-18.8			Minson and Milford, 1966
M. atropurpureum	16.2	14.3-17.2			Jones, Davies and Waite, 1967
M. atropurpureum	16.8		33.4		Milford, 1967
M. atropurpureum	14.3	8.4-19.4			van Rensburg, 1967
M. atropurpureum	18.4				Bermudez, Ceballos and Chaverra, 1968
M. atropurpureum	22.4	17.5-26.6			Andrew and Robins, 1969b

TABLE 1. **Crude protein and crude fibre percentage of tropical legumes** *(continued)*

Legume	Crude protein		Crude fibre		Reference
	Mean (%)	Range (%)	Mean (%)	Range (%)	
M. atropurpureum	24.2	23.2-25.6			Andrew and Robins, 1969b
M. atropurpureum	15.8	12.6-24.9			Blunt and Humphreys, 1970
M. atropurpureum	16.8	14.4-18.8			Catchpoole, 1970
Macroptilium lathyroides	20.5	10.4-25.9			Andrew and Norris, 1961
M. lathyroides	14.2	7.6-19.2	34.1	27.0-43.4	Milford, 1967
M. lathyroides	18.4				Bryan, 1968
M. lathyroides	19.6	14.6-26.8			Andrew and Robins, 1969b
M. lathyroides	23.8	21.2-28.0			Andrew and Robins, 1969c
Neonotonia wightii	15.5		41.0		van Wyk *et al.,* 1955
N. wightii	17.9		30.3		van Rensburg, 1956
N. wightii	21.8		27.4		van Rensburg, 1960
N. wightii	22.2	20.6-24.8			Gates, Haydock and Little, 1966
N. wightii	14.9	14.2-15.4			Neme and Nery, 1965
N. wightii	12.8		30.4		Peixoto, de Moraes and Prospero, 1965
N. wightii	16.7	12.9-20.2			Holder, 1967
N. wightii	17.0	15.5-18.6			Andrew and Robins, 1969b
Phaseolus panduratus	26.25		13.10		Otero, 1952
Pueraria phaseoloides	16.2		37.2		Reyes, 1955
P. phaseoloides	18.7		28.3		Odejar, 1949
P. phaseoloides	17.0	14.2-19.5	29.0	26.1-31.3	Guyadeen, 1951
P. phaseoloides	16.9		35.3		Loosli, Villegas and Ynalvez, 1954
P. phaseoloides	18.8		37.9		Dirven and Hart, 1959

656

TABLE 1. **Crude protein and crude fibre percentage of tropical legumes** *(continued)*

Legume	Crude protein		Crude fibre		Reference
	Mean (%)	Range (%)	Mean (%)	Range (%)	
P. phaseoloides	15.4	12.8-18.0	40.0	39.4-41.3	Dijkstra and Dirven, 1962
P. phaseoloides	20.5		37.9		Butterworth, 1963
P. phaseoloides	15.4	14.5-16.2			Neme and Nery, 1965
P. phaseoloides	19.9				Bermudez, Ceballos and Chaverra, 1968
P. phaseoloides	13.3	11.8-14.8	39.0	36.9-41.1	Blasco and Bohórquez, 1968
Pueraria thunbergiana	13.3		40.3		van Wyk *et al.*, 1955
Rhynchosia minima	23.7				Bermudez, Ceballos and Chaverra, 1958
Stizolobium atterimum	16.0	15.5-16.5	28.9	37.2-40.5	Harrison, 1942
S. atterimum	22.8		30.2		Otero, 1952
Stizolobium deeringianum	15.9	15.2-16.2	31.0	29.0-33.8	Elliott and Fokkema, 1960
S. deeringianum	11.8	9.7-14 2	29.6	28.9-30.3	Miller, Rains and Thorpe, 1964
S. deeringianum	15.6		34.5		Axtmayer, Hernandez and Cook, 1938
Stylosanthes guianensis	18.1				Scaut, 1959
S. guianensis	17.6				Nwosu, 1960
S. guianensis	15.7	14.9-17.3			Andrew and Norris, 1961
S. guianensis	12.1				Allen and Cowdry, 1961
S. guianensis	12.5				Smith, 1962
S. guianensis	13.4		32.3		Miller and Rains, 1963
S. guianensis	16.5				Hymowitz *et al.*, 1967
S. guianensis	11.9	8.5-15.7			van Rensburg, 1967
S. guianensis	11.8		37.7		Milford, 1967
S. guianensis	8.9				Newman, 1968

Legume	Crude protein		Crude fibre		Reference
	Mean (%)	Range (%)	Mean (%)	Range (%)	
S. guianensis	12.9	11.3-15.1			Blunt and Humphreys, 1970
Stylosanthes humilis	11.2	7.5-17.5			Norman, 1959
S. humilis	13.8	13.5-14.2			Fullerton, personal communication
S. humilis	10.4	9.4-11.3			Norman, 1965
S. humilis	18.1	14.8-20.5			Shaw, Gates and Wilson, 1966
S. humilis	18.2	14.4-20.9			Jones, 1968
S. humilis	5.6				Newman, 1968
S. humilis	10.0				Little, 1968
S. humilis	21.4	16.4-24.8			Andrew and Robins, 1969b
S. humilis	20.5	19.8-21.7			Andrew and Robins, 1969c
S. humilis	15.2	11.6-20.6			Fisher, 1969
Teramnus uncinatus	14.6		39.6		Otero, 1952
Teramnus volubilis	16.9				Bermudez, Ceballos and Chaverra, 1968
Vigna angularis	18.4				Bermudez, Ceballos and Chaverra, 1968
Vigna catjang	13.5	11.1-15.8			Ranjhan, Talapatra and Kala, 1967
Vigna luteola	16.9	12.1-19.9			Andrew and Robins, 1969b
Vigna marina	18.1	15.6-20.6			Jones, Davies and Waite, 1967
Vigna mungo	19.5		27.0		Harrison, 1942
Vigna repens	18.0	16.7-19.4			Jones, Davies and Waite, 1967
Vigna sinensis	23.7		20.3		Otero, 1952
V. sinensis	15.2	14.6-15.8	31.0	28.2-33.9	Elliott and Croft, 1958

TABLE 1. **Crude protein and crude fibre percentage of tropical legumes** *(concluded)*

Legume	Crude protein		Crude fibre		Reference
	Mean (%)	Range (%)	Mean (%)	Range (%)	
V. sinensis	14.3	12.5-17.5	27.1	21.6-34.8	Miller, Rains and Thorpe, 1964
V. sinensis	17.9		39.0		Majumdar, Sen and Roy, 1968
V. sinensis	16.8	12.6-20.6			Milford and Minson, 1968
Vigna vexillata	18.2	16.2-20.3	2.6	26.4-38.9	Milford, 1967
V. vexillata	18.2	14.4-22.1			van Rensburg, 1967

TABLE 2. **Phosphorus as percentage of dry matter**

Legume	Mean	Range	Reference
		(%)	
Alysicarpus rugosus	0.19		Verboom, 1965
Atylosia scarabaeoides	0.15		Singh, 1962
Cajanus cajan	0.25	0.21-0.29	Vergara, 1967
Calopogonium mucunoides	0.16		Vergara, 1967
Centrosema pubescens	0.18	0.17-0.19	Verboom, 1965
C. pubescens	0.17	0.13-0.24	Neme and Ncry, 1965
C. pubescens	0.26	0.21-0.36	Andrew and Robins, 1969a
C. pubescens	0.39	0.33-0.45	Andrew and Robins, 1969c
Desmodium intortum	0.47		Compère, 1961
(leaves)	0.28		Compère, 1961
D. intortum	0.27		Dougall and Bogdan, 1966
D. intortum	0.27	0.20-0.36	Vergara, 1967
D. intortum	0.20	0.11-0.31	Andrew and Robins, 1969a
D. intortum	0.41	0.27-0.70	Andrew and Robins, 1969c
Desmodium uncinatum	0.27		Verboom, 1965
D. uncinatum	0.33		Dougall and Bogdan, 1966
D. uncinatum	0.18	0.11-0.30	Andrew and Robins, 1969a
D. uncinatum	0.37	0.27-0.62	Andrew and Robins, 1969c
Indigofera arrecta	0.35		Dougall and Bogdan, 1966
Indigofera hirsuta	0.37		Dougall and Bogdan, 1966
Indigofera spicata	0.19		Verboom, 1965
Indigofera vohemarensis	0.37		Dougall and Bogdan, 1966
Lablab purpureus	0.21		French, 1937
L. purpureus	0.29		Dougall and Bogdan, 1966
L. purpureus	0.30	0.24-0.35	Vergara, 1967
Leucaena leucocephala	0.26	0.26-0.27	Vergara, 1967
Lotononis angolensis	0.37		Verboom, 1965
Lotonónis bainesii	0.24		Verboom, 1965
L. bainesii	0.22	0.11-0.37	Andrew and Robins, 1969a
L. bainesii	0.51	0.42-0.71	Andrew and Robins, 1969c
Macroptilium atropurpureum	0.21	0.18-0.25	Truong, Andrew and Skerman, 1967

660

TABLE 2. **Phosphorus as percentage of dry matter** *(concluded)*

Legume	Mean	Range	Reference
		(%)	
M. atropurpureum	0.19	0.09-0.30	Andrew and Robins, 1969a
M. atropurpureum	0.31	0.27-0.43	Andrew and Robins, 1969c
M. atropurpureum	0.18	0.09-0.27	White and Haydock, 1970
Macroptilium lathyroides	0.25		Verboom, 1965
M. lathyroides	0.17	0.08-0.30	Andrew and Robins, 1969a
M. lathyroides	0.36	0.30-0.58	Andrew and Robins, 1969c
Macrotyloma uniflorum	0.14		Verboom, 1965
Mimosa invisa	0.32		Dougall and Bogdan, 1966
Neonotonia wightii	0.10		van Wyk *et al.,* 1955
N. wightii	0.16		Verboom, 1965
N. wightii	0.13	0.11-0.13	Neme and Nery, 1965
N. wightii	0.13	0.10-0.30	Andrew and Robins, 1969a
Pueraria phaseoloides	0.14	0.12-0.16	Neme and Nery, 1965
P. phaseoloides	0.39		Vergara, 1967
P. phaseoloides	0.45	0.25-0.65	Blasco and Bohórquez, 1968
Pueraria thunbergiana	0.10		van Wyk *et al.,* 1955
P. thunbergiana	0.31	0.23-0.36	Abruna and Figarella, 1957
Stizolobium deeringianum	0.16	0.13-0.19	Axtmayer, Hernandez and Cook, 1938
S. deeringianum	0.15	0.12-0.18	Vergara, 1967
Stylosanthes humilis	0.08	0.05-0.11	Fullerton, 1961
S. humilis	0.08	0.05-0.10	Norman, 1965
S. humilis	0.17	0.07-0.27	Gates, Haydock and Claringbold, 1966
S. humilis	0.13	0.06-0.21	Shaw, Gates and Wilson, 1966
S. humilis	0.19	0.08-0.39	Jones, 1968
S. humilis	0.07		Little, 1968
S. humilis	0.19	0.08-0.29	Andrew and Robins, 1969a
S. humilis	0.22	0.17-0.33	Andrew and Robins, 1969c
S. humilis	0.08		Fisher, 1969
S. humilis	0.08		Playne, 1969
Stylosanthes mucronata	0.22		Verboom, 1965
Teramnus uncinatus	0.14		Verboom, 1965
Vigna luteola	0.24	0.15-0.34	Andrew and Robins, 1969a
Vigna umbellata	0.58	0.11-0.24	Vergara, 1967

661

TABLE 3. **Calcium as percentage of dry matter**

Legume	Mean	Range	Reference
		(%)	
Alysicarpus rugosus	0.82		Verboom, 1965
Atylosia scarabaeoides	1.6		Singh, 1962
Cajanus cajan	0.27	0.22-0.31	Vergara, 1967
Calopogonium mucunoides	1.00	0.66-1.22	Vergara, 1967
Centrosema pubescens	0.94		Verboom, 1965
C. pubescens	1.36	1.27-1.35	Neme and Nery, 1965
C. pubescens	0.78	0.71-0.86	Andrew and Robins, 1969b
C. pubescens	1.14	1.00-1.47	Andrew and Robins, 1969c
Desmodium intortum	0.89	0.81-0.95	Compère, 1961
(leaves)	0.99		Compère, 1961
D. intortum	0.90	0.78-1.01	Dougall and Bogdan, 1966
D. intortum	1.29	0.93.1-63	Vergara, 1967
D. intortum	0.89	0.68-1.28	Andrew and Robins, 1969b
D. intortum	1.25	0.96-1.82	Andrew and Robins, 1969c
D. intortum	0.74		Dougall and Bogdan, 1966
D. intortum	0.84	0.63-1.16	Andrew and Robins, 1969b
Desmodium uncinatum	0.78		Verboom, 1965
D. uncinatum	1.51	1.22-2.13	Andrew and Robins, 1969c
Dolichos uniflorus	0.50		Verboom, 1965
Indigofera arrecta	1.36		Dougall and Bogdan, 1966
Indigofera hirsuta	1.88		Dougall and Bogdan, 1966
Indigofera spicata	1.25		Verboom, 1965
Indigofera vohemarensis	2.24		Dougall and Bogdan, 1966
Lablab purpureus	1.19		Dougall and Bogdan, 1966
L. purpureus	0.57	0.31-0.99	Vergara, 1967
Leucaena leucocephala	0.57	0.56-0.60	Vergara, 1967
L. leucocephala	0.97	0.91-1.03	Andrew and Robins, 1969b
Lotononis angolensis	0.83		Verboom, 1965
Lotononis bainesii	0.69		Verboom, 1965
L. bainesii	0.70	0.49-1.01	Andrew and Robins, 1969b

TABLE 3. **Calcium as percentage of dry matter** *(concluded)*

Legume	Mean	Range	Reference
		(%)	
L. bainesii	1.55	1.34-1.67	Andrew and Robins, 1969c
Macroptilium atropurpureum	0.93	0.68-1.31	Andrew and Robins, 1969b
M. atropurpureum	1.74	1.71-1.93	Andrew and Robins, 1969c
Macroptilium lathyroides	1.54		Verboom, 1965
M. lathyroides	0.81	0.56-1.06	Andrew and Robins, 1969b
M. lathyroides	1.92	1.75-2.11	Andrew and Robins, 1969c
Macrotyloma uniflorum	0.14		Verboom, 1965
Mimosa invisa	1.10		Dougall and Bogdan, 1966
Neonotonia wightii	1.90		van Wyk *et al.,* 1955
N. wightii	0.88		Verboom, 1965
N. wightii	1.32	1.12-1.48	Neme and Nery, 1965
N. wightii	0.86	0.58-1.10	Andrew and Robins, 1969b
Pueraria phaseoloides	1.58	1.34-1.72	Neme and Nery, 1965
P. phaseoloides	1.11	1.05-1.16	Abruna and Figarella, 1957
P. phaseoloides	0.36		Vergara, 1967
P. phaseoloides	0.66	0.46-0.88	Blasco and Bohórquez, 1968
Stizolobium deeringianum	1.21		Axtmayer, Hernandez and Cook, 1938
S. deeringianum	1.15	0.88-1.67	Vergara, 1967
Stylosanthes humilis	0.89	0.63-1.30	Andrew and Robins, 1969b
S. humilis	2.00	1.60-2.82	Andrew and Robins, 1969c
Stylosanthes mucronata	1.32		Verboom, 1965
Teramnus uncinatus	0.68		Verboom, 1965
Vigna catjang	1.35	0.61-1.85	Ranjhan, Talapatra and Kala, 1967
Vigna luteola	1.37	1.28-1.46	Andrew and Robins, 1969b
Vigna umbellata	0.57	0.12-0.98	Vergara, 1967

663

TABLE 4. **Magnesium and sodium as percentage of dry matter**

Legume	Magnesium		Sodium		Reference
	Mean (%)	Range (%)	Mean (%)	Range (%)	
Centrosema pubescens	0.27	0.24-0.32	0.04	0.04-0.05	Andrew and Robins, 1969b
C. pubescens	0.22	0.20-0.28	0.01	0.01-0.02	Andrew and Robins, 1969c
Cyamopsis tetragonoloba	0.57				Misra, Bahn and Prasad, 1968
Desmodium intortum	0.28	0.25-0.30			Compère, 1961
D. intortum	0.53	0.35-0.70	0.03	0.02-0.04	Andrew and Robins, 1969b
D. intortum	0.31	0.25-0.47	0.01	0.01-0.01	Andrew and Robins, 1969c
Desmodium uncinatum	0.30	0.27-0.34	0.03	0.02-0.11	Andrew and Robins, 1969b
D. uncinatum	0.34	0.23-0.54	0.02	0.02-0.02	Andrew and Robins, 1969c
Lotononis bainesii	0.31	0.23-0.41	0.11	0.05-0.25	Andrew and Robins, 1969b
L. bainesii	0.35	0.25-0.45	0.12	0.05-0.20	Andrew and Robins, 1969c
Leucaena leucocephala	0.32	0.30-0.36	0.02	0.01-0.03	Andrew and Robins, 1969b
Macroptilium atropurpureum	0.73	0.64-0.89	0.03	0.02-0.04	Andrew and Robins, 1969b
M. atropurpureum	0.61	0.49-1.00	0.02	0.01-0.03	Andrew and Robins, 1969c
M. atropurpureum			0.02	0.02-0.04	Playne, 1970
Macroptilium lathyroides	0.44	0.34-0.52	0.10	0.50-0.13	Andrew and Robins, 1969b
M. lathyroides	0.50	0.41-0.75	0.05	0.03-0.08	Andrew and Robins, 1969c
Neonotonia wightii	0.36	0.32-0.40	0.03	0.02-0.06	Andrew and Robins, 1969b
Stylosanthes humilis	0.38	0.29-0.42			Andrew and Robins, 1969b
S. humilis	0.43	0.23-0.63			Andrew and Robins, 1969c
S. humilis			0.06	0.01-0.15	Playne, 1970
Vigna luteola	0.28	0.27-0.29	0.27	0.09-0.47	Andrew and Robins, 1969b

TABLE 5. Edible trees and shrubs: crude chemical composition analysis, moisture-free

Legume	Crude protein	Crude fat	Crude fibre	Nitrogen-free-extract	Ash	Ca	P	Mg	K	Cl	Reference
					(%)						
Acacia sp.	11.2	4.1	25.1	56.1	3.5	0.82	0.05				Everist, 1969
Acacia albida											
pods	10.53	0.94	27.55	57.06	3.92	0.51	0.51				Wilson and Bredon, 1963
pods	12.36	0.68	40.73	42.67	3.56						Wilson and Bredon, 1963
dry pods	11.13	0.84	32.45	46.10	9.48						Wilson and Bredon, 1963
green pods	10.19	0.49	30.36	54.94	4.02						Wilson and Bredon, 1963
whole pods	11.5	1.4	23.4	55.1	3.2						Verboom, 1965
pods husks	5.9	0.4	30.9	52.6	3.4						Verboom, 1965
seeds	22.2	2.9	7.2	60.3	3.5						Verboom, 1965
pods	10.5	1.4	26.7	57.7	3.6	0.65	0.23				West, 1950
Acacia aneura											
leaves (whipstick)	11.7	2.7	29.0	49.6	6.9	1.29	0.07				Everist, 1969
leaves (umbrella)	13.2	2.2	27.9	51.5	5.0						Everist, Harvey and Bell, 1958
leaves	12.8	1.1	32.9	47.2	6.0	1.50	0.15				Everist, Harvey and Bell, 1958
Acacia brachystachya											
leaves	10.4	5.7	41.9	38.2	4.1	0.72	0.10				Everist, 1969
Acacia brevispica											
leaves	18.79	5.67	17.99	51.01	6.54	1.86	0.31				Dougall and Bogdan, 1958
young pods	17.5	1.08	24.02	52.79	4.61	0.46	0.27				Dougall and Bogdan, 1958
Acacia cambagei											
leaves	13.3	3.1	15.9	55.2	11.0	2.29	0.07				Everist, 1969
Acacia cana											
leaves	11.3	3.2	30.2	48.6	6.8	1.09	0.05				Dougall and Bogdan, 1958

TABLE 5. **Edible trees and shrubs: crude chemical composition analysis, moisture-free** (continued)

Legume	Crude protein	Crude fat	Crude fibre	Nitrogen-free-extract	Ash	Ca	P	Mg	K	Cl	Reference
					(%)						
Acacia catenulata leaves	13.9	2.9	21.9	58.4	2.9	0.87	0.07				Everist, 1969
Acacia deanei leaves	14.3	4.8	18.9	55.6	3.4	0.89	0.06				Everist, 1969
Acacia excelsa leaves	11.8	3.6	21.4	56.7	6.5	1.79	0.06				Everist, 1969
Acacia farnesiana leaves	18.4	2.0	18.5	56.5	4.4	0.47	0.24				Everist, 1969
Acacia gerrardii leaves and young twigs	17.72	1.60	24.19	50.48	6.01	0.63	0.25				Dougall and Bogdan, 1958
Acacia giraffae leaves	12.6	1.8	34.2	47.9	3.6	0.71	0.11				West, 1950
Acacia harpophylla leaves (mature trees)	13.6	2.2	25.8	53.0	5.3	1.29	0.11				Everist, 1969
leaves (30-cm suckers)	16.2	2.1	33.9	41.6	6.2	0.94	0.19				Everist, 1969
leaves (15-30 cm suckers)	18.6	1.9	34.2	38.6	6.7	1.06	0.23				Everist, 1969
leaves (smaller than 15-cm suckers)	26.7	1.1	19.3	45.7	7.2	0.49	0.49				Everist, 1969

TABLE 5. **Edible trees and shrubs: crude chemical composition analysis, moisture-free** (continued)

Legume	Crude protein	Crude fat	Crude fibre	Nitrogen-free-extract	Ash	Ca	P	Mg	K	Cl	Reference
	(%)										
Acacia liitakunensis											
whole pods	17.3	3.1	24.8	49.2	5.7	0.79	0.34				West, 1950
pod husks	8.2	1.6	34.3	49.4	6.2	1.1	0.14				West, 1950
seeds	37.8	6.0	10.9	39.6	5.6	0.5	0.73				West, 1950
Acacia mellifera											
very young leaves	42.85	2.6	16.19	32.19	6.17	0.51	0.58				Dougall and Bogdan, 1958
Acacia nilotica											
pods	10.17	1.36	15.07	68.84	4.56						Wilson and Bredon, 1963
dry pods	9.50	1.11	14.94	70.03	4.42						Wilson and Bredon, 1963
green pods	11.87	0.85	16.19	66.13	4.96						Wilson and Bredon, 1963
dry pods	10.00	0.94	16.28	66.10	6.68						Wilson and Bredon, 1963
pods	12.4	2.7	15.6	63.2	5.1	0.51	0.14				West, 1950
pods	13.1	1.7	13.6	67.7	3.9	0.58	0.17				Everist, 1969
leaves	12.9	2.6	11.3	56.7	5.5						Everist, 1969
pods	10.44	0.82	19.76	62.43	6.55	0.53	0.16				Dougall and Bogdan, 1958
top branches, some leaves and flowers	11.92	2.18	21.35	59.68	5.47	0.87	0.13				Dougall and Bogdan, 1958
Acacia nubica											
leaves (Nov.)	32.45	1.73	15.76	41.21	8.85	1.71	0.44				Dougall and Bogdan, 1958
pods (Nov.)	15.17	1.91	37.42	38.04	7.46	1.46	0.23				Dougall and Bogdan, 1958
Acacia pendula											
leaves	13.2	3.4	29.6	46.2	7.8	0.36	0.08				Everist, 1969
Acacia polyacantha											
seeds (Sept.)	27.40	4.0	16.20	47.60	4.8						van Rensburg, 1968
pods (Sept.)	10.40	0.80	37.80	43.10	7.9						van Rensburg, 1968
leaves and twigs (Nov.)	15.50	1.54	16.88	60.46	5.62	0.78	0.21	0.28	1.73		van Rensburg, 1968

TABLE 5. **Edible trees and shrubs: crude chemical composition analysis, moisture-free** *(continued)*

Legume	Crude protein	Crude fat	Crude fibre	Nitrogen-free extract	Ash	Ca	P	Mg	K	Cl	Reference
					(%)						
Acacia senegal											
dry pods	19.65	2.14	29.64	43.26	5.31	1.40	1.78	0.09	0.73		Wilson and Bredon, 1963
green pods	22.03	0.96	39.02	30.93	7.06	1.14	2.45	0.45	2.29		Wilson and Bredon, 1963
whole	23.42	3.78	19.35	43.96	9.49	3.19	3.98	0.16	1.16		Wilson and Bredon, 1963
(kumat)	10.0		9.5	65.7	16.4	16.4	6.9	0.05	0.06		Ganguli, Kaul and Nambiar, 1964
Acacia seyal											
bark and branches											
(Aug.)	4.34	0.71	20.68	67.68	6.59	4.09	0.03				Dougall and Bogdan, 1958
(Feb.)	10.59	0.91	22.15	57.54	8.81	2.50	0.07				Dougall and Bogdan, 1958
Acacia sieberiana											
pods (Oct.)	11.62	1.43	40.24	42.69	4.00	0.10	0.14				van Rensburg, 1968
pods (Nov.)	8.30	2.10	27.80	57.00	4.80						van Rensburg, 1968
seed (Aug.)	18.70	4.30	13.90	59.20	3.90						van Rensburg, 1968
leaves and twigs	21.13	4.76	24.70	42.65	6.76	0.94	0.29	0.35	1.65		van Rensburg, 1968
pods	8.4	0.6	25.9	57.7	7.4						West, 1950
whole pods	10.3	1.0	24.8	51.9	3.4						Verboom, 1965
pod husks	9.2	0.7	27.8	50.1	3.9						Verboom, 1965
seed	18.5	2.8	13.4	52.6	4.1						Verboom, 1965
Acacia sparsiflora											
leaves	17.8	3.3	25.4	47.7	5.8	1.19	0.11				Everist, 1969
Acacia tortilis											
pods (Jan.)	17.79	1.74	17.50	54.60	8.37	1.34	0.36				Dougall and Bogdan, 1958
pods (Feb.)	10.44	0.82	19.76	62.43	6.55	0.53	0.16				Dougall and Bogdan, 1958
Adenodolichos rhomboides	14.13					0.40	0.17	0.24	1.40		Lawton, 1968
Albizia adianthifolia	29.13					0.20	0.35	0.20	1.66		Lawton, 1968

TABLE 5. **Edible trees and shrubs: crude chemical composition analysis, moisture-free** *(continued)*

Legume	Crude protein	Crude fat	Crude fibre	Nitrogen-free-extract	Ash	Ca	P	Mg	K	Cl	Reference
					(%)						
Albizia amara Feb.											
young branches & leaves	26.78	1.93	26.80	38.35	6.14	0.47	0.25				Dougall and Bogdan, 1958
flowers	26.85	1.73	21.44	42.96	7.02	1.01	0.28				Dougall and Bogdan, 1958
Albizia basaltica											
leaves	12.8	4.7	21.5	56.5	4.5	1.0	0.07				Everist, 1969
Albizia lebbek	29.2		25.3	43.8	7.5	1.8	0.2	0.5			Ganguli, Kaul and Nambiar, 1964
Albizia versicolor											
pods	12.60	1.5	47.40	31.80	6.7						van Rensburg, 1968
Baphia bequaertii											
coppice regrowth	22.38					0.17	0.31	0.28	1.07		Lawton, 1968
Baphia nitida											
leaves	21.2		26.8		5.4	0.85	0.11			0.74	Rose-Innes and Mabey, 1964
Bauhinia carronii											
leaves	13.7	1.6	13.5	64.8	6.3	0.88	0.18				Everist, 1969
young pods	11.7	2.7	25.6	55.5	4.5	0.79	0.21				Everist, 1969
mature pods	10.0	3.0	30.6	53.0	3.4	0.52	0.19				Everist, 1969
Bauhinia hookeri											
leaves	11.8	2.3	13.5	64.3	8.1	2.43	0.10				Everist, 1969
Brachystegia longifolia											
coppice regrowth	12.13					0.48	0.19	0.27	1.18		Lawton, 1968

TABLE 5. **Edible trees and shrubs: crude chemical composition analysis, moisture-free** *(continued)*

Legume	Crude protein	Crude fat	Crude fibre	Nitrogen-free-extract	Ash	Ca	P	Mg	K	Cl	Reference
	(%)										
Cajanus cajan											
before bloom	21.4	6.0	30.74	36.08	5.8	0.22	0.06				Axtmayer, Hernandez and Cook, 1938
before bloom	19.07	5.6	30.5	39.04	5.8	0.24	0.05				Otero, 1952
foliage	11.46	5.13	22.60	52.55	8.26						Otero, 1952
seeds	18.36	7.87	5.43	64.27	4.07						Otero, 1952
silage	16.42	9.6	28.4	35.7	6.9						Otero, 1952
hay	16.9	9.9	29.2	36.8	7.1						Otero, 1952
Cassia auriculata (tarward)	10.9		12.5	67.3	9.1	3.7	0.1	0.7			Ganguli, Kaul and Nambiar, 1964
Ceratonia siliqua											
pods	9.3	0.5	13.3	74.1	2.8	0.28	0.12				Everist, 1969
Cratylia floribunda	24.36	2.92	29.03	35.17	8.52	0.93	0.51				Otero, 1952
Cratylia mollis	19.30	2.92	9.36	67.39	1.03						Otero, 1952
Desmanthus virgatus											
leaves	22.40										Takahashi and Ripperton, 1949
stems	7.11										Takahashi and Ripperton, 1949
whole	12.80										Takahashi and Ripperton, 1949
Desmodium discolor											
before flowering	20.85	2.95	27.25	38.10	10.85						Compère, 1961
hay	17.04	2.44	36.10	35.00	9.42						Otero, 1952
	18.50	1.80	25.80	39.10	1.50						Otero, 1952
Desmodium molle	16.86	4.39	23.61	46.09	9.05						Otero, 1952
Desmodium pabulare	20.19	5.80	22.62	41.63	9.76						Otero, 1952

TABLE 5. Edible trees and shrubs: crude chemical composition analysis, moisture-free *(continued)*

Legume	Crude protein	Crude fat	Crude fibre	Nitrogen-free extract	Ash	Ca	P	Mg	K	Cl	Reference
					(%)						
Dichrostachys cinerea											
pods	11.40	1.20	25.60	56.40	5.50	0.51	0.20				West, 1950
leaves and shoots (Nov.)	15.00	2.08	21.16	53.43	8.33	1.53	0.18	0.46	1.22		van Rensburg, 1968
leaves	15.33	1.23	28.09	50.46	4.89						Wilson and Bredon, 1963
Eriosema engleranum	18.50					0.24	0.18	0.29	1.74		Lawton, 1968
Gleditsia triacanthos											
pods	14.30	1.90	18.40	60.80	4.70						Everist, 1969
Griffonia simplicifolia											
leaves	15.70	29.30			8.50	2.20	0.12		0.02		Rcse-Innes and Mabey, 1964
Indigofera arrecta											
young shoots	16.38	2.36	12.18	57.30	11.78	2.52	0.29	0.85	1.57		van Rensburg, 1968
Julbernadia paniculata	12.38		16.40			0.70	0.24	8.25	1.50		Lawton, 1968
Leucaena leucocephala											
meal	12.50		12.40		10.60						Farinas, 1951
	24.40	6.40	9.60								Gantt, 1953
	18.50		27.40								Loosli, Villegas and Ynalyez, 1954
	15.50										Oakes and Skov, 1967
	28.80										Hutton and Bonner, 1960
	27.50		16.40								van Rensburg, 1960
leaves	26.70										Takahashi and Ripperton, 1949
stem	9.10										Takahashi and Ripperton, 1949
whole	18.10										Takahashi and Ripperton, 1949
young shoots	19.75	5.07	16.06	49.26	9.86	2.20	0.08	0.58	1.27		van Rensburg, 1968
leaves	20.00	6.50	14.10	48.3	11.20						Everist, 1969
green tips	18.80	2.30	37.70	34.5	6.60						Everist, 1969
						0.57					Vergara, 1967
						0.97	0.26	0.32		Na0.02	Andrew and Robins, 1969b

671

TABLE 5. Edible trees and shrubs: crude chemical composition analysis, moisture-free *(concluded)*

Legume	Crude protein	Crude fat	Crude fibre	Nitrogen-free-extract	Ash	Ca	P	Mg	K	Cl	Reference
	(%)										
Millettia thonningii leaves	19.00		29.50		8.80	1.42	0.10			0.18	Rose-Innes and Mabey, 1964
Parkinsonia aculeata leaves	13.14	1.90	24.90	47.9	11.90	2.85	0.16				Everist, 1969
Poliostigma thonningii pods	6.80	2.40	23.7	63.80	4.90						West, 1950
whole pods	6.40	2.40	24.0	59.40	3.30						Verboom, 1965
pod husks	4.90	2.70	27.5	57.10	3.70						Verboom, 1965
seeds	20.30	2.20	10.0	58.20	2.80						Verboom, 1965
whole pods (Aug.)	5.10	0.90	31.70	57.40	4.90						Verboom, 1965
green leaves & twigs (Nov.)	9.88	1.85	22.15	59.06	7.06	0.70	0.23	0.20	2.45		van Rensburg, 1968
Prosopis juliflora meal	21.40		20.80	50.00	7.70	1.50	0.20	0.50			Ganguli, Kaul and Nambiar, 1964
	6.00		48.00	35.00	0.30	ppm carotene					Marion, Fisher and Robinson, 1957
Prosopis spicigera	13.90		20.30	59.20	6.50	1.90	0.20	0.50			Marion, Fisher and Robinson, 1957
Tamarindus indica leaves	13.50	6.80	18.10	52.00	9.80	3.62	0.52				Patel and Patel, 1957
leaves, summer (March-June)	13.00	7.80	16.20	53.30	9.60	2.98	0.56				Patel and Patel, 1957
leaves, monsoon (July-October)	14.00	6.00	21.00	49.70	9.20	2.56	0.49				Patel and Patel, 1957
leaves, winter (Nov.-Feb.)	13.50	7.10	17.20	53.10	9.30	3.48	0.59				Patel and Patel, 1957

Appendix 2
Sources of *Rhizobium* cultures
for tropical legumes

This list of names and addresses of organizations handling inoculants is necessarily incomplete, but supplies information as to where inoculants can be obtained.

As more effective strains are developed, the present ones will be superseded: an up-to-date list is published from time to time in the *Rhizobium Newsletter*. Copies may be obtained by writing directly to the editors:

A.H. Gibson, G.L. Turner,
   J. Brockwell
Microbiology Section
PO Box 1600
Canberra City
ACT 2601 Australia

R.A. Date
CSIRO Division of Tropical Crops
   and Pastures
The Cunningham Laboratory
Mill Road
St Lucia (Brisbane), Queensland 4067
Australia

AUSTRALIA

CSIRO Division of Tropical Crops
   and Pastures
The Cunningham Laboratory
Mill Road
St Lucia (Brisbane)
Queensland 4067

Australian Mercantile, Land and
   Finance Co. Ltd

Australian Estates Ltd
PO Box 598
Brisbane, Queensland 4001
(cultures sold under the trade name
   Tropical Inoculants)

R. Roughley
Biological and Chemical Research
   Institute
PMB 10
Tydalmere
New South Wales 2116

673

D.J. Pulsford
Agricultural Laboratories
PO Box 8
New South Wales 2143
(cultures sold under the trade name
Noculaid)

R. Daniels
Root Nodule Pty Ltd
PO Box 188
Epping, New South Wales 2121
(cultures sold under the trade names
Nitrogerm and Nitrofix)

G. Bullard
Horticultural Research Station
PO Box 720
Gosford, New South Wales 2250

Arthur Yates and Co Pty Ltd
PO Box 117
Rockhampton, Queensland 4700

The Head of the Department of Soil
Service and Plant Nutrition
Institute of Agriculture
University of Western Australia
Nedlands, Western Australia 6009

BRAZIL

J.R. Jardim Freire, Ing. Agr.
IPAGRO
CP 776
90 000 Porto Alegre, Rio Grande do
Sul

Instituto de Biología y Pesquisas
Tecnológicas

CP 357
Curitiba, Paraná

Leivas Leite S.A. Indústrias
Químicas e Biólogicas
CP 91
Pelotas, Rio Grande do Sul

or Rua Cel. Vicente, 156
Porto Alegre, Rio Grande do Sul

or Rua Monsenhor Anacleto, 86
São Paulo, S.P.

CANADA

D.C. Jordan
University of Guelph
Department of Microbiology
Guelph, Ontario N1G 2W1

CHILE

Luís S. Longeri
Universidad de Concepción
Dep. Microbiología
Casilla 272, Concepción
(cultures sold under the trade name
Nitrofix)

CZECHOSLOVAKIA

Central Institute for Plant Production Research
Microbiology Department
16106 Prague 6, Ruzyne 507

674

EGYPT

Dr Farouk M. Hammouda
Sakha Agricultural Research Station
Kafr, El Sheik

FRANCE

Lipha
R. Buquet
115, av. Lacassagne
69212 Lyon Cedex 1
(cultures sold under the trade name
    Biodoz)

M. Obaton
Laboratoire de recherches sur les
    symbiotes des racines
Institut national de la recherche
    agronomique (INRA)
9, place Viala
34060 Montpellier

N. Amarger
Laboratoire de microbiologie des
    sols
INRA
17, rue Sully
BP 1540
21034 Dijon

Institut de recherches agronomiques
    tropicales et des cultures vivrières
Laboratoire de rhizobiologie
IRAT/CERDAT - BP 5035
34032 Montpellier

Bioprox
6, rue Barbès
92305 Levallois-Perret

INDIA

Bacfil Inoculants Pvt. Ltd
23 Nawal Kishore Road
Lucknow, Uttar Pradesh

International Crops Research Insti-
    tute for the Semi-Arid Tropics
    (ICRISAT)
Hyderabad, Andhra Pradesh

Rajasthan Bio Fertilizer Industries
14/110 Market No. 3
Bhopalganj
Bhilwara - 311 001, Rajasthan

KENYA

Kenya Seed Company Ltd
PO Box 553
Kitale
Telex: 38646

S.O. Keya
Rhizobium MIRCEN
Department of Soil Science and
    Botany
University of Nairobi
PO Box 30197
Nairobi

675

MEXICO

C.C. Casas
Escuela Nacional de Ciencias
    Biológicas
IPN Apartado Postal 42-186
Mexico, D.F.
(cultures sold under the trade names
    Nitragin and Pagador)

NEW ZEALAND

Department of Scientific and Indust-
    rial Research (DSIR)
Plant Diseases Division
Private Bag, Auckland

NIGERIA

A. Ayanaba
International Institute of Tropical
    Agriculture (IIIA)
Oyo Road PMB 5320
Ibadan

PAPUA NEW GUINEA

The Director
Departement of Agriculture, Stock
    and Fisheries
Konedobu

PERU

Ing. Agric. Rodolfo Vargas
Estación Experimental Agrícola de
    La Molina

Apartado 2791, Lima
(American cultures distributed
    under the trade names Nodogen
    and Nitragin)

ROMANIA

C. Hera
Institut de recherche pour les plantes
    industrielles
Département de microbiologie
    Fundulea Jud. Calarasi

SENEGAL

Y. Dommergues/B. Dreyfus
Laboratoire de biologie des sols
Centre ORSTOM de Dakar
BP1386
Dakar

UNITED STATES

Plant Cultures
PO Box 284
Gainesville, Florida

U.M. Means
U.S. Soils Laboratory
U.S. Department of Agriculture
Beltsville, Maryland 20705

676

D.F. Weber/M.H. Keyser
Agricultural Research, Northeastern Region
Beltsville Agricultural Research Center
Cell Culture and Nitrogen Fixation Laboratory
Beltsville, Maryland 20705

A.A. Henrickson
The Nodogen Laboratories
PO Box 170
Princetown, Illinois
(cultures sold under the trade names Nodogen and Nodogen's Mr Fixo)

The Nitragin Co. Inc.
PO Box 09186
Milwaukee, Wisconsin 53209
(cultures sold under the trade name Nitragin)

University of Hawaii
College of Tropical Agriculture and Human Resources
PO Box "C"
Paia, Hawaii 96779

URUGUAY

C. Labandera
ALAR
Bulevar Artigas 3802
Montevideo

BENIN

Centre de recherches agronomiques
IRAT
Niaouli (par Altogon)

BURKINA FASO

IRAT
PO Box 596
Saria
(par Ouagadougou)

BRAZIL

Brazilsul Agropecuária Ltda
CP 1457
Porto Alegre
Rio Grande do Sul

Contibrasil Sementes Ltda
CP 81
14140 Cravinhos, São Paulo
Telex: 0166109

Departamento de Produçao Animal
Secretaria de Agricultura
São Paulo, S.P.

Estaçao Experimental de Forrages
São Gabriel, Rio Grande do Sul

Instituto de Pesquisas IRI
Campinas, São Paulo

Instituto Pesquisas Experimentos
Agropecuario Centro-Sul
Seccào de Agrostologia
Caixa Postal 28, ZC-00
Rio de Janeiro

Instituto de Zootecnica
Universidad de Cearà
Fortaleza, Cearà

COLOMBIA

Centro Nacional de Investigaciones
   Agropecuarias
"Diulio Ospina"
Medellín

Gómez Guillermo
Cra. 10 N. 11-48
Of. 201
Cali

Muñoz Tomás E.
San Diego (César)

Programa de Forrajes
Instituto Colombiano Agropecuario
   (ICA)
Bogotá

Prosecar Ltda
Calle 6 Carrera 31
Apartado Aéreo 063
Aguachica (César)

Prosemillas
Calle 8 N° 11-15
Armero (Tolima)

Semillano Ltda
Cra. 33 N° 19-63
Apartado Aéreo 2012
Villavicencio

Sociedad Agrícola Alto de Gualan-
day
Apartado Aéreo 1165

COSTA RICA

"El Semillero"
Luis Cruz B.
Apartado 783
San José

Central Agrícola de Cartago S.A.
Apartado Aéreo 38
Cartago

Instituto Interamericano de Ciencias
Agrícolas (ILCA)
Centro de Investigación y Enseñanza
Graduada
Turrialba

CÔTE D'IVOIRE

Directeur des recherches
Office de la recherche scientifique
d'Outre-Mer
BP Adiopodoumé, Abidjan

Institut d'enseignement et de
recherches tropicales
Abidjan

CUBA

Estación Experimental de Pastos y
Forrajes
"Indio Huatey"
Matanzas

GHANA

Plant Introduction and Exploration
Crop Research Institute
Ghana Academy of Sciences
Burso

GUYANA

Botanic Gardens
PO Box 256
Georgetown 7A

Central Agricultural Station
Department of Agriculture
Georgetown

INDIA

Agricultural College
Poona, Maharashtra

Agricultural Research Institute
Chotagnapar (Ranchi)
Bihar

Agricultural Research Institute
Coimbatore 3, Tamil Nadu

681

Agricultural Research Service
Tindivanam (South Arcot)
Tamil Nadu

BAI
Bangalore, Karnataka

Bangalore Nursery & Gardens
4 Krumbiegal Road
Bangalore 560 004, Karnataka

Central Arid Zone Research
    Institute
Jodhpur
Rajasthan

Deochanda Experiment Station
Deochanda

Division of Botany
Indian Agricultural Research
    Institute
New Delhi

Economic Botanist
Institute for Agricultural Research
Jaipur, Rajasthan

Indian Veterinary Research Institute
PO, Izantnagar 243122
Uttar Pradesh

Government Economic Botanist
Kanpur
Uttar Pradesh

National Seeds Corporation Ltd
F. 44 South Extension Part II
Ring Road New Delhi, 3

Tamil Nadu Agriculture Depart-
    ment
Madras, Tamil Nadu

Pratap Nursery Seed Stores
PO Premnagar
Dehra Dun 6
Uttar Pradesh

Shri Takhatmal Parasmal Jain
Tilak Road
Nandurbar (Dhulia) - 425 412
Maharashtra

INDONESIA

Institute for Agricultural Research
Bogor

ISRAEL

Hazera Seed Co
PO Box 1565
Haifa

Plant Introduction Service
The Volcani Institute of Agricultural
    Research
PO Box 6, Beit Daga

KENYA

Kenya Seed Company Ltd
PO Box 553
Kitale
Telex: 38646

682

Pasture Research Unit
National Agricultural Research
 Station
PO Box 450, Kitale

MADAGASCAR

Ferme d'Etat Omby
BP 911
Antananarivo

MALAWI

Agricultural Research Centre
PO Box 215
Lilongo

Chitedze Agricultural Research
 Station
PO Box 158
Lilongwe

Makanga Experiment Station
PO Box 20
Chirono

National Seed Co. of Malawi Ltd
PO Box 30050
Lilongwe 3

MALAYSIA

Pusat Penyelidikan Getan Tanan
 Malaysia
Petit Surat 150
Kuala Lumpur

MEXICO

Agricultores Unidos de Chihuahua
 S.A.
AP 522
CD Juárez, Chihuahua

Horizon de México S.A.
AP 474
Matamoros, Tamps.

Semillas Internacionales de Bajiío
Hidalgo 404
Salamanca, Gto.

Semillas y Agroquímicos Macor
Blvd. Zapata 2226
Culiacán, Sin.

Semisol S.A.
Calle Bustamante y 17 157
Matamoros, Tamps.

NIGERIA

Institute for Agricultural Research
Ahmadu Bello University
PO Box 116, Samaru
Sheká Baria

PAKISTAN

Agricultural College
Karachi

Ayub Agricultural Research Institute
Lyallpur

683

Plant Introduction Officer
FACP, Karachi

Punjab Agricultural College
Lyallpur

Tandojam Agricultural College
Tandojam

PANAMA

John Fraser
Edif. Cía. Panameña de Seguros
3er. piso
Calle Ricardo Arias
Apartado 4546
Panamá 5

PAPUA NEW GUINEA

Director of Agriculture
Department of Agriculture
Port Moresby

PARAGUAY

Instituto Agronómico Nacional
Ruta N° 2, Km. 18
Caacupé

PERU

Agro Perú
28 de Julio 1837
Lima

Carlesi
Av. Nicolás Arriola 2400
Lima

Casa Flor
Prolong. Italia 1708
C-7-Avia C.
Lima

Manrique
Av. Aviación 379
Lima

Universidad Agraria la Molina
Lima

PUERTO RICO

Río Piedras Estación Experimental
Mayagüez

PHILIPPINES

Bureau of Plant Industry
Dept of Agriculture and National
    Research
Manila

College of Agriculture
University of the Philippines
Manila

SINGAPORE

Hooglandt & Co
PO Box 245
Singapore 9040
Telex: 21523

Inland & Foreign Trading Co. Ltd
PO Box 2098
Singapore 9040
Telex: 25254

SRI LANKA

Agriculture Department
Peradeniya

Anderson and Co.
Colombo

H.D. Constantine and Sons
GPO Box 378
Colombo

R.R. Johnson and Co.
Colombo

SUDAN

Agricultural Research Division
Wad Medani

Range and Pasture Research Dept
Ministry of Agriculture
Khartoum

SWAZILAND

Veld and Pasture Officer
Agricultural Research Station
PO Box 4
Malkerns

THAILAND

Department of Agriculture
Srisomrong Street, Bangkok

Department of Botany
Kasetsart University
Bangkok

Pakchong Forage Crops Station
Korat

Rubber Research Institute
Haadyai

UGANDA

Makerere University College
PO Box 7062, Kampala

UNITED STATES

Agronomy and Soil Department
University of Hawaii
Honolulu, Hawaii

Bingham Seed Co.
2940 West 45 St.
Jacksonville, Florida 32201

College of Tropical Agriculture
University of Hawaii
Honolulu, Hawaii

Dixie Seed Co.
PO Drawer 10
Ochlocknee, Georgia 31773

Everglades Branch Station
Fort Lauderdale, Florida

Florida Foundation Seed Producers Inc.
Florida Agricultural Experiment Station
Gainesville, Florida

Georgia Experiment Station
Tifton, Georgia

Indian River Research Station
Fort Pierce, Florida

Hawaii Agricultural Experiment Station, RRI
Kapaa, Hawaii

Kalmbach-Buckett Co. Inc.
Shreveport, Louisiana

Louisville Seed Co.
PO Box 120
Louisville, Kentucky 40201

Segrest Feed and Seed Co.
PO Box 338
Slocomb, Alabama

USDA Crop Research Branch
Plant Introduction
Beltsville, Maryland

UNITED STATES VIRGIN ISLANDS

Kingshill St. Croix
Agricultural Institute
Virgin Islands

VENEZUELA

Centro de Investigaciones Agronómicas
Maracay

Servicio de Pastos y Forrajes
Facultad de Agronomía
Universidad del Zulia
Maracaibo

ZAIRE

Centre de recherche agronomique
Yangambi
Kinshasa

ZAMBIA

Mount Makulu Research Station
PO Box 7, Chilanga

National Agricultural Marketing Board
PO Box 30122, Lusaka

Pasture Research Officer
Nisamfu Research Station
PO Box 55
Kasama

ZIMBABWE

Farm Seeds
PO Box 653
Harare

686

Appendix 4
Common names of tropical legumes

Compiled by Camille Trentacoste

a kwakwa - *Vigna luteola*
acacia - *Albizia lebbek*
acacia bella rosa - *Leucaena leucocephala*
acacia negra - *Gleditsia triacanthos*
acacia tree - *Albizia lebbek*
alfalfa de Townsville - *Stylosanthes humilis*
alfalfa do Brasil - *Stylosanthes guianensis* cv. Schofield
alfalfa do nordeste - *Stylosanthes guianensis* cv. Schofield
alfalfa Paulista - *Teramnus uncinatus*
alfalfa selvagem - *Stylosanthes humilis*
algaroba - *Prosopis chilensis, P. pallida*
algarrobo - *Samanea saman*
algarrobo blanco - *Prosopis chilensis*
algarrobo de Chile - *Prosopis chilensis*
Alice clover - *Alysicarpus vaginalis*
amendoim de veado - *Teramnus uncinatus*
American joint vetch - *Aeschynomene americana*
amor de campo - *Desmodium triflorum*
amor seco - *Desmodium intortum*

amorsinho secco - *Desmodium triflorum*
Angola pea - *Cajanus cajan*
antejuela - *Aeschynomene americana*
arb - *Arachis glabrata*
Archer axillaris - *Macrotyloma axillare*
arhar - *Cajanus cajan*
aroma blanco - *Leucaena leucocephala*
aromo - *Acacia farnesiana*
artillery plant - *Cassia mimosoides*
arveja - *Cajanus cajan*
atro - *Macroptilium atropurpureum*
Australian shamrock - *Trigonella suavissima*
babul - *Acacia nilotica*
ban baburi - *Acacia farnesiana*
barba de burro - *Zornia diphylla*
barbadinho - *Desmodium barbatum*
bastard sensitive plant - *Aeschynomene americana*
bauhinia - *Bauhinia* sp.
bejuco de chivo - *Centrosema pubescens*
bejuco de conchitas - *Clitoria ternatea*
belalie - *Acacia stenophylla*
bendee - *Acacia catenulata*

687

Bengal bean - *Mucuna pruriens*
black gidgea - *Acacia argyrodendron*
black locust - *Robinia pseudoacacia*
black mattle - *Acacia salicina*
black sally wattle - *Acacia salicina*
black wattle - *Acacia stenophylla*
bluebell - *Centrosema virginianum*
bohemia - *Bauhinia* sp.
booyakka - *Acacia* sp.
boree - *Acacia cana*
bowyakka - *Acacia* sp.
Brazilian lucernc - *Stylosanthes guianensis* cv. Schofield
brigalow - *Acacia harpophylla*
budda pea - *Aeschynomene indica*
bullamon lucerne - *Psoralea eriantha, P. patens*
Burchell's clover - *Trifolium burchellianum*
burn mouth vine - *Rhynchosia minima*
butterfly pea - *Centrosema pubescens, Clitoria ternatea*
caballo - *Desmodium biarticulatum*
calopo - *Calopogonium mucunoides*
camel thorn - *Acacia giraffae*
campanilla - *Clitoria ternatea, Centrosema pubescens*
camwood - *Baphia nitida*
Cape clover - *Trifolium burchellianum*
Capitan Juan - *Stylosanthes scabra*
Caribbean stylo - *Stylosanthes hamata*
carpon desmodium - *Desmodium heterocarpon*
carrapicho - *Desmodium triflorum*
catjang - *Cajanus cajan*
centro - *Centrosema pubescens*

centurion - *Centrosema pascuorum*
channel clover - *Trigonella suavissima*
chickasro lima bean - *Canavalia ensiformis*
cina-cina - *Parkinsonia aculeata*
cooba - *Acacia salicina*
coolan - *Acacia salicina*
Cooper clover - *Trigonella suavissima*
cordwood wattle - *Acacia bidwillii*
corkwood wattle - *Acacia sutherlandii*
cowpea - *Vigna unguiculata*
creeping beggar weed - *Desmodium canum*
creeping vigna - *Vigna parkeri*
cresta de gallo - *Sesbania grandiflora*
cujicillo - *Aeschynomene americana*
cumandai birai - *Cajanus cajan*
Cunningham's wattle - *Acacia cunninghamii*
currawong - *Acacia burrowii*
dahl - *Cajanus cajan*
Dalrymple vigna - *Vigna luteola*
Darling clover - *Trigonella suavissima*
Darwin - *Desmodium biarticulatum*
Darwin engordo caballo - *Desmodium biarticulatum*
dead finish - *Acacia tetragonophylla, Albizia basaltica*
desert oak - *Acacia coriacea*
desmanthus - *Desmanthus virgatus*
desmodium - *Desmodium heterophyllum*
desmodium ovalifolium - *Desmodium heterocarpon* var. *ovalifolium*

688

discoloured clover - *Desmodium discolor*
dolichos lab-lab - *Lablab purpureus*
dormilonga - *Aeschynomene americana*
driedoring - *Gleditsia triacanthos*
dunthy - *Acacia stenophylla*
dwarf koa - *Desmanthus virgatus*
Ellington curse - *Acacia farnesiana*
feijao de porco - *Canavalia ensiformis*
five-leaf cassia - *Cassia mimosoides*
Florida beggarweed - *Desmodium tortuosum*
French tamarind - *Samanea saman*
frijol de arbol - *Cajanus cajan*
frijol de los arrozales - *Macroptilium lathyroides*
frijol de monte - *Macroptilium lathyroides*
frijol terciopelo - *Mucuna pruriens*
frijolillo - *Galactia striata*
frijolito - *Teramnus volubilis*
fundo-fundo - *Neonotonia wightii*
galactia - *Galactia striata*
gallito - *Sesbania grandiflora*
Georgina gidgea - *Acacia georginae*
gidga - *Acacia cambagei*
gidgea - *Acacia cambagei*
gidgee - *Acacia cambagei*
gidya - *Acacia cambagei*
gidyea - *Acacia cambagei*
gilibande - *Vigna luteola*
gliricidia - *Gliricidia sepium*
glycine - *Neonotonia wightii*
goko - *Vigna luteola*
green gram - *Vigna radiata*
green wattle - *Acacia deanei*

greenleaf desmodium - *Desmodium intortum*
guando - *Cajanus cajan*
guandul - *Cajanus cajan*
guango - *Samanea saman*
gum-arabic tree - *Acacia senegal*
gundabluey - *Acacia victoriae*
gundabluie - *Acacia victoriae*
haakdoring - *Acacia litakunensis*
haak-en-steek - *Acacia litakunensis*
hairy indigo - *Indigofera hirsuta*
hashab - *Acacia senegal*
hediondilla - *Leucaena leucocephala*
hetero - *Desmodium heterophyllum*
hickory wattle - *Acacia aulacocarpa*
hierba cuartillo - *Desmodium triflorum*
hila-hila - *Mimosa pudica*
Hindu pujali - *Desmodium triflorum*
honey locust - *Gleditsia triacanthos*
horse gram - *Canavalia ensiformis*
horse marmalade - *Desmodium discolor*
horse vine - *Teramnus labialis*
huisachee - *Acacia farnesiana*
Indian siris - *Albizia lebbek*
indolo - *Vigna luteola*
ipil-ipil - *Leucaena leucocephala*
ironwood - *Acacia excelsa*
ironwood wattle - *Acacia excelsa*
jack bean - *Canavalia ensiformis*
Jerusalem thorn - *Parkinsonia aculeata*
jetirana - *Centrosema pubescens*
joint vetch - *Aeschynomene falcata*
jongli-math - *Vigna trilobata*
jumbie bean - *Leucaena leucocephala*
kachan - *Cajanus cajan*

kaimi clover - *Desmodium canum*
kakada - *Dichrostachys cinerea*
kavuhivahi - *Vigna luteola*
Kenya white clover - *Trifolium semipilosum*
kharroub - *Bauhinia rufescens*
kiawe - *Prosopis chilensis*
kitr - *Acacia mellifera*
klein kau - *Calopogonium caeruleum*
klu - *Acacia farnesiana*
koa - *Acacia koa*
koa haole - *Leucaena leucocephala*
koemataballi - *Zornia latifolia*
Kordofan pea - *Clitoria ternatea*
kudzu - *Pueraria thunbergiana*
kuru vine - *Desmodium intortum*
lancewood - *Acacia doratoxylon, A. shirleyi*
laot - *Acacia nubica*
lead tree - *Leucaena leucocephala*
least rhynchosia - *Rhynchosia minima*
Leichhardt biflorus - *Macrotyloma uniflorum*
leucaena - *Leucaena leucocephala*
lotononis - *Lotononis bainesii*
low mulga - *Acacia aneura*
magsaysay lucerne - *Stylosanthes humilis*
Maloga bean - *Vigna lanceolata*
mangeraçiao do campo - *Stylosanthes guianensis* cv. Schofield
marmelada de cavalo - *Desmodium discolor*
masheke - *Vigna luteola*
mata-ratón - *Gliricidia sepium, Albizia lebbek*
Mendinee clover - *Trigonella suavissima*

mesquite - *Prosopis glandulosa, P. juliflora*
mgunga - *Acacia nilotica*
midger - *Acacia oswaldii*
Miles lotononis - *Lotononis bainesii*
mimosa bush - *Acacia farnesiana*
mineritchie - *Acacia cyperophylla*
minnaritchie - *Acacia cyperophylla*
monkey pod - *Samanea saman*
mother segal - *Stylosanthes hamata*
mugulula - *Vigna luteola*
mukni - *Vigna trilobata*
mulga - *Acacia aneura*
mung bean - *Vigna radiata*
munkolo - *Rhynchosia sublobata*
Murray phasey bean - *Macroptilium lathyroides*
myall - *Acacia pendula*
namie napirang - *Calopogonium caeruleum*
native verbine - *Psoralea australasica*
needle bush - *Acacia farnesiana*
nelia - *Acacia oswaldii*
ol'erbat - *Acacia nilotica*
one-eye bean - *Canavalia ensiformis*
opoponax - *Acacia farnesiana*
overlock - *Canavalia ensiformis*
Oxley fine-stem stylo - *Stylosanthes guianensis* var. *intermedia* cv. Oxley
pacpaclanhao - *Desmodium triflorum*
papito - *Clitoria ternatea*
parkinsonia - *Parkinsonia aculeata*
pata de terecay - *Stylosanthes scabra*
pea glycine - *Glycine tabacina*
pega pega - *Aeschynomene americana, Desmodium canum, D. intortum*

pega ropa - *Aeschynomene americana*
pegunny - *Bauhinia* sp.
pencil flower - *Stylosanthes hamata*
perennial soybean - *Neonotonia wightii*
phasey bean - *Macroptilium lathyroides*
phillipesara - *Vigna trilobata*
pica-pica - *Mucuna pruriens*
pico de flamenco - *Sesbania grandiflora*
pigeon pea - *Cajanus cajan*
pois d'angole - *Cajanus cajan*
pokindang - *Clitoria ternatea*
polymorphous clover - *Trifolium polymorphum*
poor man's friend - *Stylosanthes viscosa*
poroto guandul - *Cajanus cajan*
poroto paraguayo - *Cajanus cajan*
prickly acacia - *Acacia nilotica*
puero - *Pueraria phaseoloides*
Puerto Rican bean - *Cajanus cajan*
quinchoncho - *Cajanus cajan*
rabbit vine - *Teramnus labialis*
rabo de iguana - *Calopogonium mucunoides*
rain tree - *Samanea saman*
ram's-horn - *Acacia oswaldii*
red gram - *Cajanus cajan*
rhynchosia - *Rhynchosia minima*
river myall - *Acacia stenophylla*
ronte - *Aeschynomene americana*
round-leaf cassia - *Cassia rotundifolia*
Rueppell's clover - *Trifolium rueppellianum*

saca-estrepe - *Stylosanthes guianensis* cv. Schofield
sacha-cafe - *Cajanus cajan*
samaan - *Samanea saman*
saman - *Samanea saman*
sand witch - *Desmodium sandwicense*
Schofield stylo - *Stylosanthes guianensis* cv. Schofield
sebeldit - *Acacia gerrardii*
senivakacegu - *Desmodium heterophyllum*
sensitive plant - *Mimosa pudica*
shrubby stylo - *Stylosanthes scabra*
sien-sien - *Mimosa pudica*
silverleaf desmodium - *Desmodium uncinatum*
silverleaf Spanish clover - *Desmodium uncinatum*
siratro - *Macroptilium atropurpureum*
slender sweetroot - *Glycine tabacina*
soja perene - *Neonotonia wightii*
southern pea - *Vigna unguiculata*
Spanish clover - *Desmodium sandwicense*
spreading scurf pea - *Psoralea australasica*
sprinkaan boom - *Gleditsia triacanthos*
sticky stylo - *Stylosanthes viscosa*
stylo - *Stylosanthes guianensis*
sweet fenugreek - *Trigonella suavissima*
sword bean - *Canavalia ensiformis*
tall mulga - *Acacia aneura*
tamarind - *Tamarindus indica*
tan-tan - *Leucaena leucocephala*

691

tarbadillo - *Stylosanthes guianensis* cv. Schofield
tebeneque - *Stylosanthes hamata*
Temben's clover - *Trifolium tembense*
tencilla - *Zornia latifolia*
thornless mimosa - *Aeschynomene americana*
tibit tree - *Albizia lebbek*
Tinaroo glycine - *Neonotonia wightii* cv. Tinaroo
toshimbo shimbo - *Vigna luteola*
Townsville lucerne - *Stylosanthes humilis*
Townsville stylo - *Stylosanthes humilis*
trencilla - *Zornia diphylla*
trevinho de campo - *Desmodium triflorum*
trifolio - *Stylosanthes guianensis* cv. Schofield
tropical kudzu - *Pueraria phaseoloides*
tropical lucerne - *Stylosanthes guianensis* cv. Schofield
tur - *Cajanus cajan*
turpentine mulga - *Acacia brachystachya*
umbrella mulga - *Acacia aneura*
umbrella thorn - *Acacia litakunensis*
umung - *Acacia salicina*

vaivai - *Leucaena leucocephala*
vaivai vakvotona - *Acacia farnesiana*
variable glycine - *Glycine tabacina*
velvet bean - *Mucuna pruriens*
vigna - *Vigna hosei*
Virginian centro - *Centrosema virginianum*
waddy wood - *Acacia peuce*
wakutu - *Desmodium heterophyllum*
whipstick mulga - *Acacia aneura*
white thorn - *Acacia sieberiana*
wild blue vine - *Centrosema virginianum*
wild kulthi - *Atylosia scarabaeoides*
wild lucerne - *Stylosanthes humilis*
wild pea - *Centrosema virginianum*
wild pea bean - *Macroptilium lathyroides*
wild tamarind - *Leucaena leucocephala*
willow - *Acacia salicina*
woman's tongue tree - *Albizia lebbek*
woolly glycine - *Glycine tomentella*
woolly sweetroot vine - *Glycine tomentella*
woolly-flowered scurf pea - *Psoralea eriantha, P. patens*
yarran - *Acacia homalophylla*
zapatillo de la reina - *Clitoria ternatea*
zornia - *Zornia diphylla*

● **ALGÉRIE**
ENAMEP – Entreprise nationale des messageries de presse
47, rue Didouche Mourad, Alger.

● **ANGOLA**
Empresa Nacional do Disco e de Publicaçoes, ENDIPU-U.E.E.
Rua Cirilo de Conceiçao Silva, No. 7, C.P. No. 1314-C Luanda.

● **ARGENTINA**
Libreria Agropecuaria S.A.
Pasteur 743, 1028 Buenos Aires.

● **AUSTRALIA**
Hunter Publications
58A Gipps Street, Collingwood, Vic. 3066.

● **AUSTRIA**
Gerold & Co.
Graben 31, 1011 Vienna.

● **BAHRAIN**
United Schools International
PO Box 726, Manama.

● **BANGLADESH**
Association of Development Agencies in Bangladesh
1/3 Block F, Lalmatia, Dhaka 1209.

● **BELGIQUE**
M. J. De Lannoy
202, avenue du Roi, 1060 Bruxelles. CCP 000-0808993-13.

● **BOLIVIA**
Los Amigos del Libro
Perú 3712, Casilla 450, Cochabamba; Mercado 1315, La Paz.

● **BOTSWANA**
Botsalo Books (Pty) Ltd
PO Box 1532, Gaborone.

● **BRAZIL**
Fundação Getulio Vargas
Praia de Botafogo 190, C.P. 9052, Rio de Janeiro.
Livraria Canuto Ltda
Rua Consolação, 348 – 2° andar, Caixa Postal 19198, São Paulo.

● **BRUNEI-DARUSSALAM**
SST Trading Sdn. Bhd.
Bangunan Tekno No. 385, Jln 5/59, PO Box 227, Petaling Jaya, Selangor.

● **CANADA**
Renouf Publishing Co. Ltd
1294 Algoma Road, Ottawa, Ont. K1B 3W8.
Editions Renouf Ltée
route Transcanadienne, Suite 305, St-Laurent (Montrèal), Qué.
Toll free calls: Ontario, Quebec and Maritime – 1-800-267-1805; Western Provinces and Newfoundland – 1-800-267-1826.
Head Office/Siège social: 1294 Algoma Road, Ottawa, Ont.

● **CHILE**
Libreria – Oficina Regional FAO
Avda. Santa Maria 6700, Casilla 10095, Santiago.
Teléfono: 228-80-56.

● **CHINA**
China National Publications Import Corporation
PO Box 88, Beijing.

● **CONGO**
Office national des librairies populaires
B.P. 577, Brazzaville.

● **COSTA RICA**
Libreria, Imprenta y Litografia Lehmann S.A.
Apartado 10011, San José.

● **CUBA**
Ediciones Cubanas, Empresa de Comercio Exterior de Publicaciones
Obispo 461, Apartado 605, La Habana.

● **CYPRUS**
MAM
PO Box 1722, Nicosia.

● **CZECHOSLOVAKIA**
ARTIA
Ve Smeckach 30, PO Box 790, 111 27 Prague 1.

● **DENMARK**
Munksgaard Export and Subscription Service
35 Nørre Søgade, DK 1370 Copenhagen K.

● **ECUADOR**
Libri Mundi, Libreria Internacional
Juan León Mera 851, Apartado Postal 3029, Quito.
Su Libreria Cia. Ltda.
Garcia Moreno 1172 y Mejia, Apartado Postal 2556, Quito.

● **EL SALVADOR**
Libreria Cultural Salvadoreña, S.A. de C.V.
7ª Avenida Norte 121, Apartado Postal 2296, San Salvador.

● **ESPAÑA**
Mundi-Prensa Libros S.A
Castelló 37, 28001 Madrid.
Libreria Agricola
Fernando VI 2, 28004 Madrid.

● **FINLAND**
Akateeminen Kirjakauppa
PO Box 128, 00101 Helsinki 10.

● **FRANCE**
Editions A. Pedone
13, rue Soufflot, 75005 Paris.

● **GERMANY, FED. REP.**
Alexander Horn Internationale Buchhandlung
Kirchgasse 39, Postfach 3340, 6200 Wiesbaden.
UNO Verlag
Poppelsdorfer Allee 55, D-5300 Bonn 1.
Triops Verlag
Raiffeisenstr. 24, 6060 Langen.

● **GHANA**
Ghana Publishing Corporation
PO Box 4348, Accra.

● **GREECE**
G.C. Eleftheroudakis S.A.
4 Nikis Street, Athens (T-126).
John Mihalopoulos & Son S.A.
75 Hermou Street, PO Box 73, Thessaloniki.

● **GUATEMALA**
Distribuciones Culturales y Técnicas "Artemis"
5ª Avenida 12-11, Zona 1, Apartado Postal 2923, Guatemala.

● **GUINEA-BISSAU**
Conselho Nacional da Cultura
Avenida da Unidade Africana, C.P. 294, Bissau.

● **GUYANA**
Guyana National Trading Corporation Ltd
45-47 Water Street, PO Box 308, Georgetown.

● **HAÏTI**
Librairie "A la Caravelle"
26, rue Bonne Foi, B.P. 111, Port-au-Prince.

● **HONDURAS**
Escuela Agricola Panamericana, Libreria RTAC
Zamorano, Apartado 93, Tegucigalpa.
Oficina de la Escuela Agricola Panamericana en Tegucigalpa
Blvd. Morazán, Apts. Glapson, Apartado 93, Tegucigalpa.

● **HONG KONG**
Swindon Book Co.
13-15 Lock Road, Kowloon.

● **HUNGARY**
Kultura
PO Box 149, 1389 Budapest 62.

● **ICELAND**
Snaebjörn Jónsson and Co. h.f.
Hafnarstraeti 9, PO Box 1131, 101 Reykjavik.

● **INDIA**
Oxford Book and Stationery Co.
Scindia House, New Delhi 100 001; 17 Park Street, Calcutta 700 016
Oxford Subscription Agency, Institute for Development Education
1 Anasuya Ave, Kilpauk, Madras 600010.

● **INDONESIA**
P.T. Inti Buku Agung
13 Kwitang, Jakarta.

● **IRAQ**
National House for Publishing, Distributing and Advertising
Jamhuria Street, Baghdad.

● **IRELAND**
Agency Section, Publications Branch
Stationery Office, Bishop Street, Dublin 8.

WHERE TO PURCHASE FAO PUBLICATIONS LOCALLY
POINTS DE VENTE DES PUBLICATIONS DE LA FAO
PUNTOS DE VENTA DE PUBLICACIONES DE LA FAO

● **ITALY**
FAO (see last column)
Libreria Scientifica
Dott. Lucio de Biasio "Aeiou"
Via Meravigli 16, 20123 Milan.
Libreria Commissionaria
Sansoni S.p.A. "Licosa"
Via Lamarmora 45,
C.P. 552, 50121 Florence.
Libreria Internazionale Rizzoli
Galleria Colonna, Largo Chigi,
00187 Rome.

● **JAPAN**
Maruzen Company Ltd
PO Box 5050,
Tokyo International 100-31.

● **KENYA**
Text Book Centre Ltd
Kijabe Street, PO Box 47540, Nairobi.

● **KOREA, REP. OF**
Eulyoo Publishing Co. Ltd
46-1 Susong-Dong, Jongro-Gu,
PO Box 362, Kwangwha-Mun,
Seoul 110.

● **KUWAIT**
The Kuwait Bookshops Co. Ltd
PO Box 2942, Safat.

● **LUXEMBOURG**
M. J. De Lannoy
202, avenue du Roi,
1060 Bruxelles (Belgique).

● **MALAYSIA**
SST Trading Sdn. Bhd.
Bangunan Tekno No. 385, Jln 5/59,
PO Box 227, Petaling Jaya, Selangor.

● **MAROC**
Librairie "Aux Belles Images"
281, avenue Mohammed V, Rabat.

● **MAURITIUS**
Nalanda Company Limited
30 Bourbon Street, Port-Louis.

● **NETHERLANDS**
Keesing b.v.
Hogeliweg 13, 1101 CB Amsterdam.
Postbus 1118, 1000 BC Amsterdam.

● **NEW ZEALAND**
Government Printing Office
Bookshops
25 Rutland Street.
Mail orders: 85 Beach Road,
Private Bag, CPO, Auckland;
Ward Street, Hamilton;
Mulgrave Street (Head Office),
Cubacade World Trade Centre,
Wellington;
159 Hereford Street, Christchurch;
Princes Street, Dunedin.

● **NICARAGUA**
Libreria Universitaria,
Universidad Centroamericana
Apartado 69, Managua.

● **NIGERIA**
University Bookshop (Nigeria) Limited
University of Ibadan, Ibadan.

● **NORWAY**
Johan Grundt Tanum Bokhandel
Karl Johansgate 41-43,
PO Box 1177, Sentrum, Oslo 1.

● **PAKISTAN**
Mirza Book Agency
65 Shahrah-e-Quaid-e-Azam,
PO Box 729, Lahore 3.
Sasi Book Store
Zaibunnisa Street, Karachi.

● **PARAGUAY**
Agencia de Librerías Nizza S.A.
Casilla 2596, Eligio Ayala 1073,
Asunción.

● **PERU**
Libreria Distribuidora "Santa Rosa"
Jirón Apurimac 375, Casilla 4937,
Lima 1.

● **POLAND**
Ars Polona
Krakowskie Przedmiescie 7,
00-068 Warsaw.

● **PORTUGAL**
Livraria Portugal,
Dias y Andrade Ltda.
Rua do Carmo 70-74, Apartado 2681,
1117 Lisbonne Codex.

● **REPUBLICA DOMINICANA**
Editora Taller, C. por A.
Isabel la Católica 309,
Apartado de Correos 2190,
ZI Santo Domingo.
Fundación Dominicana de Desarrollo
Casa de las Gárgolas,
Mercedes 4, Apartado 857,
ZI Santo Domingo.

● **ROMANIA**
Ilexim
Calea Grivitei No 64066, Bucharest.

● **SAUDI ARABIA**
The Modern Commercial
University Bookshop
PO Box 394, Riyadh.

● **SINGAPORE**
MPH Distributors (S) Pte. Ltd
71/77 Stamford Road, Singapore 6.
Select Books Pte. Ltd
215 Tanglin Shopping Centre,
19 Tanglin Rd., Singapore 1024.

● **SOMALIA**
"Samater's"
PO Box 936, Mogadishu.

● **SRI LANKA**
M.D. Gunasena & Co. Ltd
217 Olcott Mawatha,
PO Box 246, Colombo 11.

● **SUDAN**
University Bookshop,
University of Khartoum
PO Box 321, Khartoum.

● **SUISSE**
Librairie Payot S.A.
107 Freiestrasse, 4000 Basel 10.
6, rue Grenus, 1200 Genève.
Case Postale 3212, 1002 Lausanne.
Buchhandlung und Antiquariat
Heinimann & Co.
Kirchgasse 17, 8001 Zurich.

● **SURINAME**
VACO n.v. in Suriname
Domineestraat 26,
PO Box 1841, Paramaribo.

● **SWEDEN**
Books and documents:
C.E. Fritzes Kungl. Hovbokhandel,
Regeringsgatan 12,
PO Box 16356, 103 27 Stockholm.
Subscriptions:
Vennergren-Williams AB
PO Box 30004, 104 25 Stockholm.

● **TANZANIA**
Dar-es-Salaam Bookshop
PO Box 9030, Dar-es-Salaam.
Bookshop, University
of Dar-es-Salaam
PO Box 893, Morogoro.

● **THAILAND**
Suksapan Panit
Mansion 9, Rajadamnern Avenue,
Bangkok.

● **TOGO**
Librairie du Bon Pasteur
B.P. 1164, Lomé.

● **TUNISIE**
Société tunisienne de diffusion
5, avenue de Carthage, Tunis.

● **TURKEY**
Kultur Yayinlari Is-Turk Ltd Sti.
Ataturk Bulvari No. 191,
Kat. 21, Ankara
Bookshops in Istanbul and Izmir.

● **UNITED KINGDOM**
Her Majesty's Stationery Office
49 High Holborn,
London WC1V 6HB (callers only).
HMSO Publications Centre,
Agency Section
51 Nine Elms Lane,
London SW8 5DR (trade and London
area mail orders);
13a Castle Street,
Edinburgh EH2 3AR;
80 Chichester Street,
Belfast BT1 4JY;
Brazennose Street,
Manchester M60 8AS;
258 Broad Street,
Birmingham B1 2HE;
Southey House, Wine Street,
Bristol BS1 2BQ.

● **UNITED STATES**
OF AMERICA
UNIPUB
4611/F, Assembly Drive,
Lanham, MD 20706.

● **URUGUAY**
Libreria Agropecuaria S.R.L.
Alzaibar 1328, Casilla Correo 1755,
Montevideo.

● **YUGOSLAVIA**
Jugoslovenska Knjiga, Trg.
Republike 5/8,
PO Box 36, 11001 Belgrade
Cankarjeva Zalozba
PO Box 201-IV, 61001 Ljubljana.
Prosveta
Terazije 16, Belgrade.

● **ZAMBIA**
Kingstons (Zambia) Ltd
Kingstons Building,
President Avenue, PO Box 139, Ndola.

Other Countries
Autres Pays
Otros Países

**Distribution and
Sales Section, FAO**
Via delle Terme
di Caracalla,
00100 Rome, Italy.

Tipo-lito SAGRAF - Napoli